Proceedings of the Third International Workshop on Phosphorus in Sediments

Developments in Hydrobiology 84

Series editor
H. J. Dumont

Proceedings of the Third International Workshop on Phosphorus in Sediments

Edited by

P.C.M. Boers, Th.E. Cappenberg & W. van Raaphorst

Reprinted from Hydrobiologia, vol. 253 (1993)

Springer - Science+Business Media, B.V.

Library of Congress Cataloging-in-Publication Data

```
International Workshop on Phosphorus in Sediments (3rd : 1991 : Zeist,
  Netherlands)
    Proceedings of the Third International Workshop on Phosphorus in
  Sediments / edited by P.C.M. Boers, Th. E. Cappenberg, and W. van
  Raaphorst.
        p.   cm. -- (Developments in hydrobiology ; 84)
    Held in Zeist, The Netherlands, September 30-October 3, 1991.
    ISBN 978-94-010-4696-1    ISBN 978-94-011-1598-8 (eBook)
    DOI 10.1007/978-94-011-1598-8
    1. Phosphorus cycle (Biogeochemistry)--Congresses.  2. Sediments
  (Geology)--Congresses.   I. Boers, P. C. M. (Paul C. M.)
  II. Cappenberg, Th. E. (Thomas Ernst)  III. Raaphorst, W. van (Wim)
  IV. Title.  V. Series.
  QH344.I59  1991
  551.48--dc20                                        92-46549
```

ISBN 978-94-010-4696-1

Contents

vi

Session IV: Water management measures focussed on the regulation of the release of phosphorus from the sediment, in relation to the functioning and restoration of the whole ecosystem

viii

Third International Workshop on Phosphorus in Sediments, 30 September–3 October 1991, Zeist, The Netherlands.

List of participants

Anderson, N. J., Copenhagen, Denmark
Auer, M., Houghton, USA
Beltman, B., Utrecht, The Netherlands
Bloesch, J., Dübendorf, Switzerland
Blomqvist, S., Stockholm, Sweden
Boers, P. C. M., Lelystad, The Netherlands
Brinkman, A. G., Den Burg (Texel), The Netherlands
Cappenberg, Th. E., Nieuwersluis, The Netherlands
Caraco, N. F., Millbrook, USA
Cooke, D. C., Kent, USA
Danen-Louwerse, H., Wageningen, The Netherlands
Davelaar, D., Arnhem, The Netherlands
Does, van der, J., Leiden, The Netherlands
Donze, M., Delft, The Netherlands

Eck, van, G. Th. E., Middelburg, The Netherlands
Eckerrot, Å., Uppsala, Sweden
Filip, A., Beograd, Yugoslavia
Forsgren, G., Umea, Sweden
Fox, L. E., Cambridge, USA
Gächter, R., Kastanienbaum, Switzerland
Golterman, H. L., Arles, France
Graaf, de, I. M., Leiden, The Netherlands
Groot, de, C. J., Arles, France
Gunnars, A., Stockholm, Sweden
Heller, S., Berlin, Germany
Ignatieva, N., St Petersburg, Russia
Istvánovics, V., Tihany, Hungary
Jansson, M., Umeå, Sweden
Jensen, H. S., Odense, Denmark
Jonge, de, V. N., Haren, The Netherlands
Keizer, P., Nieuwersluis, The Netherlands
Kelly, L., Sterling, UK
Kleeberg, A., Rostock, Germany

Kristensen, P., Silkeborg, Denmark
Laane, W. E. M., Lelystad, The Netherlands
Liere, van, L., Bilthoven, The Netherlands
Lijklema, L., Wageningen, The Netherlands
Löfgren, S., Uppsala, Sweden
Lopez Laseras, M., Barcelona, Spain
Lucotte, M., Montreal, Canada
Martin, A., Kent, USA
Marxsen, J., Schlitz, Germany
Meuleman, A. F. M., Utrecht, The Netherlands
Molen van der, D., Lelystad, The Netherlands
Montigny, de, C., Montreal, Canada
Mortensen, P. B., Aarhus, Denmark
Mourkides, G., Thessaloniki, Greece
Moutin, T., Montpellier, France
Nishri, A., Tiberias, Israel

Penn, M. R., Houghton, USA
Pettersson, K., Norrtälje, Sweden
Portielje, R., Wageningen, The Netherlands
Quaak, M. P., Rotterdam, The Netherlands
Raaphorst, van, W., Den Burg (Texel), The Netherlands
Rasmussen, E. K., Hørsholm, Denmark
Ripl, W., Berlin, Germany
Sinke, A. J. C., Nieuwersluis, The Netherlands
Slomp, C. P., Den Burg (Texel), The Netherlands
Smits, J. G. C., Delft, The Netherlands
Søndergaard, M., Silkeborg, Denmark
Baciu, S., Bicaz, Romania
Stone, M., Waterloo, Canada
Sundby, B., Mont-JóLi, Canada
Waara, L. T., Uppsala, Sweden
Wisniewski, R. J., Torun, Poland

Hydrobiologia **253**: xi–xviii, 1993.
P.C.M. Boers, T.E. Cappenberg & W. van Raaphorst (eds),
Proceedings of the Third International Workshop on Phosphorus in Sediments.
© 1993 *Kluwer Academic Publishers.*

The third international workshop on phosphorus in sediments
Summary and Synthesis

Paul C. M. Boers [1], Thomas E. Cappenberg [2] & Wim van Raaphorst [3]
[1] *National Institute for Inland Water Management and Waste Water Treatment, P.O. Box 17, 8200 AA
Lelystad, The Netherlands;* [2] *Netherlands Institute for Ecology, Centre for Limnology, Rijksstraatweg 6, 3631
AC Nieuwersluis, The Netherlands;* [3] *Netherlands Institute for Sea Research, P.O. Box 59, 1790 AB Den
Burg, The Netherlands*

Key words: phosphorus, sediments, chemistry, microbiology, modelling, lake restoration

Field of research

The phosphorus cycle in aquatic environments
has been a scientific topic for several decades. To
quote Hutchinson (1957, p. 727): 'Of all the elements present in living organisms, phosphorus is
likely to be the most important ecologically.' Although this statement may need some differentiation, it clearly points to the basic reason for
scientific interests. In limnology most studies focused on the mobilization and (bio)-availability of
phosphorus. The first limnologists who recognized the importance of iron-phosphorus interactions and the coupling between mobilization and
redox conditions were Einsele (1936, 1938) and
Mortimer (1941, 1942). Their observations form
the basis of many studies carried out nowadays.
Many of these studies concern the precise nature
of iron-phosphorus bindings in sediments and
suspended matter and their effect on the phosphorus buffering system, i.e. the ability of the sediments to constrain phosphate concentrations
within narrow limits (Froelich, 1988). The sorp-
tive properties of soil minerals were first investigated by agronomists and soil scientists to assess
'phosphate fixation' which makes part of the
phosphorus applied in fertilizers unavailable to
plants (e.g. Kittrick & Jackson, 1956). In sediment-water systems phosphate fixation is studied
to determine the long-term loss of phosphorus

from the ecosystem to the sediments. Soil scientists also initiated the development of reliable sequential extraction schemes to distinguish between phosphorus compounds bound to different
complexes and minerals (e.g. Chang & Jackson,
1957). Sequential extractions have proven to be
useful tools in quantifying metal-P associations as
non-mobilizable phosphorus. The fundamental
mechanism for the kinetics of phosphate sorption
on solids was already described by Carritt &
Goodgal (1954), who studied estuarine sediments. The mechanism that they proposed consists of a rapid surface adsorption reaction followed by a much slower diffusion controlled
process towards the interior of the solids. Essentially, this concept still holds and is nowadays
receiving renewed attention. Particularly the slow
step is important to predict sedimentary phosphorus binding capacity and future behaviour of
sedimentary P complexes after changing external
loadings.

All the above aspects are largely of an inorganic geochemical nature. Yet, microbiology and
organo-geochemistry also form a substantial part
of the field of research. The most obvious reason
is that mineralization of organic phosphorus is
the first and driving step in benthic P cycling
and sediment-water exchange processes. Consequently, mineralization and geochemistry of organic P compounds are important subjects. Re-

cently Gächter *et al.* (1988) identified the potential role of bacteria in directly releasing and fixating phosphates in lake sediments, likely through synthesis of poly-phosphates in the cells. This phenomenon potentially could be an important modification of the classical Mortimer-Einsele concept.

The aspects mentioned above form the body of the scientific discipline concerned with phosphorus in sediments. Like many other disciplines it contains a fundamental and an applied approach. The latter is strongly related to the productivity of aquatic ecosystems and, especially for lakes and estuaries, to eutrophication control. The main objective of the fundamental research is to identify and understand underlying mechanisms and processes, but also to quantitatively assess the importance of sedimentary P cycling within overall phosphorus budgets of aquatic systems, including streams and rivers, lakes, estuaries, shelf seas and oceans. Mathematical models are potentially valuable tools that may connect both approaches. They summarize the existing knowledge and make it available for water management purposes. Missing links in the understanding of key processes can be highlighted by modelling exercises, thus pointing to new fields of fundamental research. Also, a lack of knowledge of fundamental processes limits our ability to manage ecosystems and to solve e.g. eutrophication problems. Models in some aspects thus form a 'bridge between science and management'.

The last, but not least, branch on the tree of phosphorus in sediments is dedicated to the design and evaluation of management measures. This part of the field is not as new as it may seem at first sight. Hasler & Einsele (1948) suggested to add sulfate to lakes to increase the availability of phosphorus to primary producers by precipitating ferrous sulphide, thus decreasing the amount of Fe(III) sorption surfaces. Nowadays management is normally not aiming at fertilizing water bodies, but rather at restoring them from eutrophication. The historical example demonstrates, however, that real world application of existing knowledge and concepts may not only be useful from the management perspective, but that it may also result in new scientific insights.

History of the workshop

Traditionally the research was strongly directed towards the behaviour of phosphorus at the sediment-water interface in anoxic hypolimnia of deep, stratifying lakes. The reasons for this were that many of the lakes in the world belong to that category and that phosphorus release in such lakes is readily evident from an increase in the SRP concentration in the hypolimnion. In the sixties Han Golterman pioneered on the biological availability of sediment-bound phosphorus for algae (Golterman *et al.*, 1969). Probably he was the first to emphasize that upon a reduction of the phosphorus loading to a lake, the sediments can provide enough P to the water column to maintain a high algal biomass for many years. Golterman also brought together for the first time a number of scientists in Amsterdam, the Netherlands, in 1976 to discuss interactions between sediment and water (Golterman, 1977). Phosphate interactions formed an important topic of this symposium. The first symposium was followed by a second in 1981 in Kingston, Canada, and a third in 1984 in Geneva, Switzerland. On this last occasion, a number of scientists, working on phosphate in sediments decided to organize a special workshop dedicated to this research field.

The result was the first International Workshop on Phosphorus in Sediments held in Vienna, Austria, in 1985. It counted 25 participants (Psenner & Gunatilaka, 1988). It was followed by a second in Fyskebäckskill, Sweden, in 1988 that was visited by 42 participants (Enell *et al.*, 1989). Unfortunately, there are no proceedings of this latter workshop. From 30 September–3 October 1991 66 scientists from 16 countries met in Zeist, the Netherlands, at the Third International Workshop on Phosphorus in Sediments. The fourth workshop will be organized in 1994 in Orlando, USA by the NALMS (North American Lake Management Society).

Summary of the third workshop

More or less according to the different research orientations the plenary lectures were organized in four sessions. All sessions were introduced by an invited keynote speaker, followed by in total 45 oral presentations. In addition, 11 research posters were presented.

The four topics were:

Chemical transformations of phosphorus in sediments (21 presentations)

Keynote speaker: Dr L. E. Fox, Harvard University, USA. Fox reviewed previous work dealing with the chemistry of phosphate in turbid rivers with low Ca concentrations. However, his theories can largely be generalized for lakes, estuaries and sediments. He concluded that phosphate concentrations in such rivers are controlled by a solid ferric hydroxide-phosphate solution present in colloids or particles. Sorption of phosphate consists of a rapid, reversible uptake followed by a slow irreversible process. This second process may depend on solid phase diffusion, thus producing a distinct phase. Consequently, a semi-stable solid, intermediate between adsorption and pure solid, develops which behaves as a solid-solution. The review of Fox demonstrates the importance of iron-oxides for controlling dissolved phosphate concentrations in rivers. His model needs to be tested for sediment-pore water systems, but if applicable it can provide additional insights into the classical Mortimer-Einsele concepts.

Other contributions to this theme further elucidated the binding of phosphates to iron-oxides. Danen et al. showed that the sorption capacity of shallow lake sediments as estimated from adsorption isotherms is related to the sedimentary contents of oxalate extractable iron. Results of Jensen & Thamdrup suggest that FeOOH is not readily reduced in anoxic sediments and that amorphous Fe(III)-oxides can control pore water phosphate concentrations in anoxic marine sediments. These results point at the need for a better understanding of the cycling of iron in sediments. The close coupling between iron and phosphate probably provokes a combined research on both elements.

Portielje & Lijklema showed that iron-oxide is not the only candidate to bind phosphates. With sediments of artificial ditches they examined the fast initial as well as the slow second sorption step and concluded that Al-hydroxides provided the most important sorption sites in their system. Since Al-hydroxides are not to be reduced in sediments, systems where sorption to Al dominates P binding are expected not respond to anoxia in the same way as Fe-OOH controlled systems. In addition, Driscoll et al. clearly demonstrated the role of Ca for binding in a Ca-rich environment, probably through the coprecipitation of phosphate with calciumcarbonate. Lopez & Morgui and Stone & English found correlations between fractional composition of P and elemental composition in Spanish reservoir sediments and suspended solids in Lake Erie tributaries, respectively. The latter authors also showed that phosphate is mainly bound to clay-sized materials. Finally, Caraco et al. returned to the classical concept of Hasler & Einsele (1948) concerning the role of sulfate on phosphate dynamics. They concluded that higher sulfate concentrations in stratified lakes may have a double effect on P cycling: the benthic release is stimulated by full reduction of Fe(III) due to increased HS^- production, and the availability of P for primary producers is stimulated by a decreased amount of sorption sites due to precipitation of FeS. In conclusion, P binding in sediments is much more complicated than simple models and concepts as those of Mortimer and Einsele suggest.

The role of microorganisms in sediment phosphorus dynamics (12 presentations)

Keynote speaker: Dr R. Gächter, EAWAG, Switzerland. Gächter discussed the apparent role of microorganisms in mobilization and fixation of phosphorus in sediments. Important processes are the temporal storage of polyphosphates that are released under anoxic conditions, phosphate uptake by actively growing bacteria, release of orthophosphate by mineralization and production of refractory organic P compounds. Gächter suggested that redox conditions influence the transfer between dissolved and particulate P due to physiological changes of the bacterial population. He also concluded that sediment bacteria do not necessarily release P upon mineralization of settled organic matter. This depends on the C:P ratio of the substrate and their growth yield. Many of the bacterial processes in sediments could be similar to those during biological phosphate removal in sewage treatment plants. In these plants phosphate is taken up during the oxic period and subsequently released in the absence of oxygen. *Acinetobacter* species appear to be the dominant genus storing phosphate intracellularly as polyphosphates, but actual field data in natural ecosystems are still lacking. Bacteria may play a much more important role in P cycling in sediments than was generally assumed up to now. Sinke *et al.* studied the effect of an enhanced oxygen consumption by methane-oxidizing bacteria on the phosphate flux and found that uptake of phosphate by the growth of the methanotrophic bacteria could compensate for the decrease in chemical adsorption caused by a decrease in oxygen penetration depth. Several other authors (Istvánovics, Waara, de Montigny) also demonstrated the active role of microorganisms in the phosphorus cycling in various sediments. However, the importance of bacterial processes compared to chemical processes could not be quantified yet.

Sediment-water dynamics with special attention for functional models describing fluxes across the sediment–water interface (7 presentations)

Keynote speaker: Prof. Dr L. Lijklema, Agricultural University of Wageningen, the Netherlands. Lijklema discussed the processes involved in P dynamics at the sediment-water interface and the possibilities to model them. He stressed the need for models that can properly predict gradual changes in sediment composition after manage-

ment measures have been taken. Probably only 'simple' models will be useful for this purpose. In addition, the availability of data for calibration should be considered.

In this session most presentations dealt with transport processes and flux measurements. In lakes, core experiments appear to be the best available method to measure the phosphate flux across the sediment-water interface. Auer *et al.* demonstrated that in a statified lake this method produces data, comparable to calculations from mass balance data. Unfortunately, up to now such comparisons have been impossible in shallow lakes. In marine systems, *in situ* bell jar experiments with landers appear to be the most reliable method. However, no data produced with this method were presented at this workshop. Promising was the method presented by Kleeberg, consisting of *in situ* core experiments in a small river. Slomp *et al.* and De Jonge *et al.* presented detailed studies of phosphorus transformations in marine sediments using mesocoms and laboratory techniques, respectively. De Jonge introduced a new approach to quantify bioavailable phosphorus using bacteria instead of algae.

Only few presentations of modelling efforts were given, all from the Netherlands. Brinkman presented detailed modelling of adsorption of phosphate onto metal-oxide surfaces. The model is aimed at explaining of experimental data. Van der Molen & Smits presented the complete but simple model SWITCH that describes the behaviour of phosphate in the active sediment layer. They concluded that a basic problem is the proper assessment of a sedimentary mass balance for nutrients, including longer term processes as burial in deeper sediment layers and irreversible binding to sediment particles. Søndergaard set up sedimentary mass balances for P and Fe for lake Søbygard and concluded that full recovery of that lake would take at least another decade. These studies indicate that more information on slow processes is urgently needed to better understand eutrophication and restoration effects on a longer time scale.

Water management measures focused on the regulation of the release of phosphorus from the sediment (9 presentations).

Keynote speaker: Prof. Dr G. D. Cooke, Kent State University, USA. Cooke presented a review of phosphorus inactivation experiments in mainly the USA. He concluded that when aluminium compounds are used, the technique is reliable and safe and is effective for up to at least 9 years. He stressed the need to reduce external phosphorus loadings before phosphorus inactivation is applied. Both Keizer *et al.* and Quaak *et al.* presented results of phosphorus inactivation experiments with Fe(III) compounds. This technique works on the short term, but the effectiveness on the long term is questionable and seems to depend on the long term stability of Fe(III). Kelly presented a case study in lakes used for fish farming. Increases in available phosphorus were limited to the area near the cages. Anderson presented a method to reconstruct pre-eutrophication phosphorus concentrations from paleolimnologic data.

Conclusions and recommendations

The workshop ended with four group discussions on the subject: 'How is phosphorus removed from aquatic ecosystems?' from four different points of view, corresponding with the scopes of the four lecture sessions: chemistry, microbiology, modelling and lake restoration. The most relevant statements made by the four groups were:

Chemistry

This group focused on techniques to study processes that lead to the permanent fixation of phosphorus in sediments. It was concluded that the suitability of selective extractions is limited. Some of the many problems are related to:

- changes in ion strength.
- transformations of P to oxidized phases; no differentiation is possible between P associated

to Fe^{2+} and Fe^{3+}, because most of the iron minerals dissolve during the extractions.
- problems caused by diluted sodium hydroxide as an extraction solvent, because it induces undesirable chemical reactions like degradation of refractory organic matter and uptake of P in calcium minerals.
- similar problems caused by diluted acids, namely hydrolysis of polyphosphates and re-precipitation of mineral phases.

Suggestions for alternatives were:

- to quantify the fixation of P in sediments with mass balances of phosphorus and adsorbence for the sediments
- to gain more profit from recent developments in analytical chemistry. Recently developed or markedly improved techniques like MS-EDAX, Mössbauer spectroscopy, XRD, ^{31}P-NMR and electron microprobes might offer new perspectives.

Microbiology

Most participants doubted whether bacteria can take care of long term removal of P. It is accepted that bacteria do regulate P fluxes across the sediment-water interface through mineralization and influences on the redox potential. Particularly when local amounts of P in bacteria are relatively constant over time, the importance of bacterial pools may be questioned. At the moment, adequate experimental techniques and generally accepted theoretical concepts are most urgently needed in this research area. Definite conclusions on the role of bacteria are still premature.

Modelling

In the view of the modellers, an adequate description of processes on a long time scale poses the largest problems. Calibration of such models is difficult and small uncertainties in initial conditions may lead to large deviations. Possibly, simple mass balances can be made for P and adsorbing agents (Fe-oxides, Al-oxides, Ca), if necessary completed with a term for the loss of adsorption capacity, e.g. by sulphate reduction. The main bottleneck is considered to be a lack of knowledge of anoxic P fixation. It is certain that this occurs, but the mechanism is not well understood. Possible candidates are: burial as refractory organic P, binding to Fe(II) or even Fe(III) compounds, formation of apatite. A future shift of scientific interest from internal loading processes towards removal of P is expected.

Lake restoration

This group focused on the role of science in eutrophication control. The dominant opinion of this group was that full recovery of lakes and other ecosystems may take decades. The most important mission of science was thought to prove so and explain it. It was questioned whether methods to speed up recovery are justified, because they form just a another intervention of man in the ecosystem. It is more important to pay attention to the natural quality of the systems.

Looking back over the past three workshops, a number of trends can be discerned. The further development and evaluation of selective extraction techniques has become much less important. During the first and second workshop special sessions and discussions were dedicated to this subject, but not at this third workshop. Most researchers seemed to consider selective extractions as a tool with limited applicability. The same seems true for bio-assays. During the 70's and early 80's much effort was put in the development of this technique as well as the comparison with selective extractions. This seems to be over.

During the last workshop many questions and ideas returned to the classical concepts and potential modifications. A relatively new development is the role of bacteria and other microorganisms in the behaviour of phosphorus in sediments. The indirect role of bacteria through changes in redox conditions was recognized earlier and agrees well with the theories of Mortimer & Einsele. Gradually, the idea that microorganisms play another role in the phosphorus cycling in sediments gains. However, as pointed out by Gächter and by the participants in the final discussion, generally accepted theories are lacking. Also, reliable techniques are needed to determine the P content of bacteria as a function of varying redox conditions, to measure polyphosphates in sediments in combination with the population dynamics and activities of *Acinetobacter*-like organisms, to differentiate between refractory and bio-available P and to study production and chemical nature of refractory P under different physiological and environmental conditions. All these studies have to be performed in close connection with those on chemical adsorption and desorption phenomena under the same redox conditions.

It is getting clearer and clearer that the chemistry of phosphate in sediments is closely connected to the chemistry of iron, aluminium and to a lesser extent calcium. A better understanding of the chemistry of these metals in sediments is necessary to improve the understanding of the cycling of P. The theory that mobilization of phosphate is caused by reduction of iron(III)oxyhydroxides, is fifty years old. New contributions are the effects of organic ligands on P adsorption. The theory that sorption on oxyhydroxides occurs in a fast initial and a slow secondary phase has also been known for decades. However, the mechanisms of permanent fixation of inorganic P compounds are still largely unexplained. Adequate experimental techniques are needed.

Mathematical models may summarize scientific knowledge in such a way that it can be used for water quality management purposes. Probably quite a number of such models are used all over the world. However, usually only very few results of modelling efforts are presented at scientific congresses and workshops. The few presentations on this workshop made clear that modelling of the long term behaviour of lakes after restoration measures is still beyond our ca-

pabilities, probably because the processes leading to permanent fixation of phosphorus in sediments are still not well known.

The attention for the cycling of phosphorus in marine ecosystems is growing, reflecting the increased concern for the quality of these systems. At the preceding workshop it was concluded that the lack of comparative studies makes evaluation of similarities and differences between marine and freshwater sediments difficult and that basically phosphorus turnover in both sediments should be identical (Enell *et al.*, 1989). Such comparative studies are still lacking. The studies on basic processes in both sediments, which are increasing in number, confirm the conclusion of the previous workshop. In both systems, iron oxides form the major part of the phosphate binding capacity. Differences are probably caused by variations in chemical environment caused by salinity and trophic level.

Some studies presented concerned lakes under restoration. Although often detailed results of flux measurements and processes in the sediments were given, apparently it is not yet possible to predict how a specific lake will respond to a decrease in phosphorus loading, *i.e.* to what extent internal loading will occur and for how long. Separate mass sedimentary balances for P and binding capacity are promising. At the moment insight into crucial processes as burial is lacking. Part of the research is directed at manipulating sediment-water exchange of phosphorus, e.g. by phosphorus inactivation. The subject of the discussion is not only the merits of such techniques but also the question if it is justified to use them.

All research presented was done in systems in the temperate zone. Research in tropical or arctic lakes and estuaries was absent. Regarding the increased attention for the role of bacteria in phosphorus cycling one may question to what extent such systems differ from temperate ones. The problem was raised by Kilham & Kilham (1990) for tropical lakes but unfortunately remains answered.

Finally, it is our opinion that enough research questions remain to justify a fourth international workshop on phosphorus in sediments.

Acknowledgements

A committee consisting of Paul C. M. Boers (National Institute for Inland Water Management and Waste Water Treatment, treasurer), Thomas E. Cappenberg (Centre for Limnology/NIE, chairman), Peer Keizer (Centre for Limnology/NIE), Wim van Raaphorst (Netherlands Institute for Sea Research) and Anja J. C. Sinke (Centre for Limnology/NIE, secretary) organized the third workshop on phosphorus in sediments. Cecilia Kroon of the Centre for Limnology assisted the secretary.

Financial help from the European Environmental Research Organization, National Institute for Inland Water Management and Waste Water Treatment, Centre for Limnology/Netherlands Institute for Ecology, Ministry of Housing, Physical Planning and Environment of the Netherlands, Royal Netherlands Academy of Arts and Sciences and Netherlands Integrated Soil Research Programma made the workshop possible.

The staff of conference centre 'Woudschoten' in Zeist, the Netherlands, gave hospitality. They created the informal atmosphere necessary to have a successful workshop.

These proceedings could be published thanks to the cooperation of Kluwer Publishers. We, the editors, are highly grateful to the referees for their critical reviews and suggestions. Most of them responded very promptly, sparing us largely the need to look for alternative referees or to send reminders. All manuscripts were reviewed by at least two independent reviewers from different countries of the world. The authors too were generally very cooperative. Finally, we thank Annemieke Wagenaar-Hart for linguistically editing most of the papers.

References

Carritt, D. E. & S. Goodgal, 1954. Sorption reactions and some ecological implications. Deep-Sea Res. 1: 224–243.

Chang, S. C. & M. L. Jackson, 1957. Fractionation of soil phosphorus. Soil Sci. 84: 133–144.

Einsele, W. G., 1936. Über die Beziehungen des Eisenkreislaufs zum Phosphatkreislauf im eutrophen See. Arch. Hydrobiol. 29: 664–686.

Einsele, W., 1938. Über chemische und kolloidchemische Vorgänge in Eisen-Phosphat-Systemen unter limnochemischen und limnogeologischen Gesichtspunkten. Arch. Hydrobiol. 33: 361–387.

Enell, M., S. Fleischer & M. Jansson, 1989. Phosphorus in sediments. Ambio 18: 137–138.

Froelich, P. N., 1988. Kinetic control of dissolved phosphate in natural rivers and estuaries: A primer on the phosphate buffer mechanism. Limnol. Oceanogr. 33: 649–668.

Gächter, R., J. S. Meyer & A. Mares, 1988. Contribution of bacteria to release and fixation of phosphorus in lake sediments. Limnol. Oceanogr. 33: 1542–1558.

Golterman, H. L., C. C. Bakels & J. Jakobs- Mögelin, 1969. Availability of mud phosphates for the growth of algae. Verh. int. Ver. Limnol. 17: 467–479.

Golterman, H. L. (ed.), 1977. Interactions between sediment and water. Dr W. Junk Publishers, The Hague, The Netherlands.

Hasler, A. D. & W. G. Einsele, 1948. Fertilization for increasing productivity of natural inland waters. Trans. 13th. N. Amer. Wild L. Conf.: 527–554.

Hutchinson, G. E., 1957. A treatise on limnology. Volume I, geography, physics, and chemistry. Chapter 12, The phosphorus cycle in lakes. Wiley & Sons, New York: 727–752.

Kittrick, J. A. & M. L. Jackson, 1956. Electron microscope observations of the reaction of phosphate with minerals, leading to an unified theory of phosphate fixation in soils. J. Soil. Sci.7: 81–89.

Kilham, P. & S. S. Kilham, 1990. Endless summer: internal loading processes dominate nutrient cycling in tropical lakes. Freshwat. Biol. 23: 379–389.

Mortimer, C. H., 1941. The exchange of dissolved substances between mud and water in lakes. I. J. Ecol. 29: 280–329.

Mortimer, C. H., 1942. The exchange of dissolved substances between mud and water in lakes. II. J. Ecol. 30: 147–201.

Psenner, R. & A. Gunatilaka (eds), 1988. Proceedings of the first international workshop on sediment phosphorus. Arch. Hydrobiol. Beih. Ergebn. Limnol. 30.

Hydrobiologia **253**: 1–16, 1993.
P.C.M. Boers, T.E. Cappenberg & W. van Raaphorst (eds),
Proceedings of the Third International Workshop on Phosphorus in Sediments.
© 1993 *Kluwer Academic Publishers.*

1

The chemistry of aquatic phosphate: inorganic processes in rivers*

Lewis E. Fox
108 Pierce Hall, Harvard University, 29 Oxford St. Cambridge, MA 02138, USA

Key words: phosphate, ferric hydroxide, solution theory, solid-solution, solubility, river, colloidal suspension

Abstract

Phosphate levels in turbid rivers with low calcium concentrations are controlled by a solid ferric hydroxide-phosphate solution present in colloidal suspensions or suspended particulates. A chemical model, based on this behavior, is consistent with data from dialyzed suspensions of iron and phosphorus prepared in the laboratory as well as from the Amazon, Zaire, Orinoco, Sepik, Delaware, Hudson, Negro, and Mullica rivers. Data indicate that solid Fe/P ratios are related to solid activity coefficients by an exponential parameter, y, which represents the deviation of solid-solution from ideality. The model is mathematically consistent with Langmuir and Freundlich sorption isotherms under equilibrium conditions, and demonstrates that the isotherm parameters consist of a combination of selected constants and variables defined by solution theory. The reciprocal of the model parameter-y is shown to be equivalent to the exponential parameter in a Freundlich isotherm. The Langmuir parameter and Freundlich exponential parameter are related through the model parameter-y in systems at constant pH and ionic strength.

Introduction

Aqueous phosphate shares with nitrate the distinction of being a major growth limiting nutrient in the global biosphere. Limited availability and universal biologic demand place these nutrients at the foundation of all food chains. In agriculture, phosphate is generally less plentiful than nitrate because of its lower solubility in soils. In the

world's oceans, below 40 degrees latitude, the abundance of both nutrients in the photic zone is vanishingly small, and phytoplanktonic growth is enhanced immediately after the addition of either. However, both sources and sinks of nitrate are biologically mediated. This biologic connection means nitrate availability must increase upon biologic demand over time. Phosphate, on the other hand, has no comparable biologic source. Apatite and iron oxide weathering are its major aquatic supply. Since these processes are mostly inorganic, supply is unresponsive to biotic demand. This economic distinction between nutrient cycles suggests growth in the aquatic biosphere is limited by phosphate over geologic time.

The abundance of phosphate in the hydrosphere appears to be controlled by a combination

---

* From a presentation given at the Third International Workshop on Phosphorus in Sediments, Woudschoten/Utrecht, The Netherlands, September 30, 1991, under the auspices of: International Association of Theoretical and Applied Limnology, Limnological Institute (Royal Netherlands Academy of Arts and Sciences), Institute for Inland Water Management and Waste Water Treatment, and the Netherlands Institute for Sea Research.

of chemical and biological processes. For example, photosynthetic uptake appears to control phosphate levels in oligotrophic lakes and oceanic surface waters (Lean & Nalewajko, 1976; Lean & White, 1983; Broecker & Peng, 1982). And heterotrophically mediated release of phosphate may be a significant process in oceanic bottom waters (ref. Broecker & Peng, 1982). On the other hand, release of phosphate from sediments is due to the chemical reduction of metal hydroxides (for example see: Callender, 1982; Callender & Hammond, 1982). This manuscript concerns regulation of phosphate abundances through solubility reactions and focuses on a chemical model that relates phosphate levels to the solubilities of a solid-solution of ferric phosphate-hydroxide in colloidal suspensions or associated with suspended sediments in rivers with low calcium levels (ref. Fox, 1989).

Phosphorus exists in the surface of the earth almost exclusively in its oxidized form. As a result, the sum of the concentrations of aqueous phosphate species is frequently referred to as 'phosphate' without differentiation from the third dissociation product of phosphoric acid. To avoid confusion, the activities of specific phosphate species will be used throughout this manuscript. Activity coefficients are calculated from the ion pairing model of Millero & Schreiber (1982). The sum of oxidized, inorganic phosphorus species will continue to be called 'phosphate' or further differentiated into dissolved phosphate or suspended phosphate depending on the phases in question.

Model development

Inorganic processes dominate the regulation of phosphate levels in turbid rivers because light penetration is too shallow to support vigorous photosynthesis. In these waters, phosphate concentrations are controlled through an interaction with suspended sediments as demonstrated by the constancy of dissolved phosphate levels when suspensions of river or estuarine sediments are diluted with distilled and/or sea water, (Stephens-

en, 1949; Rochford, 1951; Carritt & Goodgal, 1954; Jitts, 1959; Gessner, 1960; Pomeroy et al., 1965; Burns and Solomon, 1969; Butler and Tibbits, 1972; Wormald and Stirling, 1979; Chase and Sayles, 1980; Fox et al., 1985, 1986). This control appears to be the result of binary exchange between phosphate and amorphous ferric hydroxide in suspension or as surface coatings on suspended sediments (Fox, 1989, 1990, 1991).

Such a reaction is consistent with the strong affinity amorphous iron and aluminum hydroxides are recognized to have for phosphate (for iron see; Swenson et al., 1949; Atkinson et al., 1967, 1972, 1974; Hingston et al., 1967; Yates & Healey, 1975; Parfitt et al., 1975; Breeuwsma & Lyklema, 1973; Lijklema, 1980; Crosby et al., 1983, 1984; for aluminum see; Bache, 1963; Hsu, 1965; 1976; Chen et al., 1975; Rajan, 1975; Bolan et al., 1985). That this reaction occurs in soils and, consequently, in fresh waters is evident from numerous observations of phosphate sorption to soils and sediments rich in these metal phases (for soils see; Russell et al., 1954; Talibudeen, 1958; Nagarajah et al., 1968; Tandon & Kurtz, 1968; Mattingly, 1975; Barrow & Shaw, 1975; Rajan & Fox, 1975; Munns & Fox, 1976; Rajan & Watkinson, 1976; Ryden & Syers, 1977; Taylor & Ellis, 1978; Barrow, 1983; for sediments see; Williams et al., 1967, 1970, 1976; Ku et al., 1978; Shukla et al., 1971; Kuo & Lotse, 1974).

In waters with low calcium concentrations, amorphous ferric hydroxide, as opposed to aluminum hydroxide, appears to comprise the controlling solid phase (Fox, 1989). This is consistent with the fact that (1) phosphate concentrations covary with ferric iron concentrations in the Sepik River (Fig. 3); (2) iron, generally, has a greater affinity for phosphate; (3) ferric salts hydrolyze more readily than aluminum salts (Hsu, 1976); and (4) rivers are near equilibrium with amorphous ferric hydroxide (Fox, 1988b), but not with amorphous aluminum hydroxide. Aluminum values seem closer to kaolinite saturation than amorphous aluminum hydroxide (Beck et al., 1974; Jones et al., 1974).

The binary exchange of ferric hydroxide and ferric phosphate can be expressed as two solubility reactions which share a common ferric ion:

$$\text{Fe(OH)}_n^{3-n}X_{(3-n)}^{-} + \text{PO}_4^{3-}$$

$$= \text{FePO}_4 + n(\text{OH}^-). \qquad (1)$$

Fox (1988a) empirically defines the composition of amorphous ferric hydroxide as $n = 2.35$ over a range of pH values between 1.5 and 7. Counterion (X) provides charge balance but does not seem to effect solubility (Fox, 1988a), perhaps because it is excluded from the polymer chain by ferric-hydroxide polycations (Music $et\ al.$, 1982).

At equilibrium, through the mass action law, equation (1) can be rewritten as

$$\frac{\alpha_1}{\alpha_2^n} = \frac{f_1 X_1 K_1}{f_2 X_2 K_2}, \qquad (2)$$

where α_1 and α_2 represents the activities of the PO$_4$ and OH respectively; X_1, f_1, K_1 and X_2, f_2, K_2 represent the mole fraction, solid activity coefficient, and solubility product of FePO$_4$ and Fe(OH)$_n$, respectively, and α_2, f_2, X_2, K_2 are not equal to zero.

Equation 2 is derived from the mass action law and is independent of any mechanistic consideration. Thus, no knowledge of the underlying process responsible for the formation of the relevant phases is necessary, only that equation 1 is at equilibrium. Constants can be obtained empirically, but the equation itself is strictly derived from chemical theory. In nonideal solutions, the solid activity coefficients are a function of the solid mole ratio. A logarithmic plot of solution activities as a function of the solid mole fractions empirically delineates this relationship. The necessary data was obtained from a series of dialysis experiments (see Appendix 1), the results of which are illustrated in Fig. 1 and formulated in equation 3:

$$\log \alpha_1 - n\log \alpha_2 = y\log \frac{X_1}{X_2} + \log K, \qquad (3)$$

where y is an empirically determined factor that relates the solid activity coefficients to the solid mole fractions. Over the experimental range of mole ratios, $y = 1.8 \pm 0.2$.

Figure 1A represents the averages of over 70 data points which have a total correlation coefficient of 0.91. The range of values plotted is indicative of variations in the elemental Fe/P ratios found in turbid surface waters. The conformance of field data to these values ($i.e.$ equation 3) is used as a measure of the extent to which this model can be applied to a specific body of water. Application to a variety of rivers is illustrated in figures 1–4 and will be discussed presently.

As a consequence of the model, the y-intercept at log X = 0, of Fig. 1, must be equal to the ratio of the logs of the solubility products of ferric phosphate and amorphous ferric hydroxide. This value, as interpolated from a linear regression of data in Fig. 1A, is equal 7.1 ± 0.5. Values from the literature for the negative logarithm of the solubility products of amorphous ferric phosphate and ferric hydroxide are 24.8 (Chang & Jackson, 1957), and 31.7 (Fox, 1988a) respectively, producing a ratio of 6.9 which is statistically equivalent to the empirically derived value. From this, it is evident that a change in slope (y) represents a phase change. This may account for the deviation in data points at Fe/P ratios greater than 80, but they more likely reflect phosphate values which are below our analytical sensitivity.

The relationship between the mole fractions of solute and solvent and corresponding rational activity coefficients in a non-electrolyte solution is essentially a measure of deviation from ideality and is explicitly given by the Gibbs-Duhem equation as follows (cf. Denbigh, 1981):

$$X_1 d\ln f_1 + X_2 d\ln f_2 = 0, \qquad (4)$$

where terms are defined in equation 2.

Since, equation 3 also expresses the solute-solvent relationship in the system specified here, it can be used to obtain an exact solution to equation 4. This exercise provides a direct demonstration of model compliance to solution theory, and its conclusion provides a convenient comparison to the definition of a regular solid solution. A similar discussion is given in Langmuir (1981).

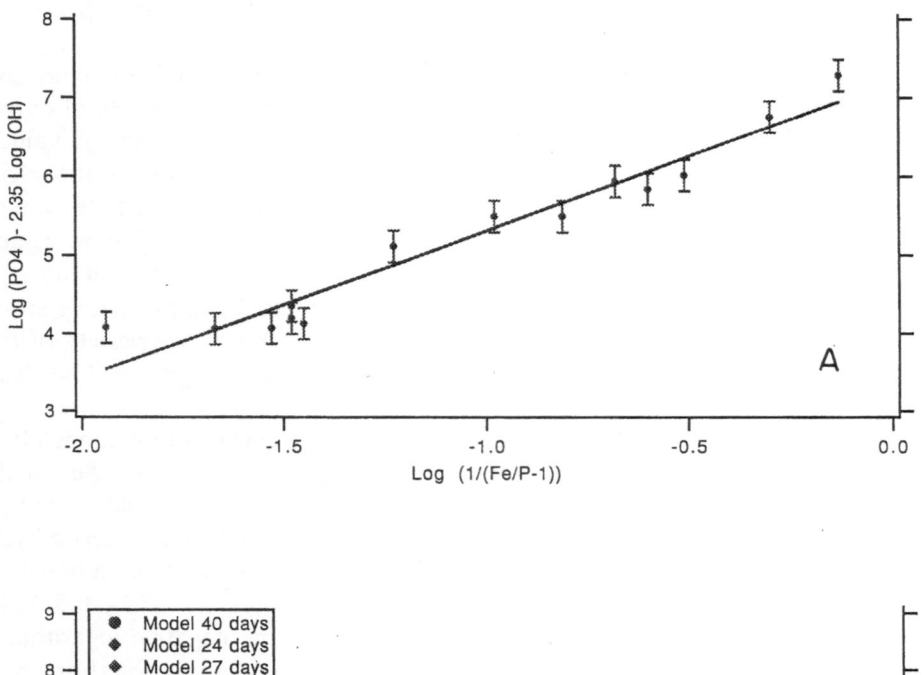

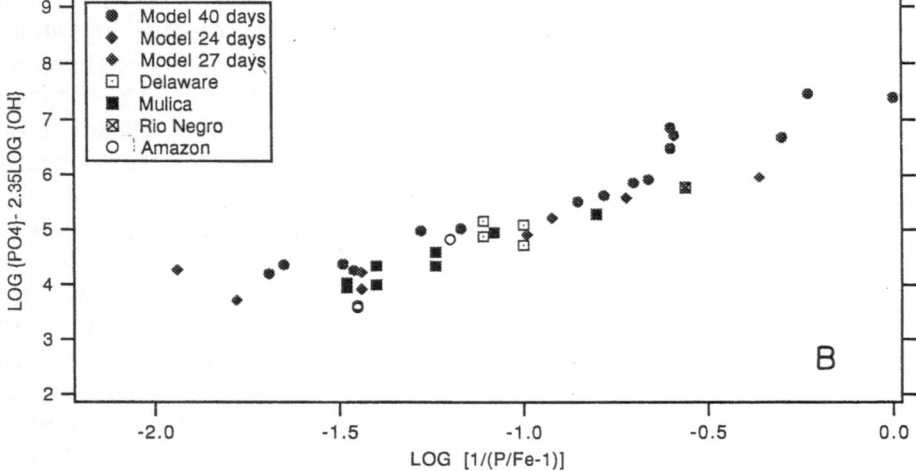

Fig. 1A and B is a plot of data from various dialysis experiments fit to equation 3. Variables plotted on the y-axis represent activities. The x-axis represents solid mole fractions in terms of solid Fe/P ratios. See Fox (1989) for the relationship between mole fraction and elemental ratios. Symbols are defined in the figure. Both model solutions consisting of inorganic ferric hydroxide – phosphate and natural suspensions from river waters are illustrated.

Combining equations 2 and 3 (in natural log form) yields,

$$\ln \frac{f_1}{f_2} = (y - 1) \ln \frac{X_1}{X_2}, \qquad (5)$$

where symbols are the same as in equation 2.

Taking the derivative with respect to X_2 and re-arranging,

$$\frac{d\ln f_1}{dX_2} - \frac{d\ln f_2}{dX_2} = (y - 1)\left[\frac{d\ln X_1}{dX_2} - \frac{d\ln X_2}{dX_2}\right]. \qquad (6)$$

Substituting $X_1 = (1 - X_2)$ into equation 6 and

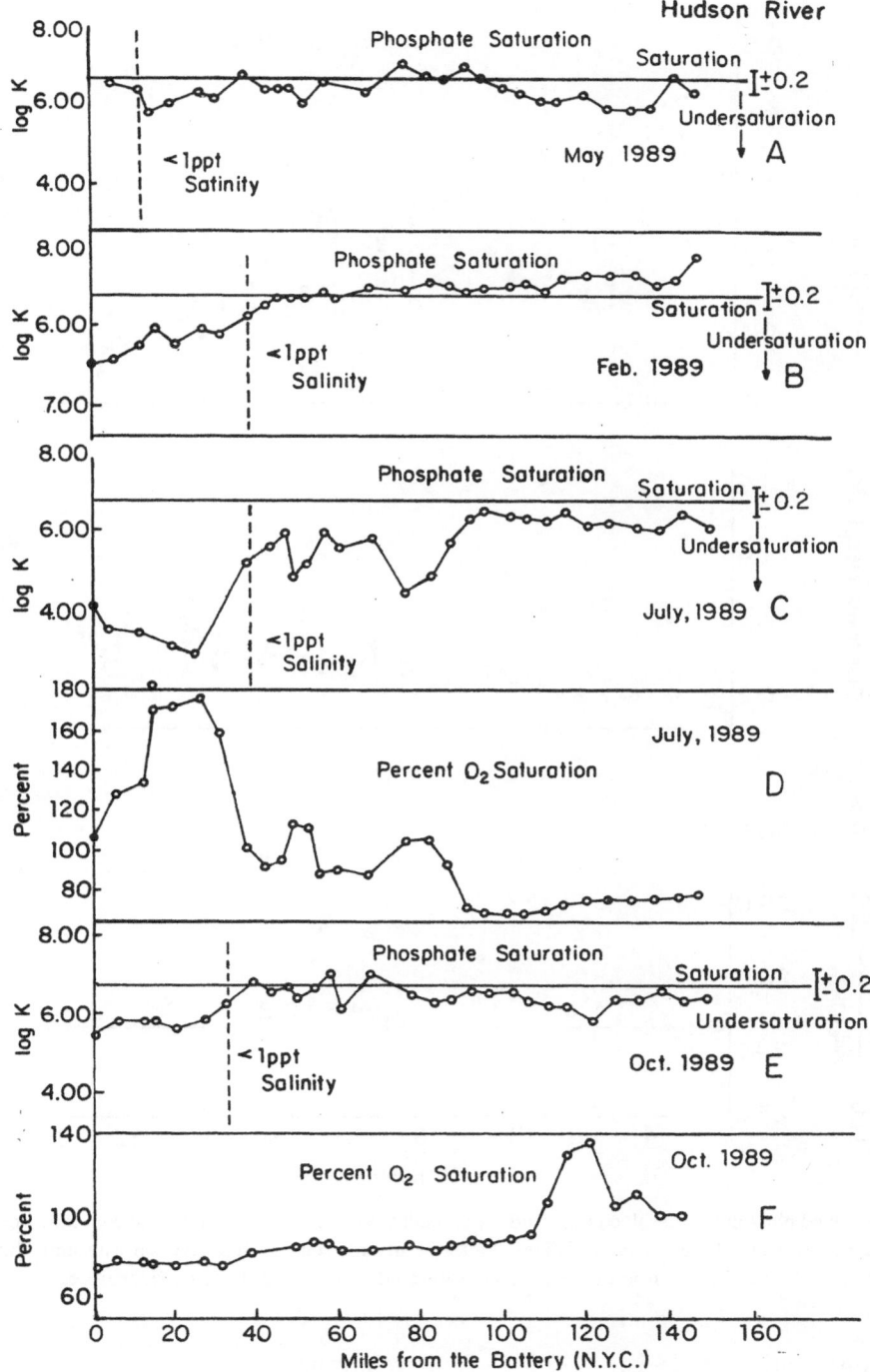

Fig. 2. demonstrates both log K from equation 3 and percent oxygen saturation plotted against distance down the Hudson River at different times of the year. Note the inverse relationship in July (panels C and D).

6

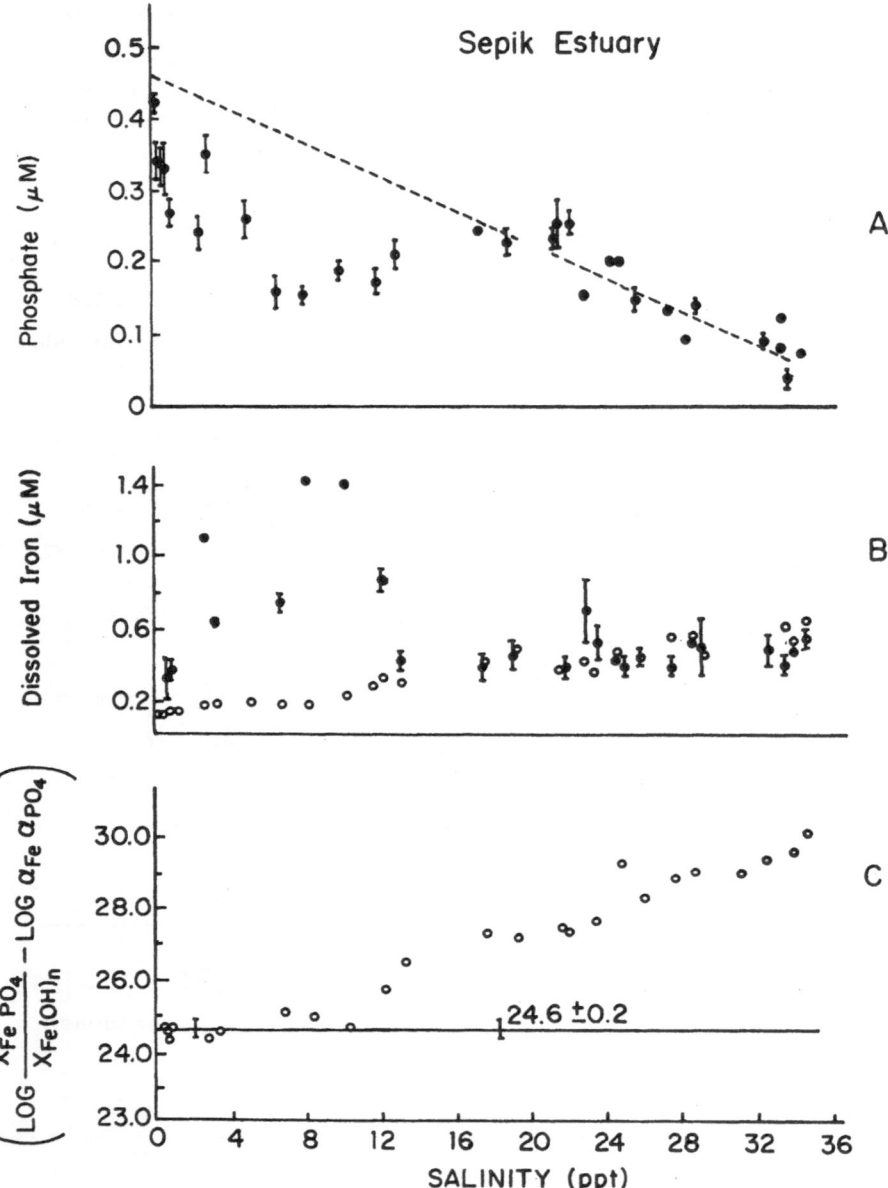

Fig. 3. illustrates dissolved phosphate, dissolved iron, and the solubility constant for the reaction in equation 1 plotted as a function of salinity in the Sepik River and Estuary, New Guinea. Note the inverse relationship between iron and phosphate. Open circles in panel B represent dissolved ferric iron concentrations saturated with respect to ferric hydroxide.

completing the differentiation yields,

$$\frac{d\ln f_1}{dX_2} = (y - 1)\left[\frac{1}{1 - X_2} - \frac{1}{X_2}\right] + \frac{d\ln f_2}{dX_2}. \quad (7)$$

Dividing equation 4 by the derivative of X_2 and

re-arranging yields

$$\frac{d\ln f_1}{dX_2} = -\frac{X_2}{1 - X_2}\frac{d\ln f_2}{dX_2}. \quad (8)$$

Combining equations 7 and 8, re-arranging,

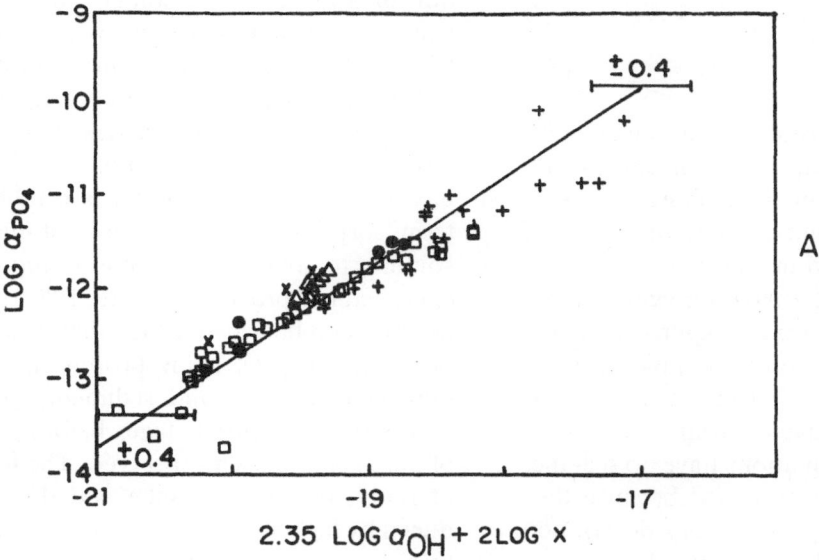

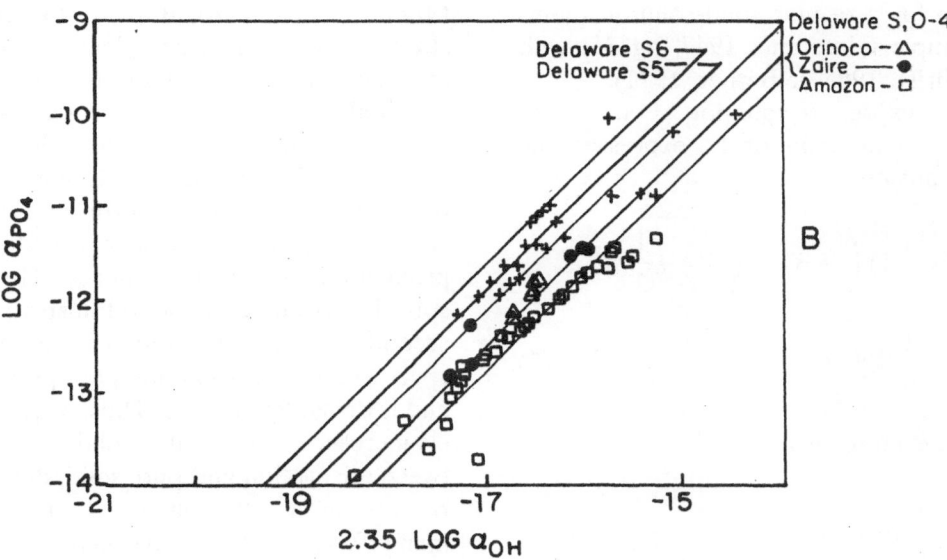

Fig. 4. represents a plot of equation 3 with and without inclusion of the solid mole fractions. Panel B represents river data which was obtained without analyses of solid elemental fractions. Panel A represents the same data after solid Fe/P ratios indicative of each watershed are included. The Sepik River, where both solid and solution parameters were measured is also included in panel A (symbol = X). The sources of the data are given in the text and symbols are defined in the figure.

and integrating over the range of interest,

$$\int_{f_2}^{1} d\ln f_2 = y - 1 \int_{x_2}^{1} \left(\frac{1}{X_2}\right) dX_2 .$$ (9)

Here, as is customary, ideal solution is chosen as the reference state, thus f_2 goes to 1 as X_2 goes to 1 in accordance with Raoult's law.

Completing the integration for f_2 and X_2, then repeating the process for f_1 and X_1 yields

$$\ln f_2 = (y - 1) \ln X_2, \tag{10}$$

$$\ln f_1 = (y - 1) \ln X_1. \tag{11}$$

Equations 10 and 11 represent a unique solution to the Gibbs-Duhem equation and clearly demonstrate that iron and phosphorous behave as a thermodynamic solution over the range of mole fractions illustrated in Fig. 1.

In 1895, Margules suggested a polynomial expansion series could be used to approximate the dependence of vapor pressures on mole fractions in a binary solution, and that the resulting coefficients could be evaluated from the Gibbs-Duhem equation. The equations have no real theoretical bases, but were proposed because they appeared to represent a satisfactory description of experimental data (cf. Glasstone, 1947; Denbigh, 1981). However, an abbreviated form of the Margules expansion series is used as a geochemical definition for a regular solid-solution (compare for example Glasstone, 1947, pg335, with Garrels & Christ, 1965 or Berner, 1971).

Equation 12 expresses equation 11 as an expansion series and equation 13 expresses the Margules expansion:

$$\ln f_1 = (y - 1)\left(-X_2 + \frac{1}{2} X_2^2 - \frac{1}{3} X_2^3 \ldots \right.$$

$$\left. + \frac{1}{n} X_2^n \right), \tag{12}$$

where n is an even integer:

$$\ln f_1 = \alpha_2 X_2 + \frac{1}{2} \beta_2 X_2^2 + \frac{1}{3} \gamma_2 X_2^3 \ldots + \frac{1}{n} k X_2^n. \tag{13}$$

(Denbigh, 1981).

When the Margules polynomials are limited to the second power of X, $\alpha_2 = 0$, and $\beta_2 = \beta$ (Denbigh, 1981). Thus, $(y - 1) = \beta$, and

$$\ln f_1 = \beta (X_2 + X_2^2), \tag{14}$$

demonstrating that the solid ferric-phosphate phase under discussion deviates from the defini-

tion of a regular solid solution by the non-zero value in the first power of X (ref. Garrels and Christ, 1965; Truesdell & Christ, 1968).

Nevertheless, interpretation of equation 1 as a solubility reaction is consistent with the characteristics of phosphate sorption in soils and sediments whenever equilibrium is established. The term 'sorption' refers to the loss of material from solution to solid phases without implying a specific chemical process (Sposito, 1984), and is used here to emphasize the fact that a similarity of product, not process, is proposed. Sorption of phosphate to soils and sediments generally involves two sequential stages before precipitation of pure solid phase is observed. The first consists of rapid, reversible uptake and the second, induced by higher levels of sorbate, is characterized by a slow irreversible process (Carritt & Goodgal, 1954; Kittrick & Jackson, 1956; Kuo & Lotse, 1974; Rajan & Fox, 1975; Barrow & Shaw, 1975; Munns & Fox, 1976; Barrow, 1983a). The first stage involves surface complexation and has been modelled using electrostatic (Barrow, 1983a) and colloidal (cf. Sposito, 1984) principles. The sorption rate of the second stage may depend on solid phase diffusion (Carritt & Goodgal, 1954; Barrow, 1983a; reviewed by Froelich, 1986), which coupled with characteristic irreversibility, implies production of a distinct phase. However, dissolved phosphate in natural suspensions of the second-stage product is under saturated with respect to the corresponding pure phosphate solid and is dependent on pH. Thus, as phosphate levels increase, a semi-stable solid, intermediate between adsorption and pure solid, develops which behaves as a solid-solution. Blanchar & Stearman (1984) use a model based on the solubility of a solid-solution of aluminum phosphate in aluminum hydroxide to approximate phosphate concentrations in soil solutions.

This concept is further supported by studies of sorption to pure substrates and used in the derivation of sorption models. In studies of cation sorption to calcite (Davis et al., 1987; Lorens, 1981) and siderite (Wersin et al., 1989) an intermediate, but stable phase, corresponding to stage 2, is identified as a solid-solution between cation-substrate and substrate. Wersin *et al.* (1989) directly observe a solid-solution of manganese carbonate in ferrous carbonate with electron-spin resonance spectroscopy. Development of a similar stable intermediate at ambient temperatures during the conversion of ferric phosphate from ferric hydroxide can be interpreted from the work of Swenson *et al.* (1949) and Haseman *et al.* (1950). The concept of surface sorption followed by solid-solution formation as part of a continuous progression toward precipitation of a pure phase is used for modelling cation sorption to metal hydroxides (Farley *et al.*, 1985) and calcite (Comans & Middelburg, 1987).

Model comparison to standard isotherms

A substantial body of literature concerned with sorption of phosphate by solid phases report results in the form of adsorption isotherms (see above references for phosphate sorption to metal hydroxide, soil, and sediments). If equation 1 represents chemistry fundamental to these observations, then it must be mathematically consistent with the general form of an adsorption isotherm wherever sorbate is equilibrated with sorbent. Adsorption isotherms are not derived from thermodynamics, but are empirically generated and originally used to describe the observed loss of gas in the presence of various solid surfaces. They are very useful for cataloging the experimental results of sorption to heterogeneous solids, such are found in soils, where stoichiometry and chemical pathways cannot be directly determined. Therefore, discussion of the relationship between equation 2–3 and standard isotherms may be useful.

Equation 1 is now rewritten in a more general form:

$$AB_n + B = AB + nC, \qquad (15)$$

where B and C are ionic activities, AC and AB are solids, and the reaction is equilibrated.

Equation 15 is rewritten in accordance with the mass action law as follows:

$$\frac{\alpha_B}{\alpha_C^n} = \frac{f_{AB} X_{AB} K_{AB}}{f_{AC} X_{AC} K_{AC}}, \qquad (16)$$

where α_x is the activity of x, X_y is the solid molar fraction of y, K_j is the solubility product of compound j, and f_i is the solid activity coefficient of species i:

Let,

$$K' = \frac{f_{AB} K_{AB}}{f_{AC} K_{AC}}. \qquad (17)$$

Combining 16 and 17 and rearranging yields

$$\frac{\alpha_B}{\alpha_C^n} = \frac{X_{AB}}{X_{AC}} K'. \qquad (18)$$

By definition $X_{AB} + X_{AC} = X_{max}$, then

$$\frac{\alpha_B}{\alpha_C^n} = \frac{X_{AB}}{X_{max} - X_{AB}} K', \qquad (19)$$

where X_{max} is the maximum solid mole ratio.

In cases were phosphate is sorbed to metal hydroxides, the anion C represents hydroxide ion. Isotherms from these experiments consist of families of curves, each curve at constant pH. Thus, $\alpha_C^n K' = K^*$. Substitution into equation 19 and rearrangement yields

$$\frac{\alpha_B X_m}{K^* + \alpha_B} = X_{AB}. \qquad (20)$$

Activities are now converted to concentrations using the activity definition, $\alpha_B = k\delta C_B$, where δ is the activity coefficient of species B, k is the dissociation constant for species B, and C is the concentration. To avoid needless complexity, it is assumed that $\alpha_B \gg \alpha_{B1} + \alpha_{B2} \cdots + \alpha_{Bn}$, where subscripts represent other species of B. Dividing by $1/K^*$, combining constants, $k\delta/K^* = K$, and rearranging equation 20 yields

$$X_{AB} = \frac{KCX_{max}}{1 + KC}. \qquad (21)$$

Equation 21 is in the form of the standard Langmuir isotherm. In this example, hydrogen ion activity, solid and solution activity coefficients, dissociation constants, and the solubility products of the pure solids are contained in the Langmuir parameter, K.

Using similar assumptions, the expression for a Freundlich isotherm can also be generated. Rewriting equation 3 in terms of equation 15 yields

$$\ln\alpha_B - n\ln\alpha_C = y(\ln X_{AB} - \ln X_{AC}) + \ln K'' \,, \quad (22)$$

where terms are the same as in equation 16.

When the solid mole fraction, X_{AB}, is very small (i.e. less than 5% of X_{AC}), X_{AC} is near 1 and relatively constant, and pH is constant, then

$$\ln\alpha_B = y(\ln X_{AB} - \ln m) + \ln K'' + n\ln k \,, \quad (23)$$

where y, m, n, k, and K'' are constants.

Activity is converted to concentration by the preceding argument, $\alpha_B = k\delta C_B$, where k is the dissociation constant of B, δ is the activity coefficient of B, C is concentration, and $\alpha_B \gg \alpha_{B1} + \alpha_{B2} \ldots + \alpha_{Bn}$. Combining constants, rearranging, and exponentiating equation 23 yields,

$$X_{AB} = AC_B^\beta \,, \quad (24)$$

where A represents the combined constants m, n, k, K, and activity coefficient δ, and $\beta = 1/y$.

Equation 24 is in the general form of a Freundlich isotherm equation. The hydrogen ion activity, mole fraction of sorbant, dissociation constants, and activity coefficients are combined in the Freundlich parameter A. The exponential parameter β is equal to the reciprocal of the slope of equation 3. Thus, in a well defined chemical system, it is a measure of the deviation from a specified reference state (i.e ideality).

The Freundlich parameter, β, has also been shown to be a measure of the sharpness of a log-normal distribution of Langmuir parameters K. The closer this term is to zero, the broader is the distribution; the closer it is to unity, the more the distribution approaches a single value (Sposito, 1980; 1984). Since the parameter K is a variable with respect to activity coefficients under the conditions imposed on equation 21, it tends to a single value as the solution tends to ideality, provided ionic strength is constant; that is to say, at

constant pH and ionic strength, K approaches a constant value as the ratio of activity coefficients approach 1. Thus, the distributional relationship between the Langmuir parameter K, and Freundlich parameter β is consistent with their common relationship to y.

This can be demonstrated directly by substituting

$$\frac{f_{AB}}{f_{AC}} = \left(\frac{X_{AB}}{X_{AC}}\right)^{y-1} \quad (25)$$

into equation 17. The Langmuir expression now becomes

$$X_{AB} = \frac{K^{\frac{1}{y}}C_B X_M}{1 + K^{\frac{1}{y}}C_B} \,, \quad (26)$$

where symbols are the same as in equation 21. As y approaches unity, K approaches a constant equal to the ratio of the solubility products times dissociation constants at constant ionic strength and pH.

Application to rivers

Conformance of both laboratory and field data to equation 3 is illustrated in figures 1–5 and reported by Fox (1989, 1990, 1991). Figure 1A and B summarizes the results of several experiments using artificially prepared suspensions of iron and phosphate and using samples taken from a variety of rivers. Suspensions of ferric hydroxide – ferric phosphate were prepared to desired elemental Fe/P ratios and dialyzed against deionized water producing the 'model solutions' which are averaged in Fig. 1A. Unfiltered waters taken from the Delaware, Mullica, Amazon, and Negro Rivers were treated similarly and are plotted with selected model solution data in Fig. 1B. In addition, Delaware and Mullica samples were perturbed by deliberate changes of pH and additions of phosphate (Fox, 1989) (see Appendix 1).

The dialysis experiments demonstrate that river waters and model solutions behave similarly and in accordance with equation 3 when incubated for several weeks. However, short residence times and numerous natural perturbations may prevent equilibration in actual rivers. Nevertheless, field measurements from the Hudson River (Fox, 1991) and the Sepik River (Fox, 1990), illustrated in Figs 2 and 3, give a good fit to equation 3 (the activity of ferric ion was used in place of hydroxide in the Sepik but both are equivalent). These results indicate that the proposed mechanism of phosphate control operates over residence times and suspended sediment levels ranging from large tropical to moderate temperate rivers.

A deviation from saturation clearly caused by intense photosynthesis is evident in the Hudson in July as illustrated by the inverse correlation between oxygen content and degree of phosphate saturation in panels E and F of Fig. 2. These observations tend to define the conditions under which inorganic phosphate control no longer dominates. Certainly it is no surprise that the low flows, low suspended sediment levels, and elevated temperatures characteristic of a moderate tidal river in mid summer would support maximum biologic activity and minimum solid phase interaction.

Figure 4 illustrates two plots of field measurements taken from the literature: Amazon (Stallard & Edmond, 1983), Zaire (Van Bennekom et al., 1978), Orinoco (Lopez, 1989), and Delaware Rivers (Culberson et al., 1982). Solid elemental ratios were measured separately from the solution parameters, except for the Sepik. Thus, panel B illustrates the solution parameters as reported. Panel A demonstrates the data in panel B after solid Fe/P ratios indicative of each river were included in accordance with equation 3. Solid ratios were either obtained from separate articles or measured in our lab (Amazon ratio = 30 (Fox, 1989); Zaire ratio = 25 (Sholkovitz et al., 1978); Orinoco ratio = 25; Delaware ratio = 10 (Fox, 1989)). With the addition of solid ratios, data from each river virtually overlaps and defines a line with slope equal 1 and scatter less than 0.4 log units. The variability among the rivers is reduced from a factor of 10 in panel 4B to a factor of 2.5 in 4A.

Figure 5 illustrates both phosphate and calcium

plotted against chloride ion from the Amazon River (Stallard & Edmond, 1983). The high solubility and low biologic affinity for chloride, make it a good tracer of physical mixing. The linear relationship between chloride and calcium and the non-linear relationship between chloride and phosphate indicates that calcium but not phosphate concentrations are controlled by dilution along the length of the Amazon River. This plot also demonstrates that the correlation between phosphate and hydroxide ion activities observed in fig. 4 is not the result of simultaneous dilution or the solubility of a solid calcium-phosphate phase.

The data illustrated in figs 1 – 5 demonstrate a wide range of river environments where phosphate concentrations are predictively related to solid Fe/P ratios by the suppositions implicit in equation 3. The data encompasses the world's three largest rivers (Amazon, Zaire & Orinoco) as well as two that are industrially impacted (Hudson & Delaware). Such a range of environments implies a global significance to the chemical processes described by equation 3 and suggests applicability to environmental issues ranging from regional water-quality management to the global phosphorus cycle.

Acknowledgements

I wish to gratefully acknowledge financial support for this work provided to Harvard University by the Cousteau Society/Foundation Cousteau, NASA grant NAG-1-55, NSF Grant DEB-79-20282, EPA grant R810219-01-0, Hudson River Foundation grant 015-88A-039.

Appendix 1: materials and methods

1. Experimental

Unfiltered river waters from the Amazon, Negro, Delaware, and Mullica rivers and model solutions consisting of various mole fractions of iron hydroxide and ferric phosphate were dialyzed

12

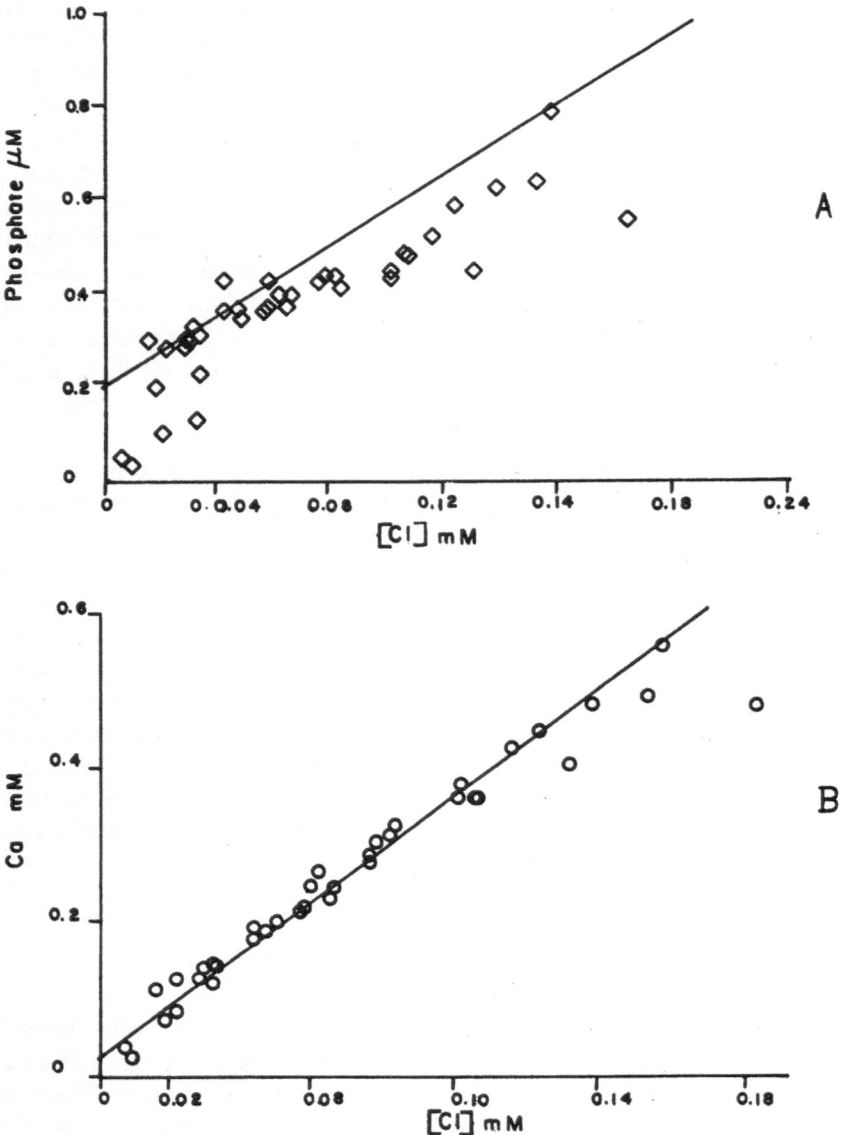

Fig. 5. illustrates mixing diagrams for phosphate and calcium in the Amazon River. Deviation from linearity with chloride ion indicates phosphate concentrations are not the result of mixing, while calcium concentrations probably are. Clearly phosphate is controlled by some process not related to calcium or dilution. Source of the data is Stallard & Edmond (1983).

against deionized water. The amorphous ferric hydroxide suspensions were prepared by hydrolysis of ferric chloride at room temperature. Suspensions of ferric hydroxide-ferric phosphate were prepared by mixing molar concentrations of ferric hydroxide with potassium phosphate at desired Fe/P molar ratios. The suspensions were

dialyzed over a period of 1 to 2 days against large volumes of deionized water in order to remove excess counter ions. Dialysis was repeated 3 times. The river suspension were treated with sodium azide to reduce biologic activity. All of the suspensions were subsequently dialyzed against deionized water. Analyses of phosphate, pH (alkalinity in sea water dialyzate), ferric and ferrous iron, and calcium were done on samples from the dialyzate on a semi-weekly to weekly basis until the solutions appeared to be equilibrated. Samples from the suspensions were taken for microprobe analysis at the beginning and end of all experiments.

Field sampling was done on a variety of rivers and estuaries. For details refer to Fox (1989, 1990, 1991).

2. Analytical

All dialysis was done with pre-washed dialysis tubing with 1000 molecular weight exclusion. Hydrogen ion activity was measured potentiometrically using a Beckman glass combination electrode with ceramic reference junction fitted to a Beckman Omega 71 pH meter. Calibration consisted of measuring standard buffer solutions of pH 4 and 7 prior to each sampling. Field measurements of pH were done immediately after sample collection.

Aqueous iron was determined by atomic adsorption spectroscopy using standard methods and a graphite furnace. Ferrous iron was determined spectrophotometrically using ferrozine (Stookey, 1970). Ferric iron is defined as the difference between total iron and ferrous iron. Dissolved iron was defined as the total aqueous iron concentration after filtration with a tangential flow filtration apparatus using 10 000 molecular weight exclusion membranes and/or the aqueous iron which past a 1000 molecular weight exclusion dialysis membrane. Ferric iron species were calculated using dissociation constants published in Stumm & Morgan (1981). The possibility was considered that monomeric, dimeric, and trimeric iron hydroxides and small organo-ferric complexes could cross the exclusion membranes. Published dissociation constants for ferric hydroxide complexes imply only monomeric species exist in significant concentrations at the range of pH values considered here. Likewise, organoferric complexes would only equal ferric hydroxide activities at pH values below 6. No special precautions to reduce CO_2 concentrations in test solutions were taken and possible ferric carbonate complexes were not considered. Results suggest that no significant levels of unidentified ferric complexes exist at pH values between 5.5 and 8.3.

Aqueous phosphorus was determined spectrophotometrically by formation of a molybdate-phosphate complex (Strickland & Parsons, 1972). Ambient phosphate concentrations in the waters studied here are well above levels where uncertainties in the molybdate method are significant (Lean & White, 1983). Dissolved phosphorus was defined, for field samples, as the total aqueous phosphorus that passes a 0.2 μm Nuclepore filter. Colloidal iron is not sufficiently separated from dissolved iron when filtered with 0.2 μm filters. But colloidal phosphate levels are 10 to 50 times lower than iron. Thus, the fraction of colloidal phosphate that slips through an 0.2 μm filter comprises an insignificant portion of the dissolved phosphate fraction.

Magnesium and calcium concentrations were determined by atomic adsorption spectroscopy using standard methods. Dissolved cations were separated form particulate by filtration through 0.2 μm Nuclepore filters in the field or dialysis in the laboratory. Alkalinity was determined by potentiometric titration with standardized HCl. Carbonate ion concentrations were calculated from the alkalinity titration results using dissociation constants given in Stumm & Morgan (1981). Activity coefficients for all dissolved species were calculated using the ion pairing model of Millero & Schreiber (1982).

Suspended solids were analyzed for iron and phosphorus with a Comeca MBX electron microprobe equipped with Tracor Northern TN5502/1310 automation using wavelength dispersive analysis. The instrument demonstrates sensitivity of 10^{-5} weight % for the corresponding oxide standards. Since only surficial amorphous material is sought and the probe's depth of measurement is about 1 μm, care was taken to focus the probe between particles where previous studies using scanning electron-microscopy show filter residues which consist of a rather uniform coating of colloidal sized material. Three to 7 replicate analysis of each sample were done. Each replicate was obtained at a different part of a sample filter so that a measure of sample uniformity over the filter surface is obtained. Replicates rarely exceeded a standard error of 20%. Blank 0.2 μm Nuclepore filters were determined to contain negligible levels of iron and phosphorus. Variations in salinity and sample volume had no effect on elemental values, and there was no significant difference between solids collected on Nuclepore or ultrafilter membranes. Samples from the field were collected on Nuclepore filters and/or ultrafilter membranes. Samples from dialysis experiments were collected either by filtering the suspensions from within the dialysis tubes through 0.2 μm Nuclepore filters or by transferring a samples of the suspension to the Nuclepore with a Q-tip.

References

Atkinson, R. J., A. M. Posner & J. P. Quirk, 1967. Adsorption of potential-determining ions at the ferric oxide-aqueous electrolyte interface. J. Phys. Chem. 71: 550–558.

Atkinson, R. J., A. M. Posner & J. P. Quirk, 1972. Kinetics of isotopic exchange of phosphate at the α-FeOOH – aqueous solution interface. J. Inorg. Nucl. Chem. 34: 2201–2211.

Atkinson, R. J., R. L. Parfitt & R. St. C. Smart, 1974. Infrared study of phosphate adsorption on goethite. J. Chem. Soc. Faraday Trans. 170: 1472–1479.

Bache, B. W., 1963. Aluminum and iron phosphate studies relating to soils. I. Solution and hydrolysis of variscite and strengite. J. Soil Sci. 14: 113–123.

Barrow, N. J. & T. C. Shaw, 1975. The slow reaction between soil and anions: 2. Effect of time and temperature on the decrease in phosphate concentration in the soil solution. Soil Sci. 119: 167–177.

Barrow, N. J., 1983. A mechanistic model for describing the sorption and desorption of phosphate by soil. J. Soil Sci. 34: 733–750.

Barrow, N. J., 1983. On the reversibility of phosphate sorption by soils. J. Soil Sci. 34: 751–758.

Beck, K. C., J. H. Rueter & E. M. Purdue, 1974. Organic and inorganic geochemistry of some coastal plain rivers of the southeastern United States. Geochim. Cosmoschim. Acta 38: 341–364.

Berner, R. A., 1971. Principles of Chemical Sedimentology. McGraw-Hill Co. 240 pp.

Blanchard, R. W. & G. K. Stearman, 1984. Ion products and solid-phase activity to describe phosphate sorption by soils. Soil Sci. Soc. Amer. J. 48: 1253–1258.

Bolan, N. S., N. J. Barrow & A. M. Posner, 1985. Describing the effect of time on sorption of phosphate by iron and aluminum hydroxides. J. Soil Sci. 36: 187–197.

Breeuwsma, A. & J. Lyklema, 1973. Physical and chemical adsorption of ions in the electrical double layers on hematite (α-Fe$_2$O$_3$). J. Colloid. Interface Sci. 44: 437–448.

Broecker, W. S. & T. -H. Peng, 1982. Tracers in the sea. Eldigio Press, New York, 690 pp.

Burns, P. A. & N. Solomon, 1969. Phosphate adsorption by Kaolin in saline environments. Proc. Nat. Shellfish Assn. 59: 121–125.

Butler, E. I. & S. Tibbitts, 1972. Chemical survey of the Tamar estuary. J. Mar. Biol. Assn. U.K. 52: 681–699.

Callender, E. & D. E. Hammond, 1982. Nutrient exchange across the sediment-water interface in the Potomac River estuary. Estuar. coast. Shelf Sci. 15: 395–413.

Callender, E., 1982. Benthic phosphorus regeneration in the Potomac river estuary. In P. G. Sly (ed.), Sediment/Freshwater Interaction. Developments in Hydrobiology 9. Dr W. Junk Publishers, The Hague: 431–446. Reprinted from Hydrobiologia 91/92.

Chang, S. C. & M. L. Jackson, 1957. Solubility product of iron phosphate. Proc. Soil Sci. Soc. Amer. 21: 265–269.

Carritt, D. E. & S. Goodgal, 1954. Sorption reactions and some ecological implications. Deep Sea Res. 1: 224–243.

Chase, E. M. & F. L. Sayles, 1980. Phosphorus in suspended sediments of the Amazon River. Estuar. Coast. Mar. Sci. 11: 383–391.

Chen, Y. S. R., J. N. Butler & W. Stumm, 1975. Kinetic study of phosphate reactions with aluminum oxide and Kaolinite. Envir. Sci. Technol. 7: 327–332.

Coman, R. N. J. & J. J. Middleburg, 1987. Sorption of trace metal on calcite: Applicability of the surface precipitation model. Geochim. Cosmochim. Acta 51: 2581–2591.

Crosby, S. A., D. R. Glasson, A. H. Cuttler, I. Butler, M. Turner, M. Whitfield & G. E. Millward, 1983. Surface areas and porosities of Fe(III)- and Fe(II)-derived oxyhydroxides. Envir. Sci. Technol. 17: 709–713.

Crosby, S. A., G. E. Millward, E. I. Butler, D. R. Turner & M. Whitfield, 1984. Kinetics of phosphate adsorption by iron oxyhydroxides in aqueous systems. Estuar. Coast. Shelf Sci. 19: 257–270.

Culberson, C. H., J. H. Sharp, T. M. Church & B. W. Lee, 1982. Data from the Salsx Cruises May 1978–July 1980: University of Delaware Oceanographic Data Report Number 2. University of Delaware, Newark, DE., 53 pp.

Davis, J. A., C. C. Fuller & A. K. Cook, 1987. A model for trace metal sorption processes at the calcite surface: Adsorption of Cd^{+2} and subsequent solid solution formation. Geochim. Cosmochim. Acta 51: 1477–1490.

Denbigh, K., 1981. The Principles of Chemical Equilibrium. Cambridge University Press, New York, 494 pp.

Farley, K. J., D. A. Dzombak & F. M. M. Morel, 1985. A

surface precipitation model for the sorption of cations on metal oxides. J. Colloid. Interf. Sci. 106: 226–242.

Fox, L. E., S. L. Sager & S. C. Wofsy, 1985. Factors controlling the concentrations of soluble phosphorus in the Mississippi estuary. Limnol. Oceanogr. 30: 826–832.

Fox, L. E., S. F. Sager & S. C. Wofsy, 1986. The chemical control of soluble phosphorus in the Amazon estuary. Geochim. Cosmochim. Acta 50: 783–794.

Fox, L. E., 1988. Solubility of colloidal ferric hydroxide. Nature 333: 442–444.

Fox, L. E., 1988. The solubility of colloidal ferric hydroxide and its relevance to iron concentrations in river water. Geochim. Cosmochim. Acta 52: 771–777.

Fox, L. E., 1989. A model for inorganic control of phosphate concentrations in river waters. Geochim. Cosmochim. Acta 53: 417–428.

Fox, L. E., 1990. Geochemistry of dissolved phosphate in the Sepik River and Estuary, Papua, New Guinea. Geochim. Cosmochim. Acta 54: 1019–1024.

Fox, L. E., 1991. Phosphorus chemistry in the tidal Hudson River. Geochim. Cosmochim. Acta 55: 1529–1538.

Froelich, P. N., 1986. Kinetic control of dissolved phosphate in natural rivers and estuaries: a primer on the phosphate buffer mechanism. In S. Nixon (ed.), Symposium on comparative ecology of freshwater and coastal marine ecosystems. Limnol. Oceanogr. (special edition).

Garrels, R. M. & C. L. Christ, 1965. Solutions, Minerals, and Equilibria. Freeman, Cooper, & Co., San Francisco, 450 pp.

Gessner, F., 1960. Untersuchungen uber den phosphathaushalt des Amazonas. Int. Revue ges. Hydrobiol. 45: 339–345.

Glasstone, S., 1947. Thermodynamics for Chemists. D. Van Nostrand Co., Inc., New York, 524 pp.

Haseman, J. F., E. H. Brown & C. D. Whitt, 1950. Some reactions of phosphate with clays and hydrous oxides of iron and aluminum. Soil Sci. 70: 257–271.

Hingston, F. J., R. J. Atkinson, A. M. Posner & J. P. Quirk, 1967. Specific adsorption of anions. Nature 215: 1459–1461.

Hsu, P. H., 1965. Fixation of phosphate by aluminum and iron in acidic soils. Soil Sci. 99: 398–402.

Hsu, P. H., 1976. Comparison of iron and aluminum in precipitation of phosphate from solution. Water Resources 10: 903–907.

Jitts, H. R., 1959. The adsorption of phosphate by estuarine bottom deposits. Aust. J. Mar. Freshwat. Res. 10: 7–21.

Jones, B. F., V. C. Kennedy & G. W. Zellweger, 1974. Comparison of observed and calculated concentrations of dissolved Al and Fe in stream water. Wat Resour. Res. 10: 790–793.

Kittrick, J. A. & M. L. Jackson, 1956. Electron-microscope observations of the reaction of phosphate with minerals, leading to a unified theory of phosphate fixation in soils. J. Soil Sci. 7: 81–89.

Ku, W. C., F. A. DiGiano & T. H. Feng, 1978. Factors affecting phosphorus adsorption equilibria in lake sediment. Wat. Res. 12: 1069–1074.

Kuo, S. & E. G. Lotse, 1974. Kinetics of phosphate adsorption and desorption by lake sediments. Soil Sci. Soc. Amer. Proc. 38: 50–54.

Langmuir, D., 1981. The power exchange function: a general model for metal adsorption onto geological materials. In D. H. Twari (ed.), Adsorption From Aqueous Solution. Plenum Press, 1–18.

Lean, D. R. S. & C. Nalewajko, 1976. Phosphate exchange and organic phosphorus excretion by freshwater algae. J. Fish. Res. Bd Can. 33: 1312–1323.

Lean, D. R. S. & E. White, 1983. Chemical and radio tracer measurements of phosphate uptake by lake plankton. Can. J. Fish. aquat. Sci. 40: 147–155.

Lijklema, L. 1980. Interaction of orthophosphate with iron (III) and aluminum hydroxide. Envir. Sci. Technol. 14: 537–541.

Lopez, J. C. S., 1989. Condiciones hidrogeoquimicas de la region estuarina-deltaica del Orinoco durante el mes de Noviembre de 1985. Ph.D. Thesis, Universidad de Oriente, Instituto Oceanografico de Venezuela, Cumana, Venezuela.

Lorens, R. B., 1981. Sr, Cd, Mn, and Co distribution coefficients in calcite as a function of calcite precipitation rate. Geochim. Cosmochim. Acta 45: 553–561.

Mattingly, G. E. G., 1975. Labile phosphate in soils. Soil Sci. 119: 369–375.

Millero, F. J. & D. R. Schreiber, 1982. Use of the ion pairing model to estimate activity coefficients of the ionic components of natural waters. Amer. J. Sci. 282: 1508–1540.

Munns, D. N. & R. L. Fox, 1976. The slow reaction which continues after phosphate adsorption: kinetics and equilibrium in some tropical soils. Soil Sci. Soc. Amer. J. 40: 46–51.

Music, S., A. Vertes, G. W. Simmons, I. Czako-Nagy, H. Leidheiser, 1982. Spectroscopic study of the formation of Fe(III) oxyhydroxides and oxides by hydrolysis of aqueous Fe(III) salt solutions. J. Colloid Interface Sci. 85: 256–266.

Nagarajah, S., A. M. Posner & J. P. Quirk, 1968. Desorption of phosphate from kaolinite by citrate and bicarbonate. Soil Sci. Soc. Amer. Proc. 32: 507–510.

Parfitt, R. L., R. J. Atkinson & R. St. C. Smart, 1975. The mechanisms of phosphate fixation by iron oxides. Soil Sci. Soc. Amer. Proc. 39: 837–841.

Pomeroy, L. R., E. E. Smith & C. M. Grant, 1965. The exchange of phosphate between estuarine water and sediments. Limnol. Oceanogr. 10: 167–172.

Rajan, S. S. S. & R. L. Fox, 1975. Phosphate adsorption by soils. II. Reactions in tropical soils. Soil Sci. Soc. Amer. Proc. 39: 846.

Rajan, S. S. S. & J. H. Watkinson, 1976. Adsorption of selenite and phosphate on an allophane clay. Soil Sci. Soc. Amer. J. 40: 51–54.

Rajan, S. S. S., 1975. Adsorption of divalent phosphate on hydrous aluminum oxide. Nature 253: 434–436.

16

Rochford, D. J., 1951. Studies in Australian estuarine hydrology. I. Introduction and comparative features. Aust. J. mar. Freshwat. Res. 2: 1–116.

Russell, R. S., J. B. Rickson & S. N. Andrew, 1954. Isotopic equilibria between phosphates in soil and their significance in the assessment of fertility by tracer methods. J. Soil Sci. 5: 85–105.

Ryden, J. C. & J. K. Syers, 1977. Origin of the labile phosphate pool in soils. Soil Sci. 123: 353–361.

Sholkovitz, E. R., R. Van Grieken & D. Eisma, 1978. The major-element composition of suspended matter in the Zaire River and Estuary. Neth. J. Sea Res. 12: 407–413.

Shukla, S. S., J. K. Syers, J. D. H. Williams, D. E. Armstrong & R. F. Harris, 1971. Sorption of inorganic phosphate by lake sediments. Soil Sci. Soc. Amer. Proc. 35: 244–249.

Sposito, G., 1980. Derivation of the Freundlich equation for ion exchange reactions in soils. Soil Sci. Soc. Am. J. 44: 652–654.

Sposito, G., 1984. The Surface Chemistry of Soils. Oxford University Press, New York, 231 pp.

Stallard, R. F. & J. M. Edmond, 1983. Geochemistry of the Amazon. 2. The influence of geology and weathering environments on the dissolved load. J. Geophys. Res. 88: 9671–9688.

Stephensen, W., 1949. Certain effects of agitation upon the release of phosphate from mud. J. Mar. Biol. Assn. 28: 371–380.

Stookey, L. L., 1970. Ferrozine – New spectrophotometric reagent for iron. Analyt Chem. 42: 779.

Strickland, J. D. H. & T. R. Parsons, 1972. A practical handbook of seawater analysis. Fish. Res. Bd Can. 310 pp.

Stumm, W. & J. J. Morgan, 1981. Aquatic Chemistry: An introduction emphasizing chemical equilibrium in natural waters. J. Wiley & Sons, New York, 780 pp.

Swenson, R. M., C. V. Cole & D. H. Sieling, 1949. Fixation of phosphate by iron and aluminum and inorganic ions. Soil Sci. 67: 3–22.

Talibudeen, O., 1958. Isotopically exchangeable phosporus in soils. III. The fractionation of soil phosphorus. J. Soil Sci. 9: 120–129.

Tandon, H. L. S. & L. T. Kurtz, 1968. Isotopic exchange characteristics of aluminum – and iron – bound fractions of soil phosphorus. Soil Sci. Soc. Amer. Proc. 32: 799–802.

Taylor, R. W. & B. G. Ellis, 1978. A mechanism of phosphate adsorption on soil and anion exchange resin surface. Soil Sci. Soc. Amer. J. 42: 432–436.

Truesdell, A. H. & C. L. Christ, 1968. Cation exchange in clays interpreted by regular solution theory. Amer. J. Sci. 266: 402–412.

Van Bennekom, A. J., G. W. Berger, W. Helder & R. T. P-. DeVries, 1978. Nutrient distribution in the Zaire Estuary and River plume. Neth. J. Sea Res. 12: 296–323.

Wersin, P., L. Charlet, R. Karthein & W. Stumm, 1989. From adsorption to precipitation: Sorption of Mn or $FeCO_3$. Geochim. Cosmochim. Acta 53: 2787–2796.

Williams, J. D. H., J. K. Syers & T. W. Walker, 1967. Fractionation of soil inorganic phosphate by a modification of Chang and Jackson's procedure. Soil Sci. Soc. Amer. Proc. 31: 736–739.

Williams, J. D. H., J. K. Syers & R. F. Harris, 1970. Adsorption and desorption of inorganic phosphorus by lake sediments in a 0.1 M NaCl system. Envir. Sci. Technol. 4: 517–519.

Williams, J. D. H., J. M. Jaquet & R. L. Thomas, 1976. Forms of phosphorus in the surficial sediments of Lake Erie. J. Fish Res. Bd Can. 33: 413–429.

Wormald, A. P. & H. P. Stirling, 1979. A preliminary investigation of nutrient enrichment in experimental sand columns and its effect on tropical intertidal bacteria and meiofauna. Estuar. Coast. Mar. Sci. 8: 441–453.

Yates, D. E. & D. W. Healey, 1975. Mechanism of anion adsorption at the ferric and chromic oxide/water interfaces. J. Colloid Interface Sci. 52: 221–228.

Hydrobiologia **253**: 17–29, 1993.
P.C.M. Boers, T.E. Cappenberg & W. van Raaphorst (eds),
Proceedings of the Third International Workshop on Phosphorus in Sediments.
© 1993 *Kluwer Academic Publishers.*

Geochemical composition, phosphorus speciation and mass transport of fine-grained sediment in two Lake Erie tributaries

M. Stone & M. C. English
Wilfrid Laurier University, Department of Geography, Waterloo, Ontario, N2L 3C5 Canada

Key words: phosphorus fractionation, grain size, geochemistry, particulate phosphorus transport, tributaries

Abstract

The concentration of major elements (Si, Al, Ca, Mg, Na, K, Fe, Ti, Mn and P), particulate phosphorus forms (NH_4Cl-RP, BD-RP, NaOH-RP, HCl-RP and $NaOH_{(85)}$-RP) and carbon content were determined in six size fractions (<8, 8–12, 12–19, 19–31, 31–42 and 42–$<60\ \mu m$) of sediment collected at gauging stations located in two Lake Erie tributaries (Big Creek and Big Otter Creek). Concentrations of major elements and phosphorus forms were remarkably similar in sediment size fractions from both rivers. Nonapatite inorganic P (NAIP) and organic P (OP) concentrations increased with decreasing grain size while apatite inorganic P (AIP) content decreased with decreasing grain size. Results of phosphorus fractionation studies were combined with historical (particle size) and hydrometric data to simulate the export of particle P on tributary sediment $<63\ \mu m$. AIP represents 67 and 70% of the calculated particulate P mass while NAIP accounts for 26 and 23% of sediment-bound P transported in Big Otter Creek and Big Creek, respectively. The $<8\ \mu m$ size fraction of tributary sediment is the most significant for the potential release of bioavailable P into the water column.

Introduction

Sediment plays a major role in the transport of phosphorus (P) in fluvial systems. In rivers where sediment transport is high, a significant proportion of the total P load can be associated with the sediment phase (LEWMS, 1975; Logan *et al.*, 1979; Logan, 1987). Several rivers in southern Ontario drain predominantly agricultural land and the majority of suspended sediment mass transported to the Great Lakes by these rivers is $<63\ \mu m$ (Armstrong *et al.*, 1979; Stone & Saunderson, 1992). Once transported into lakes, this potentially nutrient enriched fine-grained sediment can remain in suspension for extended pe-

riods of time (Armstrong *et al.*, 1979; Depinto *et al.*, 1981; Lick, 1982) and influence lake productivity by allowing maximum biological assimilation of P in suspended sediment (Thomas & Munawar, 1985; Heath & Francko, 1988; Cuker *et al.*, 1990).

The availability of particulate P on tributary sediment for lake phytoplankton depends on several physical, chemical and biological factors reported by Depinto *et al.* (1981). Among these factors, particle size, phosphorus speciation and the geochemical characteristics of the particles can influence the rate of P release from particulate matter into the water column. Although fine-grained sediment is generally considered to be the most geochemically active and important in nutrient transport (Allan, 1979, 1986; Peart & Walling, 1982; Stone & Mudroch, 1989), little information concerning the geochemistry and phosphorus fractionation by particle-size class of fine-grained sediment in Great Lake tributaries is currently available. Such information is necessary to examine the physical and chemical characteristics of external P loading to Lake Erie by fine-grained tributary sediment and to provide estimates of the potential biological availability of this particulate P source.

In this paper, chemically defined fractions of particulate tributary phosphorus in six particle size fractions ($<63 \mu m$) are examined. The geochemistry and carbon content of individual size fractions were determined to correlate elemental composition with particulate P forms and grain size. Historical sediment/particle size and hydrometric data are used in combination with P speciation data to provide a first order assessment of the export of chemically defined fractions of particulate P on sediment $<63 \mu m$ to Lake Erie during high magnitude discharge events.

Materials and methods

Study sites chosen for this work are located at two Water Survey of Canada hydrometric gauging stations in southwestern Ontario; Station No. 02GC026 on Big Otter Creek near Calton and Station No. 02GC007 on Big Creek near Walsingham (Fig. 1). The drainage basin area above each gauging station is 591 km^2 for Big Creek and 676 km^2 for Big Otter Creek. Both rivers contribute significant sediment loads to Lake Erie (Dickinson et al., 1975), drain predominantly agricultural land and have long term sediment and hydrometric records that reflect a wide range of discharge regimes, including base and stormflow.

Sediment samples were collected once during the fall of 1990 and spring of 1991 along three transects at each of the above gauging stations. Sediment collected consisted primarily of river bottom sediment because of the large mass of

sediment required for this study. Particulate P transport in fluvial systems is a complex deposition/resuspension phenomenon (Verhoff et al., 1982; Brownlee & Bird, 1988) and bed sediment at the study sites is not stable, continually moving into suspension with increasing discharge then being deposited downstream. Suspended sediment was also collected but accounted for $<5\%$ of the mass of the bulk sample. Bed and suspended sediment were sampled at four equidistant locations across each transect with a scoop sampler and an ISCO Model 2700 Sampler, respectively. Approximately 500 g of sediment from each location was combined to form a representative composite of the fluvial sediment at each transect. These composite samples are considered to approximate the physical and chemical characteristics of fluvial sediment at both gauging stations. Composite sediment samples were kept at 4 °C then wet sieved through a 64 μm sieve. Particles $<64 \mu m$ were separated mechanically into six different size fractions (<8, 8–12, 12–19, 19–31, 31–42 and 42–$<60 \mu m$) with a Warman Cyclosizer (Warman International, Ltd., Artarmon, Sydney, Australia). The median diameters of these fractions (2, 10, 15, 24, 38, 45 μm) were determined on a Sedigraph at the National Water Research Institute (NWRI) Sedimentology Laboratory, Burlington, Ontario. For the remainder of the paper, median diameters rather than size ranges will be used in tables, figures and for discussion of the results. After size fractionation, all sediment fractions were immediately freeze dried and stored for P fractionation and geochemical analyses.

Sediment samples were analyzed for major elements (Si, Al, Fe, Mg, Ca, Na, K, Ti, Mn, total P) by X-ray fluorescence (Mudroch, 1985) and the results are reported as percent dry weight. Accuracy of the analyses was checked by running Canadian Reference Standards AGV-1, MRG-1, NCM-N, GSP-1, and SY-3 and comparing the analytical results with the stated reference values for major elements. Total and organic carbon were determined with a Leco Carbon Analyzer and carbon results are accurate to 1%.

This study attempts to correlate particulate P

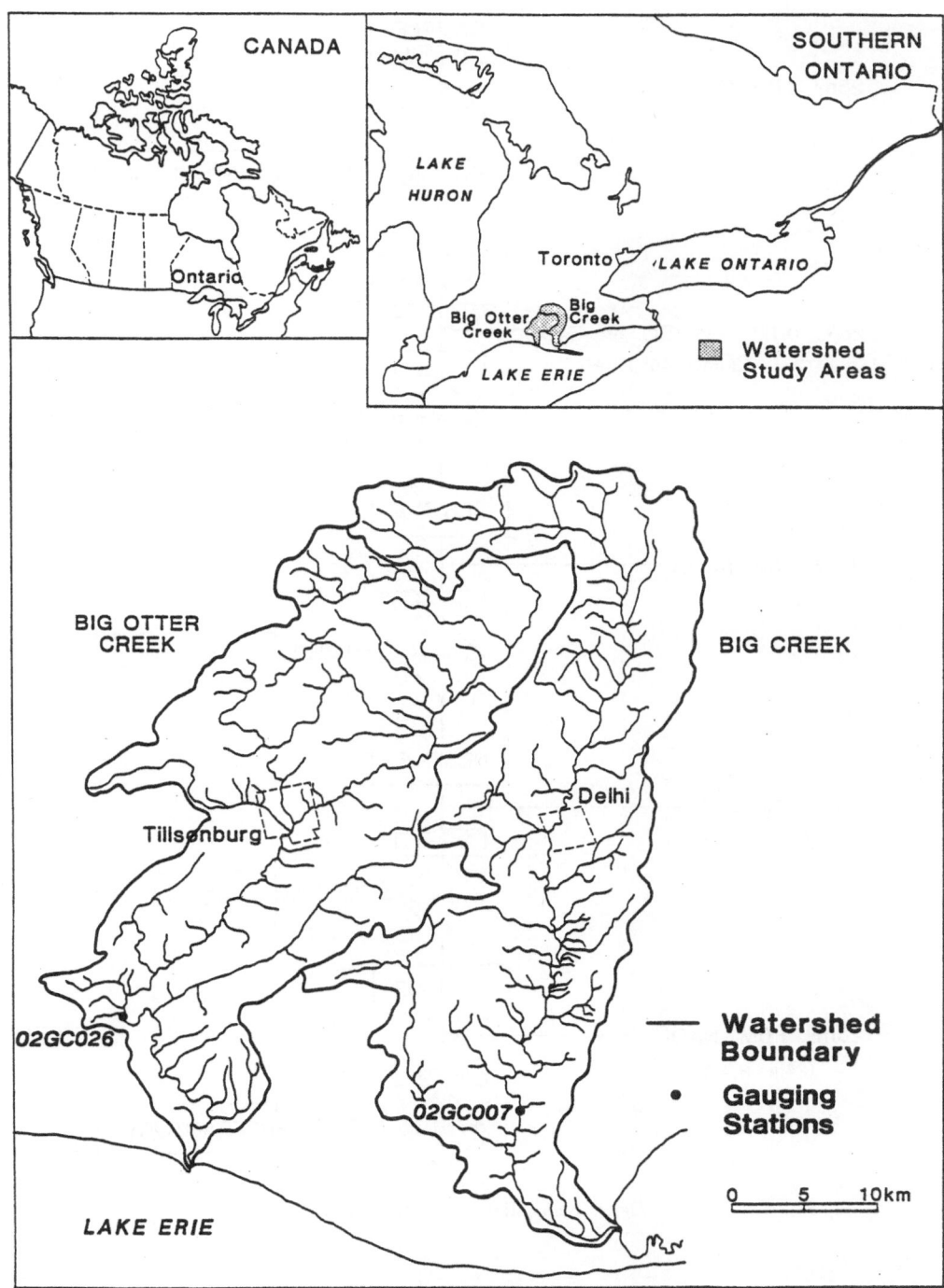

Fig. 1. Study area.

forms with other sediment characteristics such as geochemistry and grain size. The modified Psenner sequential extraction scheme reported by Petterson & Istvanovics (1988) was used to provide both quantitative and qualitative assessments of the relative fractional composition of particu-

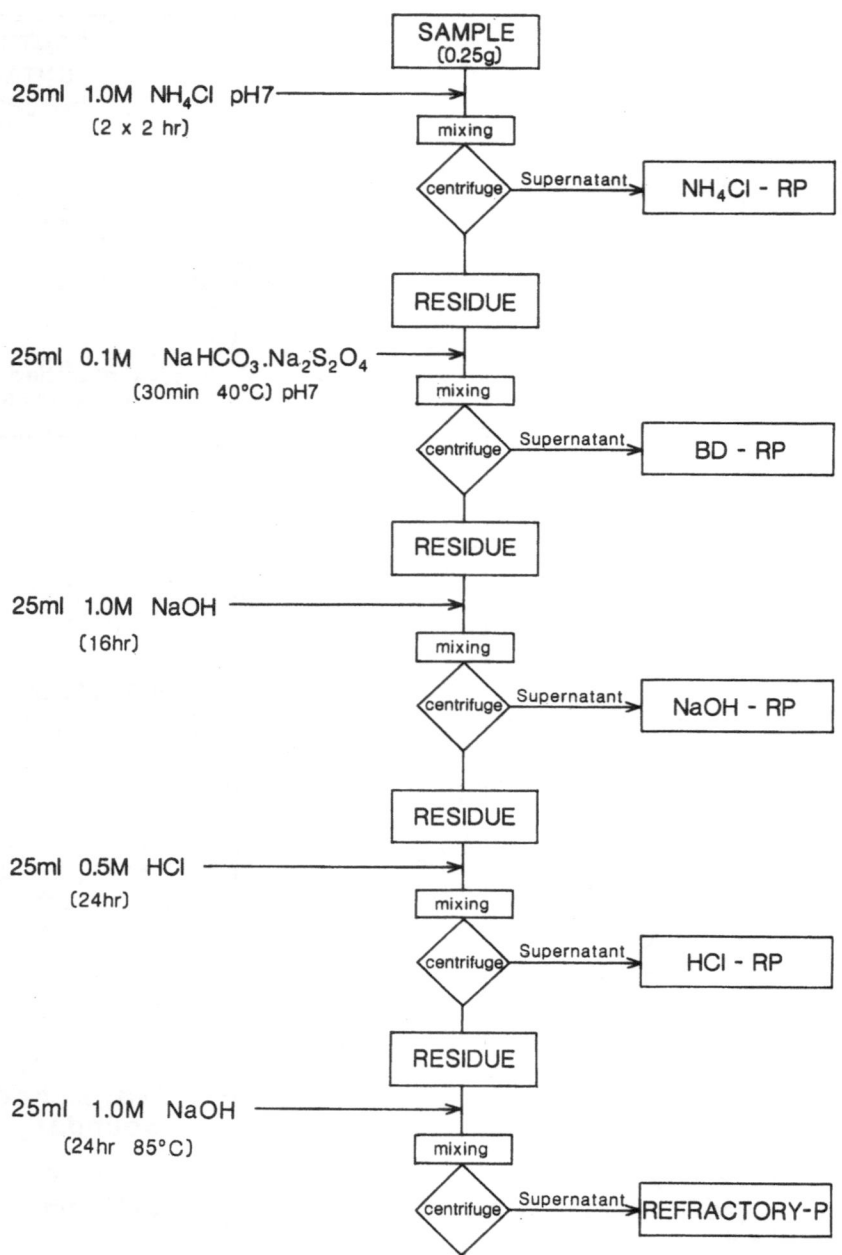

Fig. 2. Sequential extraction scheme.

late P forms in size fractions of tributary sediment. The fractionation scheme yields five chemically defined fractions of particulate P (Fig. 2). In order of extraction they are: 1) loosely sorbed P; 2) reductant soluble reactive P; 3) reactive P sorbed to metal oxides; 4) P bound to carbonates, apatite-P and P released by the dissolution of oxides; 5) non-reactive organic P extractable in hot (85 °C) NaOH. Extractions were performed in triplicate. After centrifugation, extracts were analyzed on a Technicon Autoanalyzer (Environment Canada, 1979). Detection limit of the analytical method is $1 \mu g \ P \ l^{-1}$. Variation in the analyses is reported as Standard Error of the Mean (SEM).

Previous studies have used sequential chemical extraction methods to estimate the bioavailability of sediment-phosphate (Williams *et al.*, 1976; Logan *et al.*, 1979; DePinto *et al.*, 1981; Dorich *et al.*, 1984). In this study, the bioavailable sediment-phosphate is estimated by the sum of NH_4Cl, BD and NaOH extractions and is referred to as non-apatite inorganic-P (NAIP). The HCl extraction constitutes the apatite inorganic-P (AIP) while the $NaOH_{(85)}$ extraction denotes organic-P (OP). The latter fraction is potentially available after mineralization (Blachford & Ongley, 1984).

The lack of compatible long-term sediment and water quality data increases the difficulty of assessing the transport of particulate P from river systems to receiving waters. A Water Survey of Canada hydrometric and sediment/particle size data base, which is considered to be accurate to within $\pm 5\%$ (Smith, personal communication, 1990), was combined with P fractionation data to simulate the export of different forms of particulate P to Lake Erie. The data base consists of instantaneous discharge $(m^3 \ s^{-1})$ and suspended sediment concentrations $(mg \ l^{-1})$ and depth integrating particle size data (percent finer) for a range of particle size diameters $(\leq 2–2000 \ \mu m)$. Following the method reported by Stone *et al.* (1991) the data base was used to calculate the mass of grain size fractions $(\emptyset_j)$ transported for each discharge event (Q_i) sampled. Hourly sediment loads were converted into daily loads and represent estimates of the daily mass and distribution of grain size fractions transported during 52 and 61 high magnitude discharge events in Big Creek and Big Otter Creek, respectively. Particulate P loads were calculated by multiplying the estimated mass of sediment transported for each grain size fraction by the corresponding mass of each particulate P form $(\mu g \ P \ g \ sediment^{-1})$ determined by sequential extraction.

Results and discussion

P forms and grain size

Mean concentrations of particulate P forms $(NH_4Cl-RP, BD-RP, NaOH-RP, HCl-RP$ and $NaOH_{(85)}-RP)$ in separated grain size fractions of Big Creek and Big Otter Creek sediment are presented in Fig. 3. Similar patterns between particulate P forms and grain size were observed in sediment from both tributaries. In general, concentrations of NH_4Cl-RP, BD-RP, NaOH-RP and $NaOH_{(85)}-RP$ increased with decreasing grain size while HCl-RP content decreased with decreasing grain size. These findings are in agreement with the results of Viner (1982) and Mudroch & Duncan (1986) who reported an increase in P content with decreasing grain size. However, in a study on the algal availability of phosphorus in suspended stream sediments of varying particle size, no enrichment of NAIP in small size fractions was reported (Dorich *et al.*, 1984). The authors attributed this to the low variability in the primary particle-size distribution in the sediment fractions examined. According to the data of Dorich *et al.* (1984), bioavailable particulate P will not necessarily be concentrated in the smaller size fractions in river basins where larger aggregates are primarily composed of clay and silt-sized particles.

The effect of grain size on the distribution of bioavailable particulate P forms in fine-grained tributary sediment is shown in Fig. 3. NAIP was most abundant in the smallest particle size fraction. In particular, over 50% of the particulate P in the $2 \mu m$ sediment size fraction was NAIP compared to 2% in the larger $45 \mu m$ fraction. This trend was similar for both sediments.

Variability (SEM) in the content of particulate P forms in individual sediment size fractions is presented in Fig. 4. In general, the variability in particulate P forms was remarkably low, generally less than 10%. HCl-RP was the least variable ($<5\%$) while BD-RP ranged from 13 to 27% in the $2 \mu m$ fraction.

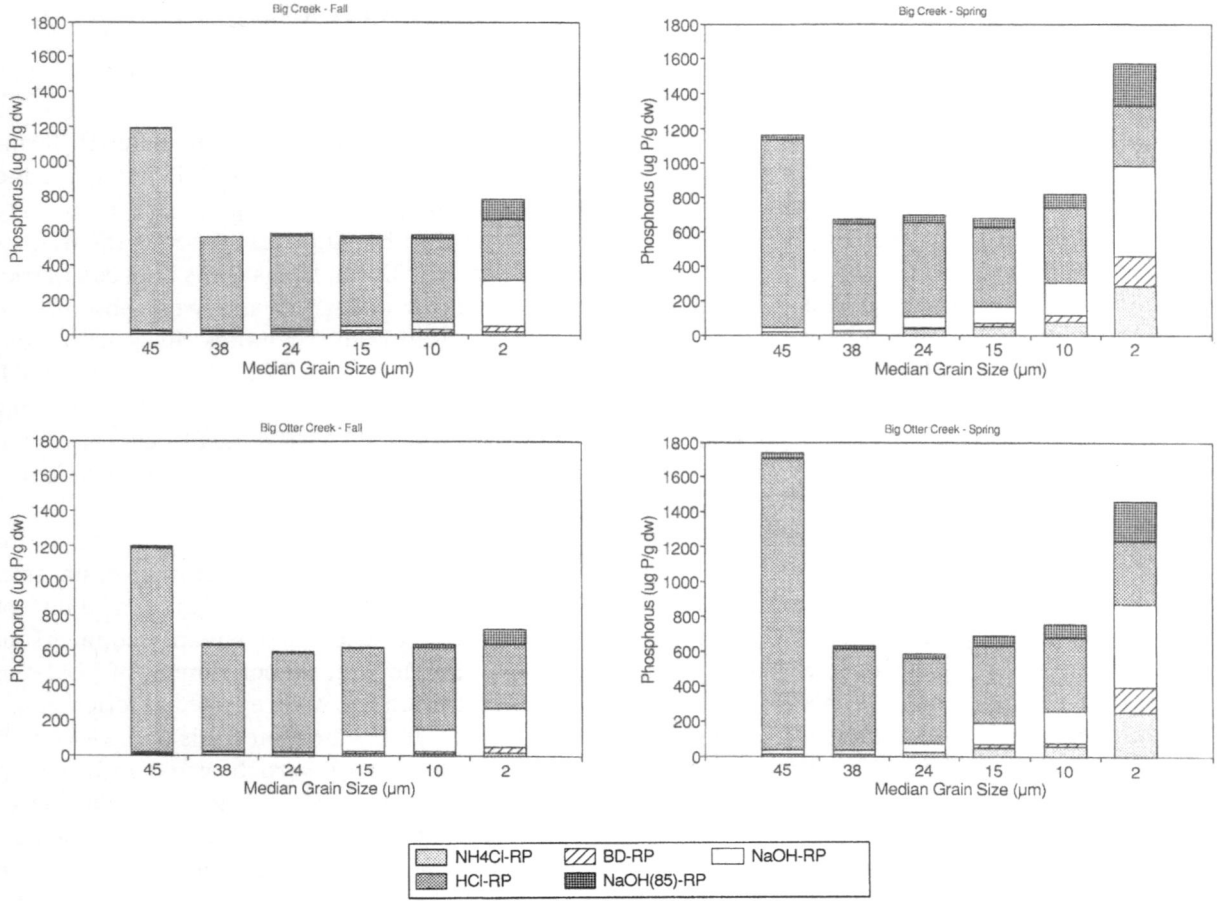

Fig. 3. Distribution of particulate phosphorus forms (mean) in sediment grain size fractions from two Lake Erie tributaries (*n* = 3).

Sediment geochemistry and P forms

The major element, total carbon and organic carbon content in separated grain size fractions for spring and fall samples are presented in Table 1. The geochemical composition of the size fractions varies but nevertheless shows definite patterns for specific elements, generally similar for both sediments.

Total P, Al, Fe, Mn and organic C content is most abundant in the 2 μm size fraction and generally decreases with increasing grain size. In this size fraction, increased concentrations of Fe, Al and Mn most likely represent oxides and hydroxides associated with the clay minerals kaolinite, illite and pyrophylite which serve to enhance P

adsorption (Stone & Mudroch, 1989). The relatively constant Ca content of particles of all sizes indicates that it is included in the particle matrix whereas increasing P content with decreasing particle size suggests it is surface bound most likely to Fe and Al oxides. In addition, Fe and Al ions probably adhere to organic matter on small size fractions and can bind phosphorus superficially in addition to organic structural phosphorus (Bostrom *et al.*, 1982).

The elemental composition in individual size fractions of sediment from Big Creek and Big Otter Creek did not vary more the 15% (Table 1). Only the content of Na varied more than 20% by grain size. The geochemical data indicate that the distribution of naturally occurring elements in the

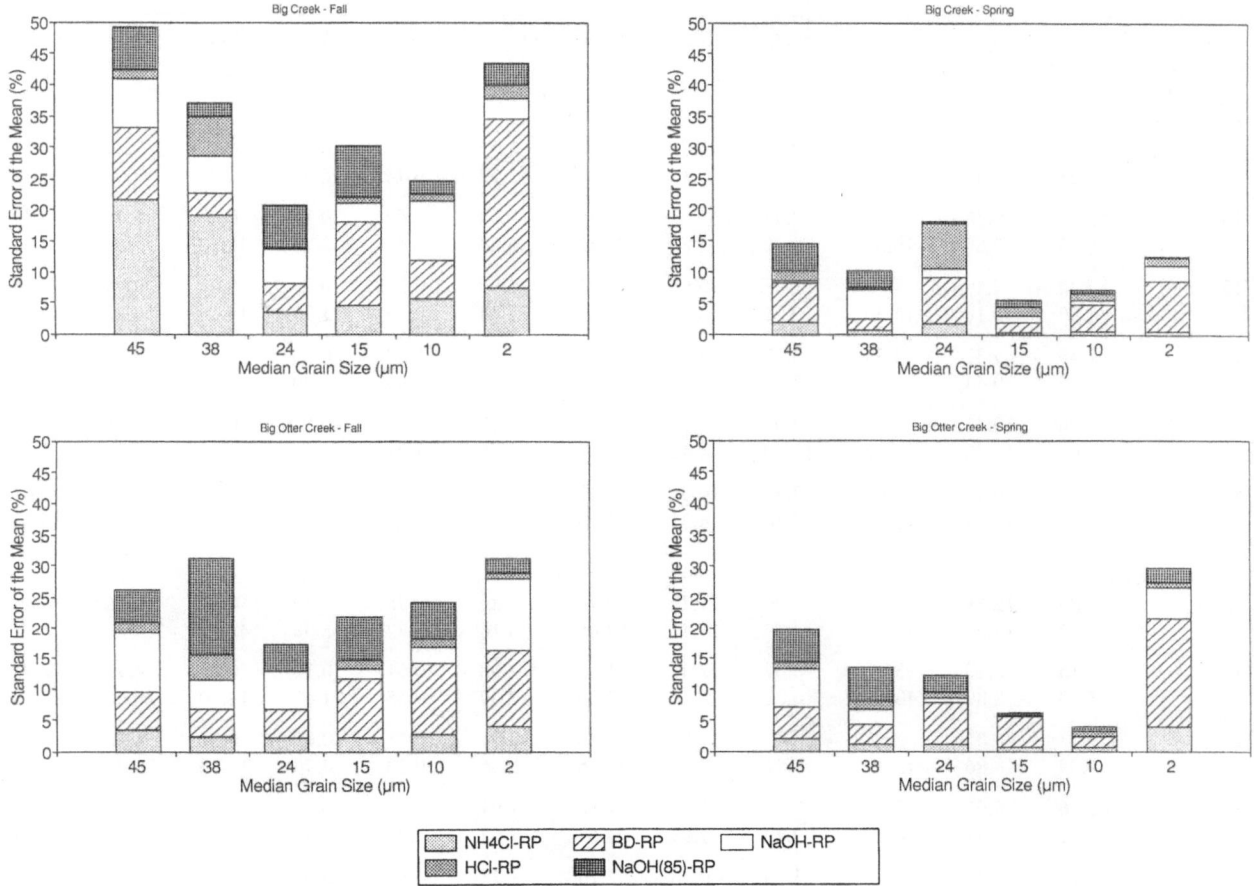

Fig. 4. Variation about the mean of particulate P forms determined by sequential extraction (*n* = 3).

sediment is remarkably consistent within these two basins.

Relationships between grain size, major elements and the various chemically defined fractions of particulate P were examined by computing Pearson product moment correlation coefficients, r (Table 2). Despite some intrabasin variability in sediment composition, similar geochemical and size fraction associations of chemically separated particulate P forms were observed. OP and all NAIP fractions were significantly related to total P (P_2O_5). In general, all three NAIP fractions are positively correlated with Al, Mn, K, and organic C while displaying a negative correlation with Si and Na. This is attributed to increased amounts of NAIP and

organic C in the 2 μm size which consists mainly of clay minerals (Stone & Mudroch, 1989). AIP in both sediments is strongly associated with Si, Ti and inversely correlated with organic C.

Due primarily to the presence of Fe and Al, organic matter (mainly humic compounds) has been shown to be important for the sorption of phosphate (Jackson & Schindler, 1975; Bostrom *et al.*, 1982). In Big Creek and Big Otter Creek sediment, OP accounted for approximately 15% of the particulate P in the 2 μm size fraction compared to 2% in the largest size fraction. Covariation between OP and metals was observed; OP was correlated with Fe, Mn and Al while correlations were also found between NAIP forms, grain size and organic C (Table 2). The observed

Table 1. Contents of major elements in separated grain size fractions of Big Creek and Big Otter Creek sediment.

Grain size (μm)	SiO$_2$	Al$_2$O$_3$	Fe$_2$O$_3$	MgO	CaO	Na$_2$O	K$_2$O	TiO$_2$	MnO	P$_2$O$_5$	TC	OC
Big Creek												
2	48.24	12.34	7.67	2.62	11.97	0.60	2.55	0.77	0.26	0.62	7.27	5.56
	5.23	7.32	7.98	2.90	8.17	62.07	3.52	6.78	6.67	26.41	*	*
10	55.26	9.36	3.93	2.88	12.09	1.15	2.03	0.62	0.15	0.24	5.44	4.47
	5.93	8.69	10.38	8.32	7.84	41.12	6.24	9.28	12.28	12.83	3.22	20.81
15	57.30	9.19	3.44	3.03	10.96	1.77	1.99	0.57	0.12	0.21	4.45	3.16
	4.80	10.16	14.39	10.75	8.11	26.07	7.84	10.16	11.71	14.89	0.22	22.15
24	61.19	9.32	2.55	3.42	10.22	2.11	2.00	0.53	0.09	0.17	3.11	2.36
	4.86	9.44	8.91	13.41	5.85	24.06	7.41	13.00	7.20	7.21	4.61	4.18
38	65.76	10.44	2.45	2.92	9.88	1.94	2.13	0.50	0.07	0.16	2.68	1.86
	7.04	11.30	8.82	13.59	4.42	25.09	9.75	11.23	6.62	4.00	6.54	26.88
45	56.42	9.43	5.60	3.74	10.06	2.13	1.77	1.38	0.14	0.24	2.79	1.64
		11.84	16.52	22.16	26.35	6.75	26.49	14.05	16.60	17.06	12.51	15.62
Big Otter Creek												
2	47.85	12.58	6.76	2.90	11.73	0.74	2.85	0.78	0.22	0.47	6.26	4.45
	7.32	11.27	17.43	4.29	13.41	53.65	6.16	13.72	5.38	43.72	*	*
10	57.15	9.10	3.37	3.07	12.33	1.45	2.08	0.61	0.12	0.22	4.90	2.93
	3.70	8.89	13.65	6.02	12.04	33.35	7.02	9.25	21.43	14.50	26.66	54.19
15	59.17	8.85	2.83	3.41	11.64	1.55	2.01	0.55	0.09	0.21	4.10	2.22
	4.09	8.86	12.38	7.72	9.01	33.55	6.56	10.20	24.70	13.33	28.05	76.52
24	61.26	8.93	2.34	3.74	11.16	1.86	1.97	0.52	0.07	0.19	2.88	1.36
	4.91	9.57	6.18	9.42	7.93	24.63	7.52	7.98	15.43	11.81	9.38	50.13
38	61.15	9.39	2.53	3.50	10.76	2.10	1.78	0.56	0.07	0.19	2.75	1.24
	7.18	11.25	9.28	12.11	1.48	16.80	14.73	5.87	6.62	15.79	2.67	60.29
45	50.86	8.81	9.54	4.41	12.15	1.89	1.48	2.11	0.18	0.26	2.67	1.35
	6.13	11.00	20.78	12.87	2.60	23.72	4.76	14.13	16.19	7.35	0.37	39.99

Mean, standard error of the mean (%). $n = 7$ for major elements; $n = 2$ for TC and OC; $n = 1$ for *.

covariation between metals and OP suggests that organic-metal oxide surfaces may influence P dynamics in Big Creek and Big Otter Creek sediment.

Seasonal variation in P composition

Earlier studies have reported temporal variation in the fractional composition of particulate P transported in fluvial systems (Ongley *et al.*, 1981, 1982; Bird, 1986; Burrus *et al.*, 1990). Seasonal relationships between the chemical forms of par-

ticulate P in size fractions of Big Creek and Big Otter Creek sediment are shown in Fig. 3. In general, both sediment types (Big Creek and Big Otter Creek) display similar trends. In sediment collected during the spring, higher contents of NH$_4$Cl-RP, BD-RP (in sediment $< 15\ \mu$m), NaOH-RP and NaOH$_{(85)}$-RP were found compared to fall sediment. The P content of the 2 μm fraction approximately doubled during spring. Both increased content and lower variability of NAIP and OP in spring samples (Fig. 4) are attributed to seasonal effects on sediment availability, source characteristics and delivery processes.

Table 2. Correlation coefficients between grain size, particulate P forms and major elements in two Lake Erie tributaries.

	Grain size	NAIP			AIP	OP	P_2O_5
		NH_4Cl-RP	NaOH-RP	BD-RP	HCl-RP	$NaOH_{(85)}$-RP	
Big Otter Creek							
Grain size		−0.66	−0.60	−0.74*	0.88**	−0.65	
SiO_2		−0.78*	−0.76	−0.81		−0.83	−0.88**
Al_2O_3		0.88**	0.91**	0.79		0.84	0.87**
Fe_2O_3					0.77		
MgO					0.73		
CaO							
Na_2O	0.76*		−0.50	−0.66		−0.61	
K_2O	−0.73*	0.85*	0.85	0.82	−0.73	0.81	0.70
TiO_2	0.70*				0.96**		
MnO		0.54	0.52	0.57		0.60	0.70
P_2O_5		0.97***	0.97***	0.94**		0.97***	
TC	−0.91**	0.70	0.64	0.81	−0.64	0.73	0.59
OC	−0.89**	0.68	0.62	0.79	−0.64	0.71	
Big Creek							
Grain size		−0.67	−0.63		0.93**	−0.66	
SiO_2	0.73*	−0.94**	−0.93**	−0.87**		−0.95**	−0.92**
Al_2O_3		0.74*	0.77*	0.85*		0.74*	0.81
Fe_2O_3		0.74*	0.76*	0.74*		0.76*	0.88
MgO		−0.60	−0.61				
CaO	−0.86*				−0.66		
Na_2O	0.91**	−0.70	−0.67	−0.54	0.72	−0.70	
K_2O		0.80	0.82	0.85		0.79	
TiO_2	0.65				0.87		
MnO		0.90	0.90**	0.82		0.91**	0.94**
P_2O_5		0.97***	0.98***	0.96**		0.97***	
TC	−0.80*	0.94**	0.92	0.84	−0.62	0.94	0.89**
OC	−0.75*	0.97***	0.97	0.89	−0.66	0.97	0.91**

Significance level (n = 6). Not marked p = 0.10. * p = 0.05. ** p = 0.01. *** p = 0.001.

In these agricultural watersheds during the spring, infiltration capacity is much reduced thus promoting Horton overland flow (Bras, 1990). The transport capacity of runoff would therefore increase the delivery of aggregates that can have higher nutrient levels than the soil (Alberts & Moldenhauer, 1981). Therefore, spring samples will reflect both the chemical characteristics of in-channel sediment and nutrient enriched eroded aggregates transported from predominantly agricultural land use.

In a study on the algal-available phosphorus in suspended sediments collected from Maumee, Sandusky and Cattaraugus rivers, DePinto *et al.* (1981) reported that the total phosphorus content varied by less than 5 percent from levels reported by an earlier investigation on the same rivers. This finding led the authors to suggest that the phosphorus content of suspended sediment may be a reproducible characteristic for a given tributary. In the present study, no significant seasonal difference (ANOVA $\alpha = 0.05$) was apparent in the AIP fraction (HCl-RP) in sediments from both tributaries (Fig. 4). The observed low seasonal variability indicates a consistency in the delivery of the AIP fraction from these rivers to Lake Erie and supports the contention of Thomas & Munawar (1985) that AIP is a constant background form in the tributaries of the lower Great Lakes. In an investigation of the seasonal delivery of particulate P forms to Lake Geneva from the upper Rhone river, Burrus *et al.* (1990) found that the content of AIP remained constant throughout the year. Given the reduced variability of AIP in Big Otter Creek and Big Creek basins, it can be expected that variation in total phosphorus export from these tributaries to Lake Erie will most likely reflect variation in the NAIP and OP fractions.

On a seasonal basis, the distribution of particulate P fractions is relatively uniform among size fractions for Big Creek and Big Otter Creek. When spring and fall P fractionation data are combined, the mean content of each particulate P form ranged (SEM) from a high of 20 to 40% for NH_4Cl-RP to a low of 1 to 7% for HClRP in both sediment types. Variation in the amount of other particulate P forms included; BD-RP (5 to 30%) NaOH-RP (14 to 28%) and $NaOH_{(85)}$-RP (15 to 28%). Therefore, simulations of particulate P export based on this data can be expected to vary within the above ranges.

Simulation of particulate P transport

Simulations of particulate P export to Lake Erie from Big Otter Creek and Big Creek are based on the assumption that results of the sequential extraction scheme represent reasonable approximations of the distribution of particulate P forms in grain size fractions of sediment in these rivers.

A quantitative assessment of the sediment transport characteristics of Big Creek and Big Otter Creek is presented in the first three columns of Tables 3 and 4. These tables report estimates of the mass of 11 size fractions transported during 52 and 61 high magnitude discharge events. Sediment ≤ 63 μm accounted for 84 and 68% of the estimated daily sediment load transported in Big Creek and Big Otter Creek, respectively. According to published data (Environment Canada, 1989), annual maximum daily suspended sediment loads have ranged from 340 to 3690 tonnes in Big Creek (Station No. 02GC007 near Walsingham) and 8340 to 34100 tonnes in Big Otter Creek (Station No. 02GC026 near Calton).

The estimated average daily sediment load was 716.1 tonnes for Big Creek (Table 3) and 5272.8 tonnes for Big Otter Creek (Table 3). These figures are lower than the annual maximum daily suspended sediment load (Environment Canada, 1989) and provide conservative estimates of suspended sediment load transported by these rivers during storm/snowmelt events.

Simulations of particulate P export to Lake Erie from Big Otter Creek and Big Creek on particle size fractions ≤ 63 μm are also presented in Tables 3 and 4. Expressed as a percentage of the total sediment-bound P load, AIP represents 67 and 70% of the calculated P mass; NAIP accounts for 26 and 23% in Big Otter Creek and Big Creek, respectively. Approximately 7% of the total calculated load was OP. Despite differences in the magnitude of sediment transported by these rivers, the relative amounts (%) of particulate P forms did not vary substantially between rivers. Therefore, variability in the total P loading can be attributed to differences in NAIP and OP. These in turn are based on seasonal variation in the particle size distribution, source characteristics of riverine sediment and land use.

The estimated mean daily load of particulate P transported during the discharge events examined was 504 kg P day^{-1} in Big Creek and 3350 kg P day^{-1} in Big Otter Creek (Tables 3 and 4). Unit area loads of 0.85 kg P ha^{-1} day^{-1} (Big Creek) and 4.9 kg P ha^{-1} day^{-1} (Big Otter Creek) for these discharge events were determined by dividing the estimated daily loads by the drainage basin area (above the gauging station). The calculated unit area loads for Big Creek and Big Otter Creek are unusually high compared to other estimates in the literature (Bird, 1986). However, these unit area loads are derived from mean daily 'event' loads and represent estimates of P transport during major storm events (52 and 61 events) rather than average daily loads based over a period of one year.

Total phosphorus (TP) concentrations in Lake Erie fluctuate seasonally. In spring and fall, TP concentrations are larger than summer concentrations due primarily to internal and external loadings (Rosa, 1987). Internal loadings in spring

Table 3. Particle size and particulate P transport characteristics in Big Creek during 52 discharge days.

Sediment transport			Estimates of P transport during 52 discharge events					
Grain size (μm)	Sediment total Wt *	% Total sediment load	NH$_4$Cl-RP (kg P)	BD-RP (kg P)	NaOH-RP (kg P)	HCl-RP (kg P)	NaOH$_{(85)}$-RP (kg P)	
2	4822	13	615	471	1824	1688	817	
4	1726	5	220	169	653	604	292	
8	4843	13	187	144	510	2257	226	
16	7286	20	187	119	412	3562	227	
31	7094	19	88	54	155	3963	114	
62	5597	15	53	34	91	6368	103	
125	3254	9						
250	1678	5						
500	653	2						
1000	228	0.6						
2000	58	0.2						
Total	37238	100	1350	990	3646	18441	1780	26207
%			5	4	14	70	7	100
Mean daily load	716		26	19	70	355	34	504

* – tonnes transported in 52 days.

Table 4. Particle size and particulate P transport characteristics in Big Otter Creek during 61 discharge days.

Sediment transport			Estimates of P transport during 52 discharge events					
Grain size (μm)	Sediment total Wt *	% Total sediment load	NH$_4$Cl-RP (kg P)	BD-RP (kg P)	NaOH-RP (kg P)	HCl-RP (kg P)	NaOH$_{(85)}$-RP (kg P)	
2	50364	16	6679	4636	17682	18677	7859	
4	19476	6	2583	1793	6838	7223	3039	
8	34087	11	1008	544	4643	15259	1484	
16	32899	10	772	483	3352	15458	1105	
31	38129	12	302	214	657	22621	515	
62	41023	13	288	192	613	57006	845	
125	51384	16						
250	35158	11						
500	11729	4						
1000	5021	2						
2000	2371	0.7						
Total	321641	100	11633	7862	33786	136244	14847	204372
%			6	4	17	67	7	100
Mean daily load	5273		191	129	554	2234	243	3350

* – tonnes transported in 61 days.

and fall isothermal periods are probably due to wave induced resuspension of sedimented particulate matter and anoxic P regeneration from reduced sediments (Rosa, 1985). However, the external loading of particulate P to Lake Erie is maximized in spring and fall (Logan, 1987). During these periods, P on fine-grained sediment, particularly the $< 8 \mu m$ size fraction, from all Lake Erie tributaries may constitute a potentially large nutrient source for planktonic and benthic microorganisms and this may substantially increase TP concentrations during these periods.

Conclusion

The present investigation provides primary field data on the form and distribution of particulate P in grain size fractions of fine-grained sediment from two Lake Erie tributaries. The study demonstrates the effect of grain size on the form of particulate P and provides additional information on the relation between particulate P form, mineralogical composition and organic C that would not be available from analysis of bulk sediment samples. Simulations of particulate P export suggest that the combined export of fine-grained sediment from all tributaries draining into Lake Erie may represent a large potential nutrient source for biotic uptake in spring and fall.

Acknowledgements

Funding from the National Water Research Institute (Burlington, Ontario) and Wilfrid Laurier University (Waterloo, Ontario) are gratefully acknowledged. Pam Schaus prepared the figures.

References

Allan, R. J., 1979. Sediment-related fluvial transmission of contaminants: some recent advances by 1979. Inland Waters Directorate, Environment Canada. Scientific Series No. 107.

Allan, R. J., 1986. The role of particulate matter in the fate of contaminants in aquatic ecosystems. Inland Waters Directorate, Environment Canada, Scientific Series No. 142.

Alberts, E. E. & W. C. Moldenhauer, 1981. Nitrogen and phosphorus transported by eroded soil aggregates. Soli Sci. Soc. Am. J. 45: 391–396.

Armstrong, D. E., J. J. Perry & D. E. Flatness, 1979. Availability of pollutants associated with suspended or settled sediments that gain access to the Great Lakes. Great Lake National Program Office. U.S. EPA-905/4-79-028.

Bird, G. A., 1986. Phosphorus dynamics in Great Lakes ecosystems. Environment Canada. 161 pp.

Blachford, D. P. & E. D. Ongley, 1984. Biogeochemical pathways of phosphorus, heavy metals and organochlorine residues in the Bow and Oldman rivers, Alberta. Inland Waters Directorate, Environment Canada, Scientific Series, 138.

Bostrom, B., M. Jansson & C. Forsberg, 1982. Phosphorus release from lake sediments. Arch. Hydrobiol. 18: 5–59.

Bras, R. L., 1990. Hydrology, an introduction to hydrological science. Addison-Wesley Co., Don Mills, Ontario, 643 p.

Brownlee, B. G. & G. A. Bird, 1988. Phosphorus dynamics in the Canagagigue Cree below the Elmira Water Pollution Control Plant. Beak Consulting Report Ref. 2397.1. Water Planning Management, Environment Canada, Burlington, Ontario.

Burrus, D., R. L. Thomas & J. Dominik, 1990. Seasonal delivery of the particulate forms of phosphorus to Lake Geneva from the upper Rhone. Aquatic Sciences 52: 221–235.

Cuker, B. E., P. T. Gama & J. M. Burkholder, 1990. Type of suspended clay influences lake productivity and phytoplankton community response to phosphorus loading. Limnol. Oceanogr. 35: 830–839.

DePinto, J. V., T. C. Young & S. C. Martin, 1981. Algal-available phosphorus in suspended sediments from lower Great Lakes tributaries. J. Great Lakes Res. 7: 311–325.

Dickinson, W. T., A. Scott & G. Wall, 1975. Fluvial sedimentation in Southern Ontario. Can. J. Earth Sci. 12: 1813–1819.

Dorich, R. A., D. W. Nelson & L. E. Sommers, 1984. Algal availability of phosphorus in suspended sediments of varying particle size. J. Envir. Qual. 13: 82–86.

Environment Canada, 1979. Analytical Methods Manual, Inland Waters Directorate, Water Quality Branch, Ottawa, Ontario, Canada.

Environment Canada, 1989. Sediment Data, Ontario. Inland Waters/Lands Directorate, Water Resources Branch, Water Survey of Canada, Ottawa, Ontario, Canada.

Heath, R. T & D. A. Francko, 1988. Comparison of phosphorus dynamics in two Oklahoma reservoirs and a natural lake varying in abiogenic turbidity. Can. J. Fish. aquat. Sci. 45: 1480–1486.

Jackson, D. A. & D. W. Schindler, 1975. The biogeochemistry of phosphorus in an experimental lake environment: evidence for the formation of humic-metal phosphate complexes. Verh. int. Ver. Limnol. 19: 211–221.

LEWMS, 1975. Preliminary Feasibility Report. Volume 1, Lake Erie Wastewater Management Study, Corps of Engineers, Buffalo District, Buffalo, N.Y.

Lick, W., 1982. Entrainment, deposition and transport of finegrained sediments in lakes. In P. G. Sly (ed.), Sediment/Freshwater Interaction. Developments in Hydrobiology 9. Dr W. Junk Publishers, The Hague: 31–40. Reprinted from Hydrobiologia 91/92.

Logan, T. J., T. O. Oloya & S. M. Yaksich, 1979. Phosphate characteristics and bioavailability of suspended sediments draining into Lake Erie. J. Great Lakes Res. 5: 112–123.

Logan, T. J., 1987. Diffuse (non-point) source loading of chemicals to Lake Erie. J. Great Lakes Res. 13: 649–658.

Mudroch, A., 1985. Geochemistry of the Detroit River sediments. J. Great Lakes Res. 11: 193–200.

Mudroch, A. & G. Duncan, 1986. Distribution of metals in different size fractions of sediment from the Niagara River. J. Great Lakes Res. 12: 117–126.

Ongley, E. D., M. C. Bynoe & J. B. Percival, 1981. Physical and geochemical characteristics of suspended solids, Wilton Creek, Ontario. Can. J. Earth Sci. 18: 1365–1379.

Ongley, E. D., 1982. Influence of season, source and distance on physical and chemical properties of suspended sediment. IAHS. Pub. No. 137: 371–383.

Peart, M. R. & D. E. Walling, 1982. Particle size characteristics of fluvial suspended sediment. IAHS Publ. No 137.

Petterson, K. & V. Istvanovics, 1988. Sediment phosphorus forms in Lake Balaton – forms and mobility. Arch. Hydrobiol. Beih. 30: 21–25.

Rosa, F., 1985. Sedimentation and sediment resuspension in Lake Ontario. J. Great Lakes Res. 11: 13–25.

Rosa, F., 1987. Lake Erie central basin total phosphorus trend analysis from 1968 to 1982. J. Great Lakes Res. 13: 667–673.

Smith, B., 1990. Water Resources Branch, Environment Canada, Guelph, Ontario.

Stone, M. & A. Mudroch, 1989. The effect of particle size, chemistry and mineralogy of river sediments on phosphate adsorption. Envir. Tech. Lett. 10: 501–510.

Stone, M., M. C. English & G. Mulamoottil, 1991. Sediment and nutrient dynamics in two tributaries of lake Erie: A numerical Model. Hydrological Processes. 5: 371–382.

Stone, M. & H. Saunderson, 1992. Particle size characteristics of suspended sediment in southern Ontario rivers tributary to the Great Lakes. Hydrological Processes. 6: 189–198.

Thomas, R. L & M. Munawar, 1985. The delivery and bioavailability of particulate bound phosphorus in Canadian rivers tributary to the Great Lakes. In: J. N. Lester & P. W. W. Kirk (eds), Proceedings of the International Conference Management Strategies for Phosphorus in the Environment: 462–469.

Verhoff, F. H., D. A. Melfi & S. M. Yaksich, 1982. An analysis of total phosphorus transport in river systems. In P. G. Sly (ed.), Sediment/Freshwater Interaction. Developments in Hydrobiology 9. Dr W. Junk Publishers, The Hague: 241–252. Reprinted from Hydrobiologia 91/92.

Viner, A. B., 1982. A qualitative assessment of the nutrient phosphate transported by particles in a tropical river. Revue Hydrobiol. trop. 15: 3–8.

Wall, G. J, W. T. Dickinson & J. P. Van Vliet, 1982. Agriculture and water quality in the Canadian Great Lakes Basin: II. Fluvial sediments. J. Envir. Qual. 11: 482–486.

Williams, J. D. H., T. P. Murphy & T. Mayer, 1976. Rates of accumulation of phosphorus forms in Lake Erie sediments. J. Fish. Res. Bd Can. 33: 430–439.

Hydrobiologia **253**: 31–45, 1993.
P.C.M. Boers, T.E. Cappenberg & W. van Raaphorst (eds),
Proceedings of the Third International Workshop on Phosphorus in Sediments.
© 1993 *Kluwer Academic Publishers.*

A double-layer model for ion adsorption onto metal oxides, applied to experimental data and to natural sediments of Lake Veluwe, The Netherlands

A. G. Brinkman
Agricultural Research Department (DLO-NL), Institute for Forestry and Nature Research (IBN-DLO), P.O. Box 167, 1790 AD Den Burg (Texel), The Netherlands

Key words: sorption model, phosphate, silicate, sediment, double-layer

Abstract

A colloid-chemical model is presented that describes sorption of ions onto metal oxide surfaces in aquatic systems. Multispecies competition for the available sorption sites and double-layer dielectric constant computation are main features of the model.

The model is used for the analysis of sorption data regarding pure crystalline sorbents such as hematite (α-Fe_2O_3) and goethite (α-FeOOH). Adsorption of potential determining ions (hydroxyl), potassium and chloride, silicate and phosphate is calculated, showing good agreement with experimental data obtained from literature.

Secondly, the model is applied to evaluate sorption data on natural sediments from Lake Veluwe, The Netherlands. The model shows a considerably improved description of sorption phenomena, compared to results from a classical (Langmuir) sorption analysis.

The research shows that a combination of model development, model system study and experimental research in natural systems is very useful for a better understanding of environmental nutrient adsorption mechanisms.

Introduction

In nature, almost any solid can adsorb ions. The importance of a solid for the binding of an ion depends on its presence, its specific surface (Table 1) and the strength of the adsorption. Clays, iron and aluminum oxides, silicates, carbonates and humic acids generally are the main constituents of soils and sediments. Although silicates and carbonates may be abundantly present, their specific surface is very small and, consequently, they play a minor role in the overall adsorption process. In humic matter, carboxyl (COOH$^-$) groups are of importance. These acid groups are negatively charged and mainly adsorb cations. The affinity for adsorption generally increases with cation charge. The humic contribution to the overall adsorption capacity in natural environments varies from almost zero in very sandy sediments to very large in peaty sediments. Clays consist of monomolecular silicate and aluminum oxide layers, and show large adsorption capacities for many ions. Under average conditions, the aluminum oxide layers are nearly neutral, and the

Table 1. Adsorption capacities of some solids.

Component	Capacity ($m^2 g^{-1}$)		Ions
Clay	Large	10–100	Mainly cations, anions to a lesser extent
Fe/Al-oxides	Large	10–100	Anions, cations to a lesser extent
		600*	
Sand	Small	≪1	Cations
Humics	(Very) large	>100	Cations
Carbonates	Small	< = 1	Both cations and anions

References: Atkinson *et al.*, 1967; Breeuwsma, 1973; Parks, 1965; Whittemore & Langmuir, 1975; Boehm; 1971; Sigg, 1979; Hingston *et al.*, 1974; Cabrera *et al.*, 1977; Parfitt *et al.*, 1975; Helyar *et al.*, 1976; Huang & Stumm, 1972; James *et al.*, 1975; Hayes *et al.*, 1991; Hiemstra *et al.*, 1989-b; De Haan, 1965; Jones & Bowser, 1978; Chen *et al.*, 1973; Griffin & Jurinak, 1973; Goujon *et al.*, 1976; Jacobson, 1978. *: for hydrous ferric oxide, Hayes *et al.*, 1988.

silicate layers are negatively charged. Adsorption of cations generally is stronger than that of anions. Most important for anion adsorption are iron(III) and aluminum oxides. These oxides are positively or neutrally charged and show high sorption affinities for a number of anions such as phosphate and silicate, and for small cations such as magnesium (Breeuwsma, 1973). They have large specific areas and, when present, they largely contribute to the anion sorption capacity of natural sediments. Thus, research into the sorption characteristics of these oxides is most relevant.

At the solid-water interface of oxides hydroxyl groups (OH^-) are formed for reasons of stability and charge neutrality. These hydroxyl groups dissociate according to

$$Me\text{-}OH_2^+ < = > Me\text{-}OH + H^+$$
$$< = > Me\text{-}O^- + 2H^+.$$

Due to these reactions, oxides are positively charged in acid media, and negatively charged in alkaline environments. The pH at which the $Me\text{-}OH_2^+$ and the $Me\text{-}O^-$ sites are balanced and the oxide is neutral, is called the zero point of charge (pH_{zpc}). This is an important characteristic of oxides, together with the number of hydroxyl groups per m^2 and the specific surface area of the oxide ($m^2 g^{-1}$).

The hydroxyl ions can be exchanged against other anions, which provides the basis for the adsorption of foreign anions such as phosphate.

One of the sorption reactions is

$$Me\text{-}OH + O\text{-}PO_3H_2^-$$
$$< = > Me\text{-}O\text{-}PO_3H_2 + OH^-.$$

The surface hydroxyl group is replaced by the phosphate ion, not just the hydroxyl's proton (Parfitt *et al.*, 1975; Sigg, 1979; Hingston *et al.*, 1974). Also, two hydroxyl ions can be replaced by one phosphate ion in which case the anion is bound to two metal atoms (Hingston *et al.*, 1974; Parfitt *et al.*, 1975; Parfitt *et al.*, 1977).

Also, other sorption reactions are possible. Aluminum and iron oxides have the general stoichiometric formula Me_2O_3, and both often have a hexagonal close packed crystal structure, with a metal atom on two-thirds of the octaeder positions. At, or close to, the surface, there is space for sorption of small cations on vacant octaeder positions, thus increasing the positive charge of the colloid. Also, Me-ions may be substituted by such small cations, generally resulting in a more negative surface charge. Under experimental conditions these sorptions or substitutions can be controlled through the ionic composition of the liquid; in nature such processes influence the anion sorption properties of these oxides.

Adsorption model

Langmuir adsorption isotherm

When phosphate (P) sorbs onto a surface, described by the reaction

$$Ads_{free} + P \Leftrightarrow Ads_{occ} , \tag{1}$$

where Ads_{free} denotes a free adsorption site and Ads_{occ} an occupied adsorption site, the equilibrium situation can be described by a Langmuir adsorption isotherm:

$$\theta = \frac{K_{ads} [P]}{1 + K_{ads} [P]} , \tag{2}$$

with

θ relative occupation of sorption sites ($-$);

$[P]$ dissolved phosphate concentration (mol l^{-1});

K_{ads} adsorption equilibrium constant (l mol^{-1});

and

$$K_{ads} = \frac{[Ads_{occ}]}{[P] [Ads_{free}]} = \exp \left(\frac{-\Delta G_{ads}}{RT} \right) , \tag{3}$$

where ΔG_{ads} is the standard free energy of adsorption (J mol^{-1}), T the absolute temperature (K) and R the gas constant (8.314 J mol^{-1} K^{-1}).

The amount of sorbed P in a solid/solution mixture then reads

$$[P_{ads}] = \theta \, Ads_{max} \quad (mol(P) \, l^{-1}) \tag{4}$$

with Ads_{max} the maximum adsorption, calculated from

$$Ads_{max} = \Gamma_{max} \, Aspec \, [solid] , \quad (mol(P) \, l^{-1}) \tag{5}$$

where

Γ_{max} = sorption site density on solid (mol m^{-2});

Aspec   = specific surface area of solid (m^2 g^{-1} (solid));

[solid]   = solid content of the mixture (g (solid) l^{-1}).

This model is valid when

(i)   each site can absorb only one ion;

(ii)   the adsorption is reversible;

(iii)   there is no interaction between adsorbing ions;

(iv)   the adsorption constant K_{ads} is independent of the absorption site.

When more ion types (i) compete for the absorption sites, the $K_{ads, i}$ values and the concentrations C_i determine the fractions (θ_i) of sites occupied by these ions. Similar to eq. 2, the Langmuir isotherm reads

$$\theta_i = \frac{K_{ads, i} \, C_i}{1 + \sum_{j=1}^{n} K_{ads, j} \, C_j} , \tag{6}$$

where n is the number of competing ion types.

When one is able to describe the mutual interactions between adsorbed ions as well as the characteristics of the adsorption sites, the assumptions iii) an iv) are not important any more. In the next section, K_{ads} will be described as a function of the type of adsorption site and the adsorbed ions.

Double-layer adsorption model

Simple schematizations of the solid/liquid interface are given in Fig. 1. These are extensions of the classical Gouy-Chapman-Stern-Grahame (GCSG-) double-layer concept (Grahame, 1947). Dehydrated ions adsorb onto the charged metal oxide surface. This type of adsorption is called chemical or specific adsorption: apart from the charge-charge interaction an extra adsorption energy gain determines the attraction. On the next plane, hydrated ions can absorb physically: only the charge-charge interaction is relevant. As a result of these adsorptions, the surface charge is largely neutralized. The remainder is neutralized by oppositely charged ions ('counterions') present in the liquid phase at larger distances from the surface.

Because charged ions absorb onto charged surfaces, it is relevant for the adsorption parameter K_{ads} (eq. 3) to distinguish between an electrostatic and a chemical attraction term. The latter, which is assumed to be constant, depends on the nature of the solid and the adsorbing ion. So:

$$\Delta G_{ads} = \Delta G_{phys} + \Delta G_{chem} \tag{7}$$

34

with

$$-\Delta G_{phys} = -z\,\Psi_j\,\frac{e}{kT} \qquad (8)$$

and

$$-\Delta G_{chem} = \phi\,\frac{e}{kT} \qquad (9)$$

with

Ψ_j = plane j potential (V)

Following eq. 3:

$$K_{ads} = \exp\!\left((-z\,\Psi_j + \phi)\,\frac{e}{kT}\right) \qquad (10)$$

ϕ = chemical or 'specific' adsorption potential (V);
k = Boltzmann constant (J K^{-1});
T = absolute temperature (K);
z = ion valence $(-)$;
e = electron charge (1.6021 10^{-19} C).

This reduces the problem of quantifying K_{ads} into quantifying the layer potentials Ψ_j and the chemical adsorption potential ϕ. ϕ is to be derived from experimental data.

The Gouy-Chapman equation describes the relationship between the charge density and the potential on plane 5 in Fig. 1. (σ_5, Ψ_5). The concentration of counterions in the liquid phase is an important parameter in that equation. In appendix I, this relationship is given.

After calculation of the potential on the outer plane, the potentials on the other planes can be found following a flat capacitor approach:

$$\Psi_{j-1} = \Psi_j + \sigma_{j-1}\,\frac{\beta_j}{\varepsilon_{r,j}\,\varepsilon_0} \qquad (11)$$

with
β_j = the thickness of layer j (m), which is situated between planes $j-1$ and j,
and

$$\frac{\varepsilon_{r,j}\,\varepsilon_0}{\beta_j} \qquad (12)$$

is the capacity $(F\,m^{-2})$ of each individual layer j.

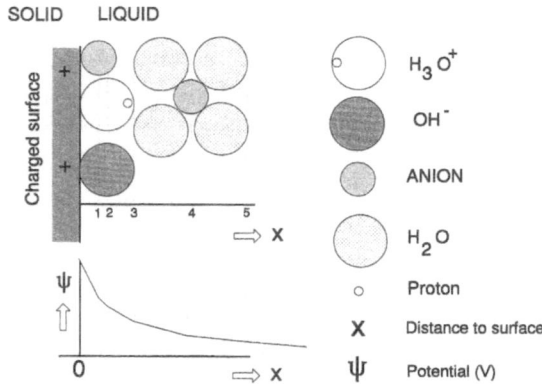

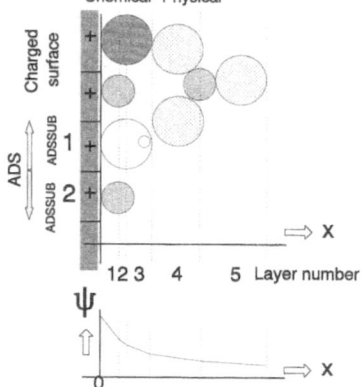

Fig. 1. a. Double layer schematization where dehydrated ions adsorb chemically onto the bare metal oxide surface, and hydrated ions adsorb physically in the next layer. b. Double layer concept from Fig. 1^a extended to a situation with (two) different adsorption sites. Surface areas of sites are ADSSUB(1) and ADSSUB(2) (m^2).

Here ε_0 is the permittivity of the vacuum $(C\,V^{-1}\,m^{-1})$ and $\varepsilon_{r,j}$ is the dielectric constant of layer j. The charge density σ_j at plane j follows from

$$\sigma_j = \sigma_0 + \sum_{l=1}^{j} \sigma_{a,l}, \qquad (13)$$

where $\sigma_{a,l}$ is the adsorbed charge density $(C\,m^{-2})$ at each plane l, σ_0 is the charge density on the bare oxide surface. This adsorbed charge density $\sigma_{a,l}$ reads

$$\sigma_{a,l} = \sum_{i=1}^{n} \theta_{i,l}\,\sigma_{max}\,z_i \qquad (14)$$

where σ_{max} is the charge density ($C\,m^{-2}$) when all sites are occupied by an ion with charge $+1$.

The calculation of the plane potentials (eq. 11) is very straightforward. However, the layer capacities are required. The crucial variable here is the dielectric constant ε_r of each layer (eq. 12). The calculation of ε_r is the most vulnerable part of the adsorption model presented in this paper. The characteristics of the adsorbing ions and the electrical field strength are major components in the description of ε_r. In appendix II, this calculation of ε_r is briefly outlined. The calculation of ε_r is tested for macroscopic model systems; the resulting description is used on the microscopic scale of the double layers. This means that the dielectric constants used in this present model are not found from adsorption data analysis, but are thought to be known according to the description as outlined in appendix II.

The position of the planes on which the adsorptions of the several ions take place has to be chosen; and the model results showed to be rather sensitive to these plane positions.

Since the potential influences adsorption and so the charge, and since the set of equations formed by the eqs. 6 to 14 is highly non-linear, an iterative procedure (Marquardt, 1963) was used for the adsorption computations.

Summarizing, the present double-layer model describes a system that

(A) has an adsorption layer where chemical adsorption takes place. In this layer:
  (i)   hydroxyl and other ions compete for the existing adsorption sites. In this concept the hydroxyl free, and positively charged bare oxide surface is considered to be the offset plane. The adsorption is reversible.
  (ii) the adsorbing ions are not hydrated.
  (iii) the number of adsorption sites is determined by the original number of hydroxyl groups.
  (iv) the plane of adsorption is that through the centre of the adsorbing ions; so, a number of adsorption planes may be distinguished within this specific adsorption layer. All these ions appear in the de-

nominator of the Langmuir adsorption isotherm (eq. 6). It is also implemented in the model that large ions that do not fit on one site may hinder neighbour ions. This is not explained in detail here, but it is relevant to, for example, chloride ions, or to phosphate ions adsorbing onto one site.
  (v)   the adsorbed charge is thought to be smeared out over the relevant adsorption plane.
  (vi) two or more adsorption site types may be distinguished, each having their own adsorption characteristics.

(B) has a physical adsorption layer, where:
  (i)   ions are reversibly bound due to electrostatic attraction.
  (ii) ions are hydrated.
  (iii) the number of adsorption sites is determined by the available area and the size of the adsorbing hydrates.
  (iv) the adsorbed charge is thought to be smeared out over the relevant adsorption plane.
  (v)   there is a diffuse outer layer that begins on the outer plane of the physical adsorption layer. The potential on this plane is calculated according to the Gouy-Chapman equation.

Application to experimental data

Data

To run the double-layer model, the number of different adsorption sites has to be known as well as the position of the adsorbing ions (site and plane of adsorption), and the extent of hindrance of neighbour ions.

The sizes of adsorbing ions can be taken from standard tables. The dielectric constant in each layer is calculated as outlined in section 2 and in appendix II. The physical adsorption potentials that partly determine the adsorption equilibrium constants K_{ads} follow from ion charges and the calculated electrostatic potentials on each plane of adsorption.

This means that three parameters are to be estimated: the specific adsorption potentials ϕ in the very first place, the position of the plane of adsorption and the extent of the hindrance of neighbour ions. The latter two are partly determined by the size of the ions and their stereometry.

The estimation of these parameters was based on experimental adsorption data published by Breeuwsma (1973) and Sigg (1979).

Breeuwsma (1973) studied adsorption of several ions onto hematite (α-Fe_2O_3): phosphate, NO_3^-, SO_4^{2-}, Cl^-, Ca^{2+}, Mg^{2+}, Li^+, Na^+, K^+. Sigg's study (1979) focused on silicate adsorption onto goethite (α-$FeOOH$). Both studies mainly employed the potentiometric titration method: artificially prepared iron oxide colloids with specific surface areas of 15–45 ($m^2\,g^{-1}$) were titrated with acid and base, pH values being recorded with an accuracy of 0.001 pH units. Ionic strength was constant during titration, determined by K^+ and Cl^- (Breeuwsma) and Na^+ and ClO_4^- (Sigg).

Due to adsorption of hydroxyl ions, the pH change during titration differs from a colloid free situation. Consequently, the pH reflects the change in adsorption in the chemical adsorption layer due to the adsorbed oxygen ($\sigma_{O^{2-}}$)- and hydroxyl (σ_{OH^-}) groups. The oxide charge σ_{ox} is defined as

$$\sigma_{ox} = \sigma_0 + \sigma_{O^{2-}} + \sigma_{OH^-} \qquad (15)$$

where σ_0 is the charge density of the bare solid surface.

When an ion, which determines ionic strength, adsorbs specifically, $\sigma_{O^{2-}}$ and σ_{OH^-} are influenced by this specific adsorption, and so is σ_{ox}. When this specific adsorption is substantial, this also affects the pH_{zpc}. If an ion does not absorb specifically, no competition between this ion and the oxygen and hydrogen groups occurs, and the pH_{zpc} does not change with ionic strength.

In the case of phosphate and silicate, direct measurements are possible because relatively large quantities of these ions adsorb onto the colloids: the difference between total ion content and ion concentration in the solution is significant and gives an accurate value for the amount of adsorbed phosphate or silicate. Breeuwsma (1973) studied phosphate adsorption onto hematite after three hours equilibration; Sigg (1979) allowed for 24–48 hours equilibration.

Calibration

The search for the best model has been performed in several steps. First a simple model (no physical adsorption layer, one type of adsorption site) was tested with only K^+ and Cl^- present as ionic strength determining ions; no specific adsorption of K^+ nor Cl^- was assumed. Results were bad, and this picture was extended several times until the model as outlined in Fig. 1[b] was reached. Two types of hydroxide sorption sites are distinguished (I, II) and one cation sorption site (III) (not mentioned in Fig. 1[b]) that is related to not occupied metal positions in the corund structure. It turned out to be necessary to assume some specific Cl^--adsorption on site I (the outermost hydroxide sorption site) and specific K^+-adsorption onto site III, not interfering with the anion adsorption sites. Results of model computations are presented in Fig. 2 for three ionic strengths.

Then, other adsorption parameters (for Ca^{2+}, SO_4^{2-}, Mg^{2+}) were tested. The results (position and specific adsorption potentials) are listed in Tables 2 and 3. Figure 3 shows calculated and

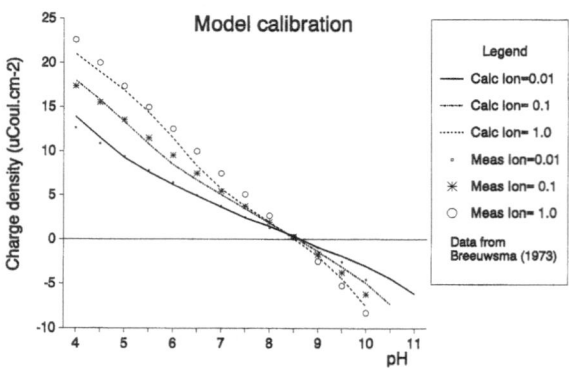

Fig. 2. Experimental (Breeuwsma, 1973) and computed data on hematite oxide charge. Parameters as listed in Tables 2 and 3. Ion = [KCl].

Table 2. Specific adsorption potentials for specific adsorption of ions onto hematite.

Ion	Site I	Site II	Site III
OH^-	0.329	0.435	
O^{2-}	0.138	0.27	
Cl^-	-0.482		
K^+			0.048
SO_4^{2-}	-0.09		
Ca^{2+}			0.11
$H_2PO_4^-$	0.00		
HPO_4^{2-}	0.22		
PO_4^{3-}	0.60		
$H_3SiO_4^-$	0.31		
$H_2SiO_4^{2-}$	0.53		

ϕ in Volts. Indexes denote: I: outer adsorption site; II: inner adsorption site; III: outer empty Fe-site. Phosphate, sulphate and silicate occupy two outer adsorption sites. Adsorption of oxygen is used to describe dissociation of a proton from an adsorbed hydroxyl ion; used as 'apparent oxygen activity' is $\{O^{2-}\} = 10^{(-20+2\,pH)}$. Since it is the product of K_{ads} and $\{O^{2-}\}$ that is important, the mentioned K_{ads}-values have only meaning in combination with this description of $\{O^{2-}\}$.

Table 3. Distances to bare surface (Å) for specific adsorption of ions onto hematite.

Ion	Site I	Site II	Site III
OH^-	1.40	1.40	
O^{2-}	1.40	1.40	
Cl^-	2.40		
K^+			1.40
SO_4^{2-}	1.75		
Ca^{2+}			1.40
$H_2PO_4^-$	1.75		
HPO_4^{2-}	1.75		
PO_4^{3-}	1.75		
$H_3SiO_4^-$	1.75		
$H_2SiO_4^{2-}$	1.75		

Distance to bare surface in Å. Sites denote: I: outer adsorption site; II: inner adsorption site; III: outer empty Fe-site. Phosphate, sulphate and silicate are large anions with smaller charge bearing groups (O-atoms). The position of these groups is indicated here.

measured (Breeuwsma, 1973) phosphate adsorption at different pH- and [P]-values; Fig. 4 shows measured (Sigg, 1979) and computed silicate adsorption.

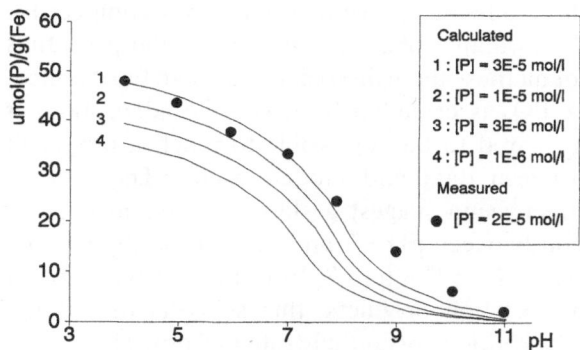

Fig. 3. Calculated phosphate sorption onto hematite. Aspec $= 30\,m^2\,g^{-1}$ (Fe). Points denote measurements as published by Breeuwsma (1973). $[P] = 2\cdot10^{-5}\,mol\cdot l^{-1}$.

For anion and cation adsorption onto hematite, the model abstraction shows reasonably good fits between data and calculations. The simpler the ions, the better the fit. A major conclusion is that (minor) specific adsorption of chloride and potassium has to be assumed. Otherwise, calculated oxide charges would not change with ionic strengths as observed by Breeuwsma. The adsorption of potassium as a positively charged ion onto the oxide surface seems somewhat abnormal; however it is often described that cations can penetrate into the oxide structure (Breeuwsma & Lyklema, 1973).

It is clear from Fig. 3 that the computation of phosphate adsorption onto hematite is of the right order, but of the wrong shape. Since not one, but

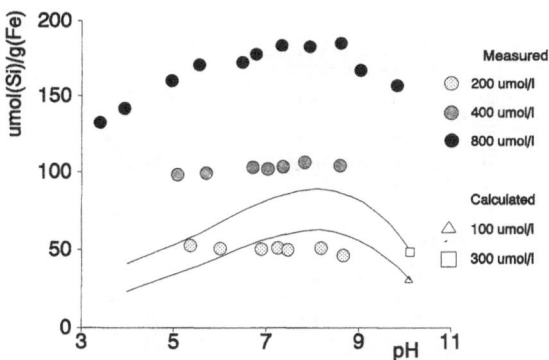

Fig. 4. Calculated and measured (published by Sigg, 1979) silicate sorption onto goethite. Ion $(KClO_4) = 0.01$. Data as in Tables 2 and 3.

38

three different phosphate ion types compete for the available sites, specific adsorption potentials for all these three dissociations are to be obtained. H_3PO_4 is excluded from specific adsorption. It appeared to be impossible to reach a proper fit between data and model results. The data of Breeuwsma suggest a slow decrease in adsorption between pH = 7 and 9.0 and hardly any decrease for pH = 6 to 7. In terms of adsorption of dissociation products, this suggests that PO_4^{3-} adsorption substantially differs from HPO_4^{2-} adsorption. The present model is unable to produce these differences. One possibility that was not studied is a change in site, or a change in occupation: one PO_4^{3-} might occupy one site while an HPO_4^{2-} might occupy two sorption sites. This could explain a relatively small decrease in adsorption when PO_4^{3-} becomes important as a sorbed anion. It might also be possible that both ions change occupation influenced by $\{OH^-\}$. An additional problem is that Breeuwsma allowed his adsorption experiments to stabilize for three hours: the observed characteristics might also be of kinetic nature.

In Fig. 4, calculated silicate sorption onto goethite is presented, using the hematite model and data as quantified in Table 2 and 3. It is clear that silicate shows an adsorption completely different from phosphate; the reason being a different dissociation behaviour. Silicate adsorption is almost as strong. The calculated data and the data published by Sigg do not fit very well, although they are of the right order and shape.

The model was further used to compute phosphate-silicate adsorption competition. In Fig. 5 P and Si adsorption onto hematite is given as a function of [Si], using the table 2 and 3 specifications. $[P] = 10^{-5}$ and $[Si] = 10^{-4}$ (mol l^{-1}) are not unusual in pore water under natural conditions. The model predicts that, at these concentrations, the oxide surface sites are occupied by silicate rather than by phosphate. Still, silicate occupies less than 10% of the total number of sites.

Figure 6 illustrates the calculated adsorption of phosphate and silicate at varying pH values. At low pH, phosphate adsorbs preferably because of the strong charge-charge interaction: the potential on the plane of adsorption is positive and the phosphate ion charge is −2 instead of −1 (silicate). At intermediate pH-values, the plane potential decreases and the silicate charge changes to −2, thus improving adsorption. At high pH-values the plane potential becomes negative, and at the same time the phosphate ion charge changes to −3, which has an opposite effect: the physical attraction term in eq. 8 changes into a physical repulsion term. Because of the ion charge, this term is less important for silicate, resulting in preferred silicate sorption. So, it is the influence of the adsorbing ions themselves on the plane potentials that largely determines the adsorption behaviour. Because of this charge influence, the

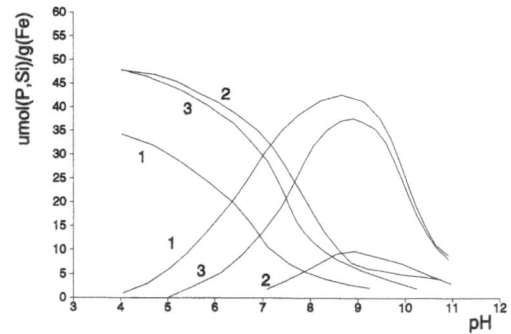

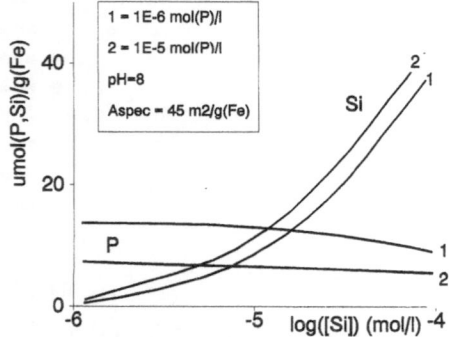

Fig. 5. Competitive silicate and phosphate sorption onto hematite, calculated according to data from Tables 2 and 3.

Fig. 6. Computed sorption of silicate and phosphate onto hematite. Characteristics in tables 2 and 3. Ionic strength (KCl) = 0.01. Aspec = 45 m^2 g^{-1} (Fe). 1: $[P] = 10^{-6}$, $[Si] = 10^{-4}$; 2: $[P] = 3 \cdot 10^{-5}$, $[Si] = 10^{-5}$; 3: $[P] = 3 \cdot 10^{-5}$, $[Si] = 10^{-4}$ (mol·l^{-1}).

sorption capacity of an oxide will never be used by phosphate or silicate completely.

Application to a natural system

Data collection

Lake Veluwe (The Netherlands, Fig. 7) is a shallow freshwater lake, formerly part off the brackish Zuiderzee (until 1932). The present state was realized in 1957. Surface area 37.8 km^2, average depth 1.26 m. Typical surface water composition is: chloride 5, sulphate 2.5, carbonate 2, sodium 6, calcium 2.5, magnesium 1, potassium 0.3, o-phosphate 0–2 10^{-3} (μmol l^{-1}).

Undisturbed sediment samples were analyzed for total-P, Fe, Si, COD, Na, Al, silt and clay content, after freeze-drying (if relevant) (Brinkman & Van Raaphorst, 1986).

Sediments ranged from sand (composition: silt <1.0%, SiO$_2$ >97%, organic matter <1%) to clay (silt 60%, SiO$_2$ 32%, organic matter 8%). Total-Fe contents ranged from less than 5 to 30 mg (Fe) g(dw)$^{-1}$, total-P contents from 0.05 to 0.8 mg(P) g(dw)$^{-1}$.

Adsorption and desorption characteristics

were studied on freeze-dried samples under aerobic conditions.

Adsorption was studied after suspension of a known amount of material (about 4 g(dw) l^{-1}) in artificial lake water (ionic strength 0.01) that contained a certain phosphate concentration, ranging from 0 to 1.5 mg(P) l^{-1}. After three days of equilibration samples were analyzed after filtration over a 0.45 μ Millipore membrane filter. Kinetics of the sorption process were also studied separately and indicated that three days of equilibration was more than sufficient (Fig. 8).

Desorption was studied by leaching a continuously stirred suspension of 4 gr(dw) l^{-1} sediment in artificial lake water (Ion = 0.01) (Fig. 9). The filtered (0.45 μ membrane filter) volume was replenished with fresh artificial lake water, at an average flow rate of 0.5 l day^{-1}. The phosphorus in the filtered water is sorbed onto an anion exchange resin (Biorad AG$_1$–X$_4$, Cl$^-$-form, 20–50 mesh size). This resin was replaced every two days, and regenerated by flushing twice with 50 ml of 3 N HCl. The exact lake water flow was measured by weighting. This procedure resulted in a 20 times increase in P-concentration, allowing accurate [P]-determination.

Desorption data revealed the amount of desorbable phosphorus initially present on the solid, and these data were added to the results of the adsorption experiments to construct the final isotherms.

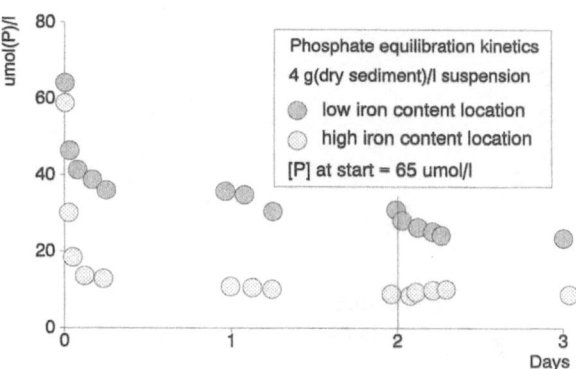

Fig. 8. Equilibration of phosphate sorption onto Lake Veluwe sediment. 4 g (dry sediment) l^{-1} suspended in artificial lake water, continuously stirred. I = 0.01. After two days the pH was raised from 7.8 to 8.8.

Fig. 7. Lake Veluwe situated in The Netherlands.

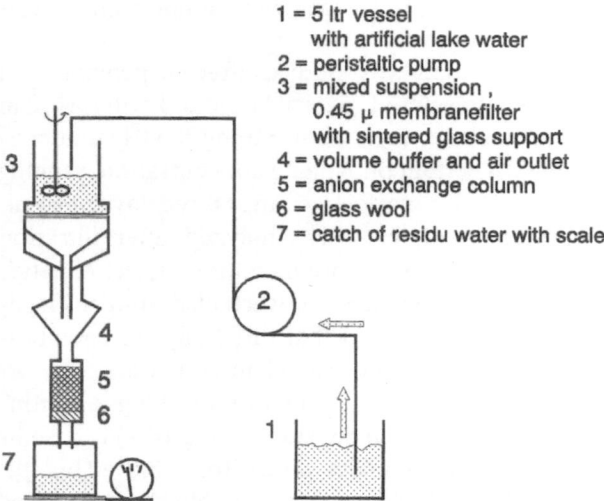

1 = 5 ltr vessel
    with artificial lake water
2 = peristaltic pump
3 = mixed suspension ,
    0.45 μ membranefilter
    with sintered glass support
4 = volume buffer and air outlet
5 = anion exchange column
6 = glass wool
7 = catch of residu water with scale

Fig. 9. Leaching experiment. Suspension in vessel 3 contains 3 g (dry sediment) l^{-1} of P-desorbing sediment.

Results

In Figs. 10 and 11 examples of Lake Veluwe adsorption isotherms are shown, for a relatively clayey area (Fig. 11) and a more sandy area (Fig. 10). Adsorption turned out to be highly related to the iron content (Brinkman & Van Raaphorst, 1986). So, all the results were related to the iron content of the samples. In Fig. 12 all the data have been grouped; the linear scale is replaced by a logarithmic one because in the desorption experiments low $[P]$ were reached (down to

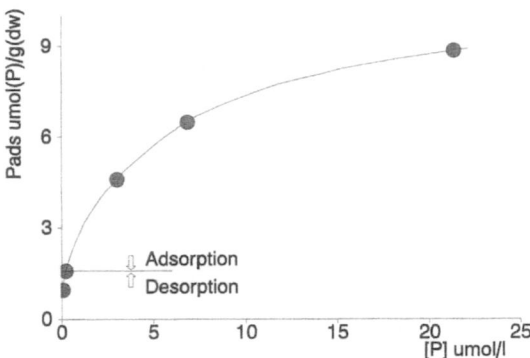

Fig. 11. Sorption isotherm for Lake Veluwe sediment with a relatively high iron content.

$0.01\ \mu mol\ l^{-1}$). Included in this figure is the estimated classical Langmuir adsorption isotherm (eqs. 2 to 5), K_{ads} being constant for all $[P]$-values.

The maximal adsorption for these samples is estimated to be $0.6\ mmol(P)\ g(Fe)^{-1}$. $K_{ads} = 2.04\ 10^5\ l\ mol^{-1}$. At low $[P]$ values, K_{ads} is too low, while K_{ads} is too large at high $[P]$-values. A quantification in terms of specific areas is not relevant in this case.

A better fit was obtained applying the double-layer model using the parameter values as listed in Tables 2 and 3 (Fig. 12).

The assumed specific surface area of the iron oxides was $358\ m^2\ g(Fe)^{-1}$ and the maximum adsorption value was estimated at $8.7\ mmol(P)\ g-$

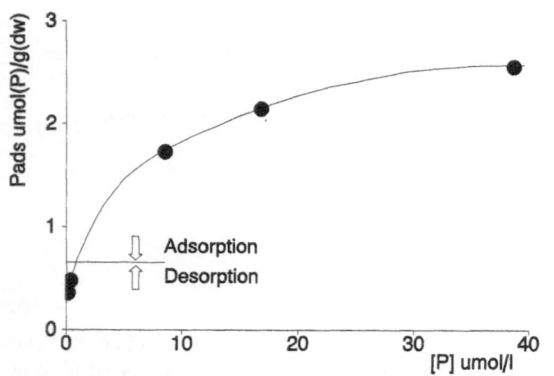

Fig. 10. Sorption isotherm for Lake Veluwe sediment with a low iron content.

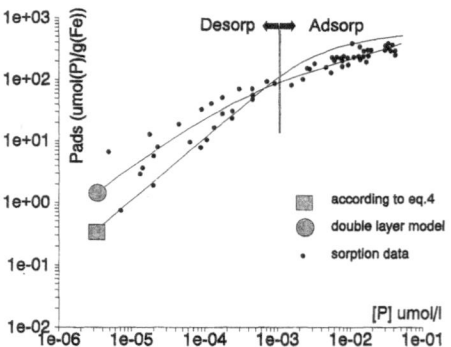

Fig. 12. Lake Veluwe sorption data and model results. Double layer computation: pH = 7.74, Aspec = 358 m^2 g^{-1} (Fe). Data in Tables 2 and 3. Eq. 4 computation: $K_{ads} = 2.04\ 10^5\ l\ mol^{-1}$, $Ads_{max} = 0.6\ mmol\ (P)\ g^{-1}\ (Fe)$.

(Fe)$^{-1}$. In Fig. 13, calculated K_{ads} values for HPO$_4^{3-}$ and PO$_4^{3-}$, the two dominant adsorbing phosphate ions, are shown. K_{ads} is high at low $[P]$ and decreases with increasing $[P]$.

Sorption may very well be described with a purely empirical formula. But, the validity of such equations is limited to the situation that is studied. On the other hand, the use of the double-layer adsorption model allows for multiple interaction calculations at different pH values, ion concentrations and ionic strengths. In Fig. 14 this is illustrated. Given the natural environment in Lake Veluwe, the extent of the sorption onto the sediment (characterized by the iron content) was computed. Sorption of K$^+$, Cl$^-$, SO$_4^{2-}$, phosphates ($= H_3PO_4 + H_2PO_4^- + HPO_4^{2-} + PO_4^{3-}$), silicates ($= H_4SiO_4 + H_3SiO_4^- + H_2SiO_4^{2-}$) and Ca^{2+} is presented as a function of pH. The ion concentrations used reflect average concentrations in the surface water of the lake. Consequently, the figure illustrates the calculated absolute participation of each of the major lake ions in the adsorption onto iron oxide at the different pH values. Since this picture is based upon the model tuning as presented here, it should be realized that the results contain considerable uncertainties. In the first place, these uncertainties concern the obtained parameters, but also the assumption that the hematite characteristics are illustrative for the natural iron oxide sorbent present in Lake Veluwe.

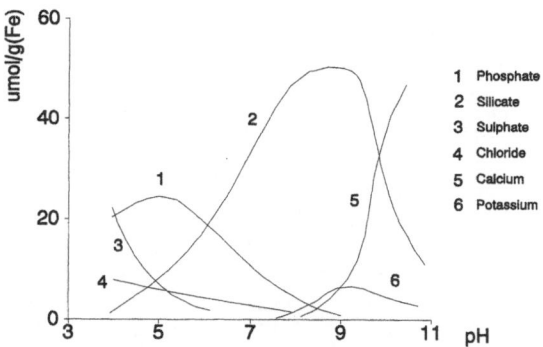

Fig. 14. Competitive sorption characteristics of Lake Veluwe sediment (defined by its iron content), as a function of pH. Double layer calculation. Sorption data as presented in Tables 2 and 3. $[P] = 10^{-6}$, $[Si] = 2\ 10^{-4}$, $[SO_4^{2-}] = 2\ 10^{-3}$, $[Cl^-] = 5\ 10^{-3}$, $[Ca^{2+}] = 2.5\ 10^{-3}$, $[K^+] = 4\ 10^{-3}$ mol l^{-1}.

From these calculations, it can, however, be concluded that in the natural situation of Lake Veluwe, silicate likely plays a dominant role in the sorption onto the iron oxides present as suspended matter. In the sediment, phosphate concentrations may be higher, and consequently, phosphate sorption may be more important.

General discussion

Double-layer model and calibration

The calibration of the double-layer model, as outlined in the previous section, was largely based on potentiometric titration data originating from literature. All the errors from the computation of the dielectric constants, from the model abstraction itself and from the experimental data result in uncertain plane positions and/or specific adsorption potentials.

Other characteristics, such as differential double-layer capacities and electrokinetic potentials, were not used here. The first can be derived from the σ_{ox}-pH curves, and contain information on the shape of these curves (Blok & De Bruyn, 1970). Electrokinetic potentials are the potentials on the final plane where ions or water molecules are bound. They determine the behaviour of a colloid, as a charged particle, in an electrophoretic

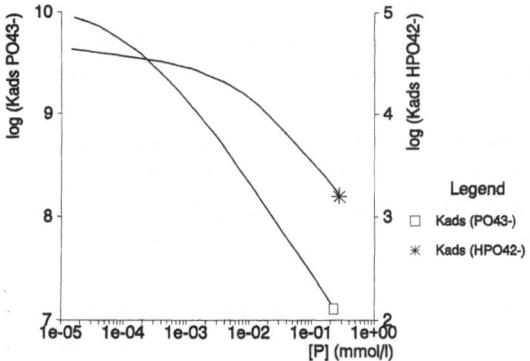

Fig. 13. Adsorption equilibrium constants for the fig. 12 isotherm (double layer computation) for HPO$_4^{2-}$ and PO$_4^{3-}$, as a function of $[P]$. pH = 7.74.

42

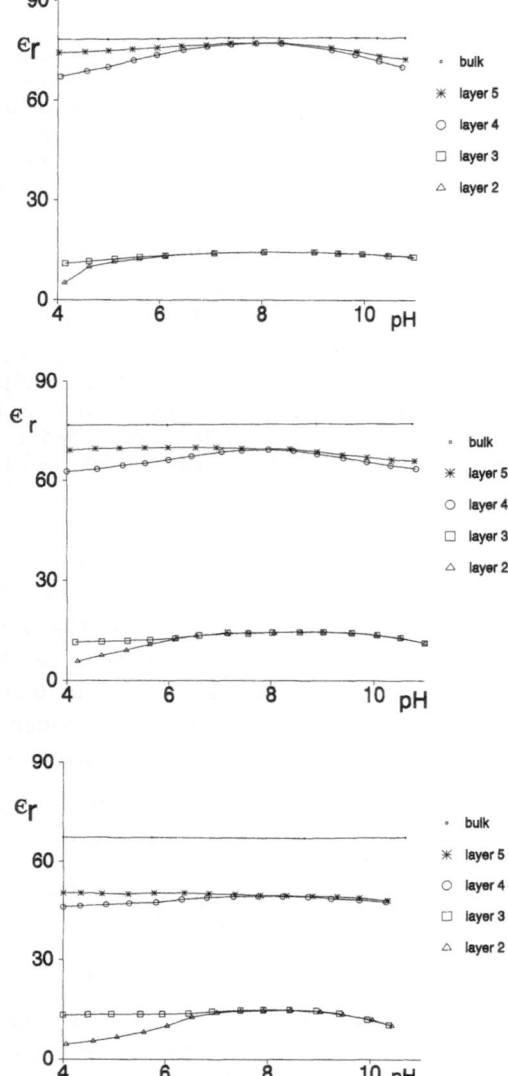

Fig. 15. Double layer dielectric constants, KCl adsorption. Sorption data listed in Tables 2 and 3. Layer numbers i denote layers between planes $(i-1)$ and i. a:[KCl] = 0.01; b: [KCl] = 0.1; c: [KCl] = 1.0 mol l^{-1}.

experiment. This electrokinetic potential is not very well defined, but it will usually not differ much from the potential at the outer plane in Fig. 1: Ψ_5. Electrokinetic data were not available for Breeuwsma's experiments.

Analyzing model adsorption and colloidal characterization data of hematite (Breeuwsma, 1973) with the presented adsorption model shows that

1. the slightly convex oxide charge – pH curves are very well simulated for I = 0.01 with the double-layer model when two different Fe-adsorption sites are assumed.
2. for a satisfactory simulation of oxide charges at all ionic strengths, a slight adsorption of Cl^- at the anion adsorption sites, and of K^+ at the empty Fe-cavities has to be assumed.
3. phosphorus adsorption and silicate adsorption are satisfactorily simulated, but not perfectly. Model adaptation might improve the results, but likely will not make the calculations fit completely to the measured data.

Most promising is the ability of calculating all kinds of interactions of adsorbing ions. Apart from steric hindrance and adsorption site occupation, these interactions mainly are electrostatic. Chemical adsorption of charged ions influences the surface charge and the double-layer properties, and through these the adsorption equilibrium constant K_{ads}. An important advantage of the double-layer model compared to classical approaches is that it explicitly takes these interactions into account.

Model translation for application to field situations

The application of the detailed double-layer model offers the possibility to simulate adsorption characteristics of natural sediments. A number of uncertainties remain like the validity of this translation of model characteristics into a natural environment where colloids never exist in a pure form but always are contaminated and do not have a well-defined structure as assumed here. These contaminations and structure deviations, however, are deviations from a general picture, and as such, this general picture certainly is of importance. Field studies have to reveal these deviations. The field study in Lake Veluwe was not detailed enough to assess all colloid-chemical adsorption properties. Consequently, the direct application of the model to the field is somewhat hazardous, but at the same time it shows that the model results are potentially useful.

Data collection

It can be concluded from the sorption computations that, when studying adsorption and desorption processes, it is very important that the concentrations of other strongly adsorbing ions are kept constant during the experiments (besides the ionic strength that should always be kept constant). This especially is the case for silicon, magnesium and calcium in the case of phosphorus sorption.

Acknowledgements

The author wishes to thank Prof. Dr C. A. Smolders and Prof. Dr L. Lijklema because a substantial part of the work has been done in their groups (General Chemistry and Environmental Engineering, respectively) at the Twente University of Technology, Faculty of Chemical Engineering, Enschede, The Netherlands.

J. de Jonge, Dr Ir. W. van Raaphorst and the author performed the field surveys and the sorption experiments. The second, presently Netherlands Institute for Sea Research, kindly commented on the text.

Appendix 1

Gouy-Chapman equation for the calculation of the potential at the outer plane

The Gouy-Chapman equation reads

$$\sigma_\delta = \sqrt{n \, \varepsilon_r \, \varepsilon_0 \, k \, T \, 2} \, (- z^- \exp(- z^+ \, y_d) +$$

$$z^+ \exp(- z^- \, y_d) - z^+ + z^-)^{0.5} \qquad (16)$$

where δ denotes the outer adsorption plane, and

$$y_d = \Psi_d \frac{e}{kT} \qquad (17)$$

where

n	=	number of equivalents per m^3
ε_0	=	permittivity of the vacuum ($C \, V^{-1} \, m^{-1}$)
ε_r	=	dielectric constant of the liquid phase ($-$)

σ_δ	=	charge density on plane δ ($C \, m^{-2}$)
z^+, z^-	=	cation and anion valence
T	=	absolute temperature (K)
Ψ_δ	=	plane δ potential (V)
k	=	Boltzmann's constant ($1.3805 \, 10^{-23} \, J \, K^{-1}$)

To calculate the layer's potential, eq. 16 has to be rearranged. With

$$A = \frac{\sigma_\delta^2}{z \, \varepsilon_0 \, \varepsilon_r \, n \, k \, T} + z^+ - z^- \qquad (18)$$

and

$$B \exp(- y_d) \gg \Psi_\delta = \frac{kT}{e} \ln(B) \qquad (19)$$

the equation

$$A = - z^- \, B^{z^+} + z^+ \, B^{z^-} \qquad (20)$$

has to be solved. Two cases exist, $z^+ = z^-$, and $z^+ \# - z^-$. In the first case, after substituting

$$D = B^{z^+} \qquad (21)$$

it follows that

$$D = \left[\frac{A}{|z|} + SIGN(- \sigma_\delta) \sqrt{\left(\frac{A}{|z|} \right)^2 - 4} \right] 0.5 \qquad (22)$$

In the second case, with

$$FI = sign(- \sigma_\delta) \, sign(z^+ + z^-) \, arccos \left[\left(\frac{A}{3} \right)^{-3/2} \right] \qquad (23)$$

it follows that

$$B = \left[- 2 \sqrt{\left(\frac{A}{3} \right)} \cos \left(120° + \frac{FI}{3} \right) \right]^{sign(z^+ + z^-)} \qquad (24)$$

The layer potential follows from eq. 19.

When Ψ_δ is small, a linearized form can be used.

This equation holds for flat particles. Sphericle particles are discussed by Loeb *et al.* (1961).

Appendix II

Calculation of the dielectric constant in the layers and calculation of the layer capacities

The calculation of the dielectric constant ε_r is the Achilles' heel of the double-layer model. The main reason for this is that there is no method to measure dielectric constants in double-layers at all, only overall capacities can be estimated from measurements. However, on a macroscale it is possible to measure the dielectric constant ε_r and to derive relationships between ε_r , ion concentrations and field strengths. These relationships are briefly outlined here; in the double-layer model these macroscale relationships are employed without microscale adaptations.   The double-layer is thought to consist of a number of flat plate capacitors. On each plane a charge density σ_j is present being the sum of the bare oxide charge σ_0 and the adsorbed charges (eq. 13). For reasons of charge neutrality, the charge density on the other plane of the capacitor is the same but opposite:

$$\sigma_j = -\sigma_{j+1} \qquad \text{(Coul m}^{-2}) \qquad (25)$$

The field strength E in a capacitor or layer reads

$$E = D\,\varepsilon_r \qquad \text{(Volt m}^{-1}) \qquad (26)$$

where

D = electric field strength in vacuum (V m^{-1})
ε_r = dielectric constant $\varepsilon_r \geq 1$ (–)

D follows from

$$D = \frac{\sigma}{\varepsilon_0} \qquad \text{(Volt m}^{-1}) \qquad (27)$$

with

ε_0 is vacuum permittivity (C V^{-1} m^{-1}) and σ the capacitor charge density (C m^{-2}).   The permittivity of solutions including ions reads $\varepsilon = \varepsilon_r\,\varepsilon_0$, ε_r is the result of a redistribution of charge. This redistribution is a result of

– dielectric polarization of molecules and ions;
– alignment of dipoles along the direction of the field.

The presence of ions influence the dielectric properties of the solution (Hasted, Ritson & Collie, 1948; Hasted & Roderich, 1958; Kraeft & Gerdes, 1965); a large electric field as it exists in double-layers results in an almost complete alignment of dipoles, and consequently a decreased extra charge compensation. This phenomenon is called dielectric saturation: ε_r approaches one (Booth, 1951; Böttcher, 1973). Both mechanisms are dealt with in the computation of ε_r.   This description is used for the double-layer calculations without any adaptation. Consequently, the flat capacitor medium is thought to be a continuum, instead of consisting of discrete particles, and the charged adsorbed onto the planes are thought to be smeared out over these planes. This, and exactly this, is the vulnerable part of the description: the macroscale interactions are assumed to be valid on the microscale of the double-layer.   In Fig. 15 double-layer dielectric constants, as computed for adsorption of KCl alone, are shown for three ionic strengths.

References

Atkinson, R. J., A. M. Posner & J. P. Quirk, 1967. Adsorption of potential determining ions at the ferric oxide-aqueous electrolyte interface. J. Chem Phys. 71: 550–558.

Beek, J. & W. H. van Riemsdijk, 1979. Interaction of orthophosphate ions with soil. In G. H. Bolt (ed.), Soil Chemistry, B. Physico-chemical models, Developments in Soil Science 5B, Elsevier, New York.

Blok, L. & P. L. de Bruyn, 1970. The ionic double-layer at the ZnO-solution interface. J. Colloid Interf. Sci. 32: 518–538.

Boehm, H. P., 1971. Acidic and basic properties of hydroxylated metal-oxide surfaces. Disc. Far Soc. 52: 264–275.

Booth, F., 1951. The dielectric constant of water and the saturation effect. J. Chem. Phys. 19: 391–394, 1327–1328, 1615.

Böttcher, C. J. F., 1973. Theory of electric polarization, I: dielectrics in static fields. 2^e edn, compl. revised by O. C. van Belle, P. Bordewijk & A. Rip. Elsevier, Amsterdam, 377 pp.

Breeuwsma, A. & J. Lyklema, 1973. Physical and chemical adsorption of ions in the electrical double-layer on hematite (α-Fe$_2$O$_3$). J. Colloid Interf. Sci. 43: 437–448.

Breeuwsma, A., 1973. Adsorption of ions on hematite α-Fe$_2$O$_3$ – a colloid chemical study. Thesis Agriculture Univ. Wageningen, 121 pp.

Brinkman, A. G. & W. van Raaphorst, 1986. De fosfaathuishouding in het Veluwemeer. Thesis Twente Univ. of Technology, Enschede (in Dutch), 700 pp.

Cabrera, F., L. Madrid & P. de Arambarri, 1977. Adsorption of phosphate by various oxides: theoretical treatment of the adsorption envelope. J. Soil Sci. 28: 306–313.

Chen, Y. S. R., J. N. Butler & W. Stumm, 1973. Adsorption of phosphate on alumina and kaolinite from dilute aqueous solutions. J. Colloid Interf. Sci 43: 421–436.

Goujon, G., J. M. Cases & B. Mufaftshiev, 1976. On the adsorption of N-dodecylammonium chloride on the surface of synthetic calcite. J. Colloid Interf. Sci. 56: 587–595.

Griffin, R. A. & J. J. Jurinak, 1973. The interaction of phosphate with calcite. Soil Sci. Soc. Am. Proc. 37: 847–850.

Grahame, D. C., 1947. The electrical double-layer and the theory of electrocapillarity. Chem. Rev. 41: 441–501.

Haan, F. A. M. de, 1965. The interaction of certain inorganic anions with clays and soils Thesis Agricultural Univ. Wageningen (Netherlands).

Hasted, J. B., D. M. Ritson & C. H. Collie, 1948. Dielectric properties of aqueous ionic solutions, part I, II. J. Chem. Phys. 16: 1–21.

Hasted, J. B. & G. W. Roderich, 1958. Dielectric properties of aqueous and alcoholic solutions. J. Chem. Phys. 29: 17–26.

Hayes, F. K., Ch. Papelis & J. O. Leckie, 1988. Modelling ionic strength effects on anion adsorption at hydrous oxide/solution interfaces. J. Colloid Interface Sci. 125: 717–726.

Hayes, F. K., G. Redden, W. Ela & J. O. Leckie, 1991. Surface complexation models: an evaluation of model parameter estimation using FITEQL and oxide mineral titration data. J. Colloid Interface Sci. 141: 448–469.

Helyar, K. R., D. N. Munss & R. G. Burau, 1976. Adsorption of phosphate by gibbsite – II: formation of a surface complex involving divalent cations. J. Soil Sci. 27: 315–323.

Hiemstra, T., J. C. M. de Wit & W. H. van Riemsdijk, 1988-b. Multisite proton adsorption modelling at the solid/solution interface of (hydr)oxides: a new approach: II. Application to various important (hydr)oxides. J. Colloid Interface Sci. 133: 105–117.

Hingston, F. J., A. M. Posner & J. P. Quirk, 1974. Anion adsorption by goethite and gibbsite – II: desorption of anions from hydrous oxide surfaces. J. Soil Sci. 25: 16–26.

Huang, C. P. & W. Stumm, 1972. The specific surface area of γ-Al_2O_3. Surface Sci. 32: 287–296.

Jacobsen, O. S., 1978. Sorption, adsorption and chemosorption of phosphate by Danish lake sediments. Freshwater Biological Laboratory, University Copenhagen Publ. 305.

James, R. O., P. J. Stiglich & T. W. Healy, 1975. Analysis of models of adsorption of metal ions at oxide/water interfaces. Disc. Faraday Soc. 59: 142–156.

Jones, B. F. & C. J. Bowser, 1978. The mineralogy and related chemistry of lake sediments. In A. Lerman (ed.), Lakes: chemistry, geology, physics. Springer Verlag, Berlin: 179–235.

Kraeft, W. D. & E. Gerdes, 1965. Ein Verfahren zur Messung der komplexen Dielektrizitätskonstanten von konzentrieten Ionlösungen mittels Hohlraumresonatoren. II: Experimentelle Durchführung und Messergebnisse. Phys. Chem. 228: 331–342.

Loeb, A. L., J. Th. Gr. Overbeek & P. H. Wiersema, 1961. The electrical double-layer around a spherical colloid particle. M.I.T. Press, Cambridge, Mass. 53 pp.

Marquardt, D. W., 1963. An algorithm for least squares estimation of non linear parameters. J. Soc. Industr. Appl. Math. 11: 431–441.

Parfitt, R. L., R. J. Atkinson & R. St. C. Smart, 1975. The mechanism of phosphate fixation by iron oxides. Soil Sci. Soc. Am. Proc. 39: 837–841.

Parfitt, R. L. & R. St. C. Smart, 1977. Infrared spectra from binuclear bridging complexes of sulphate adsorbed on goethite (α-FeOOH). J. Chem Soc. Faraday Trans. I 73: 796–802.

Parks, G. A., 1965. The iso-electric points of solid oxides, solid hydroxides and aqueous hydroxo complex systems. Chem. Rev. 65: 177–198.

Sigg, L. M., 1979. Die Wechselwirkung von Anionen und 'schwachen' Säuren met α-FeOOH (Goethit) in wässriger Lösung. Thesis Technische Hochschule, Zürich, 141 pp. (in German).

Whittemore, D. O. & D. Langmuir, 1975. The solubility of ferric oxyhydroxide in natural waters. Ground Water 13: 360–365.

Hydrobiologia **253**: 47–59, 1993.
P.C.M. Boers, T.E. Cappenberg & W. van Raaphorst (eds),
Proceedings of the Third International Workshop on Phosphorus in Sediments.
© 1993 *Kluwer Academic Publishers.*

Iron-bound phosphorus in marine sediments as measured by bicarbonate-dithionite extraction

Henning Skovgaard Jensen[1] & Bo Thamdrup[2]
[1] *Institute of Biology, Odense University, Denmark; present address: National Environmental Research Institute, Department of Freshwater Ecology, P.O. Box 314, Vejlsøvej 25, DK-8600, Denmark;* [2] *Department of Microbial Ecology, University of Aarhus, Denmark*

Abstract

A sequential five-step extraction scheme for phosphorus pools in freshwater sediment was modified for use in marine sediments. In the second step phosphate bound to reducible forms of iron and manganese ('iron-bound P') is extracted by a bicarbonate buffered dithionite solution (BD-reagent). The extraction scheme was tested on sediment from 16 m water depth in Aarhus Bay, DK and used in two other marine sediments: Kattegat at 56 m and Skagerrak at 695 m depth. By comparing the BD-extractable P-pool with both the pool of iron in the BD-fraction and the pool of oxidized, amorphous or poorly crystalline iron (am.FeOOH), highly significant correlations ($p < 0.001$) were observed in all three sediments. Thus, we conclude that the BD-reagent was very specific for iron-bound P. Further evidence for this came from two experiments: 1) Enhanced BD treatment did not result in additional phosphate extraction and 2) by sequential extraction of phosphorus pools in pure cultures of diatoms and cyanobacteria no phosphate was recovered in the BD-fraction. The pool of am.FeOOH was very important for controlling porewater phosphate concentration which was inferred from the significant inverse relationships between the two parameters ($p < 0.001$) in all sediments studied. Further, an isotopic exchange experiment with $^{32}PO_4^{3-}$ revealed that BD-extractable P was by far the most exchangeable P-pool even deep in the sediment where the pool size was small. Iron-bound P made up 33–45% of total P in the surface sediments. The ratio between iron-bound phosphate and am.FeOOH was 8–11 in Aarhus Bay and Kattegat. In Skagerrak the ratio was 17, which may indicate that the iron mineral extracted from this sediment is less capable of adsorbing phosphate or less saturated with phosphate.

Introduction

Phosphorus bound to reducible forms of iron and manganese is in general considered as a potentially mobile pool of phosphorus in sediments. It may be released from anoxic sediments upon reduction of the reactive oxidized species of the two metals. Evidence for this is mostly based on freshwater literature (e.g., Einsele, 1936; Mortimer, 1941; Stauffer, 1981; Nürnberg, 1988) but some marine studies suggest that the mechanism is valid for marine systems, too (Krom & Berner,

1981; Balzer, 1986; Yamada & Kayama, 1987; Sundby *et al.*, in press). Consequently, this phosphorus pool (normally termed 'iron-bound P') becomes a potential internal phosphorus source to water bodies which suffer from occasional oxygen depletion. Many Danish estuaries experience occasional O_2 depletion and the ability to measure the pool size of iron-bound P becomes crucial in the understanding of P cycling in these systems.

Sequential chemical extraction schemes have been widely used to distinguish between various

P-pools in freshwater sediments (e.g. reviewed by Stauffer, 1981; Boström *et al.*, 1982; Petterson *et al.*, 1988). A workshop on phosphorus in sediments held in Austria 1986 recommended the use of a five-step scheme where iron-bound P is extracted by use of a bicarbonate buffered solution of sodium-dithionite (BD-reagent) (Psenner *et al.*, 1988). The same recommendation is given by Petterson *et al.* (1988).

The BD-reagent reduces the oxidized species of iron and manganese and thereby liberates the phosphorus absorbed onto oxides/hydroxides of the two metals. Nürnberg (1988) found that iron-bound phosphorus measured by BD-extraction provided the best estimate of the internal P loading in stratified lakes with anoxic bottom water when compared to iron-bound phosphorus measured by NaOH extraction. Only few attempts have been made to apply the BD-method for marine studies. Balzer (1986) used two other reductants (hydroxylamine and oxalate) for determinations of manganese- and iron-bound P in the bay of Kiel. Lucotte (1985) examined various chemicals for simultaneously extraction of iron-bound P and oxidized iron species in suspended estuarine matter. He found the citrate dithionite bicarbonate (CDB)-reagent to be the best choice when compared to oxalate, hydroxylamine, nitriloacetic acid and acetate. Meanwhile, the specificity of the BD-reagent as well as the CDB-reagent for measuring iron-bound P is still a matter of debate. Lately, it has been questioned to which extent polyphosphates from algae and bacteria is extracted and hydrolized along with iron-bound P (R. Gächter, pers. com.).

The primary aim of the present study is to elucidate the specificity of the BD-reagent for iron-bound phosphorus in marine sediments and secondary to present some results on the entire five-step extraction scheme from different marine sediments.

We describe the slightly modified extraction scheme together with various control experiments carried out on sediment from Aarhus Bay, DK. The specificity of the BD-reagent for extracting iron-bound P is elucidated by comparing iron in the BD-extract with extractions of amorphous ferric hydroxide (am.FeOOH) (see below) and by comparing the BD extracted phosphate with the two different iron pools. This was done in three different marine sediments: Aarhus Bay at 16 m water depth, Kattegat at 56 m water depth and Skagerrak at 695 m water depth. The degree to which organic bound P would turn up as soluble reactive phosphorus (SRP) in the BD-extraction was tested by extraction of phosphorus from pure cultures of diatoms and cyanobacteria. Finally, the exchangeability of iron-bound P was studied by adding a carrier free $^{32}PO_4^{3-}$ solution to the sediment and subsequently measuring the ratio-activity in the different extracts.

Amorphous or poorly crystalline ferric iron forms, termed am.FeOOH, are much more efficient in phosphorus binding than well crystallized oxides (Jacobsen, 1978). Therefore, specific measurement of am.FeOOH is of interest, when comparing phosphorus and iron pools in sediments. Whereas the CDB extraction dissolves all free ferric oxide forms (Mehra & Jackson, 1960), schemes involving hydroxylamine, hydrochloric acid, or oxalate have been recommended for measuring amorphous iron oxides in soils (e.g. McKeague & Day, 1966; Chao & Zhou, 1983). In extractions of anoxic sediment, however, several authigenic ferrous iron compounds can contribute to the iron dissolved (Berner, 1981; Aller *et al.*, 1986). Two extraction schemes aiming at the determination of am.FeOOH in Fe(II)-containing sediments have recently been tested (Phillips & Lovley, 1987; Lovley & Phillips, 1987). In both schemes, Fe(III) is determined as the difference between total Fe and Fe(II) extracted. The anaerobic oxalate extraction (Phillips & Lovley, 1987) turned out to overestimate am.FeOOH significantly, due to dissolution of all crystalline iron oxides in the presence of catalytic amounts of Fe(II) (Fischer, 1973; Lovley, 1991). The hydroxylamine hydrochloric acid assay (Lovley & Phillips, 1987) did not dissolve crystalline iron oxides. However, when applied to marine sediments, it underestimated am.FeOOH due to reduction of Fe(III) during the Fe(II) determination step (Thamdrup, unpublished). Based on these results, another scheme for determination of am.FeOOH was developed by combination of slight modification of the two schemes mentioned. This scheme is not applicable to specific extraction of iron-bound phosphorus because of the simultaneous dissolution of apatite.

Study sites

Coring positions are given in Table 1. The sediment from Aarhus Bay on east coast of Jutland (close to st. 8 of Jørgensen *et al.* (1990)) was light brown and oxidized to 2–4 cm depth depending on the season. Below this depth, the sediment was reduced and grayish black, turning gray with depth. The infauna consisted mainly of small individuals of mussels, polychaetes and brittle stars. In the Kattegat sediment (close to st. 10 of Jørgensen *et al.* (1990)), the color zonation was similar, but here the light brown zone extended to 10–15 cm depth. Bioturbation was more intense than in Aarhus Bay due to a richer infauna dominated by brittle stars, larger polychaetes, and heart urchins. The Skagerrak station is located in the deepest part of Skagerrak. The sediment was visually quite different from the two other stations. Below the upper 5 cm, which was dark brown, the sediment was pale gray with few black spots. The infauna was dominated by small polychaetes living in slender vertical tubes in the upper 5 cm.

Methods

Sediment sampling

Sediment was sampled with use of a box-corer (40 × 40 cm, 50 cm deep) and subsampled in plexiglass tubes (inner diameter 5.2 cm). Sediment from below 40 cm depth was obtained with a piston corer.

Within a few hours of sampling, the sediment was sectioned in a glove bag under N_2-atmosphere. Sediment from at least three cores was pooled before analysis. Porewater was obtained either by N_2-pressure filtration (Kattegat and Skagerrak) or by centrifugation under N_2-atmosphere (Aarhus Bay). All chemical extractions were performed on fresh, wet sediment except for the P-pool fractionation of sediment from Kattegat and Skagerrak. In those cases, the sediment sections were frozen and stored for two weeks before analysis. As discussed later, this may cause a slight overestimation of the BD extractable SRP.

Sequential extraction of phosphorus pools

The extraction scheme (Fig. 1) is a modification of the scheme proposed for freshwater sediment by Psenner *et al.* (1988). Approximately 1 gram of wet sediment was subsampled and placed into a 50 ml polyethylene centrifuge tube.

In step one, 25 ml 0.46 M N_2-purged NaCl solution was added under N_2-atmosphere and the tube was closed by an air-tight screw cap. The centrifuge tube was shaken for 1 hour and centrifuged for 10 minutes at 3000 RPM. The supernatant was collected and another 25 ml NaCl solution was used to wash the sample and extract reabsorbed phosphate. After the second centrifugation the wash-solution was added to the first supernatant and 1 ml 1 M H_2SO_4 was added to the bottle in order to prevent co-precipitation of phosphate with iron and manganese and to preserve the sample. Step one is supposed to extract primarily the pool of loosely sorbed SRP, pore-

Table 1. Sites of sediment sampling.

Location	Position	Depth	Sediment type
Aarhus Bay	56°09,1′ N, 10°19,2′ E	16 m	Silt
Northern Kattegat	57°49,7′ N, 11°13,3′ E	56 m	Silt
Central Skagerrak	58°14,1′ N, 09°32,1′ E	695 m	Clay

50

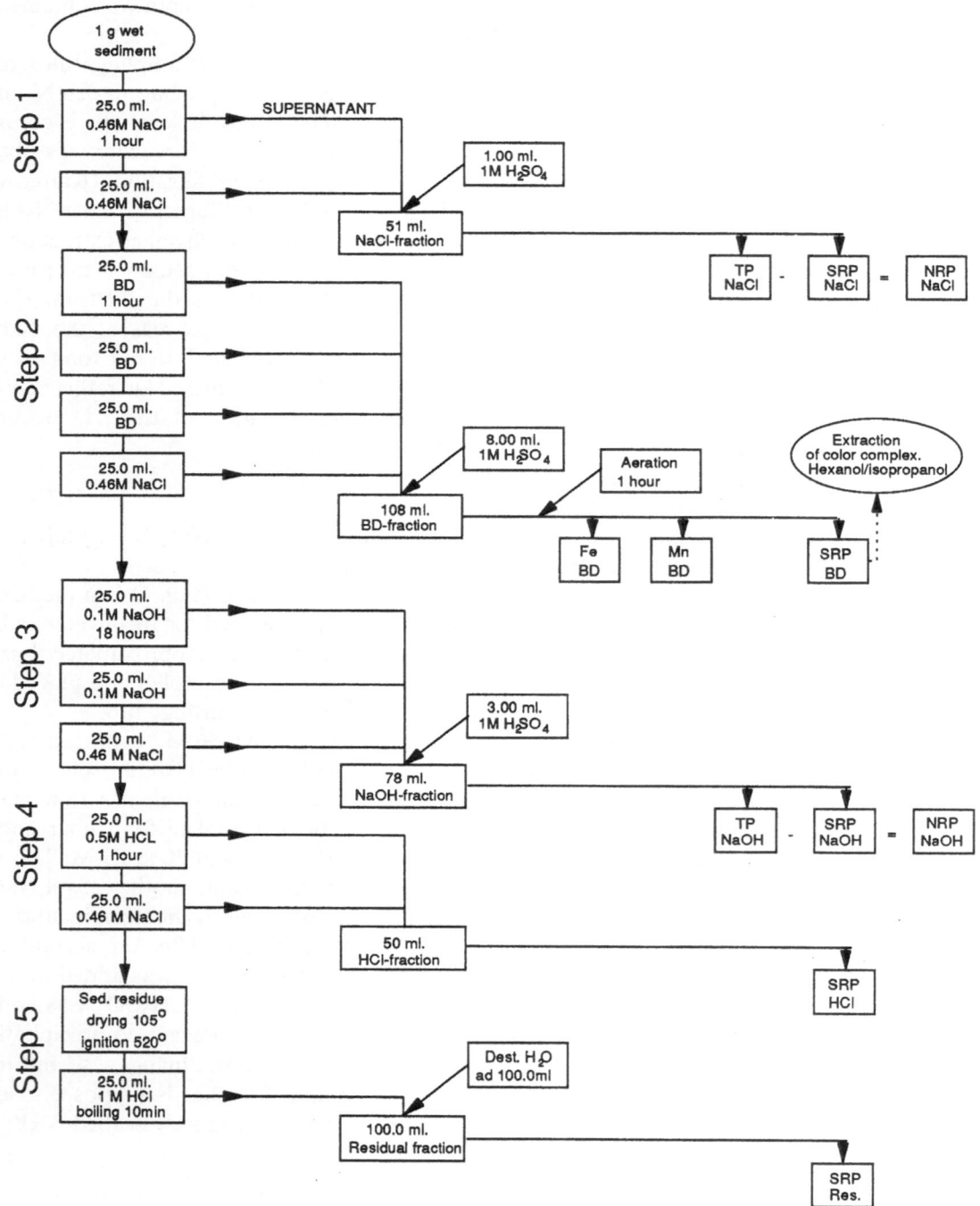

Fig. 1. Schematic presentation of the sequential extraction scheme for phosphorus pools in marine sediment. Abbreviations: TP (total phosphorus), SRP (soluble reactive phosphorus), NRP (non-reactive phosphorus), BD (Bicarbonate-dithionite). Steps 1 and 2 were performed under anaerobic conditions.

water SRP and porewater NRP (non-reactive phosphorus).

In step two, 25 ml BD-reagent (consisting of a pH 7.0 buffered solution of 0.11 M $Na_2S_2O_4$, 0.11 M $NaHCO_3$) was added to the sediment pellet. After 1 hour shaking at room temperature the extraction was followed by two BD washes and one 0.46 M NaCl wash. To the 100 ml of mixed supernatant from step 2, 8 ml 1 M H_2SO_4 were added to avoid precipitation of Fe and Mn during the following 1 hour aeration where the remaining dithionite was oxidized. The addition of acid and the aeration often led to the formation of elemental sulfur which turned the solution milky. Normally, the sulfur would precipitate in few days. Otherwise, a special technique was required for measuring phosphate as discussed later. After step 2 the sediment was handled under aerobic conditions.

In step 3, an 18 hour extraction (shaking) with 0.1 M NaOH was followed by one NaOH wash and one NaCl wash. 3 ml 1 M H_2SO_4 was added to the mixed supernatant to turn the sample acidic. This step is supposed to extract SRP from clay minerals and aluminum oxides together with much organismic phosphorus which appears as NRP in the solution. Polyphosphates e.g. will be extracted in this step (Hupfer, pers. comm.).

In step 4, a 0.5 M HCl-solution was used to extract calcium-bound phosphorus (e.g. apatite). This extraction was also followed by a NaCl wash. In the last step the sediment pellet was ignited at 520 °C for 2 hours and subsequently boiled for 10 minutes in 1 M HCl. The residual phosphorus left in the sediment was extracted. This pool is supposed to represent the more refractory organic phosphorus component.

The most important modification to the original scheme is that 0.46 M NaCl was used instead of 1 M NH_4Cl for extraction of loosely sorbed phosphorus in step 1. 1 M NH_4Cl was found to extract more than twice the amount of Ca than 0.46 M NaCl and also more SRP. This indicates a dissolution of calcium-bound P (see also Petterson et al., 1988). Moreover, a NaCl wash was added after each extraction step in order to overcome the reabsorption of phosphate. A second

NaCl wash after each step was found to extract less than 3% of the total amount of phosphorus in the respective step.

Sediment samples could be preserved in anoxic 0.46 M NaCl solution containing 4% formaldehyde or by freezing if stored under anaerobic conditions. Freezing, however, caused a 50% decrease in the pool of NRP extracted by NaOH. Some NRP was hydrolized by freezing and in the extraction scheme it showed up as BD-SRP whereby this pool was slightly overestimated.

The extraction scheme was found to be very reproducible when tested on parallel samples (see Fig. 3 and Table 2) and was applied for studying seasonal variations in sediment P-pools in Aarhus Bay (Jensen et al., unpublished).

An Aarhus Bay total P, total Fe and total Mn were measured on parallel sediment samples by HCl extraction following digestion as in step 5 (Andersen, 1976). This was done to compare the sum of P, Fe and Mn leached in the various steps with the total pools (see Fig. 2).

Analysis of phosphorus

SRP in porewater and in the various extractions was measured spectrophotometrically. Total

Table 2. Percentage of P recovery in 8 fractions from the five step sequential extraction of P in pure cultures of the diatom *Skeletonema costatum* and the blue-green algae *Synecococcus sp.* Mean values and standard deviation of three replicate samples ara given.

P-fraction	S. Costatum % of total n = 3		Synecococcus % of toal n = 3	
	Mean	S.D.	Mean	S.D.
NaCl-SRP	7.6	0.2	0.14	0.03
NaCl-NRP	58.1	1.1	0.32	0.06
BD-SRP	0	0	0	0
BD-NRP	26.5	6.4	1.14	0.17
NAOH-SRP	1.7	0.6	7.6	0.06
NAOH-NRP	4.9	0.7	72.1	1.8
HCl-SRP	0.33	0.07	8.7	0.5
HCl-NRP	0.37	0.04	0	0
Residual P	0.5	0.01	10.0	0.3

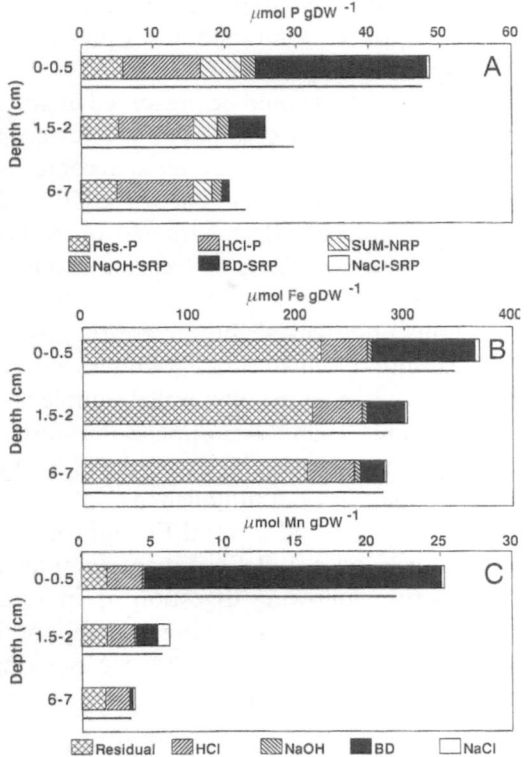

Fig. 2. Sequential extraction of phosphorus (stacked bars) in sediment from three depth in Aarhus Bay, November 1989 (A). Contemporary leaching of Fe (B) and Mn (C) in the extracts. Total P, Fe and Mn extracted from the same sediment samples are shown as solid lines below the stacked bars.

phosphorus (TP) in the extracts was determined as SRP after wet oxidation (Koroleff, 1983). NRP was calculated as the difference between TP and SRP. In step 2 it was not always possible to precipitate the elemental sulfur (or new formation of sulfur occurred upon addition of reagents). In those cases the molybdenium-blue complex was extracted by a mixture of 95% hexanol and 5% isopropanol and the absorbance was measured at 690 nm. Iron and manganese extracted together with P were measured by flame-AAS.

Extraction of am.FeOOH

The extraction scheme for am.FeOOH consisted of a 1 hour 0.5 M HCl extraction and a parallel

anaerobic oxalate extraction (Phillips & Lovley, 1987). Fe(III) was taken as the difference between HCl extractable Fe and anaerobic oxalate extractable Fe(II). We use 0.5 M HCl instead of the combined hydroxylamine hydrochloric acid regent of Lovley & Phillips (1987) (0.25 M NH_2OH in 0.5 M HCl) since the hydroxylamine was found not to affect the rate of iron dissolution at room temperature (Thamdrup, 1989). Earlier calibrations also found 0.5-1 hour extractions with dilute HCl to have nearly the same specificity for am.FeOOH as the combined reagent (Chao & Zhou, 1983; Canfield, 1988). Lovley & Phillips (1987) used the 0.5 M HCl extraction for measuring extractable Fe(II). This was not possible in our case, as we found reduction of Fe(III) when amorphous ferric hydroxide was added during the extraction of Aarhus Bay sediment. For example, addition of amorphous FeOOH (approx. 100 μmol (g sed)$^{-1}$ increased the amount of Fe(II) extracted from 78 ± 2 to 98 ± 1 μmol (g sed)$^{-1}$ with sediment from the suboxic zone (Thamdrup, 1989). With the anaerobic oxalate extraction, which we applied, no such error was found. With the combination of two different extractants in the Fe(III) determination, it is assumed that both extract the same amount of Fe(II) from the sediment. As the form of Fe(II) in sediments is largely unknown, calibration with synthetic (FeII) phases is of limited use. By applying the two extractions sequentially, the error was constrained. It was found that the scheme by underestimating Fe(II) dissolved with HCl might overestimate Fe(III) by maximally 20% (Thamdrup, 1989; Thamdrup, unpublished results).

For the HCl extraction, approx. 100 mg fresh sediment was transferred to a serum bottle containing 5 ml 0.5 M HCl and extracted on a shaker table for 1 hour. Fe in the extract was determined spectrophotometrically using reducing Ferrozine reagent (Stookey, 1970; Phillips & Lovley, 1987). For the oxalate extraction, approx. 100 mg fresh sediment was transferred under N_2 to an N_2-flushed serum bottle supplied with 5 ml N_2-purged 0.2 M ammonium oxalate/ oxalic acid at pH 3. After 16 hours of extraction

Fe^{2+} was determined with non-reducing Ferrozine.

Statistics

Significance of correlations was tested by the two-tailed t-test.

Results

Control experiments

In Aarhus Bay residual-P (Res.-P), HCl-extractable P (HCl-P) and NaOH-extractable SRP (NaOH-SRP) remained fairly constant with sediment depth while the other P-pools extracted in the three first steps (NaCl-SRP, BD-SRP and SUM-NRP) decreased with depth (Fig. 2a and Fig. 5a). SUM-NRP represents the non-reactive phosphorus extracted by NaCl and NaOH. NRP extracted by BD was negligible.

Most iron was extracted in the final step by ignition and acid boiling. Smaller amounts were extracted by BD and HCl in steps 2 and 4, respectively. Only the BD-extractable iron seems to decrease with depth (Fig. 2b). The same picture was found for manganese, however, the BD fraction was very large in the surface sediment and decreased rapidly with depth (Fig. 2c). As was the case for P extraction, the total amounts of iron and manganese extracted in the five steps agreed fairly well with parallel measurements of total pools (Fig. 2).

Prolonged BD treatment or higher temperature during BD-extraction did not extract more BD-SRP in Aarhus Bay sediment (Fig. 3). The amounts of iron, on the contrary, increased with prolonged BD-treatment and with higher temperature and in the subsequent ignition and acid extraction (step 5) iron leaching increased even more as a result of the changed BD-treatment.

In order to determine in which fraction organismic P appears, the sequential extraction was performed on pure culture of diatoms and cyanobacteria (Table 2). No SRP was detected in the

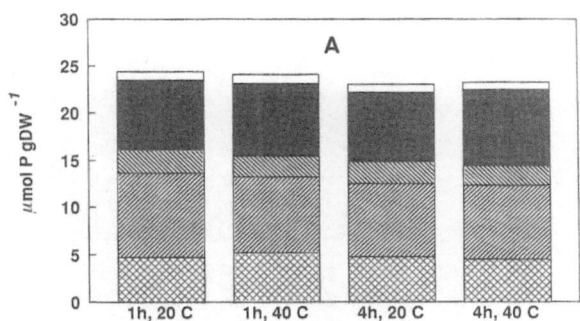

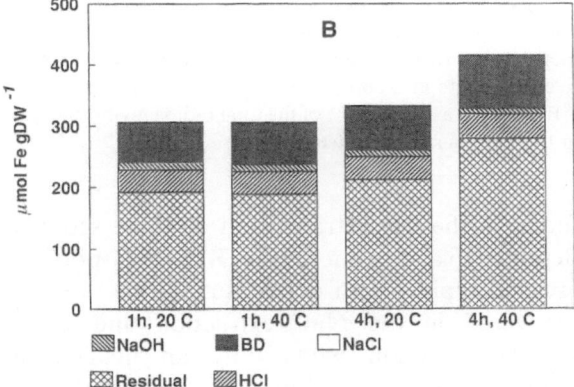

Fig. 3. Effect of increased temperature and prolonged treatment during BD-extraction (step 2) sequential extractions of P (A) and Fe (B) in the five steps with sediment from Aarhus Bay.

BD fraction even if the cultures were grown at surplus P which favours the accumulation of polyphosphates. In the first three steps 90% and 80% of the total P was recovered as NRP in *Skeletonema* and *Synecococcus*, respectively. In *Synecococcus* about 19% was recovered in the two final steps while in *Skeletonema* those fractions were insignificant.

In order to determine which fractions were important with respect to exchange of phosphate with porewater, a carrier-free solution of $^{32}PO_4^{3-}$ was injected in three depths in a sediment core from Aarhus Bay (a parallel core to the one in Fig. 2). After two days the sediment core was sectioned, the various P-pools (except Res.-P which was left out for safety reasons) were extracted and the radioactivity measured. More than 80% of the radiolabeled P recovered was

54

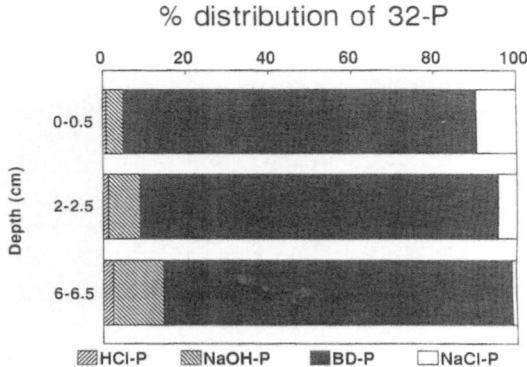

% distribution of 32-P

Fig. 4. Recovery of ^{32}P in the first four extracts in three sediment depths two days after injection of carrier-free radioactive phosphate in a core from Aarhus Bay. The amounts of
[32]P are given as percentage of the total radioactivity recovered
in the first four extracts (step 1–4).

found in the BD extraction at all three sediment
depths (Fig. 4) even if the BD-SRP pool decreased rapidly with depth (Fig. 2). 5–10% was
recovered in the NaCl-extraction and in the
NaOH-extraction while only an insignificant
amount was recovered in the HCl-extraction.

Results from the three stations

In Aarhus Bay, measurements of P-pools, Fe-
pools and porewater SRP were performed on
parallel cores at three occasions in spring 1991.
The three measurements were very similar and
only one dataset is shown in Fig. 5. Both BD-
SRP, BD-Mn, BD-Fe and am.FeOOH decreased

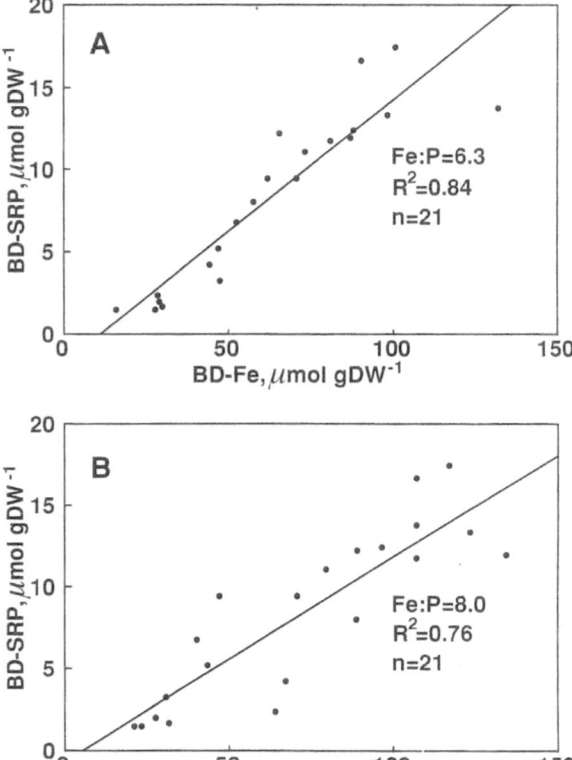

Fig. 6. BD-SRP plotted against BD-Fe (A) and am.FeOOH
(B) from 7 depths and three sampling dates (spring 1991) in
Aarhus Bay. The solid lines represent the best linear fit. Fe:P
is the molar ratio between BD-Fe and BD-P (A) and am-
.FeOOH and BD-P (B) calculated from the slope of the re-
gression lines.

with depth, the latter two being nearly identical.
The pool of BD-Mn was low compared with the

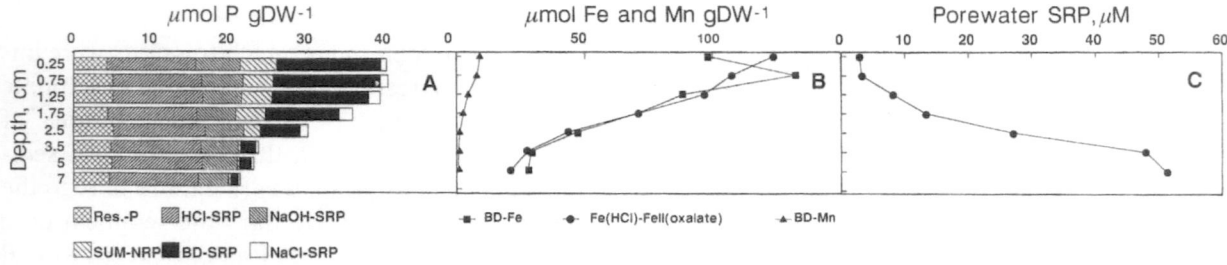

Fig. 5. Sequential extraction of P in 8 sediment depths from Aarhus Bay, February 1991 (water depth 16 m) (A). Amounts of
iron (BD-Fe) and manganese (BD-Mn) leached in the BD extraction and amounts of am.FeOOH measured in a parallel sediment core (B). Porewater concentration of SRP measured in a parallel sediment core (C).

iron (5b). Porewater SRP increased with depth from 3–5 μM in surface sediment to about 50 μM at 7 cm depth. Strong correlations were observed between BD-Fe and BD-SRP and between am.FeOOH and BD-SRP ($p < 0.001$) (Table 3 and Fig. 6). When calculated from the slope of the regression lines in Fig. 6, the molar ratio between BD-Fe and BD-SRP was 6.3 and between am.FeOOH and BD-P 8.0. The molar ratios in surface sediment varied between 6.5 and 11 for both iron extraction methods. Porewater SRP correlated negatively with am.FeOOH ($r^2 = 0.86$; $p < 0.001$) and with BD-P ($r^2 = 0.79$) (Table 3).

A similar pattern was observed in the Kattegat sediment (Fig. 7). However, BD-SRP remained a quantitatively significant P-pool even down to 30 cm and BD-Fe was considerably higher than am.FeOOH, especially in the deeper sediment layers. Porewater SRP increased from between 3 to 8 μM in the uppermost 4 cm to a maximum

value of about 60 μM at 9 cm depth. BD-SRP correlated positively with BD-Fe, am.FeOOH and BD-Mn ($p < 0.001$) (Table 3). Molar Fe:P ratios in the extracts were 10 and 15.9 for BD-Fe:BD-SRP and am.FeOOH:BD-SRP, respectively, when calculated from the slope of regression lines. The corresponding values for the surface sediment were 11–12 for BD-Fe:BD-SRP and 8–9 for am.FeOOH:BD-SRP. Porewater SRP correlated negatively with am.FeOOH ($r^2 = 0.62$, $p < 0.001$) and BD-Fe ($r^2 = 0.56$; $p < 0.001$).

In Skagerrak BD-SRP was found in considerable amounts even in the deepest sediment layers (Fig. 8a). Similarly large pools of BD-Fe, am.FeOOH and BD-Mn were determined at greater depth. In the surface sediment very high concentrations of BD-Mn (about 400 μmol gDW^{-1}) and BD-Fe (more than 600 μmol gDW^{-1}) were recorded. Am.FeOOH was only about 200 μmol

Table 3. Product-moment correlation coefficients (r) between porewater SRP and extracted amounts of iron and manganese in sediment from Aarhus Bay, Kattegat and Skagarrak. The data on BD-P, BD-Fe and am.FeOOH are idential with the data in Figs 6, 7 and 8, respectively.

Aarhus Bay, three sampling dates in spring 1991 ($n = 21$)

	BD-P	BD-Fe	BD-Mn	am.FeOOH
Porewater-SRP	−0.89	−0.87	−0.68	−0.93
BD-P	1	0.92	0.65	0.87
BD-Fe	0.92	1	0.79	0.85
BD-Mn	0.65	0.79	1	0.88

Kattegat, May 1990 ($n = 20$)

	BD-P	BD-Fe	BD-Mn	am.FeOOH
Porewater-SRP	−0.49	−0.75	−0.62	−0.79
BD-P	1	0.88	0.68	0.70
BD-Fe	0.88	1	0.84	0.86
BD-Mn	0.68	0.84	1	0.88

Skagerrak, May 1990 ($n = 24$)

	BD-P	BD-Fe	BD-Mn	am.FeOOH
Porewater-SRP	−0.74	−0.59	−0.47	−0.63
BD-P	1	0.94	0.90	0.79
BD-Fe	0.94	1	0.95	0.79
BD-Mn	0.90	0.95	1	0.69

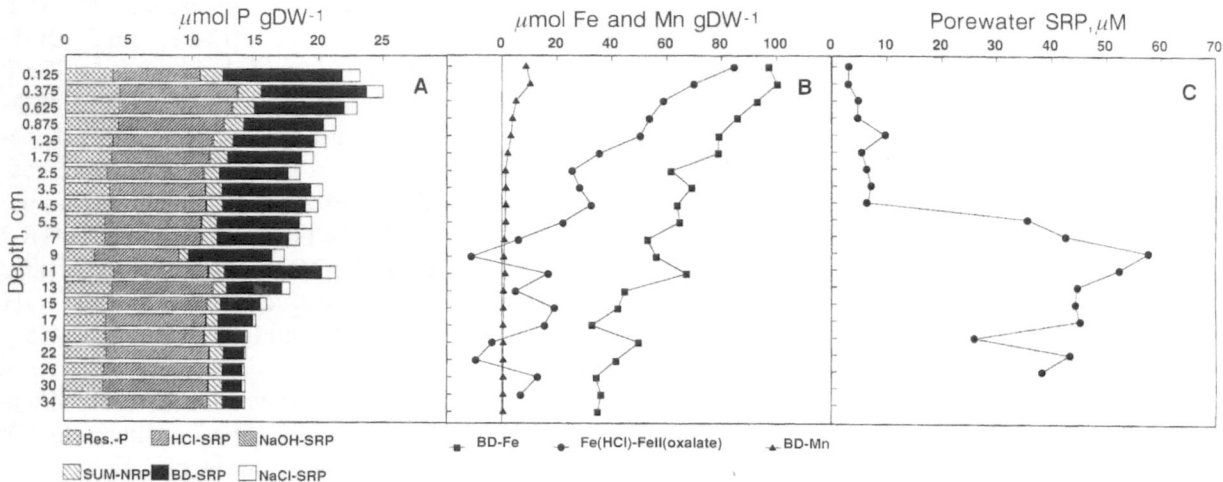

Fig. 7. Sequential extraction of P in sediment from Kattegat, May 1990 (water depth 56 m) (A). Amounts of iron (BD-Fe) and manganese (BD-Mn) leached in the BD extraction and amounts of am.FeOOH in the same sediment sample (B). Porewater concentration of SRP measured in a parallel sediment core (C).

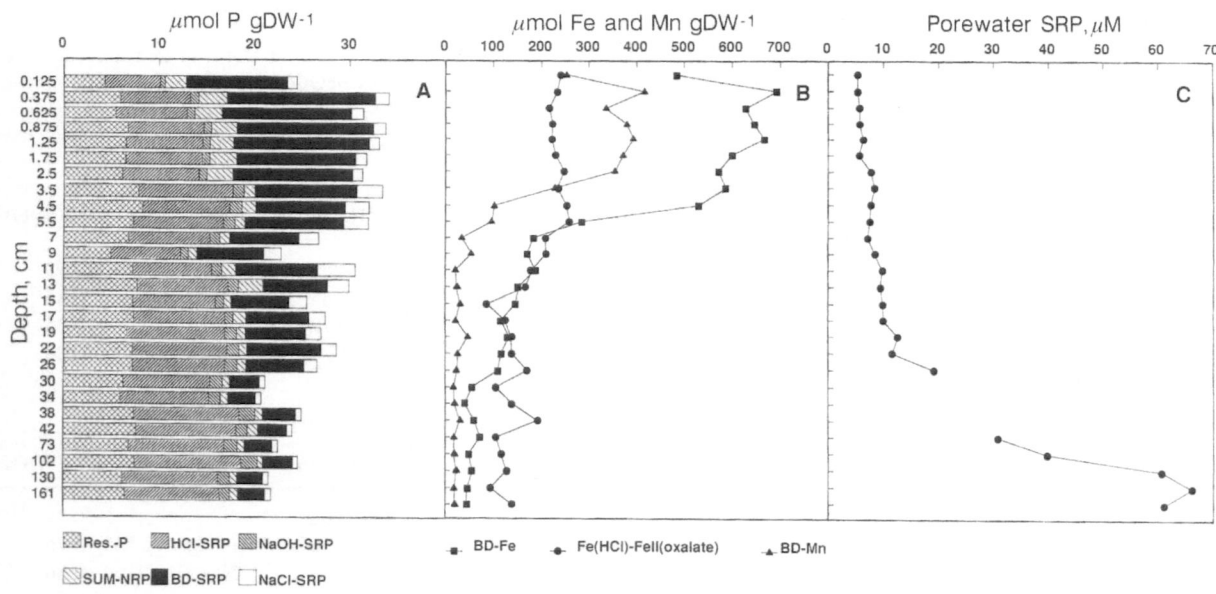

Fig. 8. Sequential extraction of P in sediment from Skagerrak, May 1990 (water depth 695 m) (A). Amounts of iron (BD-Fe) and manganese (BD-Mn) leached in the BD extraction and amounts of am.FeOOH in the same sediment sample (B). Porewater concentration of SRP measured in a parallel sediment core (C).

gDW^{-1} in the surface sediment. BD-Mn decreased rapidly at 2–3 cm depth and BD-Fe decreased rapidly at 4–5 cm. A third decrease in BD-Fe was seen at 26–30 cm depth. A parallel

decline in BD-SRP was observed in all three depths. A less dramatic decrease in am.Fe.OOH was observed (Fig. 8b). The porewater SRP profile showed an inverse tendency with smaller increases at 2 and 6 cm depth but remained low (about 10 μM) down to 20 cm depth. A SRP maximum of 60 μM was recorded at 130 cm depth (Fig. 8c). BD-SRP correlated strongly with BD-Fe ($r^2 = 0.88$) and BD-Mn ($r^2 = 0.81$) and less significantly with am.FeOOH ($r^2 = 0.62$) (Table 3). Molar Fe:P ratios in the extracts for BD-Fe:BD-SRP and am.FeOOH:BD-SRP were 63 and 16.9, respectively, when calculated from the slope of regression lines. The corresponding values for the surface sediment were 45 for BD-Fe:BD-SRP and 15–16 for am.FeOOH:BD-SRP. Porewater SRP correlated negatively with BD-SRP ($r^2 = 0.35$; $p < 0.001$) and with am.FeOOH ($r^2 = 0.40$; $p < 0.001$).

Discussion

The results indicate that the SRP extracted by the BD-treatment represents the pool of phosphate absorbed onto reducible forms of Fe and Mn. This conclusion is primarily based on three observations: 1) the close correlations between BD-SRP and BD-Fe, am.FeOOH and to some extent BD-Mn, which were observed in the three sediments studied; 2) the fact that no additional SRP could be extracted by prolonged BD treatment; 3) the fact that no organismic P from the algae or the bacteria cultures showed up as BD-SRP. This may also be inferred from the experiment where the BD-treatment was performed at higher temperatures. If any organic P was extracted as SRP in the BD step, then higher temperatures probably would have increased that process.

In addition, molar Fe:P ratios of 6.5 to 16 between BD-SRP and BD-Fe or am.FeOOH were recorded for sediment from Aarhus Bay and Kattegat. These ratios are similar to the molar ratios between FeOOH and adsorbed PO_4^{3-} found in experiments with pure crystals (slightly aged) at pH 7 (e.g. Lijklema, 1980) or the ratio in freshly precipitates in lake water (Einsele, 1936). Jensen *et al.* (1992) found that freshwater sediment with molar Fe:P ratios above 8.5 were capable of retaining phosphate in the oxidized surface layer while phosphate leached at a rate independently of the Fe:P ratio when this was below 8.5. This may indicate that the oxidized iron pool in the sediments from Aarhus Bay and Kattegat was nearly saturated with phosphate while sediment from Skagerrak was not. However, it is questionable whether the BD-Fe extracted from Skagerrak sediment came from a pool of am.FeOOH or from less reactive iron species as discussed below.

In Aarhus Bay and Kattegat, the BD-treatment obviously extracted more Fe than the pool responsible for phosphate adsorbtion. This can be inferred both from the positive X-intercept (11 and 10.6 μmol Fe gDW^{-1}) of the regression lines (BD-Fe, BD-SRP) for the two stations and from the comparison with the am.FeOOH profile in Kattegat. The am.FeOOH:BD-SRP ratio was in general smaller (and closer to the values found by Einsele and Lijklema) than the BD-Fe:BD-SRP ratio in all three sediments. Am.FeOOH is therefore more likely to represent the iron pool responsible for phosphate adsorption than BD-Fe. BD-treatment may extract some reduced iron components and some oxidized iron species which are less reactive towards phosphate e.g. aged goethite, hematite and magnetite (e.g. Mehra & Jackson, 1960). The enchanced leaching of Fe in the fifth extraction step by prolonged BD-treatment probably indicates a mobilization of oxidized iron from e.g. silicates or clay minerals.

The exchangeability of phosphate in the various P-pools studied by the $^{32}PO_4^{3-}$ experiment revealed that iron-bound P is the most important exchangeable P-pool in the sediment even in deeper layers where the pool size is rather insignificant. This points to the importance of am.FeOOH in controlling porewater SRP which was confirmed by the good correlations between SRP and am.FeOOH found in all three sediments. Also, Psenner & Pucsko (1988) found that iron hydroxides were most important in controlling porewater P concentrations of a softwater lake and Golterman (1988) demonstrated that in sediments iron hydroxides will extract phosphate from calcite and thereby control the equilibrium between adsorbed SRP and porewater SRP.

Other studies in marine sediments indicate the importance of formation and dissolution of oxidized iron for controlling porewater SRP (e.g. Krom & Berner, 1981; Yamada & Kayama, 1987; Sundby *et al.*, in press). Less information is available on adsorption of phosphate to oxidized species of manganese. Meanwhile, in another study we found that amorphous manganese was less capable of absorbing phosphate than am.FeOOH (Jensen & Thamdrup, unpublished). Given the relatively smaller amounts of oxidized manganese found in the sediments we conclude that manganese is of minor importance as a phosphate adsorbent when compared with oxidized iron.

Profiles of total phosphorus in marine sediment generally show maximum values in the bioturbated surface sediment declining to a stable value in deeper sediment layers. Results from this study support the general assumption that rather immobile P-pools dominate in deeper sediments and that the surface maximum consists of mobile P-pools like loosely sorbed SRP, iron-bound P and fresh organic P. The large exchangeability of iron-bound P indicates that the pool is highly mobile and stresses the importance of proper methods for measuring iron bound P. With the reservation that only three sediment types were studied, it may be concluded that the extraction scheme used in this study provides such a specific measurement of iron-bound P in marine sediments.

Acknowledgements

We are indebted to Kitte Gerlich for carrying out chemical analysis with great skill. The paper was markedly improved by the criticism from Drs N. Caraco, R. Psenner, P. Sampou and B. Sundby. We thank the crew and scientific staff on the April 1990 cruise of R.V. *Gunnar Thorson* for helping with the sediment sampling. The cruise was arranged by Dr B. B. Jørgensen and supported by Danish Agency of Environmental Protection. The study was financed by the Danish Agency of Environmental Protection, the 'Marine Research Programme 90'.

References

Aller, R. C., J. E. Mackin & R. T. Cox, 1986. Diagenesis of Fe and S in Amazon inner shelf muds: apparent dominance of Fe reduction and implications for the genesis of ironstones. Continent. Shelf Res. 6: 263–289.

Andersen, J. M., 1976. An ignition method for determination of total phosphorus in lake sediments. Wat. Res. 10: 329–331.

Balzer, W., 1986. Forms of phosphorus and its accumulation in coastal sediments of Kieler Bucht. Ophelia 26: 19–35.

Berner, R. A., 1980. Early diagenesis. Princeton University Press, N.Y.

Berner, R. A., 1981. Authegenic mineral formation resulting from organic matter decomposition in modern sediments. Forschr. Minre. 59: 117–135.

Canfield, D. C., 1988. Sulfate reduction and the diagenesis of iron in anoxic marine sediments. Ph. D. diss. thesis, Yale University, 248 p.

Chao, T. T. & Zhou, 1983. Extraction techniques for selective dissolution of amorphous iron oxides from soils and sediments. Soil Sci. Soc. Am. J. 47: 225–232.

Einsele, W., 1936. Über die beziehungen der Eisenkreislaufes zum Phosphorkreislauf im eutrophen See. Arch. Hydrobiol. 29: 664–686. (In German).

Fischer, W. R., 1973. Die Wirkung von zweiwertigem Eisen auf Lösung und Umwandlung von Eisen(III)-hydroxiden. In: Schlichting E. & U. Schwertmann (eds), Pseudogley and gley: Genesis and use of hydromorphic soils, Weinheim, Bergstraße, 37–44.

Golterman, H. L., 1988. The calcium- and iron bound phosphate diagram. Hydrobiologia 59: 149–151.

Hieltjes, A. H. M. & L. Lijklema, 1980. Fractionation of inorganic phosphates in calcareous sediments. J. Envir. Qual. 9: 405–407.

Jacobsen, O. S., 1978. Sorption, adsorption and chemosorption of phosphate by Danish lake sediment. Vatten 4: 230–243.

Jensen, H. S., P. Kristensen, E. Jeppesen & A. Skytthe, 1992. Iron:Phosphorus ratio in surface sediment as an indicator of phosphate release from aerobic sediments in shallow lakes. In B. T. Hart & P. G. Sly (eds), Sediment/Water Interactions. Developments in Hydrobiology 75. Kluwer Academic Publishers, Dordrecht: 731–743. Reprinted from Hydrobiologia 235/236.

Jørgensen, B. B., M. Bang & T. Henry Blackburn, 1990. Anaerobic mineralization in marine sediments from the Baltic Sea-North Sea transition. Mar. Ecol. Prog. Ser. 59: 39–54.

Koroleff, F., 1983. Determination of nutrients. In: Grasshof, K., M. Ehrhardt & K. Kremling (eds), Methods of seawater analysis. Verlag Chemie.

Krom, M. D. & R. A. Berner, 1981. The diagenesis of phosphorus in a nearshore marine sediment. Geochemica et Cosmochimica 45: 207–216.

Lijklema, L., 1980. Interaction of orthophosphate with iron(III) and aluminum hydroxides. Envir. Sci. Technol. 14: 537.

Lovley, D. R. & E. J. P. Phillips, 1987. Rapid assay for microbially reducible ferric iron in aquatic sediments. Appl. envir. Microbiol. 53: 1536–1540.

Lovley, D. R., 1991. Dissimilatory Fe(III) and Mn(IV) reduction. Microbiol. Rev. 55: 259–287.

Lucotte, M. & B. d'Anglejan, 1985. A comparison of several methods for the determination of iron hydroxides and associated orthophosphates in estuarine particulate matter. Chemical Geology 48: 257–264.

McKeague, J. A. & J. H. Day, 1966. Dithionite and oxalate-extratable Fe and Al as aids in differentiating various classes of soils; Can. J. Soil. Sci. 46: 13–22.

Mehra, O. P. & M. L. Jackson, 1960. Iron oxide removal from soils and clays by a dithionite-citrate system buffered with sodium bicarbonate. Clays Clay Minerals 7: 317–327.

Nürnberg, G. K., 1988. Prediction of phosphorus release rates from total and reductant-soluble phosphorus in anoxic lake sediment. Can. J. Fish. aquat. Sci. 45: 453–462.

Pettersson, K., B. Boström & O.-S. Jacobsen, 1988. Phosphorus in sediments – speciation and analysis. In G. Persson & M. Jansson (eds), Phosphorus in Freshwater Ecosystems. Developments in Hydrobiology 48. Kluwer Academic Publishers, Dordrecht: 91–101. Reprinted from Hydrobiolgia 170.

Phillips, E. J. P. & D. R. Lovley, 1987. Determination of Fe(III) and Fe(II) in oxalate extracts of sediment. Soil. Sci. Soc. Am. J. 51: 938–941.

Psenner, R. & R. Pucsko, 1988. Phosphorus fractionation: advantages and limits of the method for the study of sediment P origins and interactions. Arch. Hydrobiol. Beih. 30: 43–59.

Psenner, R., B. Boström, M. Dinka, K. Petterson, R. Pucsko & M. Sager, 1988. Fractionation of phosphorus in suspended matter and sediment. Arch. Hydrobiol. Beih. 30: 98–110.

Stauffer, R. E., 1981. Sampling strategies for estimating the magnitude and importance of internal phosphorus supplies in lakes. US EPA Rep. 60013-81-015, Corvallis.

Stookey, L. L., 1970. Ferrozine – a new spectrophotometric reagent for iron. Anal. Chem. 42: 779–781.

Sundby, B., C. Gobeil, N. Silverberg & A. Mucci, in press. The phosphorus cycle in coastal marine sediments. Limnol. Oceanogr.

Thamdrup, B., 1989. Chemical extractions for separate determination of easily extractable Fe(II) and Fe(III) applied to a coastal marine sediment. Master thesis, University of Aarhus.

Yamada, H. & M. Kayama, 1987. Distribution and dissolution of several forms of phosphorus in coastal marine sediments. Oceanol. Acta 10: 311–321.

Hydrobiologia **253**: 61–72, 1993.
P.C.M. Boers, T.E. Cappenberg & W. van Raaphorst (eds),
Proceedings of the Third International Workshop on Phosphorus in Sediments.
© 1993 *Kluwer Academic Publishers.*

Supply of phosphorus to the water column of a productive hardwater lake: controlling mechanisms and management considerations

Charles T. Driscoll[1], Steven W. Effler[2], Martin T. Auer[3], Susan M. Doerr[2] & Michael R. Penn[3]
[1] *Department of Civil and Environmental Engineering, Syracuse University, Syracuse, NY 13210, USA;*
[2] *Upstate Freshwater Institute, Syracuse, NY 13214, USA;* [3] *Department of Civil and Environmental Engineering, Michigan Technological University, Houghton, MI 49931, USA*

Key words: calcium, lake management, phosphorus, sediments

Abstract

Onondaga Lake is a hypereutrophic, industrially polluted lake located in Syracuse, NY. High hypolimnetic concentrations of H_2S that develop after anoxia restrict the accumulation of total Fe^{2+} due to the formation of FeS, and may limit $Fe-PO_4$ interactions. High water column concentrations of Ca^{2+} and high rates of $CaCO_3$ deposition occur due to inputs of Ca^{2+} from an adjacent soda ash manufacturing facility. Patterns of P concentration and other water chemistry parameters in the lower waters, and results from chemical equilibrium calculations, suggest that $Ca-PO_4$ minerals may regulate the supply of P from sediments to the water column in Onondaga Lake. These findings have important management implications for Onondaga Lake. First, declines in water column Ca^{2+} concentrations due to reductions in industrial $CaCl_2$ input may result in conditions of undersaturation with respect to $Ca-PO_4$ mineral solubility and increases in the release of P from sediments to the water column. Second, introduction of O_2 from hypolimnetic oxygenation, as a lake remediation initiative, may enhance P supply from sediments, because of increased solubility of $Ca-PO_4$ minerals at lower pH.

Introduction

Phosphorus has long been recognized as the most critical nutrient limiting phytoplankton growth in lakes (Hutchinson, 1973; Schindler, 1977). Efforts to reclaim culturally eutrophic lakes have focused primarily on reduction of external P loads (Uttormark, 1978). However, recently it has been recognized that in certain systems management steps should first be considered to reduce or prevent internal loading from sediments (Caraco *et al.*, 1991). The failure, or delay, of lakes to recover following major reductions in external P loading has been attributed to continued high rates of P release from sediments (e.g., Ahlgren, 1977; Larsen *et al.*, 1981; Welch *et al.*, 1986).

It is essential to understand and quantify the processes regulating sediment release, in order to effectively support (usually costly) management decisions to control P loading to lake systems. A number of mechanisms have been identified as potentially regulating the supply of P from sediments, including: redox (oxidation-reduction) reactions, adsorption, mineral phase solubility, mineralization of organic matter, and turbulence (e.g., Böstrom *et al.*, 1988; Caraco *et al.*, 1991). Related factors that have been proposed as controlling sediment P release include: O_2 (e.g., Einsele, 1936; Mortimer, 1941, 1942), SO_4^{2-} concentration (e.g., Caraco et al., 1989, 1991), pH (e.g., Andersen, 1975; Stauffer, 1985), and sediment Fe content (e.g., Baccini, 1985).

Redox conditions at the sediment-water interface are widely accepted as the principal factor regulating sediment P release. Einsele (1936) and Mortimer (1941; 1942) demonstrated that P is immobilized in oxygenated sediments by retention with Fe(III). Under reducing conditions P is released from sediments following reduction of Fe(III) to Fe(II) and the associated disruption of solid Fe(III)-PO_4 forms. This scenario has been described as the Einsele/Mortimer model (Böstrom et al., 1988) and observations consistent with this model have been documented widely. According to the Einsele/Mortimer model, maintenance of oxygenated conditions at the sediment-water interface is critical to minimize or eliminate P release. Thus artificial oxygenation of anoxic hypolimnia (as a lake management option) should decrease the supply of P from sediments to the water column (Cooke et al., 1986; Gächter, 1987).

In this study, the processes regulating the supply of P from sediments in a calcareous hypereutrophic lake were evaluated by analysis of water column chemistry. Three different processes are considered: regulation by redox (O_2/Fe (the Einsele (1936)/Mortimer (1941; 1942)) model), mineralization of organic matter, and Ca-PO_4 mineral interactions. Moreover, the results of this work are interpreted with regard to management alternatives currently being considered for remediation of the lake.

Materials and methods

Site description

Onondaga Lake is located in metropolitan Syracuse, New York, U.S.A. The lake is dimictic, has a surface area of 11.7 km^2, a mean depth of 12 m and a maximum depth of 20 m. The lake has a high flushing rate (4 times y^{-1}), and thus responds rapidly to remedial measures (Devan & Effler, 1984). It is a hardwater system (Effler & Driscoll, 1985), naturally enriched with SO_4^{2-} (1.6 mmol

l^{-1}). Onondaga Lake has received municipal effluent and industrial waste for more than a century.

High productivity is a major water quality problem of Onondaga Lake (Auer et al., 1990). A number of costly measures have been taken to remediate this problem. A ten-fold reduction in P loading was achieved from 1970–1981 (Devan & Effler, 1984). Moreover, an additional two-fold reduction has been attained since 1981. The current loading of P is still about 0.27 mol m^{-2} y^{-1}, approximately 11.5 times greater than the level considered as acceptable by Vollenweider (1975). The Metropolitan Syracuse Treatment Plant (METRO) presently contributes more than 55% of the annual total external P load to the lake. Remediation alternatives presently under consideration to further reduce P loading include additional treatment at METRO (advanced treatment) and diversion of the METRO effluent to a nearby river.

Manifestations of the continuing hypereutrophic state of the lake include: 1) high concentrations of phytoplankton biomass and low transparency (Auer et al., 1990), 2) extended periods of hypolimnetic anoxia (Effler et al., 1986), and 3) lake-wide depletion of O_2 during the fall mixing period (Effler et al., 1988). The large internal loading of P from lake sediments during stratification has recently been quantified by Auer et al. (1993). The relative importance of this input would increase substantially if external loads are further reduced through remediation efforts. Hypolimnetic oxygenation is presently being considered as a possible component of the overall lake remediation plan, primarily to improve O_2 resources. A secondary goal of the proposed hypolimnetic oxygenation program is to reduce internal P loading by decreasing the supply of P from sediments.

Onondaga Lake has also been severely polluted with inputs of Ca^{2+}. Soda ash was produced by the Solvay Process at an adjoining facility from 1885–1986. The wastewaters produced from this process resulted in stoichiometric inputs of Ca^{2+}, Na^+ and Cl^- to Onondaga Lake, and elevated concentrations in the water column.

With the closure of the facility in 1986 the Ca^{2+} concentration in the lake has decreased from about 16 to 4.3 mmol l^{-1}. However, about 40% of the current input of Ca^{2+} to the lake is received from a land disposal area associated with the chemical manufacturing facility (Effler *et al.*, 1991). The entire water column of the lake was (Effler & Driscoll, 1985), and continues to be (subsequently presented herein), oversaturated with respect to the solubility of calcite ($CaCO_3(s)$).

These conditions have resulted in a high rate of $CaCO_3(s)$ precipitation and deposition. Calcium carbonate was the dominant component of the particle assemblage of the upper waters of Onondaga Lake in the early 1980's (Yin & Johnson, 1984), and continues to be significant since closure of the soda ash facility (Johnson *et al.*, 1991). The mean $CaCO_3(s)$ content of the surficial sediments of the lake is 60% ($n = 70$), ranging from 30 to 92% of dry weight (unpublished data). Selective extraction (Williams *et al.*, 1976) of sediment trap collections suggests that P co-precipitates with $CaCO_3(s)$ in the water column. Approximately 30% of the depositing P during May–November 1980 was found to be inorganic;

56% of the P in the surficial sediments of a single deep water location was determined to be inorganic (Wodka *et al.*, 1985).

Field and laboratory methods

Water samples were collected at a depth of 19 m, with a submersible pump, from April through October of 1989 and 1990, at a semi-permanent (buoyed) station located in the southern basin of the lake. This site has been found to be representative of overall lake conditions (Field, 1980). The sampling depth was usually within 50 cm of the sediment-water interface. Water chemistry parameters which show marked temporal variation were sampled at least once per week; other more invariant parameters were monitored less frequently (Table 1). Temperature (Montedoro-Whitney, Model TC-5) and O_2 (Yellow Springs Instruments, Model 54) were measured in the field; O_2 determinations were also frequently checked by Winkler titration (APHA, 1985). Comparable water chemistry data are available for earlier years (i.e., 1980, 1981) and were used to interpret changes in lake chemistry.

Table 1. Laboratory chemical analyses, methods and frequency; and specification of average concentrations for chemical equilibrium modeling. The average concentration at 18 m was obtained for 1989.

Chemical constituent	Method*	Frequency	Average 18 m conc. (mmol l^{-1})
SRP	Colorimetric, ascorbic acid	$\geq$ 1/wk	$1.5 \cdot 10^{-1}$
pH	–	$\geq$ 1/wk	7.6 (units)
Alkalinity	Titration, H_2SO_4 (pH of 4.3)	$\geq$ 1/wk	4.2
Total Fe^{2+}	Colorimetric, 2,2'bipyridyl (Heaney and Davidson, 1977)	$\geq$ 1/wk	$1.3 \cdot 10^{-2}$
Total H_2S	Titration, thiosulfate	$\geq$ 1/wk	$1.2 \cdot 10^{-1}$
SO_4^{2-}	Turbidimetric, $BaSO_4$	$\geq$ 1/wk	1.7
Cl^-	Titration, $Hg(NO_3)_2$	$\geq$ 1/wk	14.4
Ca^{2+}	Atomic absorption spectrophotometry	Biweekly	4.4
Na^+	Atomic absorption spectrophotometry	Biweekly	9.7
Mg^{2+}	Atomic absorption spectrophotometry	(Effler and Driscoll, 1985)	$9.9 \cdot 10^{-1}$
K^+	Atomic absorption spectrophotometry		$4.4 \cdot 10^{-1}$
F^-	Ion selective electrode	(Effler and Driscoll, 1985)	$2.4 \cdot 10^{-1}$
Temperature			9.1 °C

* All according to Standard Methods (APHP, 1985), except Fe^{2+}.

The major chemical constituents of the lake, soluble reactive P (SRP), and species expected to influence the cycling P, were measured using the procedures indicated in Table 1. All analyses were performed on unfiltered samples, with the exception of SRP (filtered through a 0.45 μm pore size filter). Sample handling and analyses were conducted according to the procedures described in Standard Methods (APHA, 1985). Analyses of reactive components were usually completed within 3 hours of collection.

Calculations

To evaluate mineralization of sediment organic matter as a control on the internal supply of P to the water column, the approach of Caraco *et al.* (1991) was used. The release of P and dissolved inorganic C (DIC) was calculated by subtracting the lower water (19 m) concentrations of SRP and DIC at a given time from the value measured at spring turnover. Values of DIC were not measured directly but were calculated from measured values of alkalinity and pH (Stumm & Morgan, 1981). Caraco *et al.* (1991) have indicated that in hard water lakes this approach may result in an error in the estimate of DIC release due to loss/ release of DIC associated with $CaCO_3(s)$ precipitation/dissolution. They suggest that for hardwater systems, the sum of electron acceptors used in mineralization of sediment organic matter (e.g., O_2, NO_3^-, Fe(III), SO_4^{2-}, CO_2) be used as an alternative. We found good agreement using both approaches.

A series of chemical equilibrium calculations were conducted with the chemical equilibrium model MINEQL (Westall *et al.*, 1976). The thermodynamic data used in this analysis were obtained from Ball *et al.* (1980). Two types of chemical equilibrium calculations were conducted. First, calculations were made to evaluate water column monitoring data. Water chemistry data obtained for a given date were entered in MINEQL and the chemical speciation was calculated for fixed (ambient) pH conditions in which chemical precipitation was not allowed to occur.

From these calculations, saturation indexes (SI) with to the solubility of various mineral phases were determined (eqn. 1),

$$SI = \log Q_p/K_p \qquad (1)$$

where: Q_p is the ion activity product of the mineral of interest, and K_p is the thermodynamic solubility product of the mineral. For this analysis the minerals examined included, calcite ($CaCO_3(s) = Ca^{2+} + CO_3^{2-}$; $pK_{so} = 8.475$), whitlockite ($Ca_3(PO_4)_2(s) = 3Ca^{2+} + PO_4^{3-}$; $pK_{so} = 28.92$) and iron sulfide ($FeS(s) + H^+ = Fe^{2+} + HS^-$; $pK_{so} = 3.915$).

The second type of calculation included a series of chemical 'titrations' in which changes in the chemistry of the lower waters of Onondaga Lake were examined over a range of conditions. The initial conditions for the MINEQL titrations were average lower water concentrations of Onondaga Lake for 1989 (Table 1). Initially the influence of variations in redox conditions on solution control and concentrations of total PO_4^{3-} were examined by varying pϵ ($-\log$ (electron activity)) from oxidizing conditions ($+10$) to reducing conditions (-5), with the pH fixed at 8.0. In this simulation $CaCO_3(s)$ was considered as a potential control of Ca^{2+}, $Fe_2O_3(s)$ (hematite) and FeS were considered as potential controls on total Fe and $Ca_3(PO_4)_2(s)$, $Fe_3(PO_4)_2(s)$ (vivianite) and $FePO_3(s)$ (strengite) were considered as potential control of total PO_4^{3-}. These minerals were included in model simulations as dissolved solids (i.e., minerals were allowed to precipitate when the solution was oversaturated with respect to their solubility).

In addition, two other 'titrations' were conducted to evaluate controls on water column total PO_4^{3-} and implications of these mechanisms for lake management issues. In the first of these, the partial pressure of CO_2 (P_{co_2}) was varied from $10^{-3.14}$ to $10^{-1.44}$ atm to evaluate changes in solution chemistry following mineralization of sediment organic matter. The initial value represents water column conditions during spring turnover, while the final value represents lower water conditions in late stratification during 1980 when lake productivity was very high. In the second titration, Ca^{2+}, added as a Cl^- salt, was varied from 1.25 to 22.4 mmol l^{-1} to assess changes in water column chemistry that occurred due to changes in Ca loading to the system. The upper range of Ca^{2+} concentrations is indicative of values in 1980 when the soda ash manufacturing facility was in operation. The lower range of Ca^{2+} is more representative of conditions following closure of the industrial facility. The titrations were conducted by assuming that $CaCO_3$, FeS and $FePO_4$ occur as dissolved solids. Two alternate scenarios were considered for solution control by Ca-PO$_4$ minerals. Whitlockite ($Ca_3(PO_4)_2$) was considered to occur as either a fixed (activity) or dissolved solid in chemical titrations.

Results and discussion

Testing the Einsele/Mortimer model

Oxygen was depleted rapidly from the bottom waters of Onondaga Lake following the onset of stratification in early May; anoxia developed within 5–6 weeks (Fig. 1a). Total Fe^{2+} was evident (Fig. 1b) within about 1 week of the onset of anoxia, and soon after release of total H_2S release was observed (Fig. 1c). The short time interval between the initial detection of the byproducts of Fe(III) and SO_4^{2-} reduction, despite the rather significant range in the redox potentials of these processes (Stumm & Morgan, 1981; Böstrom et al., 1988), reflects the very large demand for electron acceptors in the lower waters due to the high level of productivity and subsequent deposition of carbon. The concentrations of total Fe^{2+} were considerably lower than values typically reported during anoxia for systems following the Einsele/Mortimer model (e.g., Mortimer, 1941; 1942; Larsen et al., 1981). The molar ratio of total Fe^{2+} to SRP was much less than 1 for most of the stratification period (mean ratio $\pm$ std.dev. = 0.52 ± 0.31). The relatively low total Fe^{2+} concentrations were almost certainly a result of control by the very high total H_2S concentrations. Molar ratios of total Fe^{2+} to total H_2S were less than 1, and for most observations were less than

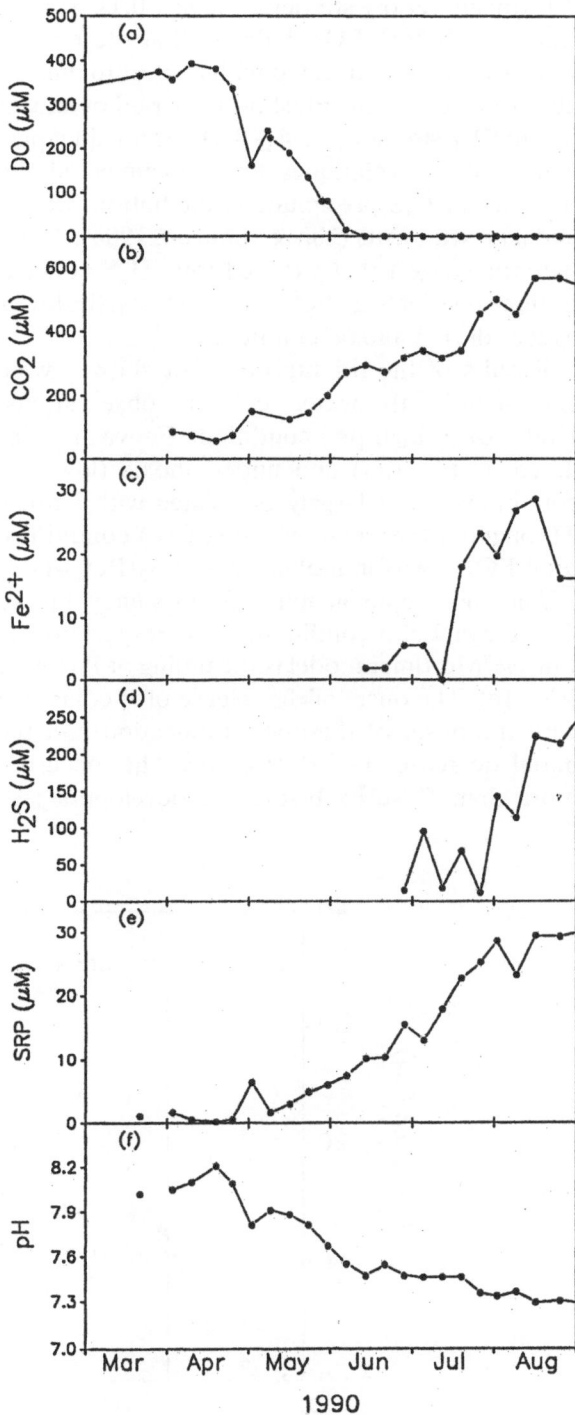

Fig. 1. Concentrations of water chemistry parameters in the lower waters (19 m) of Onondaga Lake, March–August 1990. DO is dissolved oxygen and SRP is soluble reactive phosphorus.

0.1 (mean ratio $\pm$ std.dev. $= 0.096 \pm 0.1$). Calculations with MINEQL indicate that the bottom waters were oversaturated with respect to the solubility of FeS during most of the period of anoxia (mean SI $\pm$ std. dev. $= 1.4 \pm 0.41$). Individual particle analysis techniques have documented the presence of FeS precipitate in the bottom waters and lake sediments (Yin & Johnson, 1984). These patterns suggest that elevated total H_2S concentrations restrict total Fe^{2+} availability in the lower waters during anoxic conditions.

Results of the $p\epsilon$ titration (not shown) were consistent with water column observations. Under oxic (high $p\epsilon$) conditions Fe was precipitated as $Fe_2O_3(s)$ and under anoxic (low $p\epsilon$) conditions it was largely associated with FeS(s). Throughout the range of redox ($p\epsilon$) conditions, total PO_4^{3-} was immobilized as $Ca_3(PO_4)_2(s)$.

The most conspicuous inconsistency in the Onondaga Lake conditions with respect to the Einsele/Mortimer model is the timing of P release (Fig. 1e). The onset of the release of P coincided with the onset of thermal stratification and the initial decreases in pH (Fig. 1f). This occurred more than 5 weeks before the development of anoxia and more than 6 weeks before the appearance of measurable quantities of total Fe^{2+}. Clearly Onondaga Lake does not fit the Einsele/Mortimer model; other processes apparently are important in regulating the mobilization of P from the sediments of Onondaga Lake. Following turnover the pH of the lower waters was near 8.0. During stratification the release of CO_2 results in pH decreases to near 7.3. The coincidence of the decrease in pH and release of P is consistent with the influence of pH on the solubility of Ca-P minerals (Stumm & Leckie, 1971; Staudinger et al., 1990).

Evaluating the role of mineralization

Decomposition processes can result in the mobilization of P incorporated in deposited organic material. Caraco et al. (1989; 1991) developed an approach to evaluate the potential importance of this recycle pathway. Depositing organic particles in most lakes have P: organic carbon (P:C) ratios between 3 and 10 μmol P mmol C^{-1} (Caraco et al., 1991). Available sediment trap data indi-

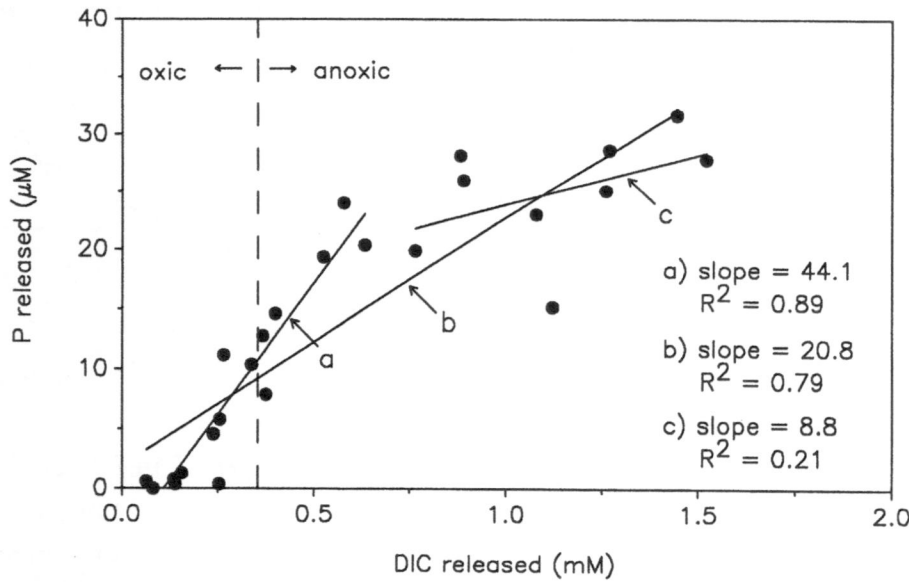

Fig. 2. Concentrations of P released (as SRP) as a function of dissolved inorganic carbon (DIC) release in the lower waters (19 m) of Onondaga Lake, March–August, 1990.

cate Onondaga Lake conditions probably fall at the upper bound of this range. When the ratio of SRP release to DIC release falls in this range, mineralization reactions alone may explain P release from sediments (Caraco *et al.*, 1989). Release ratios (e.g., SRP: DIC) less than 3 μmol P mmol C^{-1} suggest that sediment immobilization is significant, whereas ratios above this range indicate P release from sediments that was previously incorporated (Caraco *et al.*, 1989; e.g., adsorption, mineral formation, etc.).

The release of P relative to the release of DIC over the entire stratification period was 21 μmol P mmol C^{-1} (Fig. 2), considerably greater than the upper limit of values expected from mineralization of algal biomass (10 μmol P mmol C^{-1}). A very large release of P occurred initially relative to DIC release (44 μmol P mmol C^{-1}). The initial release of SRP occurred coincident with a decline in water column pH (Fig. 1f). Later in the stratification period the rate of P release declined relative to DIC release. During this later period the stoichiometry of P to DIC release was comparable to values expected from decomposition of cell biomass (8.8 μmol P mmol C^{-1}). These

trends may suggest initially a large supply of P to the water column from mineral pools. Later during stratification mineral pools become exhausted and release of P occurs from decomposition reactions (although there is considerable scatter in the stoichiometry data late during stratification (Fig. 2)). The analysis and system description presented heretofore suggest that, rather than control by redox processes, sedimentary P release in Onondaga Lake is likely the result of dissolution of, or desorption from, Ca-PO_4 minerals, in combination with decomposition of organic P.

Chemical equilibrium calculations: Ca-PO_4 mineral control

Chemical equilibrium calculations indicate that the lower waters of Onondaga Lake were oversaturated with respect to the solubility of $CaCO_3$ (mean SI $\pm$ std. dev. $= 0.41 \pm 0.24$). This pattern is consistent with water column conditions observed by Effler & Driscoll (1985) prior to the closure of the chemical facility. There are many Ca-PO_4 minerals which can potentially form (e.g.,

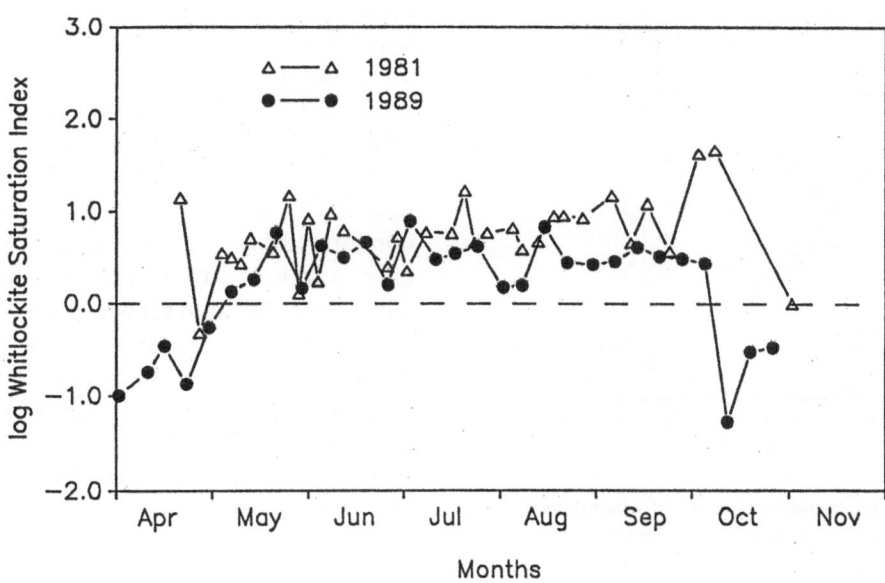

Fig. 3. Values of saturation indexes (SI) with respect to the solubility of whitlockite ($Ca_3(PO_4)_2$) for the lower waters (19 m) of Onondaga Lake during the summer stratification period, 1981 and 1989.

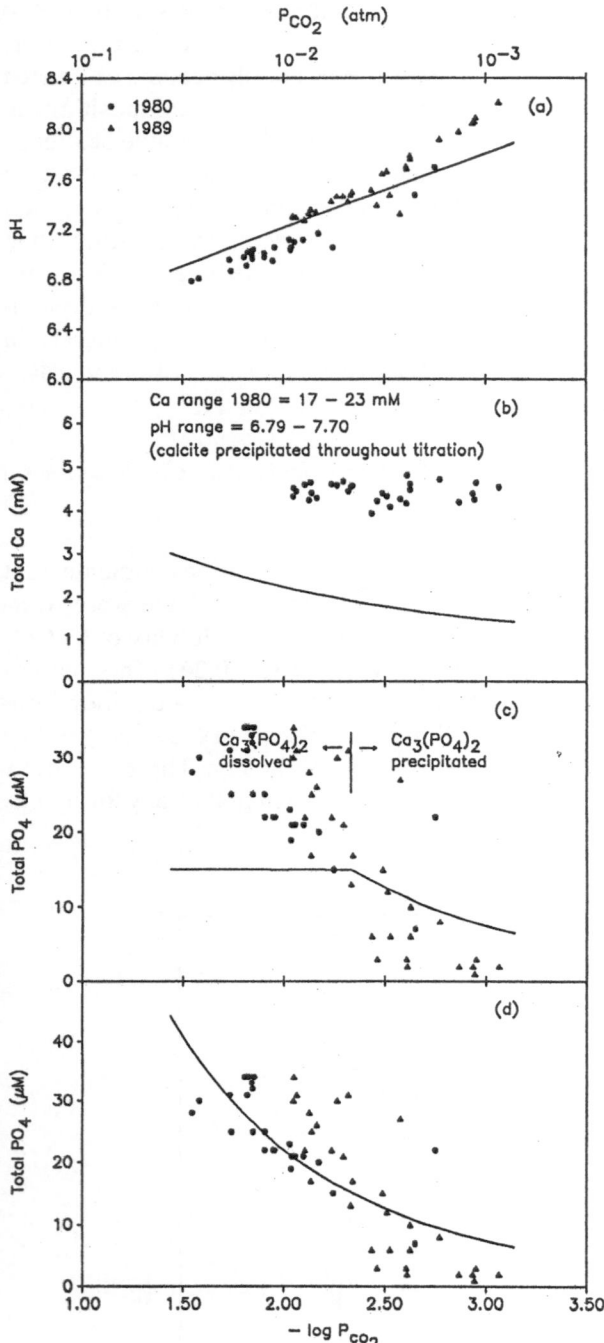

Fig. 4. Results of the chemical titration of Onondaga Lake waters to incremental addition of CO_2. Initial conditions are representative of 1989 water chemistry (Table 1). Chemical equilibrium calculations assumed that $CaCO_3$ occurred as a dissolved solid (i.e., precipitates under conditions of oversaturation). Two scenarios for Whitlockite ($Ca_2(PO_4)_3$) solubil-

monetite, brushite, octacalcium phosphate, whitlockite, hydroxyapatite, fluorapatite). In addition, total PO_4^{3-} may be regulated by adsorption on calcite particles (Leckie & Stumm, 1971) or coprecipitation with calcite. There is no direct evidence of specific Ca-PO_4 mineral formation in Onondaga Lake. However, using individual particle analysis, Honstein (1982) found that P in Onondaga Lake sediments was largely associated with Ca. Chemical equilibrium calculations with MINEQL showed conditions of oversaturation with respect to a number of relatively insoluble Ca-PO_4 minerals (e.g., hydroxyapatite, fluorapatite). Waters of Onondaga Lake were generally slightly oversaturated with respect to the solubility of whitlockite (Fig. 3; mean SI $\pm$ std. dev. for the period from early May to early October = 0.48 ± 0.20). Conditions of oversaturation were evident in recent years as well as prior to the closure to the industrial facility (1981 shown as an example year). As a result, whitlockite was used in our analysis as a reference mineral, to assess the solubility of total PO_4 from $CaPO_4$ minerals.

Two chemical equilibrium 'titrations' were conducted to evaluate the potential response of Onondaga Lake waters to perturbations. The first titration involved variable addition of CO_2 to the hypolimnion of Onondaga Lake (Fig. 4). As Pco_2 increased from values representative of spring turnover ($10^{-3.14}$ atm) to conditions of late stratification ($10^{-1.44}$ atm), the simulated pH declined from near 8.0 to 6.8 (Fig. 4a). The range of simulated pH values is similar to observed values for corresponding Pco_2 conditions. As waters were oversaturated with respect to the solubility of $CaCO_3(s)$, this mineral was precipitated throughout the titration. Increases in Ca^{2+} were evident due to increased solubility of $CaCO_3(s)$ with decreases in pH (Fig. 4b). Under low Pco_2 and high pH conditions, solutions were oversaturated with respect to the solubility of $Ca_3(PO_4)_2(s)$

ity were assumed: that the mineral occurs as a dissolved solid (4c) and as a fixed (activity) solid (4d). Measured values for two years are shown.

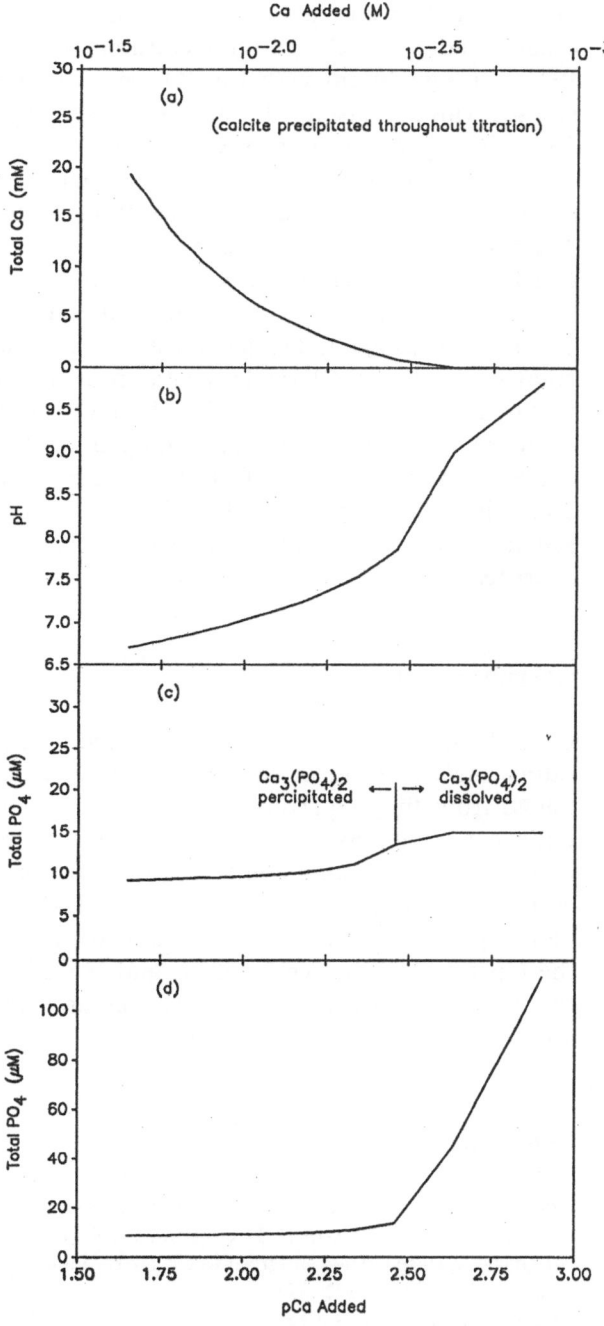

Fig. 5. Results of the chemical titration of Onondaga Lake waters to incremental addition of $CaCl_2$. Initial conditions are representative of 1989 water chemistry (Table 1). Chemical equilibrium calculations assumed that $CaCO_3$ occurred as a dissolved solid (i.e., precipitates under conditions of oversaturation). Two scenarios for Whitlockite ($Ca_2(PO_4)_3$) solubility were assumed: that the mineral occurs as a dissolved solid (5c) and a fixed (activity) solid (5d).

(Fig. 4c,d). Predictions of total PO_4^{3-} showed increasing concentrations with increasing Pco_2 and deceasing pH (Fig. 4c,d). If a finite pool of $Ca_3(PO_4)_2(s)$ exists (i.e., dissolved solid), the extent of total PO_4^{3-} release with decreasing pH is limited by the quantity of that pool (Fig. 4c). If an infinite pool of $Ca_3(PO_4)_2(s)$ is available (fixed solid activity), the release of total PO_4^{3-} increases exponentially with decreases in pH (Fig. 4d). The magnitude of total PO_4^{3-} increase with increasing Pco_2 is comparable to water column concentrations. Note that the pattern of increases in total PO_4^{3-} for the scenario in which Pco_2 is initially low and an infinite pool of $Ca_3(PO_4)_2(s)$ occurs as the fixed solid (Fig. 4d) is similar to observations of the lower waters of Onondaga Lake.

The $CaCl_2$ titration was conducted to simulate the response of Onondaga Lake to variations in Ca^{2+} loading due to changes in industrial discharge (Fig. 5). The Ca^{2+} concentrations observed in the chemical equilibrium titration spanned from lower concentrations observed in recent years following the closure of the industrial facility (e.g., 1989) to very high concentrations typical of lake conditions when the industrial facility was in full operation (e.g., 1980). Calcium concentrations increased with addition of $CaCl_2$. Due to saturation of the system with respect to $CaCO_3(s)$, most of the added Ca^{2+} precipitated as $CaCO_3(s)$ (Fig. 5a). The increase in Ca^{2+} concentrations is coincident with decreases in pH and increased solubility of $CaCO_3(s)$. Acid production occurs when inputs of $CaCl_2$ are lost from solution by $CaCO_3(s)$ precipitation (eqn. 2).

$$Ca^{2+} + 2Cl^- + HCO_3^- = CaCO_3(s)$$
$$+ 2Cl^- + H^+ \qquad (2)$$

The response of total PO_4^{3-} to changes in added Ca is interesting and may provide insight into the patterns in SRP in the lower waters of Onondaga Lake (Fig. 5c,d). Total PO_4^{3-} showed little change in concentration over the high range of the $CaCl_2$ titration (added Ca from 4 to 22.4 mmol 1^{-1}), despite a wide variation in Ca^{2+} and pH. This response is due to conditions of oversaturation and precipitation of $Ca_3(PO_4)_2(s)$. This pattern is consistent with observations that SRP concentrations released from Onondaga Lake sediments presently (Fig. 1e) are comparable to values in the early 1980's (e.g., 1980); although P inputs to the lake, productivity and P deposition to sediments have decreased in recent years. The similarity in the concentrations of SRP released from sediments under highly varying loading of P is suggestive of Ca-PO$_4$ mineral control. At lower inputs of CaCl$_2$ (somewhat lower than currently experienced by Onondaga Lake) the solution becomes undersaturated with respect to the solubility of Ca$_3$(PO$_4$)$_2$(s) and concentrations of total P O$_4^{3-}$ increase. The extent of this increase depends on the pool of Ca$_3$(PO$_4$)$_2$(s) available for dissolution. If the pool is limited, release of total PO$_4^{3-}$ will be limited by the quantity available (e.g., dissolved solid Ca$_3$(PO$_4$)$_2$(s) titration; Fig. 5c). If the pool is large (e.g., fixed solid activity Ca$_3$(PO$_4$)$_2$(s) titration) then a large increase in total PO$_4^{3-}$ will occur (Fig. 5d).

Management implications

The results of this work have management implications for Onondaga Lake. Based on P deposition from sediment traps (Wodka *et al.*, 1985), patterns of SRP release from sediments (Fig. 1,2), individual particle analysis of sediments (Honstein, 1981), mineral phase SI values (Fig. 3) and results from chemical equilibrium titrations (Fig. 5), it appears that Ca-PO$_4$ minerals play an important role in the regulation of SRP release from sediments in Onondaga Lake. Indeed if Ca-PO$_4$ minerals are important in the supply of SRP from sediments to the water column, the extent of SRP release could be affected by continued declines in Ca^{2+}. Results from the CaCl$_2$ chemical equilibrium titration suggest that if Ca^{2+} continues to decline, waters may become undersaturated with respect to Ca-PO$_4$ minerals (e.g., Ca$_3$(PO$_4$)$_2$(s)) and increases in total PO$_4^{3-}$ will occur.

Hypolimnetic oxygenation could alter P cycling from the sediment, as introduction of O$_2$ to the lower waters could result in a decrease in pH due to enhanced aerobic mineralization of organic matter to CO$_2$. The concentration of total PO$_4^{3-}$ in equilibrium with Ca-PO$_4$ minerals increases with increasing CO$_2$ and decreasing pH, for a given concentration of Ca^{2+} (Fig. 4). Preliminary results of SRP release from intact sediment cores are consistent with this observation. The rate of SRP release has been found to be considerably greater under low pH conditions than under higher pH conditions (unpublished data). In Onondaga Lake, due to Ca-PO$_4$ mineral control of sediment SRP release, it appears likely that hypolimnetic oxygenation would not result a decrease in internal cycling of P. Indeed if water column pH values decline from the treatment, it seems likely that the rate of SRP supply from sediments would be enhanced, at least over the short-term.

Acknowledgements

Carol Brooks and Jeff Addess performed laboratory analysis. Calcium and Na$^+$ data were obtained from the Department of Sanitation, Onondaga County, Syracuse, NY. Bruce Wagner and MaryGail Perkins conducted the field program. Bill Schecher advised us on the MINEQL modeling efforts. This is Contribution No. 117 of the Upstate Freshwater Institute, and No. 22 of the New York State Fresh Water Research Institute.

References

Ahlgren, I., 1977. Role of sediments in the process of recovery of a eutrophic lake. In H. L. Gotterman (ed.), Interactions between sediments and fresh water. Dr W. Junk Publishers, The Hague, 372–377.

Anderson, J. M., 1975. Influence of pH on release of phosphorus from lake sediments. Arch. Hydrobiol. 76: 411–419.

APHA (American Public Health Association), 1985. Standard Methods for the Examination of Water and Wastewater, 16th Edition. American Public Health Association, Washington, DC, 1268 pp.

Auer, M. T., N. A. Johnson, M. R. Penn & S. W. Effler, 1993. Measurement and verification of rates of sediment phosphorus release for a hypereutrophic urban lake. In P. C. M. Boers, T. E. Cappenberg & W. van Raaphorst (eds), Proceedings of the Third International Workshop on Phosphorus in Sediments. Developments in Hydrobiology 84. Kluwer Academic Publishers, Dordrecht: 301–309. Reprinted from Hydrobiologia 253.

Auer, M. T., M. L. Storey, S. W. Effler, N. A. Auer & P. Sze, 1990. Zooplankton impacts on chlorophyll and transparency in Onondaga Lake, New York, USA. In R. D. Gulati, E. H. R. R. Lammens, M.-L. Meijer & E. van Donk (eds), Fluxes between Trophic Levels and through the Water–Sediment Interface. Developments in Hydrobiology 61. Kluwer Academic Publishers, Dordrecht: 603–617. Reprinted from Hydrobiologia 200/201.

Baccini, P., 1985. Phosphate interactions at the sediment–water interface. In W. Stumm (ed.), Chemical processes in lakes. Wiley-Interscience, New York: 189–205.

Ball, J. W., D. K. Nordstrom & E. A. Jenne, 1980. Additional and revised thermodynamic data for WATEQ-2 computerized model for trace and major element speciation and mineral equilibrium of natural waters, U.S., Geol. Surv. Water Resource Investigations, Menlo Park, California.

Böstrom, B., J. M. Andersen, S. Flerscher & M. Jansson, 1988. Exchange of phosphorus across the sediment–water interface. In G. Persson & M. Jansson (eds), Phosphorus in Freshwater Ecosystems. Developments in Hydrobiology 48. Kluwer Academic Publishers, Dordrecht: 229–244. Reprinted from Hydrobiologia 170.

Caraco, N. F., J. J. Cole & G. E. Likens, 1989. Evidence for sulfate-controlled P release from sediments of aquatic systems. Nature (London) 341: 316–318.

Caraco, N. F., J. J. Cole & G. E. Likens, 1991. A cross-system study of phosphorus release from Lake sediments. In J. Cole, G. Lovett and S. Findlay (eds), Comparative analyses of ecosystems: Patterns, mechanisms, and theories. Springer-Verlag, New York: 241–258.

Cooke, G. D., E. B. Welch, S. A. Perterson & P. R. Newroth, 1986. Lake and reservoir restoration. Butterworth, Boston.

Devan, S. P. & S. W. Effler, 1984. History of phosphorus loading to Onondaga Lake. J. Envir. Eng. Div., ASCE 110: 93–109.

Effler, S. W. & C. T. Driscoll, 1985. Calcium chemistry and deposition in ionically enriched Onondaga Lake, New York. Envir. Sci. Technol. 19: 716–720.

Effler, S. W., J. P. Hassett, M. T. Auer & N. Johnson, 1988. Depletion of epilimnetic oxygen and accumulation of hydrogen sulfide in the hypolimnion of ionically enriched Onondaga Lake, NY, USA. Wat. Air Soil Pollut. 39: 59–74.

Effler, S. W., M. G. Perkins & C. Brooks, 1986. The oxygen resources of the hypolimnion of ionically enriched Onondaga Lake, NY, USA. Wat. Air Soil Pollut. 29: 93–108.

Effler, S. W., C. M. Brooks, J. M. Addess, S. M. Doerr, M. L. Storey & B. A. Wagner, 1991. Pollutant loadings from Solvay Waste beds to lower Ninemile Creek, New York. Wat. Air Soil Pollut. 55: 427–444.

Einsele, W., 1936. Über die Beziehungen des Eisenkreislaufs

Zum Phosphatkreislauf im eutrophen See. Arch. Hydrobiol. 29: 664–686.

Field, S. D., 1980. Nutrient-saturated algal growth in hypereutrophic Onondaga Lake, Syracuse, New York. Thesis. Syracuse University. Syracuse, NY.

Gächter, R., 1987. Lake restoration. Why oxygenation and artificial mixing cannot substitute for a decrease in the external phosphorus loading. Schweiz. Z. Hydrol. 49: 170–185.

Heaney, S. I. & W. Davison, 1977. The determination of ferrous iron in natural waters with 2,2′ bipyridyl. Limnol. Oceanogr. 22: 753–759.

Honstein, R. L., 1981. An assessment of mechanisms by which phosphorus may be regulated within the sediments of Onondaga Lake, New York. Thesis, Syracuse University Syracuse, NY.

Hutchinson, G. E., 1973. Eutrophication. The scientific background of a contemporary practical problem. Am. Sci. 61: 269–279.

Johnson, D. L., J. Jiao, S. G. DosSantos & S. W. Effler, 1991. Individual particle analysis of suspended materials in Onondaga Lake, New York. Envir. Sci. Technol. 25: 736–744.

Larsen, D. P., D. W. Schultz & K. W. Malereg, 1981. Summer internal phosphorus supplies in Shagawa Lake, Minnesota. Limnol. Oceanogr. 26: 740–753.

Mortimer, C. J., 1941. The exchange of dissolved substances between mud and water in lakes (Parts I and II). J. Ecol. 29: 280–329.

Mortimer, C. H., 1942. The exchange of dissolved substances between mud and water in lakes (Parts III and IV). J. Ecol. 30: 147–201.

Schindler, D. W., 1977. Evolution of phosphorus limitation in lakes. Science 195: 260–262.

Staudinger, B., S. Peiffer, Y. Avnimelech & T. Berman, 1990. Phosphorus mobility in interstitial waters of sediments in Lake Kinneret, Israel. Hydrobiologia 207: 167–177.

Stauffer, R. E., 1985. Relationships between phosphorus loading and trophic state in calcareous lakes of southeast Wisconsin. Limnol. Oceanogr. 30: 123–145.

Stumm, W. & J. O. Leckie, 1971. Phosphate exchange with sediments: its role in the productivity of surface waters. Adv. Wat. Pollut. Res. 5; 1970, III-26: 1–16.

Stumm, W. & J. J. Morgan, 1981. Aquatic chemistry, 2nd edn. Wiley Interscience, New York.

Uttormark, F. D., 1978. General concepts of lake degradation and lake restoration. Environmental Protection Agency. EPA-740/5-79-001, pp. 65–70.

Vollenweider, R. A., 1975. Input-output models, with special reference to the phosphorus loading concept in limnology. Schweiz. Z. Hydrol. 37: 53–84.

Welch, E. B., D. E. Spyridakis, J. I. Shuster & R. R. Horner, 1986. Declining lake sediment phosphorus release and oxygen deficit following wastewater diversion. J. Wat. Pollut. Contr. Fed. 58: 92–96.

Westall, J. C., J. L. Zachary & F. M. M. Morel, 1976. Technical note 18, R. M. Parsons Laboratory for Water Resources and Hydrodynamics, Massachusetts Institute of Technology, Cambridge.

Williams, J. D., J. M. Jaquet & R. L. Thomas, 1976. Forms of phosphorus in the surficial sediments of Lake Erie. J. Fish. Res. Bd. Can. 33: 413–429.

Wodka, M. C., S. W. Effler & C. T. Driscoll, 1985. Phosphorus desposition from the epilimnion of Onondaga Lake. Limnol. Oceanogr. 30: 833–843.

Yin, C. & D. L. Johnson, 1984. An individual particle analysis and budget study of Onondaga Lake sediments. Limnol. Oceanogr. 29: 1193–1201.

Hydrobiologia **253**: 73–82, 1993.
P.C.M. Boers, T.E. Cappenberg & W. van Raaphorst (eds),
Proceedings of the Third International Workshop on Phosphorus in Sediments.
© 1993 *Kluwer Academic Publishers.*

Factors influencing fractional phosphorus composition in sediments of Spanish reservoirs

P. Lopez & J. A. Morgui
Dep. Ecology, F. Biology, Avgda, Diagonal 645, 08028 Barcelona, Spain

Key words: sediment, phosphorus fractions, reservoirs, trophic status

Abstract

Total phosphorus in sediment (P_{sed}) and its fractional composition (reactive phosphate extracted with NaOH, NaOH-RP, reactive phosphate extracted with HCl, HCl-RP, and residual phosphate, residual-P) have been determined in superficial sediments of 43 Spanish reservoirs located in different limnological regions and with different trophic states. Data were evaluated by statistical analysis to examine the influence of regional distribution and trophic status. Relations with calcium, manganese, iron and aluminium contents have also been studied.

In the western part of Spain, reservoirs presented the highest values on average of P_{sed}, NaOH-RP and residual-P (1296, 328 and 877 $\mu g \ g^{-1}$ dw., respectively) and the lowest values of HCl-RP (91.0 $\mu g \ g^{-1}$ dw.). The main phosphorus fractions were residual-P ($>50\%$) and NaOH-RP ($>10\%$). In the eastern area, P_{sed} NaOH-RP and residual-P attained the lowest values on average (502, 4 and 330 $\mu g \ g^{-1}$ dw., respectively), whereas HCl-RP presented the highest values (167 $\mu g \ g^{-1}$ dw.). The main fractions were residual-P ($>50\%$) and HCl-RP ($>25\%$).

Trophic status seemed to be a secondary factor controlling P_{sed}. The highest contents of P_{sed} were found in eutrophic reservoirs, but only when those of the same region were compared, and the statistical significance (ANOVA F test) of the observed differences was very small ($p < 0.057$).

Introduction

Sediment characteristics may be involved in many processes controlling phosphorus chemistry in lakes and reservoirs. A large fraction (50–100%) of the external phosphorus loading may be retained by sediments depending on characteristics of catchments and water bodies (Lijklema, 1980). Numerous studies have focused on possible relationships between total phosphorus contents of upper sediment layers and other lake characteristics such as trophic status (Golachowska, 1984; Wisnievsky & Planter, 1985; Flannery *et al.*, 1982), sediment composition (Brunskill *et al.*,

1971; Allan & Brunskill, 1977; Lopez, 1986) or internal phosphorus loading (Boström, 1984; Nürnberg *et al.*,1986; Nürnberg, 1988). Fractional phosphorus composition has been studied in relation to internal phosphorus loading, but it has seldom been compared with other sediment characteristics. Notable exceptions are the works of Williams *et al.* (1976), Ostrofsky (1987) and Ostroksky *et al.* (1989). However, the influence of regional variations (due to differences in geological and/or climatological characteristics) on phosphorus contents in sediment and fractional phosphorus composition has not been taken into account in these studies, because only small groups of lakes or lakes with similar characteristics were considered. Consequently, many of the relationships described cannot explain the variation observed between lakes from different regions.

Spanish reservoirs represent a wide range of lake types. An extensive survey of one hundred reservoirs all over Spain established a typology based on major ionic composition and salinity of water (Armengol et al., 1991). This classification reflects geological, climatological and geographical factors and it allows us to define four limnological regions in the Iberian Peninsula. The same reservoirs were also classified according to their concentrations of chlorophyll a (Cla) and soluble reactive phosphorus (SRP), giving an estimation of the trophic status of reservoirs. Four categories were defined: oligomesotrophic (Cla < 2 mg m^{-3} and SRP < 0.2 μmol l^{-1}) meso-eutrophic (Cla from 2 to 5 mg m^{-3} or SRP from 0.2 to 0.5 μmol l^{-1}) eutrophic (Cla from 5 to 50 mg m^{-3} or SRP from 0.5 to 1.0 μmol l^{-1}) and hypereutrophic (Cla > 50 mg m^{-3} or SRP > 1.0 μmol l^{-1}) (Morgui et al., 1990; Morgui, 1991).

In the present study, we examined the contents of total phosphorus (P$_{sed}$) and its fractions NaOH-RP, HCl-RP and residual-P, as well as total iron (Fe$_{sed}$), total calcium (Ca$_{sed}$), total aluminium (Al$_{sed}$) and total manganese (Mn$_{sed}$) in the sediments of 43 selected reservoirs. We searched for trends in variations of P$_{sed}$ and its fractional composition related to regional distribution of the reservoirs or trophic characteristics. We also studied the relationships between phosphorus species and other elements. This may provide a global approach to the influence of regional and trophic factors on phosphorus composition in sediments of Spanish reservoirs.

Study area

The Iberian Peninsula can be divided into four limnological regions defined by two intersecting gradients: from west to east because of the transition from eruptive to sedimentary rocks and from north to south because of a decreasing rainfall gradient (Margalef, 1975; Estrada, 1975; Margalef et al., 1976; Armengol et al., 1991). The four regions present differences in geological and climatological characteristics, which are reflected in the major ionic composition of water. In the western region (I) (Fig. 1), reservoirs have catchments on siliceous rocks (mainly basalt and granite) and waters have low concentrations of calcium, bicarbonate and dissolved salts. Because of its Atlantic climate, mean rainfall is moderate or high and this area is sometimes called 'the wet Spain'. In the eastern region, reservoirs have catchments on sedimentary rocks and waters have high concentrations of calcium, bicarbonate and dissolved salts. Differences in the geological nature of watersheds and rainfall determine three subregions within this area: Region II includes reservoirs mainly situated in North and Northeastern Spain (although some reservoirs located in special areas of the south are included in this area). They have catchments on calcareous rocks and rainfall is similar to that of the western area. Bicarbonate and calcium are the major ions in water. Region III: includes reservoirs located in endorrheic areas of Central, South and East Spain. Calcium and sulphate are the dominant ions. Because of its low rainfall, this area is called 'the dry Spain'. Region IV includes a small number of reservoirs located in the southernmost part of Spain that are characterized by inputs of water coming from areas with accumulation of evaporites. Chloride and sodium are the dominant ions in the water, which has a high concentration of dissolved salts.

The location of the 43 reservoirs studied in this paper and the region they belong to are given in Fig. 1.

Methods

Sediment samples were obtained from a single location in the deepest area of the reservoir. 21 reservoirs were sampled in December 1987/February 1988, 13 reservoirs in June/September 1988 and 9 were sampled twice, in winter and summer. Samples were collected from the upper

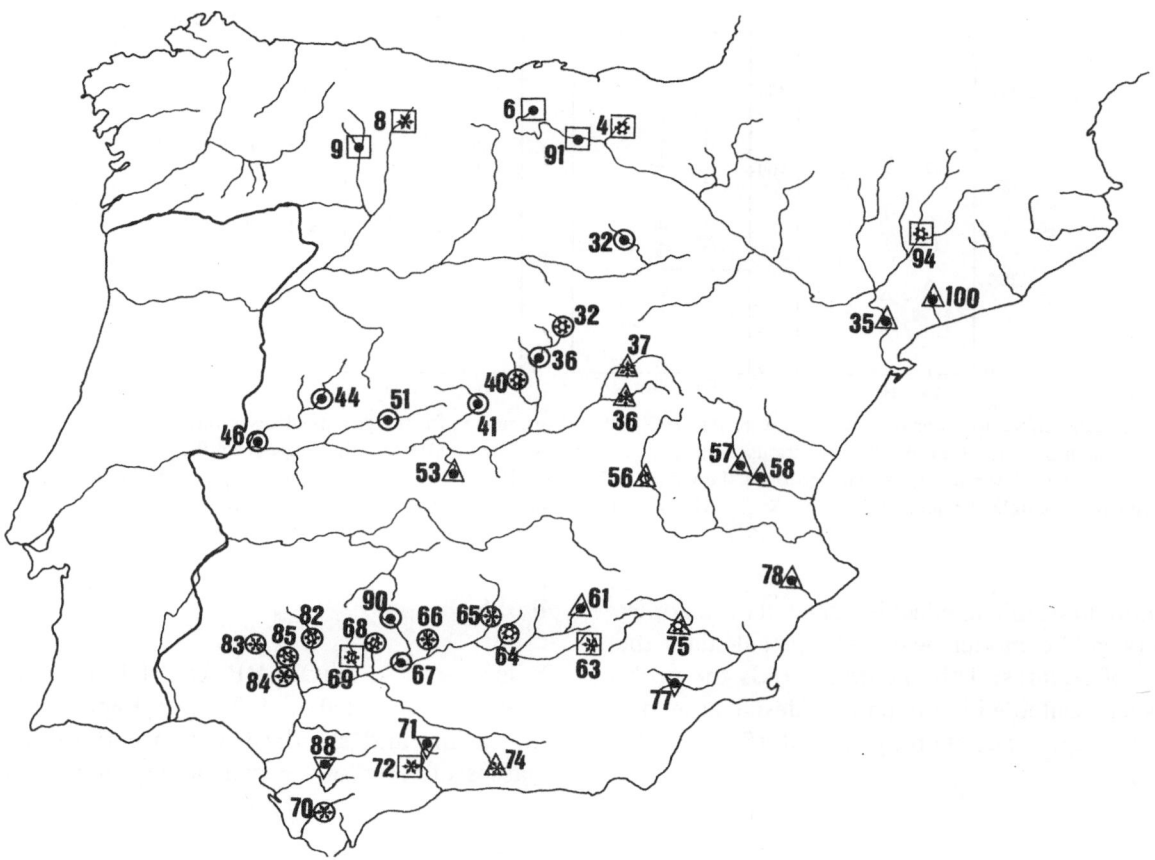

Fig. 1. Geographical distribution of the reservoirs studied. Circles: Region I; Squares: Region II; Triangles: Region III. Inverse triangles: Region IV. ⊛: oligo-mesotrophic; ▲: meso-eutrophic; ●: eutrophic.

4 cm of sediment with a modified Eckman grab and frozen for subsequent analyses.

Phosphorus fractions were determined using the sequential extraction procedure of Hieltjes & Lijklema (1980), but using distilled water instead of NH_4Cl to remove interstitial and weakly-adsorbed phosphate. Previous experiments showed that NaOH-RP extraction was not affected by contents of 15% Ca_{sed} and that values lower than $1-2 \mu g \, g^{-1}$ of NaOH-RP had poor reproducibility (Lopez & Morgui, in press). P_{sed}, Fe_{sed}, Ca_{sed}, Al_{sed} and Mn_{sed} were determined by X-ray fluorescence spectrometry. Residual-P was calculated as the difference between total phosphorus and the sum of NaOH-RP and HCl-RP.

The SPSS.PC computer software system was used for statistical analysis. The Cochrans-C and Barlett-Box F test were used to check for homogeneity of variances and Box's M test for the homogeneity of variance-covariance matrix (SPSS.INC, 1988). Data were logarithmically transformed so as not to violate homogeneity of variance-covariance and multivariate normality assumptions. The variations observed in reservoirs sampled twice did not show seasonal trends, the average value for each reservoir being used for statistical analyses. Sample sizes for each group were: Region I: oligomesotrophic $N = 6$; meso-eutrophic $N = 5$; eutrophic $N = 8$; Region II: oligo-mesotrophic $N = 3$; meso-eutrophic $N = 3$; eutrophic $N = 4$; Region III: oligo-mesotrophic $N = 3$; meso-eutrophic $N = 2$; eutrophic $N = 6$: Region IV eutrophic $N = 3$. For the multivariate analysis of variance (MANOVA), the regression

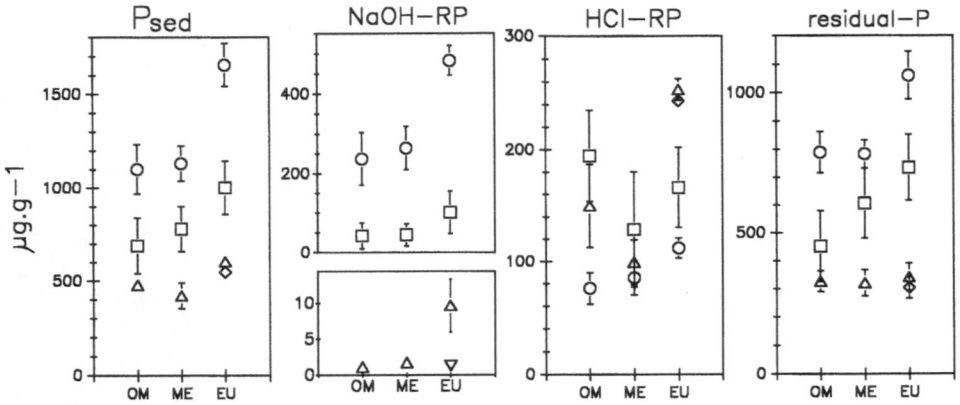

Fig. 2. Mean and standard error of P$_{sed}$, NaOH-RP, HCl-RP and residual-P, for the groups of reservoirs established according to their regional distribution and trophic status. Values in μg g^{-1} dw. For NaOH-RP, the range from 0 to 15 μg g^{-1} has been expanded in the lower part of the figure to show the values in Regions III and IV. OM: oligo-mesotrophic; ME: meso-eutrophic; EU: eutrophic. Circles: Region I; Squares: Region II; Triangles: Region III. Inverse triangles: Region IV.

method, in which an effect is adjusted for all other effects in the model, was used to calculate the sums of squares. This method avoids the problems associated with unbalanced designs (i.e. different sample sizes of groups tested) (SPSS.INC, 1988).

Results

Values of P$_{sed}$, NaOH-RP, HCl-RP and residual-P are given in Tables 1, 2, and 3. Figure 2 shows the means and standard error for the different classes of reservoirs based on regional distribu-

Table 1. Contents (in μg g^{-1}) of P$_{sed}$, NaOH-RP, HCl-RP and residual-P in the reservoirs studied in Region I. Values in brackets correspond to the percentage of P$_{sed}$.

	NaOH-RP	HCl-RP	Residual-P	P$_{sed}$
84	175.3 (18.1)	32.8 (3.4)	762.4 (78.6)	970.5
65	457.7 (31.0)	90.4 (6.1)	928.6 (62.9)	1476.8
70	35.1 (5.2)	125.4 (18.6)	514.6 (76.2)	675.1
83	352.6 (25.3)	55.1 (4.0)	984.7 (70.7)	1392.4
66	163.0 (16.4)	75.4 (7.6)	753.1 (75.9)	991.5
64	256.6 (23.9)	166.2 (15.4)	653.1 (60.7)	1075.9
38	165.2 (19.6)	63.1 (7.5)	615.5 (72.9)	843.9
82	227.2 (20.7)	83.1 (7.6)	786.7 (71.7)	1097.0
68	190.0 (17.0)	82.3 (7.4)	845.9 (75.7)	1118.1
85	191.9 (18.2)	53.7 (5.1)	809.2 (76.7)	1054.8
40	554.9 (34.6)	63.0 (3.9)	985.4 (61.5)	1603.4
67	110.7 (12.8)	69.6 (8.1)	684.6 (79.1)	865.0
32	164.2 (21.6)	53.8 (7.1)	541.5 (71.3)	759.5
90	282.1 (20.3)	56.5 (4.1)	1053.8 (75.7)	1392.4
41	1015.5 (47.7)	130.8 (6.1)	984.5 (46.2)	2130.8
60	599.0 (34.6)	95.8 (5.5)	1035.1 (59.8)	1729.9
46	901.1 (23.0)	142.9 (3.6)	2880.1 (73.4)	3924.0
44	285.6 (29.4)	49.0 (5.1)	635.8 (65.5)	970.5
51	507.0 (34.3)	296.7 (20.1)	673.1 (45.6)	1476.8

Table 2. Contents (in $\mu g \, g^{-1}$) of P_{sed}, NaOH-RP, HCl-RP and residual-P in the reservoirs studied in Region II. Values in brackets correspond to the percentage of P_{sed}.

	NaOP-RP	HCl-RP	Residual-P	P_{sed}
72	1.1 (0.2)	293.8 (43.5)	380.3 (56.3)	675.1
63	2.7 (0.7)	151.1 (39.8)	226.0 (59.4)	379.7
8	122.1 (12.1)	138.6 (13.7)	751.9 (74.3)	1012.6
69	112.4 (10.4)	64.3 (6.0)	899.2 (83.6)	1075.9
94	2.2 (0.3)	255.1 (40.3)	375.7 (59.4)	632.9
4	19.3 (3.0)	66.5 (10.5)	547.1 (86.4)	632.9
91	7.6 (1.3)	161.8 (27.4)	421.3 (71.3)	590.7
53	48.7 (5.0)	203.6 (21.0)	718.2 (74.0)	970.5
9	64.5 (4.6)	246.2 (17.7)	1081.6 (77.7)	1392.4
6	284.6 (27.0)	52.8 (5.0)	717.5 (68.0)	1054.8

tion and trophic status. Mean values of P_{sed}, NaOH-RP and residual-P increased from Region III to Region I and from oligo-mesotrophic to eutrophic reservoirs. Mean contents of HCl-RP increased from Region I to Region III, increasing from oligo-mesotrophic to eutrophic reservoirs in Region I, whereas in the other areas it presented the lowest average values in mesoeutrophic reservoirs.

To test for the significance of the differences observed, a two-way analysis of variance (ANOVA), with regional distribution and trophic status as factors, was performed for P_{sed} (Table 4). Data from Region IV were excluded from the analysis, because the reservoirs included in this area were all eutrophic. The inclusion of these data in the ANOVA would lead to the generation of empty groups (*i.e.* oligo-mesotrophic and meso-eutrophic in Region IV) which would make the results difficult to interpret.

P_{sed} was significantly related to regional distribution and only weakly related to trophic status. To identify which of the regions presented significantly different means compared with the other

Table 3. Contents (in $\mu g \, g^{-1}$) of P_{sed}, NaOH-RP, HC-RP and residual-P in the reservoirs studied in Regions III and IV. Values in brackets correspond to the percentage of P_{sed}.

	NaOH-RP	HCl-RP	Residual-P	P_{sed}
Region III				
36	0.5 (0.1)	120.7 (26.0)	343.0 (73.9)	464.1
74	1.9 (0.4)	239.1 (47.2)	265.4 (52.4)	506.3
37	1.0 (0.2)	90.0 (19.4)	373.1 (80.4)	464.1
75	2.1 (0.4)	149.7 (25.3)	438.9 (74.3)	590.7
56	1.3 (0.5)	48.8 (19.3)	203.0 (80.2)	253.2
57	3.3 (0.5)	250.1 (39.5)	379.5 (60.0)	632.9
78	0.5 (0.1)	261.2 (51.6)	244.7 (48.3)	506.3
61	26.1 (3.6)	206.9 (28.8)	484.3 (67.5)	717.3
100	9.1 (1.3)	252.6 (37.4)	413.5 (61.2)	675.1
58	2.3 (0.3)	268.3 (39.7)	404.5 (59.9)	675.1
35	16.7 (4.0)	280.5 (66.5)	124.7 (29.5)	421.9
Region IV				
77	0.1 (0.0)	249.4 (53.7)	214.6 (46.2)	464.1
88	1.6 (0.3)	226.4 (41.3)	320.5 (58.4)	548.5
71	2.1 (0.3)	254.1 (40.1)	376.7 (59.5)	632.9

Table 4. Results of the two-ways ANOVA and Tukey's (HSD) test for P_{sed}.

Source of variation	Two-ways ANOVA				
	SS	DF	Ms	F	Sig. of F
Within cells	0.81	31	0.03		
Region	0.97	2	0.49	18.74	0.000
T	0.16	2	0.08	3.14	0.057
Region by T	0.02	4	0.00	0.18	0.945
	Tukey's (HSD) test				
	Region II		Region III		
Region I	*		*		
Region II			*		

* Indicate $p < 0.05$.

Table 5. Results of the multivariate analysis of variance for fractional phosphorus composition. P_{sed} acts as covariate (regression factor). Univariate F-test for significant effects in multivariate analysis and regression parameters are also given. T: trophic status; SS: sum of squares; MS: mean squares. DF: degrees of freedom; MR: multiple regression coefficient.

Effect	A. Multivariate tests of significance (Pillais test)				
	Value	F hypoth.	DF	Error DF	Sig.
Regression	0.912	96.760	3.00	28.00	0.000
Region by T	0.332	0.933	12.00	90.00	0.518
T	0.291	1.647	6.00	58.00	0.151
Region	0.582	3.968	6.00	58.00	0.002

Effect: regression	B. Univariate test of significance					
Variable	MR^2	Adj. R^2	MS	Error MS	F	Sig.
NaOH-RP	0.366	0.344	3.309	0.191	17.297	0.000
HCl-RP	0.053	0.022	0.079	0.052	1.692	0.203
Residual-P	0.799	0.792	0.836	0.007	118.961	0.000

Effect: regional distribution						
Variable	SS	Error SS	MS	Error MS	F	Sig.
NaOH-RP	4.990	5.739	2.495	0.191	13.041	0.000
HCl-RP	0.570	1.558	0.285	0.052	5.490	0.009
Residual-P	0.003	0.211	0.002	0.007	0.247	0.783

	Regression parameters			
	B	Beta	Std. err.	Confidence interval (P = 95%)
NaOH-RP	2.027	0.605	0.487	1.032–3.023
Residual-P	1.019	0.893	0.093	0.828–1.210

regions, we used a multiple comparison test, Tukey's Honest Significant Difference (HSD) (Table 4). As can be observed, each area presented different values of P_{sed} as compared to those of the other areas.

Relationships between fractional composition, regional distribution and trophic status were studied using a multivariate analysis of variance (MANOVA) (Table 5). This statistical procedure takes into account the relationships among dependent variables, being more adequate than one single ANOVA for each dependent variable when these are correlated. The fractional composition was considered as a complex variable described by NaOH-RP, HCl-RP and residual-P. The multivariate test (Table 5A) evaluates the hypothesis that the population means for these three variables as a whole are the same for the different categories defined by factors: regional distribution and trophic status. The influence of P_{sed} on the fractional composition was also considered by using P_{sed} as covariate in the MANOVA.

Results showed that fractional composition was significantly associated with P_{sed} (significance for regression factor 0.000) and regional distribution (significance for region factor 0.002), but not related to trophic status. Only when the multivariate test is significant can the univariate test help to determine which variables contribute to overall differences (Table 5B). NaOH-RP was associated with P_{sed} and regional distribution, HCl-RP with regional distribution and residual-P only with P_{sed}. For NaOH-RP the means adjusted for covariance are compared with observed means in Fig. 3. In Region I values observed were higher than those expected from P_{sed} contents, whereas in Region III values observed were lower.

Because regional differences in NaOH-RP and HCl-RP could not be explained only from P_{sed} differences, the relative contribution of these fractions to the phosphorus content should also present regional trends. The relative fractional composition of phosphorus in sediments of the reservoirs studied is showed in Fig. 4. Regions I and III could be clearly distinguished: reservoirs of the first area presented almost constant per-

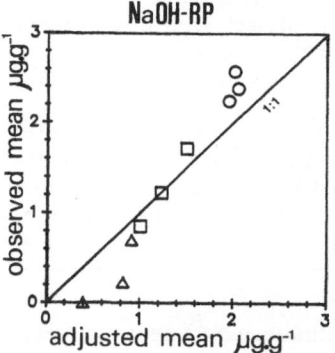

Fig. 3. Observed means of NaOH-RP (y-axis) versus adjusted means for covariance with P_{sed} (x-axis). Symbols as used in Fig. 2.

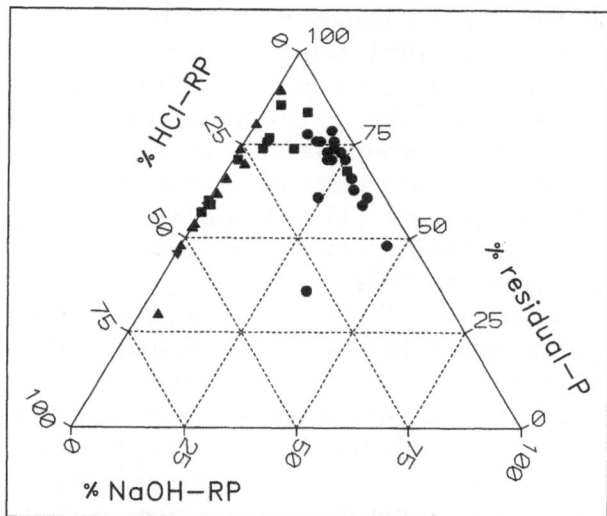

Fig. 4. Position, on a triangular diagram, of the reservoirs studied according to the relative fractional phosphorus composition. Symbols as in Fig. 2.

centages of HCl-RP, which was the least abundant fraction (5%); in Regions III and IV NaOH-RP was always lower than 1%, HCl-RP and residual-P being the most abundant fractions with a wide range of variation.

We also examined the correlations between phosphorus species and total contents of iron, calcium, aluminium and manganese for the whole set of data. The matrix of correlation coefficients is given in Table 6. P_{sed}, NaOH-RP and residual-P were positively correlated with iron, alumin-

Table 6. Correlation coefficients among phosphorus fractions and elements of sediment. Data were logarithmically transformed. (*: $p < 0.01$; **: $p < 0.001$; $N = 42$).

	Fe_{sed}	Al_{sed}	Ca_{sed}	Mn_{sed}
NaOH-RP	0.7011**	0.7207**	− 0.8905**	0.6393**
HCl-RP	− 0.2903	− 0.2660	0.6449**	− 0.3066
Residual-P	0.7159**	0.7104**	− 0.7485**	0.6571**
P_{sed}	0.6859**	0.7012**	− 0.7136**	0.6430**

ium and manganese and negatively with calcium. HCl-RP was only correlated with calcium.

Discussion

In the reservoirs studied, P_{sed} was clearly related to regional distribution, the mean values observed increasing from east (Region III) to west (Region I). The regions differ greatly in the geological nature of their catchments: siliceous rocks in the western area and calcareous rocks in the eastern area (Margalef *et al.*, 1976; Armengol *et al.*, 1991). The values of P_{sed} reported for calcareous sediments are usually lower than those observed for non-calcareous (Bortlesson & Lee, 1972, 1974, 1975; Allan & Brunskill, 1977). Estrada (1978) suggested that the high concentrations of bicarbonate and calcium in eastern reservoirs could prevent eutrophication by removing phosphate from water as calcium-phosphate precipitates. However, P_{sed} contents in eastern reservoirs (Regions II, III and IV) were lower than those in western reservoirs. This seems to indicate that mechanisms other than calcium-phosphate precipitation were probably more important in controlling phosphorus transport to sediments in the reservoirs studied.

Siliceous rocks usually have higher contents of iron and aluminium and lower contents of calcium than sedimentary rocks (Forstner, 1977). This may explain the correlations observed between P_{sed}, Fe_{sed}, Al_{sed}, Mn_{sed} and Ca_{sed}. The role of iron and aluminium in phosphorus dynamics in sediment is well documented. Chemical association of these variables frequently leads to strong correlation between phosphorus and iron

and aluminium (Brünskill *et al.*, 1971; Allan & Brünskill, 1977; Ostrofsky, 1987). However, these correlations may not be found when other factors also have a strong influence, (Premazzi & Ravera, 1977; Cross & Rigler, 1983, Flannery *et al.*, 1982, Lopez, 1986). Whether the correlations observed for the reservoirs studied were a consequence of chemical association between phosphorus and metals, or caused by another factor, which acted simultaneously on phosphorus and metals, cannot be deduced from these data.

Trophic status seems to be a minor factor controlling P_{sed} in the reservoirs studied. Literature data on the relationship between P_{sed} and trophic status are not conclusive. Some authors have reported higher contents of P_{sed} in eutrophic than in oligotrophic sediments in local groups of lakes (Flannery *et al.*, 1982; Wisnievsky & Planter, 1985). Others could not establish this relationship for lakes from different regions (Golachowska, 1984; Lopez, 1991). We found the highest contents of P_{sed} in eutrophic reservoirs, but only when those of the same region were compared, and the statistical significance of the differences observed was very small ($p < 0.057$).

We also studied P_{sed} from a fractional composition approach. The relationship between P_{sed} and fractional composition can be considered from two different points of view. On the one hand, P_{sed} corresponds to the sum of the three fractions and so changes in P_{sed} are due to changes in absolute values of any fraction. Thus P_{sed} is dependent on the individual fractions. On the other hand, chemical processes in sediments are controlled, among other factors, by the amount of reactants, so changes in one fraction may be due to the total amount of phosphate (P_{sed}). Fractions are therefore dependent on P_{sed}. The latter approach was used in the multivariate analysis of variance, where P_{sed} was considered as a covariate, because analytically residual-P was calculated from P_{sed} and not as an independent variable.

As a whole, fractional composition was related to P_{sed} and regional distribution but not affected by trophic status. However, the relationships observed were different for each fraction.

Regional differences in NaOH-RP were not only a consequence of regional differences in P_{sed}. In Region I, NaOH-RP content was higher than that expected from P_{sed}, whereas in Region III it was lower. Ostrofsky (1987) suggested constant relationships between NaOH-RP and P_{sed} for 66 North American lakes, but our data indicate that this relationship may be influenced by regional distribution of the reservoirs: in the western area, NaOH-RP contributed to P_{sed} to a greater extent than in eastern areas. This may be explained by the higher contents of iron and aluminium in the western area, compared to those of the eastern areas. The extremely low values of NaOH-RP in Regions III and IV should be considered with some caution, because they are in the lowest range of determination of the analytical procedure (see Methods).

HCl-RP was not related to P_{sed} and it was strongly influenced by regional distribution. This fraction has been reported as an allochthonous compound and mainly related to erosion rates (Williams et al., 1976; Jones & Bowser, 1978; Armengol et al., 1984), although precipitation of calcium phosphates can also occur in calcareous waters (Jacobsen, 1978; Avnimelech, 1983). The values of HCl-RP observed in Regions II and III, which attained 50% of total P_{sed}, were in agreement with the calcareous nature of the catchments, and they could be due to allochthonous input or calcium-phosphate precipitation. However, the lack of correlation between HCl-RP and P_{sed} suggests that erosion rates and apatite precipitation did not exert significant control over P_{sed} for the whole set of reservoirs.

The regional differences observed for residual-P were associated with regional differences in P_{sed}. Residual-P has been related to organic phosphorus, but it can also include inorganic refractory phosphates (Hieltjes & Lijklema, 1980). No regional differences were found for organic phosphorus, as could be expected. The lack of association between organic phosphorus and trophic status is more difficult to explain. A better distinction between organic and inorganic refractory phosphates is needed to explain the variations observed for residual-P.

The correlations observed among the three fractions and other elements in sediments were in agreement with their chemical composition and their association with P_{sed}. NaOH-RP and residual-P were correlated positively with iron and aluminium and negatively with Ca_{sed} as P_{sed}. HCl-RP was only correlated with Ca_{sed}, as could be expected from its chemical composition.

Acknowledgements

We thank the 'Serveis Generals de l'Universitat de Barcelona' for helping with the X-ray fluorescence analysis and Robin Rycroft for improving the language of this text. The research was supported by a grant from 'Caixa de Barcelona'.

References

Allan, R. J. & G. J. Brunskill, 1977. Relative atomic variation (RAV) of elements in lake sediments: Lake Winnipeg and other Canadian lakes. In: H. L. Golterman (ed.), Interactions between sediments and fresh water. Dr W. Junk Publishers, The Hague, 108–120.

Armengol, J., M. Crespo & J. A. Morgui, 1984. Phosphorus compounds in the sediment of the Sau reservoir (Barcelona, N.E. Spain) throughout its twenty-year existence. Verh. int. Ver. Limnol. 22: 1536–1540.

Armengol, J., J. L. Riera & J. A. Morgui, 1991. Major ionic composition in the Spanish reservoirs. Verh. int. Ver. Limnol. 24: 1363–1366.

Avnimelech, Y., 1983. Phosphate and calcium carbonate solubilities in Lake Kinneret. Limnol. Oceanogr. 28: 640–645.

Bortlesson, G. C. & G. F. Lee, 1972. Recent sedimentary history of lake Mendota. Environmental, 6: 799–808.

Bortlesson, G. C. & G. F. Lee, 1974. Phosphorus, iron and manganese distribution in sediment cores of six Wisconsin Lakes. Limnol. Oceanogr. 19: 794–801.

Bortlesson, G. C. & G. F. Lee, 1975. Recent sedimentary history of lake Monona, Wisconsin. Wat. Air Soil Poll. 4: 89–98.

Boström, B., 1984. Potential mobility of phosphorus in different types of lake sediments. Int. Revue ges. Hydrobiol. 69: 457–474.

Brunskill, G. J., D. Povoledo, B. W. Graham & M. P. Stainton, 1971. Chemistry of surface sediments of sixteen lakes in E.L.A., N.O.. J. Fish. Res. Bd Can. 28: 277–294.

Cross, P. M. & F. H. Rigler, 1983. Phosphorus and iron retention in sediments measured by mass budget calculations and directly. Can. J. Fish. aquat. Sci. 40: 1589–1597.

82

Estrada, M., 1975. Statistical considerations of some limnological parameters in Spanish reservoirs. Verh. int. Ver. Limnol. 19: 1849–1859.

Estrada, M., 1978. Relationship among biological and physicochemical parameters in Spanish reservoirs. Verh. int. Ver. Limnol. 20: 1642–1646.

Flannery, M. S., R. D. Snodgrass & T. J. Whitmore, 1982. Deep water sediments and trophic conditions in Florida lakes. In P. G. Sly (ed.), Sediment/Freshwater Interaction. Developments in Hydrobiology 9. Dr W. Junk Publishers, The Hague: 597–602. Reprinted from Hydrobiologia 91/92.

Forstner, U. 1977. Metal concentrations in freshwater sediments -natural background and cultural effects. In: H. L. Golterman (ed.), Interactions between sediments and fresh water. Dr W. Junk Publishers, The Hague: 94–103.

Golachowska J., 1984. Phosphorus in the bottom sediments of some lakes of the world. Pol. Arch. Hydrobiol. 31: 175–205.

Hieltjes, A. H. & L. Lijklema, 1980. Fractionation of inorganic phosphates in calcareous sediments. J. environ. Qual. 9: 405–407.

Jacobsen, O. S., 1978. A description model for phosphate sorption by lake sediments. Proceedings of Interactions between sediment and water 6th Nordic Symposium on sediments. 9–12.3. 1978. Hurdal Norwey: 127–136.

Jones, B. F. & C. J. Bowser, 1978. The mineralogy and related chemistry of lake sediments. In A. Lerman (ed.), Lakes: chemistry, geology and physics. Springer-Verlag, New York: 179–235.

Lijklema, L. 1980. Eutrophication; the role of sediments. Hydrobiol. Bull. 14: 98–105.

Lopez, P. 1986. Composición del sedimento en sistemas acuáticos del litoral mediterraneo español. Limnetica 2: 11–18.

Lopez, P., 1991. Aspects of sedimentary phosphorus dynamics in epicontinental systems: a literature review. Oecol. aquat. 10: 113–125.

Lopez, P. & J. A. Morgui, in press. Efecto de la matriz mineral sobre la determinacion de las fracciones de fósforo en

sedimento. Actas del VI Congreso Español de Limnología. Granada Septiembre 1991.

Margalef, R., D. Planas, J. Armengol, A. Vidal, N. Prat, A. Guiset, J. Toja & M. Estrada, 1976. Limnology of the Spanish reservoirs. Vols. I & II. Ministerio de Obras Publicas. Madrid. 453 and 85 pp.

Morgui, J. A., 1991. Eutrofización: Situación del problema en España. Ing. quim. Ag. 1991: 114–119.

Morgui, J. A., J. Armengol & J. L. Riera. 1990. Evaluación limnológica del estado de los embalses españoles: Composición iónica y nutrientes. Comunicaciones de las Terceras Jornadas Españolas de Presas. Comité Nacional Español de grandes Presas. Barcelona. Junio 1990: 652–668.

Nürnberg, G. K., 1988. Prediction of phosphorus release rates from total and reductant-soluble phosphorus in anoxic lake sediments. Can. J. Fish. aquat. Sci. 45: 453–462.

Nürnberg, G. K., M. Shaw, P. J. Dillon & D. J. McQueen, 1986. Internal phosphorus loading in an oligotrophic Precambrian Shield Lake with an anoxic hypolimnion. Can. J. Fish. aquat. Sci. 43: 574–580.

Ostrofsky, M. L., 1987. Phosphorus species in the surficial sediments of lakes of Eastern North America. Can. J. Fish. aquat. Sci. 44: 960–966.

Ostrofsky, M. L., D. A. Osborne & T. J. Zebulske, 1989. Relationship between anaerobic sediment phosphorus release rates and sedimentary phosphorus species. Can. J. Fish. aquat. Sci. 46: 416–419.

Premazzi, G. & O. Ravera, 1977. Chemical characteristics of lake Lugano sediments. In: H. L. Golterman (ed.), Interactions between sediments and fresh water. Dr W. Junk Publishers, The Hague, 121–124.

SPSS. Inc, 1988. SPSS/PC+ Advanced Statistics v.2.0. SPSS. Inc, Chicago. 242 pp.

Williams, J. D. H., J. M. Jaquet & R. L. Thomas, 1976. Forms of phosphorus in the surficial sediments of lake Eire. J. Fish. Res. Bd Can. 33: 413–429.

Wisniewski, R. J. & M. Planter, 1985. Exchange of phosphorus across sediment-water interface (with special attention to the influence of biotic factors) in several lakes of different trophic status. Verh. Int. Ver. Limnol. 22: 3345–3349.

Hydrobiologia **253**: 83–98, 1993.
P.C.M. Boers, T.E. Cappenberg & W. van Raaphorst (eds),
Proceedings of the Third International Workshop on Phosphorus in Sediments.
© 1993 *Kluwer Academic Publishers.*

The effect of deposition of organic matter on phosphorus dynamics in experimental marine sediment systems*

C. P. Slomp[1], W. Van Raaphorst[1], J. F. P. Malschaert[1], A. Kok[1] & A. J. J. Sandee[2]
[1] *Netherlands Institute for Sea Research, P.O. Box 59, 1790 AB Den Burg (Texel), The Netherlands;*
[2] *Netherlands Institute of Ecology – Center for Estuarine and Coastal Ecology (NIOO-CEMO), Vierstraat 28, 4401 EA Yerseke, The Netherlands*

Key words: organic matter, phosphorus, sandy marine sediment, macrofauna, boxcosms

Abstract

The effect of deposition of organic matter on phosphorus dynamics in sandy marine sediments was evaluated using an experimental system (boxcosms) and three different strategies: (1) no supply (2) one single addition (3) weekly additions of a suspension of algal cells (*Phaeocystis spec.*). Macrofauna (3 species, 6 individuals of each) were added to half of the boxes. Both in the case of the single and weekly additions a clear effect of increased organic matter loading on phosphorus dynamics was found. Following the organic matter addition, porewater phosphate concentrations in the upper sediment layer increased, phosphate release rates from the sediment increased by a factor 3–5 and in the boxes to which a single addition was applied NaOH-extractable phosphorus increased substantially. The increase in phosphate release rates from the sediment was attributed to mineralization of the added material and to direct release from the algal cells. No clear effect of the presence of macrofauna on sediment-water exchange of phosphate could be discovered. The macrofauna were very effective at reworking the sediment, however, as illustrated by the organic carbon profiles. It is hypothesized that the sediment-water exchange rates of phosphate were regulated by the layer of algal material which was present on the sediment surface in the fed boxes. In the boxes to which the single addition was applied porewater phosphate concentrations were lower and NaOH-extractable phosphorus was higher in the presence of macrofauna, suggesting that macrofauna can stimulate phosphate binding in the sediment.

Introduction

Benthic phosphorus regeneration may strongly influence water column chemistry in shallow marine systems (e.g. Balzer, 1984; Callender & Hammond, 1982; Fisher *et al.*, 1982; Hopkinson, 1987; Klump & Martens, 1981 & 1987; Rutgers van der Loeff, 1980). Therefore, the role of sediments in phosphorus recycling and eutrophica-

tion of these systems (e.g. the North Sea, Brockman *et al.*, 1988 & 1990) is of major importance, even though phosphorus generally does not limit primary production (Peeters & Peperzak, 1990; Riegman *et al.*, 1990).

Phosphorus cycling has mostly been studied in organic-rich, high porosity, fine-grained sediments (e.g. Froelich *et al.*, 1988; Klump & Martens, 1981 & 1987; Krom & Berner, 1980 & 1981; Martens *et al.*, 1978). Much less information is available on organic-poor, low porosity, sandy sediments (e.g. Hopkinson, 1987; Rutgers van

─────────
*Publication no. 40 of the project Applied Scientific Research Netherlands Institute for Sea Research (BEWON)

der Loeff, 1980; van Raaphorst *et al.*, 1990) which can be found in a major part of the North Sea (Eisma, 1990). In view of the general concern about increased eutrophication of the North Sea (Postma, 1985), presumably resulting in increased algal blooms (Cadée, 1990) and oxygen deficiency in certain areas (Westernhagen *et al.*, 1986), it is important to obtain more quantitative information on the processes controlling phosphorus dynamics in sandy sediments.

Early diagenesis in marine sediments largely depends on the supply of organic carbon (Berner, 1980; Billen *et al.*, 1990; Klump & Martens, 1987). Although a significant correlation between the amount of fine particles and of organic matter in sediments can often be found (Creutzberg *et al.*, 1984; Billen *et al.*, 1990) deposition of organic matter is not limited to fine-grained sediments. Jenness & Duineveld (1985) have shown that considerable amounts of phytoplanktonic material can be – at least temporarily – buried in sandy sediments down to a depth of 5 cm following deposition in periods of slack tidal current.

Binding of phosphorus in the sediment may cause a time lag between organic matter mineralization in the sediment and actual regeneration of phosphorus to the water column. Sorption to iron and aluminum oxides and precipitation processes (Lijklema, 1977; Martens *et al.*, 1978; Froelich, 1988; Froelich *et al.*, 1982) may substantially reduce regeneration to the overlying water. Furthermore, uptake of phosphorus by microorganisms, not only from the organic substrate but also from the porewater, may play an important role. This latter process obviously depends on the quality (e.g. C:P ratio) of the available organic matter (Billen *et al.*, 1990; Gächter *et al.*, 1988 & 1992). Under anoxic conditions chemically bound phosphorus may be released due to reduction of iron oxides (Mortimer, 1941). According to Gächter *et al.* (1988) polyphosphates which have accumulated in bacterial cells during oxic conditions may then be released as well.

The presence of macrofauna can stimulate mineralization of organic matter and uptake of phosphorus by microorganisms through reworking of the sediment. Furthermore, sediment-water

exchange rates of phosphorus can be enhanced, mostly due to bioirrigation activity (e.g. Aller, 1982; Hüttel, 1990; Hylleberg & Hendriksen, 1982; Yingst & Rhoads, 1980).

In this study the effect of deposition of organic matter on phosphorus dynamics in a sandy marine sediment is evaluated. Furthermore, the role of macrofauna is discussed. The system was a modification of the boxcosms described by van Raaphorst *et al.* (1992). This research was part of a larger study on North Sea sediment eutrophication of which further results will be published elsewhere.

Materials and methods

Boxcosms

Sediment with a median grain size of 125–160 μm and a content of particles $< 50 \mu$m of *ca* 2–5% was obtained from a station in the southern North Sea (Zeegat van Texel: 52° 53′ N, 4° 34′ E; depth: 17 m). The sediment was stored in large covered containers at outdoor (winter) temperatures for *ca* 4 months. Before use, the sediment was sieved (< 0.5 cm) and homogenized in a cement mill. The boxcosm experiments were performed in 26 cylindrical polypropylene boxes with an inner diameter and height of 30 and 35 cm, respectively. The boxes were filled with sediment up to 10 cm from the rim, resulting in a sediment depth of 25 cm in each box. Incorporation of air bubbles while filling the boxes was avoided as much as possible by adding seawater simultaneously. The thin layer (*ca* 5 mm) of fine particles which subsequently developed on top of the sediment was carefully removed.

The boxes were distributed over 2 separate basins, in order to be able to maintain the 'starved' and 'fed' boxcosms spatially apart thus avoiding mutual contamination. No communication existed between the boxes. To each box *ca* 10 cm of overlying water was added, which was continuously replaced by filtered (over sand beds, grainsize 1–1.4 mm), aged (for several weeks in 2 large containers) North Sea water of constant salinity (29‰), an average dissolved organic carbon

(DOC) content of 2.1 ± 0.4 mg l^{-1} and the following average ($n = 16$) nutrient concentrations: $PO_4 = 2.7 \pm 0.9$ μmol l^{-1}; $NO_3 + NO_2 = 54.3 \pm 7.9$ μmol l^{-1}; $Si = 17.2 \pm 3.3$ μmol l^{-1}. The NH_4 concentration in the inflow was ca 1.0 μmol l^{-1} at the beginning, increased to ca 7 μmol l^{-1} at day 14 and subsequently decreased to values < 0.8 μmol l^{-1} remaining at this level from day 24 onwards. The DOC present in the inflow water probably consisted of refractory components (Laane, 1980). The inflow rate of 10.4 ml min^{-1} resulted in a residence time of the overlying water of ca 11 hours. Outflow took place by overflow over the rim of the boxes into the basin. Constant bubbling of air was performed to keep the water column in the boxes well-mixed and saturated with respect to oxygen. The boxcosms were kept in the dark at a temperature of 11.8 ± 0.5 °C.

One week after installation each box was supplied with micro- and meiofauna through a 250 ml sediment sample consisting of the 2.5 cm surface layer of freshly collected boxcores. Two weeks later three species of macrofauna (*Tellina fabula*, *Nephtys hombergii*, *Echinocardium cordatum*; 6 of each, resulting in a total density of 255 ind. per m²) were added to 13 of the boxes. Dead individuals visible at the sediment surface were replaced on a weekly basis.

Three different strategies were used to study organic matter deposition: (1) no supply ('starved'; 8 boxes), (2) one single addition ('fed'; 10 boxes), (3) weekly additions ('fed'; 8 boxes). The organic matter consisted of a suspension of *Phaeocystis spec.*, a common alga in coastal areas of the North Sea (e.g. Cadée, 1990; Lancelot *et al.*, 1987), which was collected in the Schulpengat south-west of Texel with 50 μm plankton nets during the 1990 spring bloom. The material was homogenized by stirring with a paddle in large containers, divided into equal portions and subsequently stored at -20 °C until use (ca 4 weeks for the first addition). 16 days after the introduction of the macrofauna and 37 days after the installation of the boxes the first portion of (thawed) organic matter was added (day 0). The organic matter supply to the boxcosms resulted in loadings of ca 8 g C m^{-2} and 6.3 mmol P m^{-2} for the

weekly additions (during 19 weeks, resulting in a total of 152 g cm^{-2} and 120 mmol P m^{-2}), and 24 g C m^{-2} and 19 mmol P m^{-2} for the single additions. The amount of carbon supplied with the single addition is approximately equivalent to the annual metabolic loss of sandy North Sea sediments as estimated by de Wilde *et al.* (1984) and Cramer (1991). Although the water circulation was stopped for 24 h following each addition not all of the algal material settled on the sediment surface within this period, resulting in a loss of organic matter due to outflow from the boxes. This especially was a problem in the boxcosms to which the single addition of organic matter was applied. Therefore, the actual carbon loading in these boxes was somewhat lower than 24 g C m^{-2}. At each sampling event either intact boxes were used for the measurements (sediment-water exchange rates, oxygen respiration rates and penetration depth) or boxes were 'sacrificed' (porewater, sediment composition).

Sediment-water exchange rates

Sediment-water exchange rates of phosphate were measured in single boxes which were temporarily disconnected from the water supply. 500 ml of the overlying water was carefully removed and stored in a jar. At fixed time intervals 25 ml of sample was taken both from the overlying water and from the jar. The samples were filtered (0.45 μm cellulose acetate) and analyzed for phosphate. At the end of each experiment – which never took more than eight hours – the water supply was reconnected.

The fluxes were calculated from the concentration change in time in the overlying water of the boxcosm corrected for the consumption or production of phosphate in the jar and the decreasing depth of the overlying water due to sampling:

$$dC_o/dt = J * h^{-1} - R, \qquad (1)$$

where

C_o = concentration of the overlying water (mol m^{-3});

t = time (s);

J = sediment-water exchange rate (mol m^{-2} s^{-1};

h = the depth of the overlying water which decreases in time due to sampling (m);

R = change of the phosphate concentration in the overlying water (mol m^{-3} s^{-1}) due to production/consumption in the water column (jars).

Oxygen respiration and penetration

Benthic oxygen consumption was measured using the method described by Cramer (1990). The boxcosms were covered with a plexiglass lid in which a stirring device, O_2 electrodes (YSI 5739) and a temperature electrode were fitted. Respiration was calculated from the change in the O_2 concentration in the chamber during incubation. Oxygen concentrations in the pore water were measured with an O_2 micro-electrode (Helder & Bakker, 1985) at 0.2 mm depth intervals using a micromanipulator. The oxygen penetration depth is defined as the depth at which zero oxygen concentrations or constant and low readings were obtained.

Porewater

The boxcosms were sampled with acrylic liners (i.d. 5.2 cm, length 30 cm) which were sliced into 5, 10 and 20 mm segments (depending on sediment depth). Interstitial water was obtained by squeezing under N_2 pressure using Reeburgh-type squeezers (Reeburgh, 1967) fitted with 0.45 μm cellulose-acetate filters. In all cases segments of three cores were pooled.

Sediment composition

The boxcosms were sampled with PVC tubes (i.d. 4.5 cm) and sliced into segments of 5, 10, 20 and 40 mm (depending on sediment depth). Three cores were pooled each time. Organic-C contents

were measured on a Carlo Erba NA 1500-2 elemental analyzer following the procedure of Verardo *et al.* (1990).

The phosphorus speciation was determined using the sequential extraction scheme described by Hieltjes & Lijklema (1980). 50 mg of wet sediment was extracted sequentially with 2×50 ml of 1 M NH_4Cl, pH = 7 (2×2 hours), 50 ml of 0.1 M NaOH (17 hours) and 50 ml of 0.5 M HCl (24 hours). These fractions presumably represent the loosely bound and exchangeable fraction, the fraction bound by iron and aluminum oxides and the calcium bound fraction, respectively. A shaking table was used for continuous agitation of the suspensions. After each extraction step the suspensions were filtered (0.45 μm cellulose-acetate), the filtrate was stored at -20 °C until analysis and the filter with the sediment was added to the next extraction solution in the sequential procedure. The organic carbon content and phosphorus speciation were only determined for the sediments of the starved boxes and those that were fed once.

Easily exchangeable Fe and Mn was determined through an extraction with 0.1 M HCl (suprapur). It was assumed that most of the reactive iron and manganese oxides were released by this method. 0.1 gr of dried (60 °C) and homogenized (through grinding in an agate mortar) sediment was leached with 50 ml of HCl for 18 hours, followed by filtration over a pre-acid cleaned 0.45 μm cellulose nitrate filter (Duinker *et al.*, 1974).

Analytical procedures

Phosphate concentrations (analytical precision ± 0.03 μmol l^{-1} at a concentration of 1 μmol l^{-1}) were determined on a Technicon AA II autoanalyzer (fluxes, porewater) and on a Shimadzu Double beam Spectrophotometer W-150-02 (sediment phosphorus) following the method of Strickland and Parsons (1972). The Fe and Mn content of the HCl-leachate was determined with a Perkin Elmer 5100 PC Atomic Absorption Spectrophotometer using the standard addition method for calibration (analytical precision for Fe and Mn: ± 1 μmol l^{-1} and ± 0.5 μmol l^{-1} at a concentration of 18 μmol l^{-1}).

Results

Sediment–water exchange rates

Figure 1 shows the concentration change with time in the overlying water of the weekly fed boxcosms with macrofauna and in the jars on day −8 (no jar measurement), 2, 4 and 102. Calculated phosphate release rates from such data (assuming a linear relationship between the phosphate concentration and time) are given in Fig. 2. Error bars indicate the standard error of the calculated flux. Deviations from a straight line, as found for example on day 4 (Fig. 1), resulted in large standard errors for the estimated fluxes due to the small number of samples ($n = 4$–7). Phosphate release rates were generally low in the starved boxcosms with the exception of the high initial release rates in the boxes with macrofauna. Following deposition of organic matter (day 0) an increase in phosphate release rates from the sediment was found within 4 days in the case of the single additions, followed by a period of very low phosphate release from day 10 (with macrofauna) or 15 (without macrofauna) onwards. The interpretation of the results for the weekly fed boxes is hampered by the limited amount of measurements and the large errors in the estimated fluxes. From Fig. 2 it can be observed, however, that phosphate release rates increased within 2 days after the first organic matter addition in the boxes with macrofauna and within 4 days after the second addition in the boxes without macrofauna. The maximum phosphate release rates were 2–3 times higher in the boxes fed only once compared to the weekly fed ones. Apart from a slightly higher maximum phosphate release rate in the presence of macrofauna in the boxes which were fed once, no clear effect of the presence of macrofauna on the phosphate fluxes was observed.

The phosphate concentration in the overlying water was generally higher than in the inflow water, particularly in the fed boxcosms. In the case of the single addition the phosphate concentration in the overlying water increased from ca 3 to ca 10 μmol l^{-1} immediately following the food supply. This was followed by a decrease to 2–3 μmol l^{-1} within 2 days. The same pattern was observed in the case of the weekly additions, corresponding values being ca 3, 6 and 3–4 μmol l^{-1}, respectively.

Porewater profiles

In the starved boxcosms the porewater concentration of phosphate (Fig. 3) slowly decreased in time, to a concentration of less than 5 μmol l^{-1} in the upper sediment layer both with and without macrofauna. When the boxcosms were fed only once an immediate increase of the porewater phosphate concentration was found (> 20 μmol l^{-1}) in the upper 30–40 mm of the sediment, followed by a rapid decrease, especially in the presence of macrofauna (Fig. 3b). In the case of weekly additions of organic matter only a minor increase (Figure 3b, with macrofauna) or even a decrease (Fig. 3a, without macrofauna) of the phosphate concentration in the porewater of the upper sediment layer could be detected following the addition of organic matter. Both in the presence and absence of macrofauna the phosphate concentration subsequently decreased rapidly, even though organic matter additions continued. In all boxes the porewater phosphate concentrations measured in the upper 5 mm were higher than those of the overlying water.

The porewater phosphate concentration declined in all of the boxes during the course of the experiments, apart from the initial increase due to the organic matter additions found in the fed boxcosms. The phosphate concentrations found at the start of the measurements, however, were very high: 15–25 μmol l^{-1}. Presumably phosphate was released from the sediment during the period prior to the first measurements either due to mineralisation of organic matter and/or due to desorption from binding sites.

88

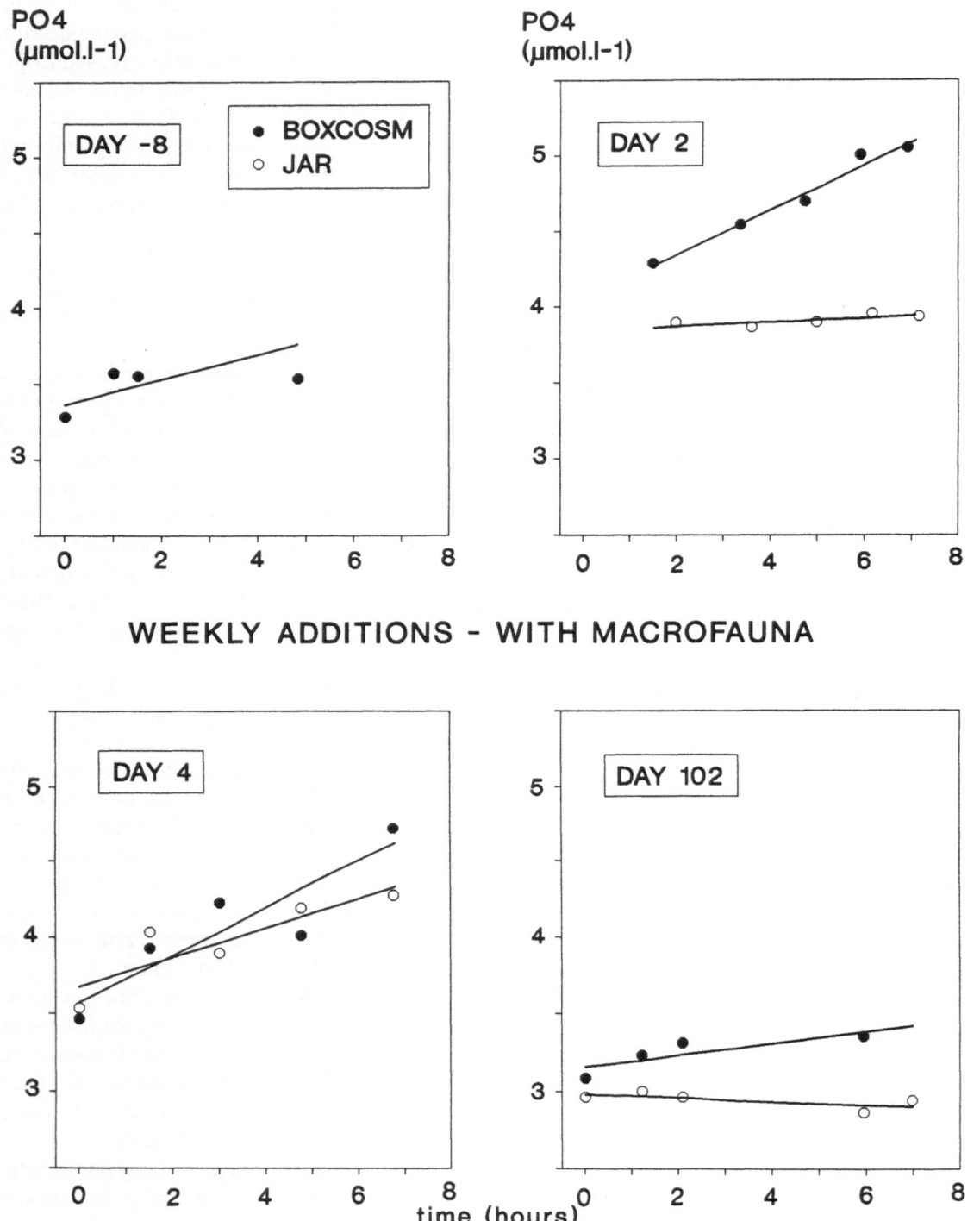

Fig. 1. Change in the phosphate concentration ($\mu mol\,l^{-1}$) with time in the overlying water of weekly fed boxcosms with macrofauna and the matching jars during fluxexperiments on day −8, 2, 4 and 102. Solid lines were obtained through linear regression.

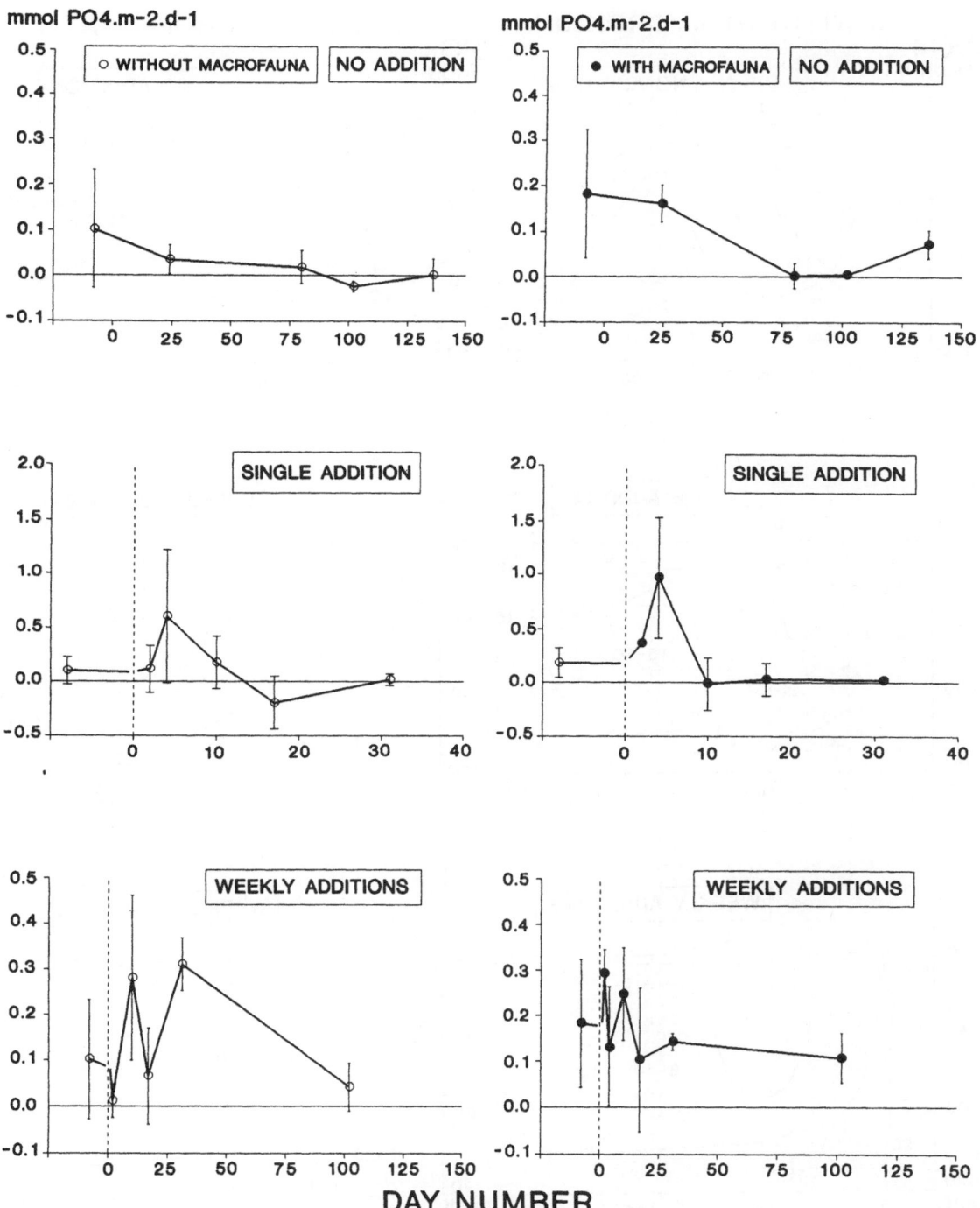

Fig. 2. Sediment-water exchange rates of PO$_4$ (mmol m^{-2} d^{-1}) measured in the boxcosms. Error bars indicate the standard error of the estimated flux.

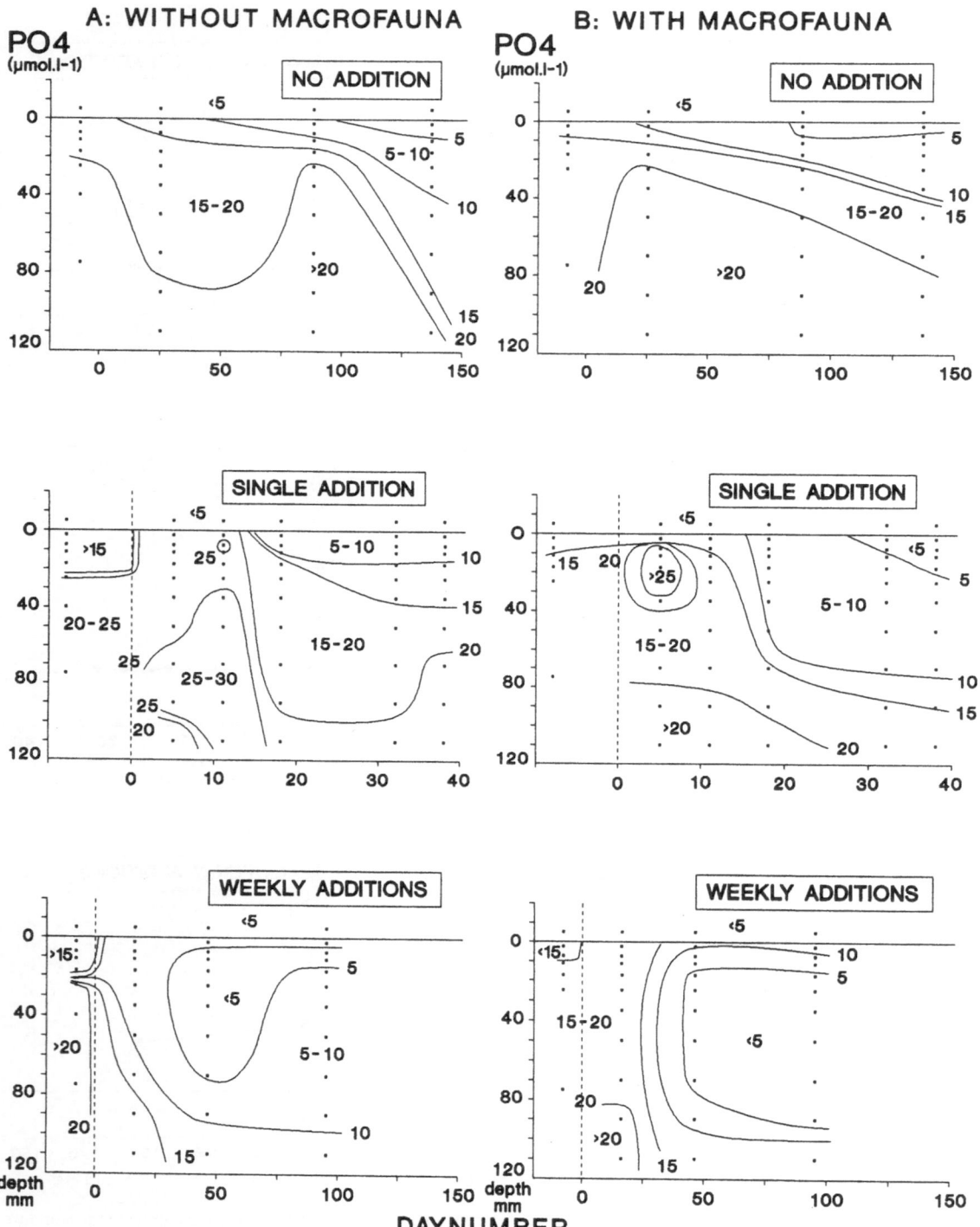

Fig. 3. PO$_4$ concentrations (μmol l^{-1}) in the porewater of the boxcosm sediment a. without macrofauna b. with macrofauna.

Oxygen respiration

Deposition of organic matter caused the benthic oxygen consumption to increase substantially (Fig. 4a). After a single organic matter addition oxygen respiration increased to *ca* 20 and 30 mmol O_2 m^{-2} d^{-1} in the boxes with and without macrofauna, respectively. Oxygen respiration rates subsequently decreased to *ca* 10 mmol O_2 m^{-2} d^{-1} within 30 days, slightly higher than the original rates (*ca* 8 mmol O_2 m^{-2} d^{-1}). In the case of weekly supply, oxygen respiration rates increased from approximately 10 to a maximum of 40 mmol O_2 m^{-2} d^{-1}. After an initial almost linear increase the oxygen respiration rates seemed to stabilize around 30–35 mmol O_2 m^{-2} d^{-1}. No clear effect of the presence of macrofauna on oxygen respiration could be discovered in the fed boxcosms.

Oxygen penetration

The addition of organic matter generally caused the oxygen penetration depth to decrease (Fig. 4b). Especially the single addition had a very clear effect on the oxygen penetration depth both in the boxes with and without macrofauna. In these boxes oxygen penetration gradually decreased to depths <1 mm 10 days after the addition, and subsequently increased again to *ca* 5–15 mm. In the weekly fed boxcosms with macrofauna oxygen penetration depths decreased from *ca* 30 to 2 mm during the experiment. In the weekly fed boxes without macrofauna large oscillations in the oxygen penetration depth were found, but overall, a decrease from 20 mm to 4 mm was observed. The oxygen penetration depth in the fed boxcosms with macrofauna was generally smaller than in the boxcosms without macrofauna (Wilcoxon's test; $n = 21$; $p < 0.05$).

Sediment composition

In the boxcosms to which no organic matter was added the carbon content remained very low,

ranging from 0.01 to 0.03% (Fig. 5). After a single addition of organic matter the carbon content rapidly rose to 0.04–0.09% in the upper sediment layer, both with (Fig. 5b) and without macrofauna (Fig. 5a). When organic matter was added weekly and macrofauna were present, this increase of the carbon content was not limited to the upper sediment layer but extended down to 50 mm in the boxcosms, indicating substantial sediment mixing.

The leachable Fe- and Mn-contents of the sediment were very low: 0.03–0.04% (5.4–7.2 μmol Fe g^{-1}) and 0.002% (0.4 μmol Mn g^{-1}), respectively. NH_4Cl-, NaOH- and HCl-extractable phosphorus amounted to 0.01–0.03, 0.02–0.05 and 0.05–0.11 μmol P g^{-1} sediment, respectively. The phosphorus content determined with this extraction was lower than found at a comparable sandy station in the North Sea where values of 0.05, 0.25 and 2.2 μmol P g^{-1} were found for the NH_4Cl, NaOH and HCl fraction, respectively (unpublished results). The low values found here can probably be explained by the removal of a part of the fine sediment fraction after filling of the boxes. No clear reaction to the single addition of organic matter could be discovered in the NH_4Cl and HCl fractions. Only NaOH-extractable phosphorus showed a clear response (Fig. 6) with the largest increase occurring in the presence of macrofauna.

Discussion

Effect of organic matter additions

Both in the case of the single and weekly additions a clear effect of increased organic matter loading on phosphorus dynamics was found. Following deposition of organic matter porewater phosphorus concentrations in the upper sediment layer increased, phosphate release rates showed a 3–5 fold increase and NaOH-extractable phosphorus increased substantially in the boxes to which a single addition was applied. Furthermore, oxygen respiration rates showed an immediate response. In the case of the single additions of

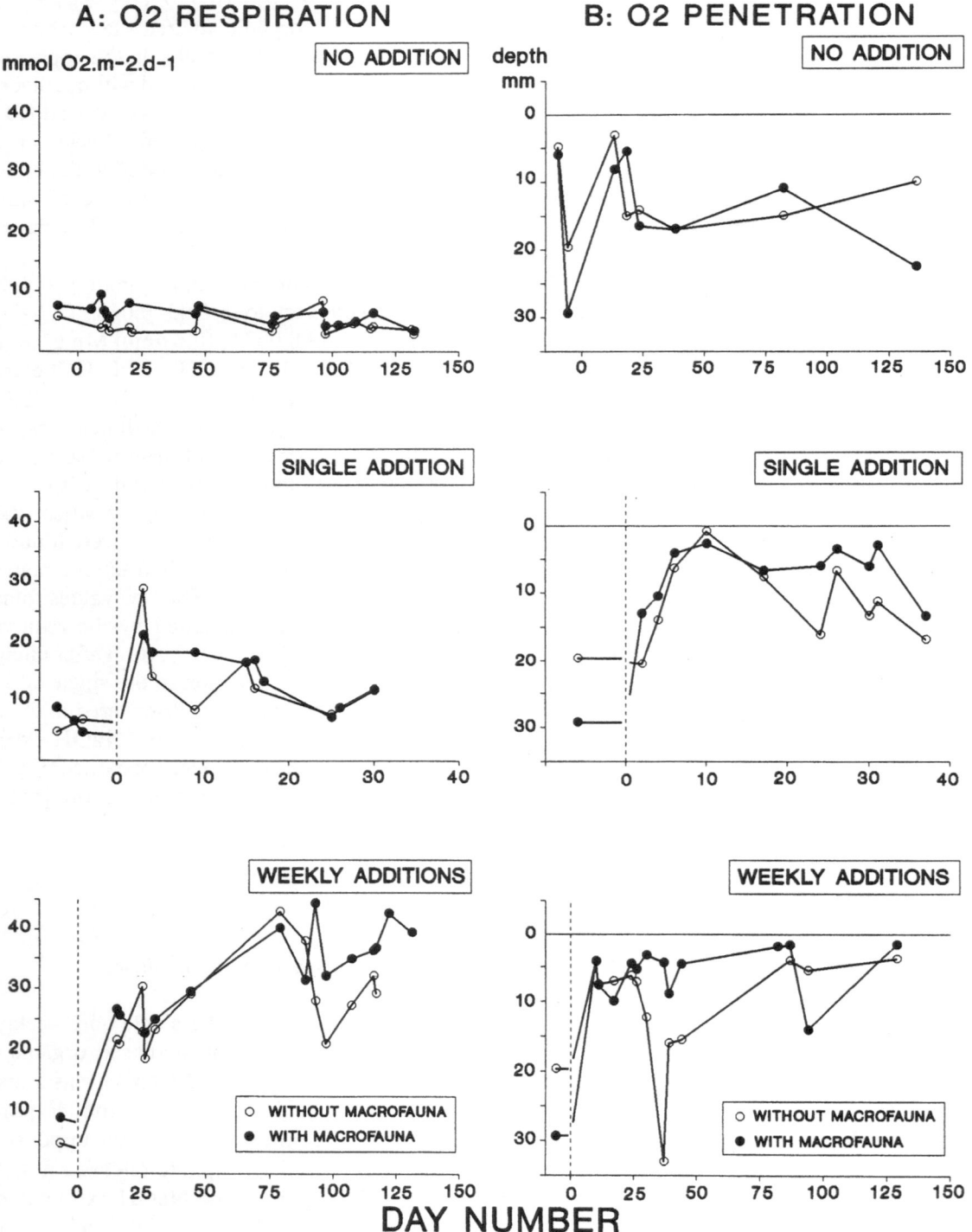

Fig. 4. a. Oxygen respiration rates (mmol O_2 m^{-2} d^{-1}) b. Oxygen penetration depths (mm) in the boxcosms.

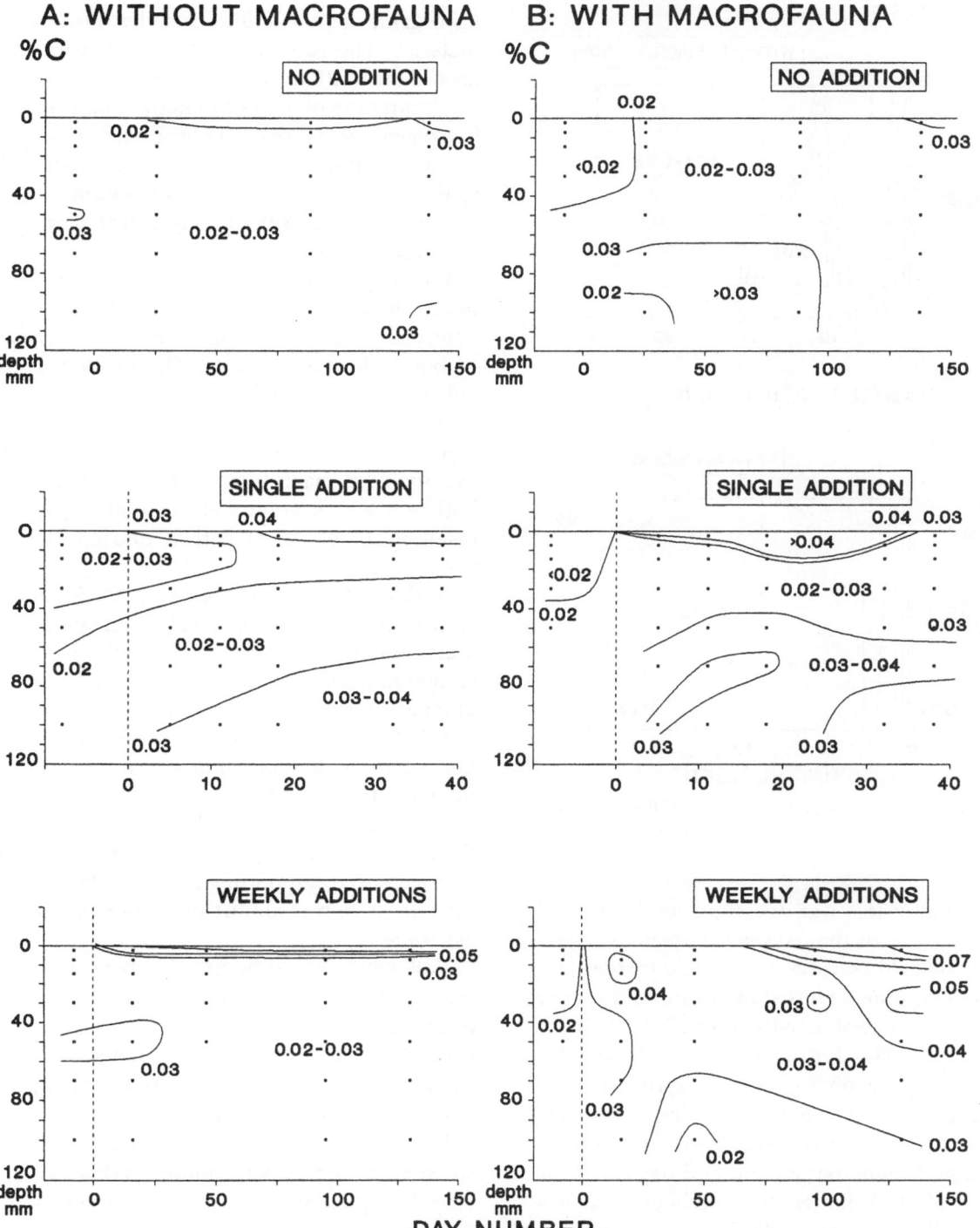

Fig. 5. Organic carbon content (%C) of the sediment in the boxcosms a. without macrofauna b. with macrofauna.

94

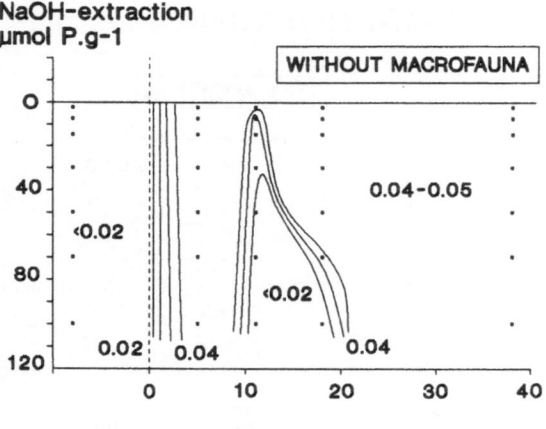

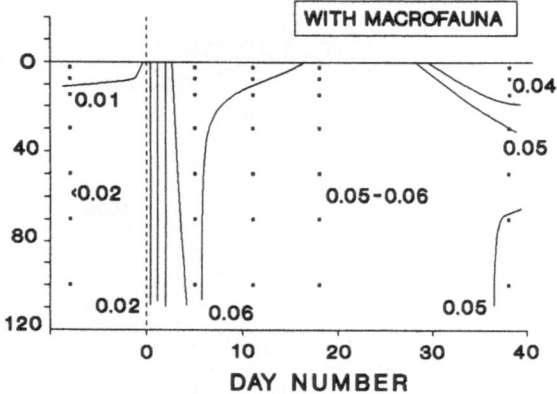

Fig. 6. NaOH-extractable phosphorus (μmol P gr^{-1}) in the sediment from the boxcosms which were fed once.

organic matter this was accompanied by a rapid initial decrease of the oxygen penetration depth.

Previous field research on organic matter deposition on sediments has shown a rapid response of benthic microbial activity (Graf, 1982 & 1989; Meyer-Reil, 1983; Jensen *et al.*, 1990) following deposition of a phytoplankton spring bloom. During laboratory experiments similar to ours using intact cores from the field, Graf (1987) found a maximum oxygen uptake 3 days after the addition of algal matter. In experimental microcosms Kelly & Nixon (1984) observed a time lag of 2–20 hours between an organic matter addition and maximum ammonium release rates. Enoksson (1987) found maximum oxygen consumption

and phosphate release rates *ca* 6 days following an organic matter addition in the form of algal material. The results of our experiments are in accordance with these observations: following the single addition of organic material a large increase in oxygen respiration and phosphate release rates occurred within 2 and 4 days, respectively. Due to the relatively large intervals between sampling events it is impossible to say when the maxima occurred exactly.

When studying phosphorus dynamics it is of major importance to know what mechanism is controlling whether phosphorus is being released or bound in the sediment. The mechanisms involved can be of a chemical (adsorption/desorption and precipitation/dissolution) or biological nature (uptake or release by bacteria, excretion by macrofauna) or a combination of both (e.g. anoxic conditions mediated by bacteria resulting in release of sorbed phosphorus from iron oxides).

In the starved boxcosms a relatively large release of phosphorus, especially in the presence of macrofauna, was observed compared to the corresponding oxygen uptake. The average O_2-uptake/P-release atomic ratio in the starved boxes was low, *ca* 55, indicating that oxic mineralization was not the dominating process. Excretion by macrofauna (Nixon *et al.*, 1980) may explain part of these results. Initial porewater concentrations in the sediment were high and only gradually decreased, supporting the observed phosphorus release during almost the entire length of the experiment.

Following the addition of organic matter increased phosphate release occurred in the fed boxcosms. This can be attributed to (1) mineralization of the added organic matter, (2) release from iron oxides due to reduction and (3) direct release from algal cells due to cell lysis.

In the same set-up, van Duyl *et al.* (1992) measured increased bacterial numbers (from *ca* 0.5 to 1.5×10^9 bacteria cm^{-3}) and bacterial production rates (from *ca* 7 to 140 mg C m^{-2} d^{-1}; methyl-^{3}H-thymidine incorporation method) on day 5 compared to day -4 in the 0–3 mm sediment layer of the boxes which were fed once. In combination with the higher oxygen respiration rates from day 2 onwards, this indicates the potential importance of oxic mineralization for the phosphate fluxes.

If reduction of iron oxides controls the phosphate release, a decrease in the oxygen penetration depth would be expected during this period. Figure 4b shows that oxygen penetration decreased to depths <1 mm on day 10 in the case of the single additions. Although sulfate reduction may have taken place in locally reduced spots, diagenesis probably did not proceed beyond nitrate reduction as nitrate was generally still present in the porewater (not shown). Furthermore, only a relatively small amount of phosphorus was present in the NaOH-extractable phosphorus fraction at the start of the experiment and this fraction was found to increase from ca 0.02 to 0.05 μmol P g^{-1} during the period of maximum phosphate release. Apparently, the adsorption capacity of the sediment was not eliminated, making a chemical control of the increased phosphate release improbable.

Preferential P-release due to cell lysis is known to occur rapidly on death of algal cells (Balzer, 1984; Garber, 1984; Krom & Berner, 1981) and certainly may have occurred in the *Phaeocystis* suspensions.

In the boxes to which a single addition was applied the sediment-water exchange rates of phosphate were very low from day 10–15 onwards, coinciding with a gradual decrease in the oxygen consumption rate. When the organic matter was applied weekly, however, substantial phosphate release continued to occur. Furthermore, the oxygen respiration rates roughly stabilized and further depletion of the porewater did not occur. It is unlikely that a steady state situation was reached, however, as the sediment organic carbon content still continued to increase.

A tentative budget for the fate of the phosphorus added to the boxcosms through the organic matter additions is presented in Table 1. As the various input and output terms could not be quantified accurately, a great deal of assumptions were necessary. Each output/storage term in Table 1 is the highest value of the estimates with and

Table 1. A tentative budget for the fate of the phosphorus added to the 'fed' boxcosms (in mmolP.m^{-2}; n.d. = not determined)

	Single addition (mmol P $\cdot$ m^{-2})	Weekly addition
Input	+ 15	+ 95
Output/storage		
P flux	− 4	− 29
Organic P	− 0.2	− 3
NaOH-extr. P	− 5	n.d.
Bacterial P	n.d.	− 4
Unaccounted for	+ 5.8 (39%)	+ 59 (62%)

without macrofauna. A loss of 20% of the added organic matter due to outflow over the rim was assumed for both feeding regimes. The release of phosphate from the sediment was estimated from the area under the curves in Fig. 2. The amount of added phosphorus which was bound in sediment organic matter and in the NaOH-extractable fraction of the sediment was calculated from the Figures 5 and 6 (assuming a sediment density of 2.65 g cm^{-3}, an average porosity of 0.40, a sediment depth of 10 cm and a C:P ratio for the organic matter of 106; Redfield *et al.*, 1963). An estimate of the amount of phosphorus bound in bacteria was obtained from the increase in bacterial biomass in the weekly fed boxes with macrofauna integrated over 63 mm depth during 130 days (2.4 g C m^{-2}; van Duyl, 1992) assuming a C:P weight ratio of 20 for the bacteria (Gächter *et al.*, 1992). *Ca* 40–60% of the phosphorus added through the organic matter is not accounted for in Table 1. Neither the phosphorus bound in bacteria in the case of the single additions nor the NaOH-extractable phosphorus in the sediment of the weekly fed boxes can account for this difference.

During the experiment total amounts of *ca* 0.26 and 3.3 mol O$_2$ m^{-2} (calculated from the area under the curve in Fig. 4 and corrected for the respiration in the starved boxes) were consumed in the boxes with single and weekly additions, respectively. This corresponds to a carbon respiration of *ca* 18–36% of the added material for both treatments when assuming a respiration quotient of 0.85 (Hargrave, 1973) and a loss due to outflow of 20%. These results suggest that for both feeding regimes a major part of the added organic matter was not mineralized during the experiment. This is in accordance with the fact that a layer of algal material was present on the sediment surface in most of the fed boxes during the entire experiment.

Effect of macrofauna

The macrofauna added to the boxcosms consisted of sub-surface and surface deposit feeders generally found in sandy sediments in the North Sea (Creutzberg *et al.*, 1984). Apart from the sessile bivalve *Tellina*, of which up to 5 individuals per boxcosm had to be replaced during the experiment, the macrofauna generally had a low mortality and were very effective at reworking the sediment, as illustrated by the organic carbon profiles. Only *Echinocardium* reached high growth rates (Duinveld, personal communication). In some of the sampled boxes, however, the macrofauna (*Echinocardium* in particular) were completely inactive. Further details on the macrofauna in this study will be published elsewhere.

Previous research has shown that macrofauna can directly increase sediment-water exchange rates of oxygen and nutrients due to bioirrigation activity (Aller & Yingst, 1985; Hylleberg & Hendriksen, 1980; Kristensen & Blackburn, 1987) and indirectly due to the fact that macrofaunal feeding and burrowing can stimulate microbial activity in the sediment (Aller, 1982; Aller & Yingst, 1985; Kristensen & Blackburn, 1987; Yingst & Rhoads, 1980).

Both in the case of the weekly and single additions of organic matter no clear effect of the presence of macrofauna on sediment-water exchange of phosphate and oxygen respiration rates could be discovered. The role of the burrowers *Echinocardium* and *Nephtys* was most likely limited to reworking of the sediment. Therefore, only indirect effects of macrofauna on solute transport would be expected.

The increase in substrate availability (higher organic carbon contents) at greater depths probably resulted in increased mineralization (van Duyl *et al.*, 1992) below the uppermost sediment layer explaining the smaller oxygen penetration depth in the presence of macrofauna. The increase of oxic mineralization below the upper sediment layer was apparently too small to have a substantial effect on the total oxygen respiration rates. The organic carbon profiles show that mixing of the food into the sediment took several weeks, thus causing a major portion to remain at the sediment surface. In any case, this holds for the most labile, freshly deposited material in the weekly fed boxcosms. Consequently, most of the organic matter mineralization probably took place in the algal layer on the sediment surface and the processes in this layer most likely determined the phosphate and oxygen fluxes. Apparently, the processes in this layer were not substantially affected by the presence of macrofauna.

In the weekly fed boxes with macrofauna higher phosphate concentrations were observed in the upper cm's of the sediment in the second half of the experiment, corresponding to the higher organic carbon contents in these boxes. In the case of the single additions this was not observed, probably because time was too short to mix a substantial amount of carbon deeper into the sediment. In these boxes porewater phosphate concentrations were lower and NaOH-extractable phosphorus was higher in the presence of macrofauna This suggests that macrofauna can stimulate phosphate binding in the sediment.

Measurements in experimental systems of this type are generally associated with a great number of problems. Although the influence of sediment heterogeneity is limited when using boxes filled with homogenized sediment instead of intact sediment cores, large differences between simularly treated boxes can still occur. Due to the manipulation of the sediment, processes may take place in the boxes which do not occur in natural sediments. Furthermore, the number of sampling events, different treatments and variables to be measured are limited by the maximum number of boxes that can be handled. Despite the drawbacks, this study illustrates that interesting results on the effect of organic matter deposition on phosphorus dynamics in sediments can be obtained in this type of system.

References

Aller, R. C., 1982. The effects of macrobenthos on chemical properties of marine sediment and overlying water. p 53–102. In: P. L. McCall & M. J. S. Tevesz (eds), Animal-sediment relations. Plenum Publishing Corporation. New York.

Aller, R. C. & J. Y. Yingst, 1985. Effect of the marine deposit-feeders *Heteromastus filiformis* (Polychaeta) *Macoma balthica* (Bivalvia) and *Tellina texana* (Bivalvia) on averaged sedimentary solute transport, reaction rates and microbial distributions. J. mar. Res. 43: 615–645.

Balzer, W., 1984. Organic matter degradation and biogenic element cycling in a nearshore sediment (Kiel Bight). Limnol. Oceanogr. 29: 1231–1246.

Berner, R. A., 1980. Early diagenesis. A theoretical approach. Princeton University press. Princeton. 241 pp.

Billen, G., C. Joiris, L. Meyer-Reil & H. Lindeboom, 1990. Role of bacteria in the North Sea Ecosystem. Neth. J. Sea Res. 26: 265–293.

Brockmann U. H., G. Billen & W. W. C. Gieskes, 1988. North Sea nutrients and eutrophication. In: W. Salomons, B. L. Bayne, E. K. Duursma & U. Förstner (eds), Pollution of the North Sea. An assessment. Springer Verlag, Berlin: pp. 348–383.

Brockmann U. H., R. W. P. M. Laane & H. Postma, 1990. Cycling of nutrient elements in the North Sea. Neth. J. Sea. Res. 26: 239–264.

Cadée, G. C., 1990. Increased bloom. Nature 346: 418.

Callender, E. & D. E. Hammond., 1982. Nutrient exchange across the sediment-water interface in the Potomac river estuary. Estuar. coast. mar. Sci. 15: 395–413.

Cramer, A., 1990. Seasonal variation in the benthic metabolic activity in a frontal system in the North Sea. In: M. Barnes & R. N. Gibson (eds), Trophic relationships in the marine environment. Proc. 24th Europ. Mar. Biol. Symp. Aberdeen. Univ. Press: pp 54–76.

Cramer, A. 1991. Benthic metabolic activity at frontal systems in the North Sea. Dissertation. University of Amsterdam, The Netherlands, 93 pp.

Creutzberg F., P. Wapenaar, G. Duineveld & N. Lopez Lopez, 1984. Distribution and density of the benthic fauna in the southern North Sea in relation to bottom characteristics and hydrographic conditions. Rapp. P. v. Réun. Cons. int. Explor. Mer. 183: 101–110.

De Wilde, P. A. W. J., E. M. Berghuis & A. Kok, 1984. Structure and energy demand of the benthic community of the Oyster Ground, Central North Sea. Neth. J. Sea Res. 18: 143–159.

Duinker, J. C., G. T. M. Van Eck & R. F. Nolting, 1974. On the behaviour of copper, zinc, iron and manganese and evidence for mobilization processes in the Dutch Wadden Sea. Neth. J. Sea Res. 8: 214–239.

Enoksson, V., 1987. Nitrogen flux between sediment and water and its regulatory factors in coastal areas. Dissertation. Department of Marine Microbiology. University of Göteborg, Sweden.

Eisma, D., 1990. Transport and deposition of suspended matter in the North Sea and the relation to coastal siltation, pollution and bottom fauna distribution. Reviews in Aquatic Sciences 3: 181–216.

Fisher, T. R., P. R. Carlson & R. T. Barber, 1982. Sediment nutrient regeneration in three North Carolina estuaries. Estuar. coast. Shelf Sci. 14: 101–116.

Froelich, P. N., M. L. Bender, N. A. Luedtke, G. R. Heath & T. DeVries, 1982. The marine phosphorus cycle. Am. J. Sci. 282: 474–511.

Froelich, P. N., 1988. Kinetic control of dissolved phosphate in natural rivers and estuaries: a primer on the phosphate buffer mechanism. Limnol. oceanogr. 33: 649–668.

Froelich, P. N., M. A. Arthur, W. C. Burnett, M. Deakin, V. Hensley, R. Jahnke, L. Kaul, K. H. Kim, K. Roe, A. Soutar & C. Vathakanon, 1988. Early diagenesis of organic matter in Peru continental margin sediments: phosphorite precipitation. Mar. Geol. 80: 309–343.

Gächter, R., J. S. Meyer & A. Mares, 1988. Contribution of bacteria to release and fixation of phosphorus in lake sediments. Limnol. Oceanogr. 33: 1542–1558.

Gächter, R. & J. S. Meyer, 1993. The role of microorganisms in mobilization and fixation of phosphorus in sediments. In P. C. M. Boers, T. E. Cappenberg & W. van Raaphorst (eds), Proceedings of the Third International Workshop on Phosphorus in Sediments. Developments in Hydrobiology 84. Kluwer Academic Publishers, Dordrecht: 103–121. Reprinted from Hydrobiologia 253.

Garber, J. H, 1984. Laboratory study of nitrogen and phosphorus remineralization during the decomposition of coastal plankton and seston. Estuar. Coast. Shelf Sci. 18: 685–702.

Graf, G. 1982. Benthic response to sedimentation of a spring phytoplankton bloom process and budget. Mar. Biol 67: 201–208.

Graf, G., 1987. Benthic energy flow during a simulated autumn bloom sedimentation. Mar. Ecol. Prog. Ser. 39: 23–29.

Graf, G., 1989. Benthic-pelagic coupling in a deep-sea benthic community. Nature 341: 437–439.

Hargrave, B. T., 1973. Coupling carbon flow through some pelagic and benthic communities. J. Fish Res. Bd Can. 35: 1317–1326.

Helder W. & J. F. Bakker, 1985. Shipboard comparison of micro- and minielectrodes for measuring oxygen distribution in marine sediments. Limnol. Oceanogr. 30: 1106–1109.

Hieltjes A. H. M. & L. Lijklema, 1980. Fractionation of in-

98

organic phosphates in calcareous sediments. J. Envir. Qual. 9: 405–407.

Hopkinson, Jr. C. S., 1987. Nutrient regeneration in shallow-water sediments of the estuarine plume region of the near-shore Georgia Bight, U.S.A. Mar. Biol. 94: 127–142.

Hüttel, M., 1990. Influence of the lugworm Arenicola marina on porewater nutrient profiles of sand flat sediments. Mar. Ecol. Prog. Ser. 62: 241–248.

Hylleberg, J. & K. Hendriksen, 1982. The central role of bio-turbation in sediment mineralization and element recycling. Ophelia 1: 1–16.

Jenness, M. I. & G. C. A. Duineveld, 1985. Effects of tidal currents on chlorophyll-a content of sandy sediments in the southern North Sea. Mar. Ecol. Prog. Ser. 21: 283–287.

Jensen, M. H., E. Lomstein & J. Sorensen, 1990. Benthic NH_4^+ and NO_3^- flux following sedimentation of a spring phytoplankton bloom in Aarhus Bight, Denmark. Mar. Ecol. Progr. Ser. 61: 87–96.

Kelly, J. R. & S. W. Nixon, 1984. Experimental studies of the effect of organic deposition on the metabolism of a coastal marine bottom community. Mar. Ecol. Prog. Ser. 17: 157–169.

Klump, J. V. & C. S. Martens, 1981. Biogeochemical cycling in an organic-rich coastal marine basin. II. Nutrient sediment-water exchange processes. Geochim. Cosmochim. Acta 45: 101–121.

Klump, J. V. & C. S. Martens, 1987. Biogeochemical cycling in an organic-rich coastal marine basin. 5. Sedimentary nitrogen and phosphorus budgets based upon kinetic models, mass balances and the stoichiometry of nutrient regeneration. Geochim. Cosmochim. Acta 51: 1161–1173.

Kristensen, E. & T. H. Blackburn, 1987. The fate of organic carbon and nitrogen in experimental marine sediment systems: influence of bioturbation and anoxia. J. mar. Res. 45: 231–257.

Krom, M. D. & R. A. Berner, 1980. Adsorption of phosphate in anoxic marine sediments. Limnol. Oceanogr. 25: 797–806.

Krom M. D. & R. A. Berner, 1981. The diagenesis of phosphorus in a nearshore marine sediment. Geochim. Cosmochim. Acta 45: 207–216.

Laane, R. W. P. M., 1980. Conservative behaviour of dissolved organic carbon in the Ems-Dollart Estuary and the Western Wadden Sea. Neth. J. Sea Res. 14: 192–199.

Lancelot, C., G. Billen, A. Sournia, T. Weisse, F. Colijn, M. J. W. Veldhuis, A. Davies & P. Wasman, 1987. Phaeocystis blooms and nutrient enrichment in the continental coastal zones of the North Sea. Ambio 16: 38–46.

Lijklema, L., 1977. The role of iron in the exchange of phosphate between water and sediments. In: H. L. Golterman (ed.), Interactions between sediments and freshwater. Dr W. Junk Publishers, The Hague: 313–317.

Martens, C. S., R. A. Berner & J. K. Rosenfeld, 1978. Interstitial water chemistry of anoxic Long Island sound sediments. 2. Nutrient regeneration and phosphate removal. Limnol. Oceanogr. 23: 605–617.

Meyer-Reil, L. A., 1983. Benthic response to sedimentation events during autumn to spring at a shallow water station in the western Kiel-Bight. II. Analysis of benthic bacterial populations. Mar. Biol. 77: 247–256.

Mortimer, C. H., 1941. The exchange of dissolved substances between mud and water in lakes. J. Ecol. 30: 280–329.

Nixon, S. W., J. R. Kelly, B. N. Furnas, C. A. Oviatt & S. S. Hale, 1980. Phosphorus regeneration and the metabolism of coastal marine benthic communities. In: K. R. Tenore & B. C. Coull (eds), Marine benthic dynamics. Univ. of South Carolina Press, Columbia: 219–242.

Peeters, J. C. H. & L. Peperzak, 1990. Nutrient limitation in the North Sea: a bioassay approach. Neth. J. Sea Res. 26: 61–73.

Postma, H., 1985. Eutrophication of Dutch Coastal waters. Neth. J. Zool. 35: 348–359.

Redfield, A. C., B. Ketchum & F. Richard, 1963. The influence of organisms on the composition of seawater. In: M. Hill (ed.). The Sea. Vol. 2. Interscience, New York, pp. 26–77.

Reeburgh, W. S., 1967. An improved interstitial water sampler. Limnol. Oceanogr. 12: 163–170.

Riegman, R., F. Colijn, J. F. P. Malschaert, H. T. Klooster-huis & G. C. Cadée, 1990. Assessment of growth rate limiting nutrients in the North Sea by the use of nutrient uptake kinetics. Neth. J. Sea Res. 26: 53–60.

Rutgers van der Loeff, M. M., 1980. Nutrients in the interstitial water of the Southern Bight of the North Sea. Neth. J. Sea Res. 14: 144–171.

Strickland, J. D. H. & T. R. Parsons, 1972. A practical handbook of seawater analysis. 2nd edn. Bull. Fish. Res. Bd Can. 167: 1–311.

Van Duyl, F. C., A. J. Kop, A. Kok & A. J. J. Sandee, 1992. The impact of organic matter and macrozoobenthos on bacterial and oxygen variables in marine sediment boxcosms. Neth. J. Sea Res. 29: 343–355.

Van Raaphorst, W., H. T. Kloosterhuis, A. Cramer & K. J. M. Bakker, 1990. Nutrient early diagenesis in the sandy sediments of the Doggerbank area, North Sea: pore water results. Neth. J. Sea Res. 26: 25–52.

Van Raaphorst, W., H. T. Kloosterhuis, E. M. Berghuis, A. J. Gieles, J. F. P. Malschaert & G. J. Van Noort, 1992. Nitrogen cycling in two sediments of the southern North Sea (Frisian Front, Broad Fourteens): Field data and mesocosm results. Neth. J. Sea Res. 28: 293–316.

Verardo, D. J., P. N. Froelich & A. McIntyre, 1990. Determination of organic carbon and nitrogen in sediments using the Carlo Erba Na-1500 analyzer. Deep Sea Res. 37: 157–165.

Westernhagen, H. v., W. Hickel, E. Bauerfeind, U. Niermann & J. Kröncke, 1986. Sources and effects of oxygen deficiencies in the south-eastern North Sea. Ophelia: 26: 457–473.

Yingst, J. Y. & D. C. Rhoads, 1980. The role of bioturbation in the enhancement of bacterial growth rates in marine sediments. In: K. R. Tenore & B. C. Coull (eds), Marine benthic dynamics. Univ. South Carolina Press, Columbia: 407–421.

Hydrobiologia **253**: 99–102, 1993.
P.C.M. Boers, T.E. Cappenberg & W. van Raaphorst (eds),
Proceedings of the Third International Workshop on Phosphorus in Sediments.
© 1993 *Kluwer Academic Publishers.*

Abstracts

Relationship between iron loading and phosphorus retention in shallow Danish lakes

P. Kristensen, M. Søndergaard, E. Jeppesen & H. S. Jensen
National Environmental Research Institute, Division of Freshwater Ecology, Lysbrogade 52, DK-8600 Silkeborg, Denmark

Recent studies have shown the importance of iron as a regulator of the phosphorus content and aerobic phosphorus release from the sediment of shallow lakes. Nevertheless, little information exists on iron loading, iron retention and factors determining the iron dynamics in lakes.

Accordingly, factors important for iron retention are discussed on basis of mass balances from 20 shallow Danish lakes. The iron loading varied between 0.1 and 160 g Fe m^{-2} y^{-1}, averaging 22 g Fe m^{-2} y^{-1}. In all lakes, except one, a net iron retention was observed. Mean retention was 12.6 g Fe m^{-2} y^{-1} corresponding to 55 percent of the loading.

In general, a higher percentage of iron retention than phosphorus retention was observed and consequently the iron:phosphorus ratio was higher in the sediment than in loading. Lakes with a high iron loading and high inlet iron concentration showed a higher percentage of phosphorus retention than in comparable lakes with low iron loading.

Emphirical models describing the relationship between iron loading and iron retention as well as relationship between iron loading and phosphorus retention are presented and discussed.

Phosphate compounds in sediments
1. Inorganic and biological aspects

H. L. Golterman,* I. M. de Graaf # & C. J. de Groot*
* Station Biologique de la Tour du Valat, Arles, 13200 France
 # Populatie Biologie, c/o LEPS, Schelpenkade, 14a, 2313 Leiden, The Netherlands

The chemical adsorption of phosphate onto sediments may be caused by $Fe(OOH)$ and/or $CaCO_3$. Mathematically the adsorption can be described by a Langmuir or a Freundlich isotherm. Analyzing several chemical reaction mechanisms for the equilibrium between $(o\text{-}P)_w$ and P_{sed} leads, under certain conditions, to the Freundlich isotherm, and the best description that can be proposed at the moment is $P_{sed} = A*(o\text{-}P)_w^{0.33}$ with A being a negative, nearly linear function of the pH.

The constant 'A' does not depend only on the pH; compounds like $NaCl$, $MgCl_2$ and $CaCl_2$ etc. influence 'A' strongly. We think that ion activities cause this effect through the electric double layer of the $Fe(OOH)$. A different effect is caused by the presence of S^{2-}, which renders the $Fe(OOH)$ inactive. This inactivation can be undone by denitrification.

The adsorption onto $CaCO_3$ can be explained by the soluble Ca^{2+} concentration, which causes the product formed (probably apatite) to be co-precipitated with $CaCO_3$. A range of values for the solubility product of apatite is proposed. Data from the rivers Rhine and Rhone suggest a value near 10^{-50}.

100

The Ca and Fe phosphate adsorption mechanisms can be combined to a solubility diagram, of which a new numerical version will be presented.

Sediments from about 10 lakes have been analyzed for $FeOOH \approx P$ and $CaCO_3 \approx P$ using sequential extractions with Ca-NTA and Na-EDTA. Furthermore the bioavailability of the P_{sed} for the alga Scenedesmus spec. was measured in bioassays. Bioavailability was compared with the different fractions. It appeared that especially the sum of the $FeOOH \approx P$ and $CaCO_3 \approx P$ fractions was available, while 'org-P' was not available.

Phosphate compounds in sediments
2. Organic aspects

C. J. de Groot* & H. L. Golterman[#]
*Station Biologique de la Tour du Valat, Arles, 13200 France
[#]Populatie Biologie, c/o LEPS, Schelpenkade, 14a, 2313 Leiden.

Although org-P accounts for a large part of the P_{sed}, little is known about its chemical and biological properties. In most studies investigating the chemical nature of org-P_{sed} compounds, only a small percentage could be characterized. Of the compounds found phytic acid and related compounds were quantitatively the most important.

In this study three different approaches have been combined, to study the properties of org-P_{sed}:

(1) The complex and adsorption chemistry of phytic acid and its bioavailability were studied in laboratory experiments.
(2) Algae were killed and left to mineralize in an 'artificial' sediment. The properties of the remaining organic phosphate compounds were investigated.
(3) After removal of $Fe(OOH) \approx P$ and $CaCO_3 \approx P$, the remaining org-P was studied on Camargue sediments. An attempt was made to characterize the nature of ASOP (Acid Soluble Org-P).

Phytic acid was found to form complexes with Ca^{2+}, Mg^{2+}, Mn^{2+}, Fe^{2+} and Fe^{3+}. The Fe complexes were found to be the most stable in the presence of chelating agents. All complexes dissolved in diluted acid.

Early diagenesis of phosphorus in recent sediments of the estuary and the gulf of St Lawrence

Marc Lucotte
GEOTOP, Université du Québec à Montréal, c.p. 8888, suc.A., Montréal, P.Q., H3C3P8, Canada

A sequential extraction method is used to differentiate between four phases of particulate phosphorus in recent sediment and in suspended particulate matter of the estuary and the gulf of St Lawrence. The four P phases are extracted in the following order: (1) the exchangeable P fraction (P_{ex}) after washing the solid matter in ammonium acetate, (2) the inorganic P fraction adsorbed onto iron-oxides (P_{cdb}) using a citrate-dithionite-bicarbonate buffered reagent ('cdb'), (3) the apatite P fraction (P_{ap}) dissolved in HCl, and (4) the organic P fraction (P_{org}) released after ashing and HCl treatment. The simultaneaous determination of the atomic ratio of the iron oxyhydroxide (Fe_{cdb}) to the P_{cdb} serves as an indicator of the level of orthophosphate adsorption in various redox conditions of the environment.

The fresh to salt water transition of the suspended particulate matter is characterized by a marked Fe flocculation and a strong P_{cdb} adsorption. In the rapidly accumulating sediments of the lower estuary (sedimentation rates > 2 mm/a), up to 50% of the P_{cdb} initially buried is released in the interstitial waters during the oxic to anoxic transition and diffuses back to the water column thanks to the bioturbation. In the presence of a noticeable bacterial activity, responding to a higher marine carbon content in the sediments, an additional 20 to 30% P_{cdb} is regenerated below the oxic layer, and appears to correspond to an equivalent formation of authigenic P_{ap}. In the sediments of the Gulf, characterized by lower sedimentation rates (< 2 mm/a), the regeneration of P_{cdb} reaches 60 to 80%. Again, the sediments containing higher fractions of marine organic carbon sustain a bacterial activity which seems to be responsible for the mineralization of a fraction of the released P_{cdb} as P_{ap}.

Combining the particulate P partitioning with sedimentation rates and organic matter burial provides the establishment of early diagenetic P regional budgets. The highest P accumulation ($P_{tot} \approx 1400$ to 5500 mgP/m^2/a) in the St Lawrence system is located in the lower estuary and arises from the high fluxes of P_{ap}-rich terrigenous particulate matter ($\approx 70\%$ of P_{tot}), from the moderate remobilization of the sedimented P_{cdb} and P_{org} fractions and from the small authigenic P_{ap} precipitation ($\approx 10\% \ P_{tot}$). In the gulf, the P accumulation drops rapidly ($P_{tot} \approx 200$ to 850 mgP/m^2/a) because of the low fluxes of terrigenous P_{ap}-rich particles and because of the almost total remobilization of P_{cdb}, and yet in spite of a partial remineralization of the remobilized P_{cdb} as authigenic P_{ap} when bacterial activity is high. Both detrital and authigenic P_{ap} ($\approx 60\%$ and $\approx 20\%$ respectively) appear to form the major P sink in the sediments of the gulf, whereas P_{cdb} and P_{org} constitute $\approx 20\%$ of the total accumulated P.

Seasonal variation in P-pools, porewater SRP and P-release in a coastal marine sediment

P. B. Mortensen[1], H. S. Jensen[2], E. K. Rasmussen[1] & B. Thamdrup[3]
[1] Water Quality Insitute. [2] Odense University. [3] Aarhus University. Denmark

In 1990 variation in P-pools and porewater SRP was studied in a coastal marine sediment from Aarhus Bight, Denmark. Sediment P-release and oxygen uptake was measured in a continuous flow system using undisturbed sediment cores collected by a box-corer. Water depth at the station was 16 meters, which was usually below the halocline.

P-release rates were low in the spring (less than 30 umoles P m^{-2} d^{-1}) except in a short period after the spring bloom of diatoms where a value of 60.5 was recorded. At this occasion NRP made up half of the released P. During summer P-release rates reached values of 160 umoles P m^{-2} d^{-1}, and in September the maximum rate of 550 was recorded. The peak value was recorded in a period of relatively high water temperature and low oxygen. In summer and autumn all P was released as SRP. Seasonal variation in P-release rates correlated negatively with oxygen tension in the bottom water ($r^2 = 0.69$; $p < 0.001$) which indicates that redox dependent P-release is an important mechanism in this sediment. P bound to reducible forms of iron and manganese (BD-SRP) made up 40–50% of total P in the mixed layer of the sediment (0–2 cm depth). This poolsize varied between 100 and 150 moles P m^{-2}. Seasonal variation in BD-SRP in seven depth intervals correlated positively with pools of reactive Fe^{3+} and maganese ($r^2 = 0.52$; $p < 0.001$).

Seasonal variations of P compounds and their concentrations in two coastal lagoons

T. Moutin, B. Picot, M. C. Ximenes* & J. Bontoux
*D.S.E.S.P., Faculté de Pharmacie, Montpellier; *C.E.M.A.G.R.E.F. Montpellier, France*

The accumulation of nutrients, with its short or long term consequences (eutrophication, dystrophic crisis), is a major problem in coastal lagoons. During 1990 a study was carried out at three sampling sites in two coastal lagoons near Montpellier in order to estimate the quantities present and the exchanges between compartments: variables were measured simultaneously in the overlying water, the interstitial water and the sediments. Monthly analyses of the overlying water (temperature, salinity, dissolved oxygen, orthophosphate, ammonium, nitrate and chlorophyll a) allow a general understanding of the functioning of these lagoons and a characterization of each site. The orthophosphate concentration is distinctly higher in summer.

Analysis of the interstitial water (ammonium, orthophosphate), carried out every three months, allows the determination of diffusion fluxes. The method (peeper technique) and the validity of the results are discussed. A strong correlation between the ammonium and orthophosphate concentrations is observed. The seasonal variations in the estimated fluxes do not appear to be related to the concentrations in the overlying water; the considerable phosphate release observed in summer might be caused by the transition from the oxidized to the reduced state of the iron in the upper sediment layer.

Analysis of the sediments, particularly phosphate fractionation according to the method of Golterman and Booman (1988), was also carried out at different depths in each season. While total phosphate is essentially constant over time, considerable variations in the fractions appear, in particular in the iron bound phosphate fractions. A gradient from surface to bottom is observed throughout the year, and a distinct decrease in summer and autumn.

Phosphate release appears to be important but is difficult to calculate. Diffusion into the sediments takes place throughout the year, but transition from the upper sediment layer to the overlying water seems to depend very much on the oxidation state of the iron. The quantity of iron bound phosphate dissolving from a 1 mm thick oxidized sediment layer appears to be roughly equivalent to 150 days' release by the upward diffusion.

Hydrobiologia **253**: 103–121, 1993.
P.C.M. Boers, T.E. Cappenberg & W. van Raaphorst (eds),
Proceedings of the Third International Workshop on Phosphorus in Sediments.
© *1993 Kluwer Academic Publishers.*

The role of microorganisms in mobilization and fixation of phosphorus in sediments

René Gächter[1] & Joseph S. Meyer[2]
[1]*Institute of Aquatic Sciences (EAWAG), Swiss Federal Institute of Technology (ETH), CH-6047 Kastanienbaum, Switzerland;* [2]*Department of Fisheries, Humboldt State University, Arcata, CA 95521, USA*

Key words: bacteria, phosphorus, phosphorus cycling in lakes, sediments

Abstract

Cycling of phosphorus (P) at the sediment/water interface is generally considered to be an abiotic process. Sediment bacteria are assumed to play only an indirect role by accelerating the transfer of electron from electron donors to electron acceptors, thus providing the necessary conditions for redox- and pH-dependent, abiotic sorption/desorption or precipitation/dissolution reactions.

  Results summarized in this review suggest that

(1) in eutrophic lakes, sediment bacteria contain as much P as settles with organic detritus during one year,

(2) in oligotrophic lakes, P incorporated in benthic bacterial biomass may exceed the yearly deposition of bioavailable P several times,

(3) storage and release of P by sediment bacteria are redox-dependent processes,

(4) an appreciable amount of P buried in the sediment is associated with the organic fraction,

(5) sediment bacteria not only regenerate PO_4, they also contribute to the production of refractory, organic P compounds, and

(6) in oligotrophic lakes, a larger fraction of the P settled with organic detritus is converted to refractory organic compounds by benthic microorganisms than in eutrophic lakes.

From this we conclude that benthic bacteria do more than just mineralize organic P compounds. Especially in oligotrophic lakes, they also may regulate the flux of P across the sediment/water interface and contribute to its terminal burial by the production of refractory organic P compounds.

Introduction

While, scattered information indicates that bacteria might directly be involved in the control of P flux across the sediment/water interface, an overwhelming number of case studies have described P cycling at the sediment/water interface ignoring bacteria or by considering them only as incorporeal catalyzers of mineralization as sketched in Fig. 1a.

This paper, critically reviews, reevaluates and analyzes reported results on the contribution of bacteria to the phosphorus (P) cycling across the sediment/water interface. It emphasizes the state of knowledge, defines gaps, reassesses research strategies, and addresses the following two questions:

1. Under what environmental conditions do bacteria need to be considered as organisms in

order to understand P cycling across the sediment/water interface, and under what conditions can they be neglected?

2 What environmental conditions favour the production of refractory organic P compounds, and what conditions favour regeneration and recycling of bioavailable P?

Classical view of phosphorus cycling in lakes

According to the classical view of P cycling in lakes (Fig. 1a), phosphate (PO_4) flushed into a lake by its tributaries or released from the sediments is taken up from the water by living and non-living suspended particles and lost to the lake bottom in proportion to the settling velocities of the particles. In the sediment, organic P is liberated as soluble PO_4 to solution during decomposition by bacteria, or it becomes buried as refractory organic P. The liberated PO_4 may be adsorbed to inorganic surfaces, complexed by refractory organic materials, or precipitated as apatite or vivianite. Part of it may remain in solution or be recycled into the overlying water. In many cases, the flux of PO_4 from the sediment to the water is largely controlled by the prevailing redox conditions at the sediment/water interface. An oxidized sediment surface often prevents or strongly diminishes release of PO_4 from the sediment to the water. It forms an efficient trap for dissolved iron (Fe), manganese (Mn) and PO_4 moving the interstitial water from reduced sediments to the oxic surface. When the hypolimnion and, hence, the sediment surface become anoxic, this barrier disappears and release rates of Fe, Mn and PO_4 often increase markedly, indicating reduction of ferric hydroxide complexes and subsequent release of ferric iron and adsorbed PO_4 (Mortimer, 1941, 1942, 1971; Einsele, 1936; Einsele & Vetter, 1938). In lakes with too low a Fe:P ratio or with too high a pH, release and binding of inorganic P cannot be controlled exclusively by iron. Boström *et al.* (1982) discussed additional chemical, biological and physical mechanisms (e.g., sorption of P to clay minerals and humic substances, bacterial activity, bioturbation by benthic organisms, mobilization of P by rooted macrophytes, wind-induced turbulence in the sediment overlying water, resuspension, gas ebullition and temperature) controlling the cycling of P at the sediment/water interface.

Can bacteria be neglected as a transient P pool?

In the scenario described above, bacteria are treated only as 'catalysts' that accelerate solubilization of PO_4, by oxidation of organic detritus and reduction of various oxidants. Consuming O_2, NO_3^- and SO_4^{2-}, they provide the necessary conditions for abiotic or biotic reduction of ferric iron (Fe^{3+}), subsequent release of PO_4 and precipitation of iron sulfide (FeS). The concentration of dissolved PO_4 in the interstitial water is assumed to be controlled by mineralization, by abiotic physical/chemical equilibria of precipitation/dissolution or sorption/desorption, and by diffusion.

By definition, catalysts are neither produced nor consumed during a chemical reaction, and their composition and pool size are constant. Thus, it is assumed that the amount of P bound to bacteria does not vary (i.e., uptake is always compensated by release, independent of varying environmental conditions). As a consequence, uptake and release of PO_4 by bacteria and, hence, their potential role in accelerating or retarding P exchange across the sediment/water interface, often is disregarded or considered to be negligible in comparison to mineralization rate and chemical equilibria.

Bacteria are not, however, only incorporeal sites of enzyme production (Fig. 1b). More likely, as other organisms, they depend on P as a nutrient; hence, they are able to take it up from the organic substrate or from the water. Net release seems to be controlled by their demand for P. For example, Gächter & Mares (1985) and others (cited in their paper) observed that during decomposition of settling detritus, no PO_4 was released. They suggested that heterotrophic organisms colonizing and decomposing P-deficient settling organic material may be a sink, rather than a source,

Traditional model

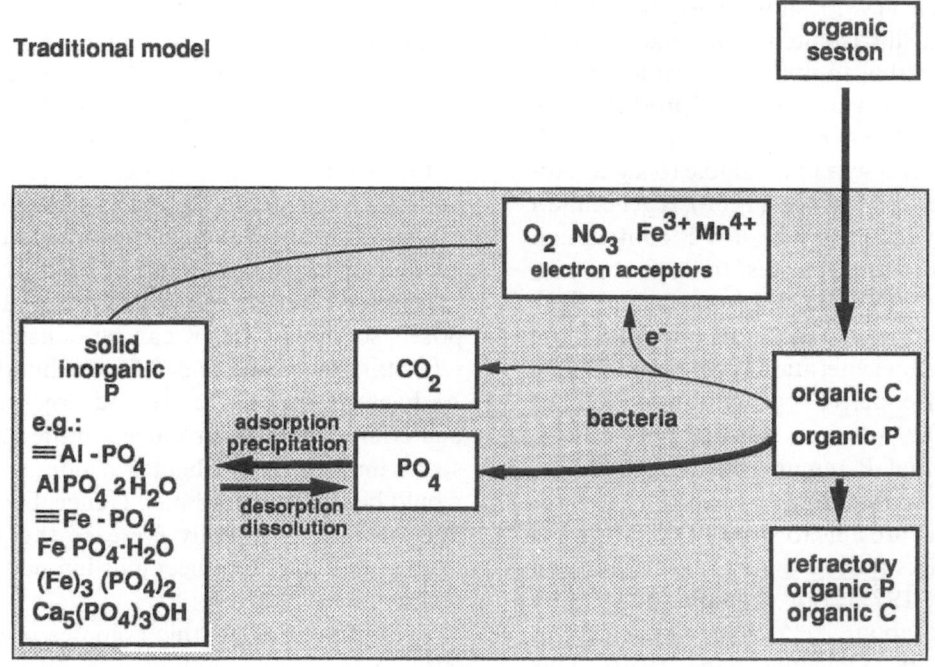

Revised model

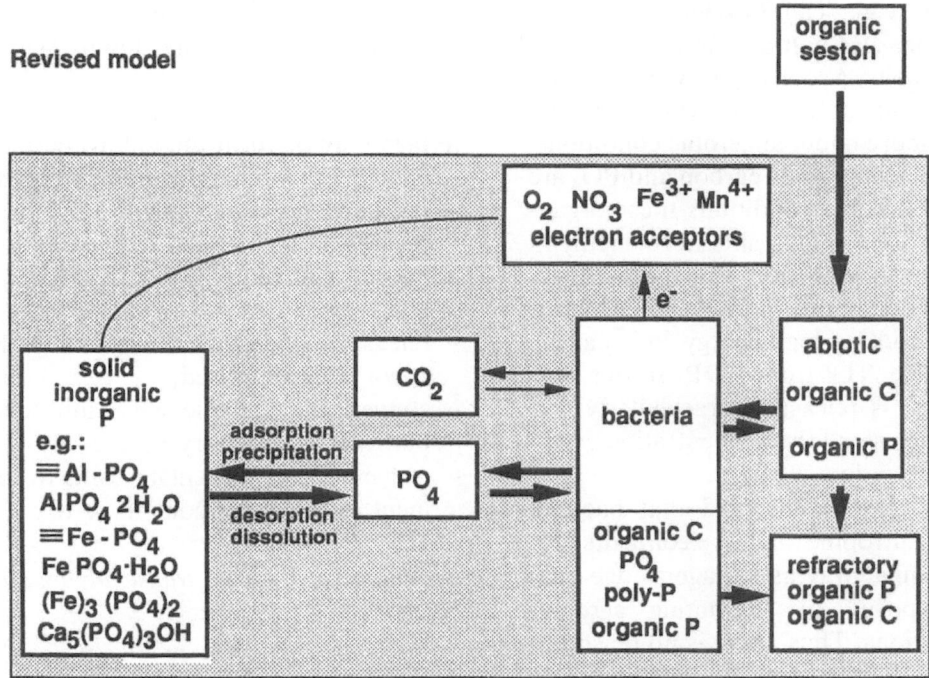

Fig. 1. Classical model (Fig. 1a) and a more conceptual model (Fig. 1b) of bacterial constribution to nutrient cycling.

for PO_4. Stöckli (1985) showed experimentally that microorganisms isolated from lakewater did not release but rather took up PO_4 while decomposing dissolved organic matter produced by *Chlamydomonas rheinhardii;* half of the organic P excreted by bacteria was unavailable to algae. According to Jewell & McCarty (1968), regeneration of P from decaying algae with low P content can be subdivided into three phases: In the first phase, soluble reactive P (SRP) is taken up by the algae or by their decomposers; in the second phase, the apparent nutrient regeneration is zero; and finally, in the third phase, active regeneration of P takes place. When this nutrient regeneration ceases, some of the initial P remains in the particulate refractory organic material.

Many bacteria are able to store PO_4 in the form of polyphosphates (poly-P), if PO_4 is available in excess (Kulaev, 1979). For example, poly-P can constitute up to about 20% of the dry weight in *Acinetobacter* spp. (Deinema *et al.*, 1980). This phenomenon has been studied extensively because it offers a possibility to remove PO_4 from wastewater (see e.g., Van Groenestijn, 1988). These bacteria are also common in aquifers, surface waters and soils. According to Wentzel *et al.* (1986), they are not able to gain energy by fermentation of glucose under anaerobic conditions. However, if sufficient organic carbon and PO_4 are available, under aerobic conditions these organisms polymerize excess PO_4 and store energy in the resulting energy rich poly-P. Under subsequent anaerobic conditions, poly-P can be hydrolyzed and the resulting high-energy PO_4 can be used to synthesize ATP from ADP. In this process, inorganic P is released intracellularly and eventually diffuses along the concentration gradient into the surrounding medium.

Gächter *et al.* (1988) observed that bacteria isolated from eutrophic lakes sediments responded in a similar way as sewage sludge bacteria, when exposed to alternating aerobic/anaerobic conditions. Thus, there is no reason to assume *a priori* that the size of the P pool in sediment bacteria would be constant under changing environmental conditions (e.g., changing redox potential, or changing supply and composition of organic material). However, it remains to be seen whether variation in their pool size could significantly affect PO_4 concentrations in the pore water and, hence, the exchange of PO_4 across the sediment/water interface.

From a thermodynamic perspective, all organic material is eventually decomposable if redox conditions are favourable. However, concentration profiles of organic material in sediments suggest that part of the settled organic material is decomposed so slowly that it can be considered to be refractory on a time scale of some hundred years. As bacteria decompose settled organic material, bacterial biomass is produced in the sediment. It seems unlikely that all bacterial biomass produced would be easily degradable. Hence, bacterial production may eventually increase the amount of refractory debris that becomes buried in the sediment.

Recruitment of benthic colonies of cyanobacteria to the water column is another biological mechanism that might increase the flux of P from sediments to lake water, as suggested by Boström (1985) for *Microcystis* and observed by Osgood (1988) for *Aphanizomenon* and by Wolf (1986) for *Oscillatoria*.

In summary, bacteria at the sediment/water interface may perform the following functions:
– catalyze the mineralization of organic P;
– take up and transiently store P in their biomass;
– release P if redox conditions are not favourable;
– release P when they die and thus, themselves, become decomposed;
– convert bioavailable SRP into dissolved or particulate refractory P; and
– transfer P in particulate form from the sediment to the water column or *vice versa*.

Estimation of P bound transiently by sediment bacteria

The P/biovolume ratio in bacteria
In addition to bacterial P, sediments contain many other sorts of detrital organic and inorganic P compounds. Since most bacteria are closely

associated with such abiotic solid phases, determination of the bacterial P pool size requires a quantitative physical separation of bacteria from other solids. To our knowledge, no such technique is currently available. Hence, P bound to bacteria cannot be measured directly. Indirect estimations require information about bacterial biovolume and the P content of bacterial cells.

The biovolume can be obtained from direct microscopic inspection of stained bacteria, yielding number of cells and average cell volume.

Determination of bacterial P content requires a bacterial culture that contains no particulate phosphorus (PP) other than bacteria. In addition, environmental conditions of the culture should mimic those of the investigated sediment as closely as possible. Cultivation of bacteria in interstitial water might be a suitable method, especially if a technique could be found to dissolve (e.g., by blending, grinding or ultrasonifying) at least part of the decomposable particulate organic material into the interstitial water before separating interstitial water from the sediment. Formation of iron oxyhydroxides sorbing P could be prevented by keeping the culture anoxic or by adding a chelator, that is not easily biologically degradable (e.g. EDTA).

As an alternative, Gächter et al. (1988) sampled bacteria grown in a benthic chamber in an-oxic water overlying the sediment of eutrophic Lake Sempach; that water contained no settling detritus and no particulate Fe. Assuming that all PP was incorporated in bacteria, they estimated that bacterial biomass contained 15 μg P mm^{-3} of cell volume. This value lies only slightly above the range of 3 to 12 μg P mm^{-3} that Shuter (1978) and Vadstein et al. (1988) reported as a minimum cell quota of P. Despite the large variability of the results gathered in Table 1, we estimate that bacteria not limited by P might accumulate about 11 μ g mm^{-3}, even under anoxic conditions. In sediments where bacteria accumulate poly-P (e.g., Uhlmann & Bauer, 1988), 11 μg mm^{-3} likely is a conservative estimate for the bacterial P content.

The C/P ratio in bacteria

The particulate organic carbon (POC): bacterial volume ratio is affected by nutritional status and community structure (Bratbak & Dundas, 1984). According to Riemann & Bell (1990), values vary from 0.12–0.58 mg POC mm^{-3}. If we accept 0.22 mg POC mm^{-3} as an average value (Bratbak, 1985), assume that POC is 50% of the dry weight of the particulate organic matter and add some weight for inorganic salts, then the ratio of dry weight/biovolume must be close to 0.5 mg mm^{-3}. Bornsheim et al. (1990) report a dry

Table 1. Estimated P content of microbial biomass reported in literature. [1] Converted from % P of dry weight (DW), assuming a DW/biovolume ratio of 0.5 g m^{-3} (s. page 8).

Microbes	P content (μg P mm^{-3})	Reference
Mixed hypolimnic bacteria	15	Gächter et al. (1988)
Mixed limnetic bacteia	6–17	Vadstein et al. (1988)
Acinetobacter calcoaceticus	16–33[1]	Hoffmeister et al. (1990)
Acinetobacter wolffii	32–37[1]	Hoffmeister et al. (1990)
Sewage bacteria	4–42[1]	Kämpfer et al. (1990)
Planktonic bacteria	13	Borsheim et al. (1990)
Minimum cell quota		
Bacillus subtilis	12	Shuter (1978)
Pseudomonas aeruginosa	7	Shuter (1978)
Corynebacterium bovis	8	Shuter (1978)
Nitrosomonas europea	4	Shuter (1978)
Escherichia coli	6	Shuter (1978)

weight/biovolume ratio of 0.6 mg mm^{-3}. Hence, this ratio is much larger for bacteria than for phytoplankton, where often a ratio of 0.2 mg mm^{-3} is reported (Vollenweider, 1974).

The C:P ratio of bacteria is affected by the redox condition and the C:P ratio of the substrate. Fenchel & Blackburn (1979) and Gächter *et al.* (1988) reported C:P (weight) ratios for bacteria of 18.4 and 14.9, respectively. Vadstein *et al.* (1988) observed values varying between 3 and 29, with a median value of 11. Combination of the widely accepted conversion factor of 220 μg C mm^{-3} (Bratbak, 1985) and a assumed average of C:P ratio of 20 yields an average P content of 11 μg mm^{-3} cell volume. This value lies at the lower end of the values gathered for bacteria not limited by P (Table 1), indicating that for bacteria not limited by P, the C:P ratio might even be < 20. Definitely, the C:P ratio of bacteria is lower than the C:P ratio of settling seston which was reported to vary between 40 in eutrophic and 200 in oligotrophic lakes (Gächter *et al.*, 1985).

P in sediment bacteria of lakes of different trophic states

The number of bacteria generally increases about 3 to 5 orders of magnitude from the water to the sediment surface and then decreases rapidly

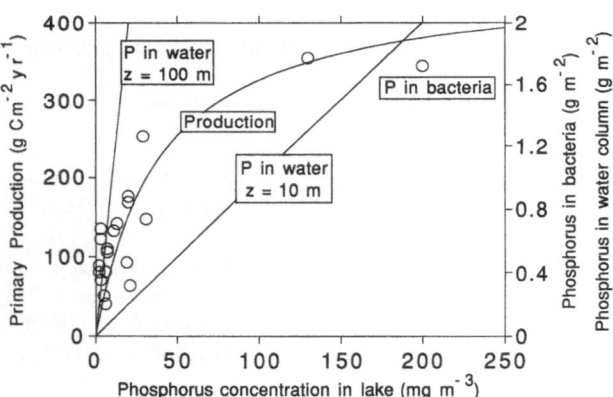

Fig. 2. Areal primary production (curve), benthic bacterial P content (circles) and areal P content of the water column for average depths of 10 and 100 m (dashed lines) as a function of P concentration at spring overturn. Primary production = 456 [P]/([P] + 40) (Fricker, 1980); data for P content of benthic bacteria from Table 2.

within the sediment at greater depth. It also seems to vary horizontally over the lake basin, with higher numbers in the littoral zone covered by macrophytes and lower numbers in profundal sediments (Wetzel, 1983).

Figure 2 shows the relationship between yearly primary production rate (Prod) and a lake's P concentration at spring overturn, as suggested by Fricker (1980). It also includes data on P fixed in benthic bacteria (P_{bact}) (derived from Table 2) and on the P content of lakes, both related to the lakes' P concentrations. Comparison of the two axes for Prod and P_{bact} suggests, as a first approximation a linear relationship between average benthic bacterial P content and primary production:

$$P_{bact} = k \cdot Prod. \qquad (1)$$

with

$$k = 0.005 \text{ yr gP/gC}$$

Applying a C:P ratio of 20, equation 1 suggests that organic C incorporated in benthic bacteria is equivalent to about 10% of the yearly primary production. Since a large fraction (normally > 80%) of C assimilated by primary production is recycled in the water column and since part of the organic material reaching the sediment surface is refractory, only about 10% of the assimilated organic C might be mineralized in the sediment. Hence, the amount of organic C fixed in the benthic bacterial biomass is approximately equal to the amount of organic C that it dissimilates per year. This high ratio between standing crop of bacterial C and metabolized organic C suggests a very low metabolic activity of benthic bacteria. This conclusion is supported by Boström & Törnblom (1990), who showed that at low temperatures average bacterial doubling times were as high as 2000–3000 days in marine sediments and higher than 1000 days in the sediment of the highly eutrophic Lake Vallentunasjön. Such low average growth rates indicate that a large fraction of sediment bacteria might be in a 'dormant' stage, not growing at all.

At steady state, bacterial production (BP), par-

Table 2. Estimated biomass and P content of benthic bacteia in various lakes. BM = biomass, BV = biovolume, CN = cell number, DW = dry weight, POC = particulate organic carbon, WW = wet weight.

Lake	Bacterial numbers (10^6 cells g^{-1})	Biomass (g C m^{-2})	P content (mg P m^{-2})	P conc. (μg P $liter^{-1}$)	Sediment layer (cm)
Beloe (Kosino)	2326[a]	12.8–25.6[b]	640–1280[c]		0–5
B. Medvezh'e	1905[a]	10.5–21.0[b]	525–1050[c]		0–5
Chernoe (Kosino)	1285[a]	7.1–14.1[b]	355–705[c]		0–5
M. Medvezh'e	1624[a]	8.9–17.9[b]	445–875[c]		0–5
Sviatoe (Kosino)	922[a]	5.1–10.1[b]	255–505[c]		0–5
Krugloe	1110[a]	6.1–12.2[b]	305–610[c]		0–5
Gab (Karelian region)	1883[a]	10.4–20.7[b]	520–1035[c]		0–5
Lake Sempach[d]		25.5	1770	130	0–5
Mirror Lake[e]		4.0	200[c]	6	0–10
Vallentunasjön[f]		34.0	1720[c]	200	0–10
Lake Luzern[g]		7.5–25.4[i]	375–1270[c]	20	0–5
Wastwater[h]		6.8–9.3[l]	340–465[c]	2	0–10[k]
Ennerdale[h]		7.6–10.2[l]	380–510[c]	2	0–10[k]
Buttermere[h]		4.2–5.9[l]	210–295[c]	5	0–10[k]
Crummock[k]		5.1–9.0[l]	255–450[c]	3	0–10[k]
Coniston[h]		10.2–16.9[l]	510–845[c]	3	0–10[k]
Thirimere[h]		11.0–13.5[l]	550–675[c]	3	0–10[k]
Windermere (N)[h]		12.7–15.2[l]	635–785[c]	13	0–10[k]
Ullswater[h]		7.7–15.7[l]	330–770[c]	7	0–10[k]
Derwentwater[h]		6.8–8.5[l]	385–425[c]	6	0–10[k]
Bassenthwaite[h]		5.9–6.8[l]	295–340[c]	21	0–10[k]
Windermere (S)[h]		9.3–24.5[l]	465–1225[c]	20	0–10[k]
Loweswater[h]		10.2–11.0[l]	510–550[c]	7	0–10[k]
Rydal Water[h]		13.1–13.5[l]	655–675[c]	11	0–10[k]
Grasmere[h]		6.8–11.8[l]	340–590[c]	19	0–10[k]
Esthwaite Water[h]		11.8–17.8[l]	590–890[c]	31	0–10[k]
Blelham Tarn[h]		16.1–34.3[l]	805–1720[c]	29	0–10[k]

[a] Wetzel (1983: 595).

[b] BM = [CN g^{-1} WW]·[1.05 g WW cm^{-3}]·[5.10^4 cm^3 m^{-2}]·[0.5 to 1 μm^3 $cell^{-1}$]·[10^{-9} mm^3 μm^{-3}]·[0.22 mg C mm^{-3}]·[10^{-3} g mg^{-1}].

[c] mg P m^{-2} = [POC in BM]·50 mg P g^{-1}C.

[d] Gächter *et al.* (1988).

[e] Jordan *et al.* (1985).

[f] Boström *et al.* (1985).

[g] Laczko (1988).

[h] Jones *et al.* (1979).

[i] BM = [BV g^{-1} DW]·[0.1 g DW cm^{-1}]·[0.22 mg C mm^{-3} BV]·[10^4 cm^3 m^{-2}].

[k] Assumption: total areal BM = 2·areal BM of top cm.

[l] BM = 2·[CN g^{-1} DW]·[0.1 g DW cm^{-3}]·[0.5 μm^{-3} $cell^{-1}$]·[0.22 mg C mm^{-3}]·[10^{-9} mm^3 μm^{-3}]·[10^4 cm^3 m^{-2}]·[10^{-3} g mg^{-1}].

ticulate organic carbon in bacterial biomass (POC_{bact}), substrate consumption (CS), specific growth rate (μ) and growth yield (Y) are related as follows:

$$BP = \mu \cdot (POC_{bact}) = Y \cdot (CS). \qquad (2)$$

Solving equation (2) for POC_{bact} allows one to estimate bacterial POC as a function of μ, Y and

CS. We estimated that yearly substrate consumption might equal 10% of primary production. According to Boström & Törnblom (1990), growth yield of bacterial communities varied between 0.17 and 0.4 (average of about 0.3), and average growth rate of sediment bacteria at temperatures between 4 °C and 5 °C was about 0.0005 day^{-1} or 18 yr^{-1}. From this it follows that bacterial biomass can be expressed in g C m^{-2} as

$$POC_{bact} = Y \cdot (CS)/\mu$$
$$= 0.3 \cdot 0.1 \cdot Prod./0.18$$
$$= 0.17 \; yr \cdot Prod.$$

Assuming again a C:P ratio of 20, the P fixed in bacterial biomass (P_{bact}) can be estimated as

$$P_{bact} = POC_{bact}/20$$
$$= 0.0085 \; yr \; gP/gC \cdot Prod. \qquad (3)$$

Taking into account all uncertainties related to the various assumptions, equation (3) agrees well with equation (1). Thus, the two independent approaches predict a similar relationship between P_{bact} and primary production.

P_{bact} an also be related to P that settled to the lake bottom with organic detritus ($P_{settled}$). If 20% of organic C assimilated settles to the lake bottom, then $P_{settled}$ can be estimated as

$$P_{settled} = 0.2 \cdot Prod. \cdot (1/C:P) \qquad (4)$$

where C:P is the C to P weight ratio of the settled seston. Combining (1) and (3) with (4) yields

$$P_{bact}/P_{settled} = (0.025 \; to \; 0.043) \cdot (C:P) \qquad (5)$$

In oligotrophic lakes, the C:P ratio of settling detritus can be as high as 200; whereas in eutrophic lakes, it is close to 40 (Gächter & Bloesch, 1985). Thus, in oligotrophic lakes, the P fixed in benthic bacteria might exceed the annual flux of P to the sediment 5- to 9-fold; whereas in eutrophic lakes, P_{bact} is expected to equal 1 to 2 times the annual flux of P to the sediment. Since bacteria are concentrated in the sediment close to the surface, these high values suggest that in this

layer a substantial part of the P might be incorporated in bacterial biomass.

At spring overturn, the areal P content of a lake can be obtained by multiplying its average concentration times its average depth. As shown in Fig. 2, in shallow lakes the estimated P incorporated in sediment bacteria might exceed the amount present in the water. Thus, a partial release of P incorporated in benthic bacteria could lead to a significant increase of the PO_4 pool available for primary producers, if the released P does not become sorbed to abiotic surfaces or is precipitated in the sediment. On the other hand, it seems unlikely that seasonal changes in bacterial biomass or changes of its P content could significantly alter the P content of the water column in deep lakes.

Evidence for bacterial P from sequential P extraction

Boström *et al.* (1985) extracted various forms of sedimentary P from Lake Vallentunasjön. They distinguished between loosely bound P, Fe and bound P, Ca bound P and residual P consisting mainly of organic P but also including the inert fraction. They classified about 75% of total P in surface sediments (0–2 cm) of Lake Vallentunasjön as residual P (Fig. 3). About 50% of this fraction was suddenly released from the sediment when the O_2 concentration in the hypolimnion

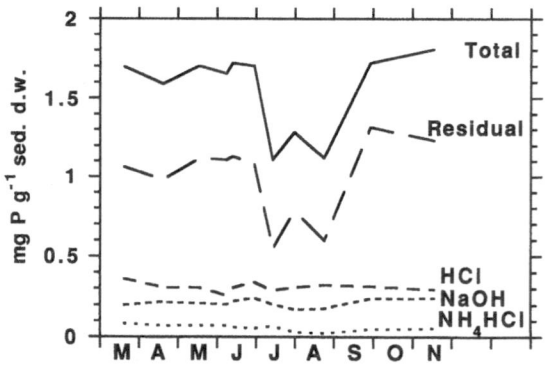

Fig. 3. Seasonal variation in concentration of total P and different P fractions in surface sediments (0–2 cm) of Lake Vallentunasjön; redrawn from Boström *et al.* (1985).

decreased from about 10 mg liter^{-1} to 2 mg liter^{-1} in July. It increased rapidly to its original concentration when the O_2 concentration increased, indicating that release and uptake were reversible, redox-dependent processes. The redox-dependent dissolution and formation of solid P could be explained by dissolution and formation of inorganic phosphate sorbed to iron oxyhydroxides. However, in the extraction scheme applied by these investigators, inorganic P adsorbed to Fe would have been extracted as SRP in the NaOH-P fraction, a fraction that was rather constant throughout the observed period. Boström et al. (1985) suggested that an increase in mineralization rate, lysis of bacteria and other organisms, or release of P from bacteria under anoxic conditions could have contributed to the observed loss of P from the sediment.

Recently, Uhlmann et al. (1990) extracted P from activated sludge according to the extraction scheme of Psenner et al. (1988) and in a separate extraction with hot water. They analyzed NaOH extracts and hot water extracts using NMR spectroscopy, a technique that allows poly-P to be identified. Using the method of Psenner et al. (1988), the major part of the total P was recovered as non-reactive P (NRP) in the NaOH fraction ('residual P' in the terminology of Boström et al., 1985). Most of this P was identified as inorganic poly-P. These results suggest that the 'residual P' observed in the sediment of Lake Valentunasjön was not necessarily of organic nature. Changes in the residual P could well be interpreted as hydrolysis of inorganic poly-P under anaerobic conditions and reformation of poly-P when O_2 concentrations close to the lake bottom (4 m) again reached 10 mg liter^{-1}.

Since production and dissolution of poly-P could be the major processes that contribute to transient storage of P in a nearly constant microbial benthic biomass, there is an obvious need to have a method available to quantify poly-P concentrations in sediments. As long as poly-P cannot be measured directly in sediments, the quantitative contribution of benthic bacteria to sudden redox-dependent release and fixation of P remains speculative and obscure.

Uhlmann et al. (1990) suggested that the non-reactive P in the NaOH fraction was a good measure of poly-P. This conclusion needs to be further tested. It might be valid for 'young' activated sludge with a P content as high as 10% of dry weight. In sediments with a higher content of organic compounds and a lower P content, a substantial part of the NRP in the NaOH fraction might, however, be incorporated in the refractory organic fraction.

Contribution of bacteria to production of refractory organic P compounds

Theoretical considerations

Although bacteria require a minimum amount of P and can store it excessively if supply exceeds their demand and redox conditions are favourable, they are only a transient sink for P. When they die, decay and become decomposed by those which survive, part of the previously fixed P might be assimilated by their decomposers, part may be released as inorganic P, and part might eventually become dissolved or particulate refractory organic P compounds.

Generally, in deep lakes about 80% of C assimilated by primary producers is recycled in the epilimnion. About 20% settles across the thermocline and reaches the sediment. There, about half of it becomes mineralized within a few years. The remaining 10% is buried in the sediment as refractory organic carbon compounds. Table 3 exemplifies mineralization of settled organic material and production of refractory organic material in lake sediments. As it illustrates, decomposition of bacterial biomass is always coupled with production of new biomass and, hence, with the production of additional refractory organic carbon. Thus, after complete mineralization of the bacterial biomass about 16% of the initial bacterial C remains as refractory C.

Of course, a 50% mineralization could also be explained if we had chosen other assumptions regarding the content of refractory organic POC in settled seston and benthic bacteria. Possible

Table 3. Scheme exemplifying production of refractory organic material in lake sediments. Assumptions: 20% of C assimilated by primary producers settles to the sediment. 9% of C assimilated by primary producers and by bacteria are refractory. Bacteria mineralize organic C with an efficiency of 50%.

POC to be mineralized	Refractory POC	POC in bacteria	CO_2
20	9	5.5	5.5
5.5	0.495	2.5025	2.5025
2.5025	0.22523	1.1386	1.1386
1.1386	0.10248	0.51806	0.51806
0.51806	0.04663	0.23572	0.23572
0.23572	0.021215	0.10725	0.10725
0.10725	0.009653	0.048799	0.048799
0.048799	0.004392	0.022204	0.022204
0.022204	0.001998	0.010103	0.010103
Sum	9.90659		10.08324

extreme, but unrealistic assumptions would be:
– 50% of the settled POC is refractory, and bacteria produce no refractory POC; or
– all sestonic POC is decomposable, but bacterial POC is completely refractory.

As more precise information is lacking, however, it seems reasonable to assume equal percentages of refractory organic material in the biomass of all trophic levels.

As long as we are not able to measure production of refractory organic material by bacteria, this discussion is mainly of didactical value. However, it shows that bacteria contribute to the production of refractory organic material. Even if we know that bacterial biomass and activity might change seasonally, it has to be in cyclic steady state in a lake of constant trophic state. Accordingly, regardless of the bacterial pool size, on a yearly basis, uptake rates of all nutrients need to be compensated by loss rates. Thus, contrary to mineral sites, bacteria can never act as a terminal sink for P. However, because they produce some refractory organic P compounds they contribute not only to the recycling, but also to the removal of bioavailable P from aquatic systems. Let us assume (as sketches in Fig. 4) that at steady state, bacteria consume per year an amount l_C of organic C and an amount of I_P of P (i.e., the C:P ratio of its substrate equals $l_C/I_P = a$). Because at

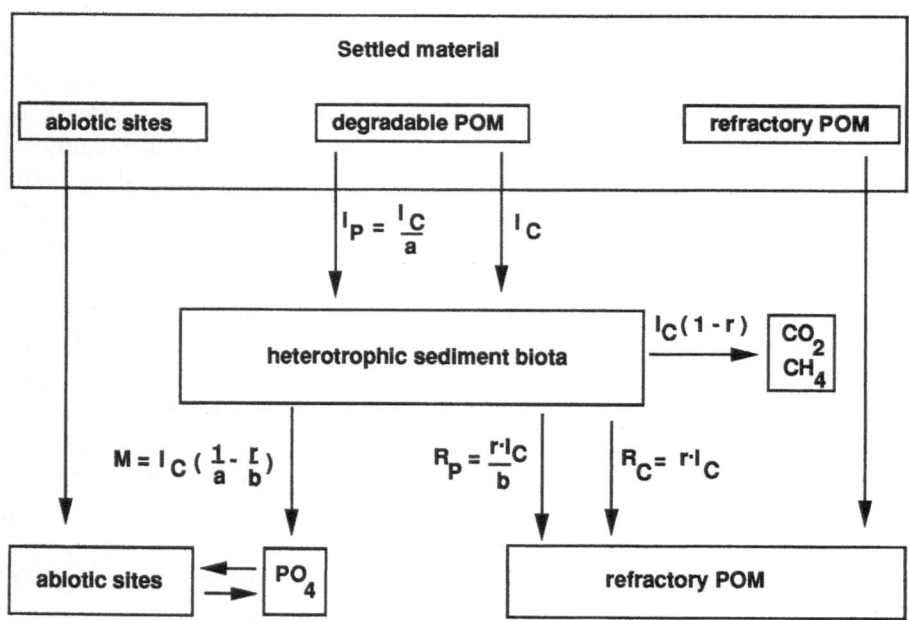

Fig. 4. Release of PO_4 and production of refractory organic P compounds by sediment bacteria (see text for explanation of symbols).

steady state the pool size of bacteria does not change, net production of the bacterial community equals zero. It transforms bioavailable organic material into CO_2 (CH_4), PO_4 and refractory organic material. Hence, production of refractory organic carbon (C_r) can be formally expressed as $r \cdot I_C$ where r defines the fraction of the total bioavailable organic C that is transformed by bacteria to refractory organic carbon. Production of refractory organic P equals $r \cdot I_C/b$, where b is the C:P ratio in the refractory organic material produced by the bacteria.

At steady state the amount P 'mineralized' and released as inorganic PO_4 (M) must equal $I_P - r \cdot I_C/b$. Relating mineralization to input yields

$$M/lP = 1 - r \cdot a/b \qquad (6)$$

indicating that bacteria mineralize P more efficiently at lower values of r, lower C:P ratios of the substrate, and higher C:P ratios of the produced refractory material. The absolute values of r, a and b are currently unknown.

For the following reasons, the values of a and b cannot be determined simply by measuring the C:P ratio of the settling material and of refractory organic material in deep sediments, respectively:
(1) settled organic material is an unknown composition of degradable and refractory organic material with unknown C:P ratios, and
(2) refractory organic C pool in deep sediments

is likely dominated by refractory organic C originating from settled organic material.

Even if the absolute values of a, b and r are not known, relative release rates can be predicted from equation (6). In oligotrophic lakes, sediment bacteria mineralize organic P compounds less efficiently than in eutrophic lakes, because in oligotrophic lakes a larger fraction of the settled organic material is decomposed aerobically (causing a higher growth yield and, hence, larger values of r and because the settling material has a higher C:P ratio suggesting a higher value of a than in eutrophic lakes.

Sediments will release no P when the integrated rate of bacterial refractory P production plus abiotic and biotic immobilization equal the supply rate of labile P to the sediment surface. Thus, very likely, release rates of PO_4 from sediments in oligotrophic lakes are small not only because little organic material is decomposed and the PO_4 sorbed to inorganic sites is small in comparison to their sorption capacity, but also because bacteria mineralize organic P compounds less efficiently than in eutrophic lakes.

Refractory organic P compounds in lake sediments

Refractory organic material includes refractory P compounds, partly as cell remains but also as

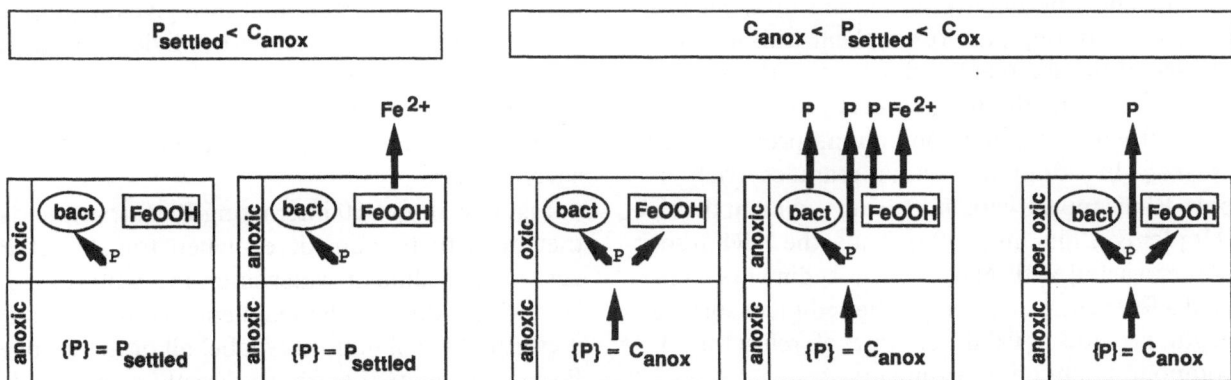

Fig. 5. Schematic presentation of P sinks and sources in the sediments of oligotrophic and eutrophic lakes under oxic and anoxic conditions. {P} is total P concentration (particulate + dissolved).

high molecular weight organic compounds either dissolved in interstitial water or adsorbed to solid surfaces. Lean (1973a, b), Lean & Rigler (1974) and Rigler (1973) identified in epilimnetic waters four P compartments: PO_4, low molecular weight organic P (XP), a colloidal high molecular weight P fraction (HMP) and particulate P incorporated in seston (PP). XP was excreted by planktonic microorganisms and, after polycondensation formed HMP, which was no longer bioavailable to phytoplankton. Hence, HMP might be called refractory. Filtration as well as poisoning of the water with formaldehyde (Jackson & Schindler, 1975; Brassard & Auclair, 1984) prevented labelling of XP and HMP when ^{32}P was added as PO_4, indicating that the formation of XP and HMP was mediated by microorganisms. There is no reason why sediment bacteria should behave differently. Thus, they as well probably produce not only solid refractory P compounds as cell remains but also HMP compounds, which may either be dissolved in the interstitial water or adsorbed to organic and inorganic surfaces. However, it needs to be demonstrated that these compounds are 'refractory' on the time scale of years.

In Psenner's extraction scheme (Psenner et al., 1984; Psenner & Pucsko, 1988; Psenner et al., 1988), most organic P compounds are recovered as NRP in the NaOH fraction. Of course, this fractionation does not discriminate between degradable and refractory compounds and, as demonstrated by Uhlmann et al. (1990), it may include inorganic poly-P as well. However, if we assume that the quality of settling material did not change during past years, then, with increasing age of the sediment and, hence, with increasing sediment depth, the ratio between degradable and refractory organic compounds probably decreases. In addition, volutin granules (rich in poly-P) disappear with increasing sediment depth (Hupfer & Uhlmann, 1990). Thus, the NRP fraction extracted with NaOH from sediments sampled a few centimetres below the sediment surface might provide a valid estimate of refractory P compounds buried in sediments.

From results obtained by Williams et al. (1976) in Lake Erie, it can be deduced that in sediment layers that were deposited 23 to 36 years before sampling, 30% to >50% of the buried P was organic refractory P, if the allochthonous apatite-P was disregarded. In Lakes Skaha and Wood, 15 to 32% and 25 to 70% of the non-apatite P sampled deeper than 10 cm below the sediment surface was organic P (Williams, 1973). In Kleiner Montigglersee, 40 to 60% of the P was of organic nature in the top 10 cm of the sediment (Psenner et al., 1984); and in the oligotrophic but meromictic Piburger See, 75 to >90% of the total P was organic in the top 8 cm of the sediment.

Obviously, in many cases, after initial mineralization of the autochthonously formed particulate P, a large fraction is buried as organic P in sediment. Thus, bacterial mineralization of organic P is not very efficient, or, as discussed above, bacteria may even contribute to the formation of such compounds.

Experimental evidence of biological control of phosphorus fluxes across the sediment/water interface

In order to test whether the transfer of P across the sediment/water interface is controlled not only by inorganic chemical equilibria but also by bacteria, bacterial activity could be suppressed by heat, by gamma irradiation or by addition of antibiotics or poisons. If bacteria play an important role, then inactivation of bacteria will markedly alter P fluxes. However, it must be kept in mind that such experiments can at best give qualitative results, because they compare a living sediment with one that has been poisoned just recently. A true comparison would require a control with sterile deposition of settling seston during at least several years.

Figure 6 shows 10 interacting P compartments that need to be considered when transfer of P across the sediment/water interface is discussed. If $^{32}PO_4$ is added to the sediment-overlying water, it eventually will uniformly label all of the mobile P compartments (i.e., those directly or indirectly in exchange with the PO_4 pool). Bacterial activity controls the labelling of particulate P incorpo-

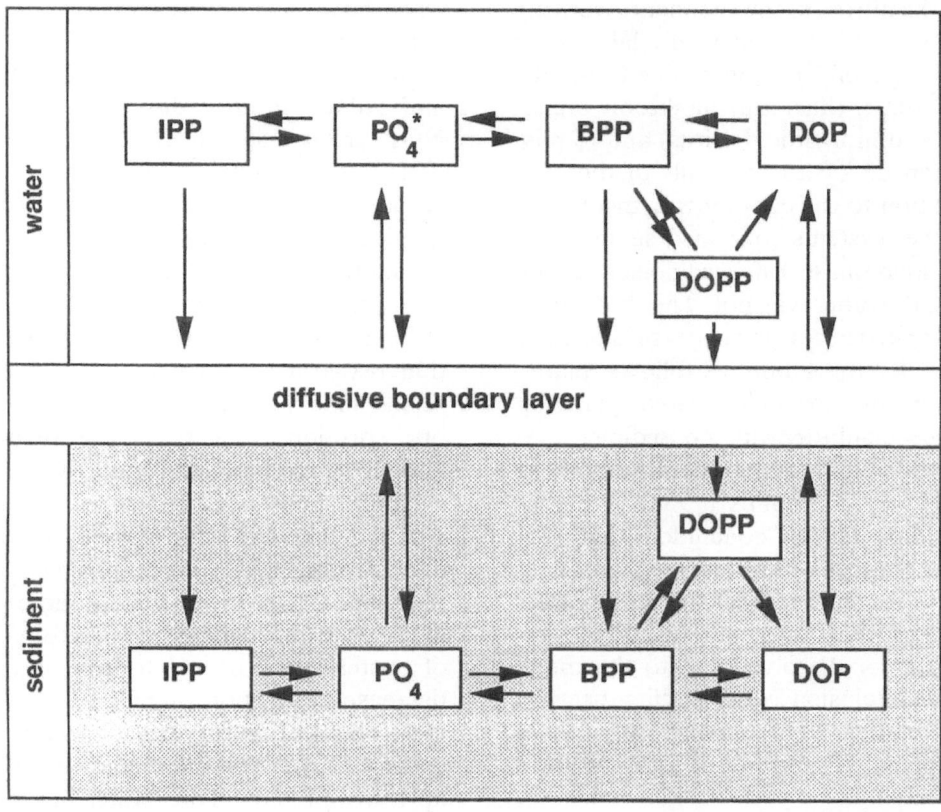

Fig. 6. P compartments involved in the exchange of $^{32}PO_4$ (PO_4^*) across the sediment/water interface: IPP = inorganic particulate P, PO_4 = ortho-P, BPP = particulate P in biomass, DOP = dissolved organic P, DOPP = detrital organic particulate P.

rated in bacteria (BPP), dissolved organic P (DOP) and detrital bacterial P (DPP). ^{32}P is transferred to the sediment as PO_4, as dissolved organic P (DOP), as inorganic particulate or colloidal P (IPP), as particulate P incorporated in bacteria (BPP) or in organic detritus (DPP). Transport of dissolved species (PO_4, DOP) is controlled by molecular diffusion through the diffusive boundary layer, separating the well mixed water from the sediment. Transport of particulate species (IPP, BPP and DPP) is controlled by their settling velocities.

In the sediment, inactivation of bacteria retards or even prevents the labelling of BPP, DOP and DPP. As a consequence, the specific activity of the PO_4 pool of the pore water increases faster if sediment bacteria are inactive. Since diffusion of $^{32}PO_4$ from the water to the sediment depends on its concentration gradient across the interface,

inactivation of sediment bacteria likely results in a lower uptake rate of ^{32}P in a later phase of an experiment and in a higher steady-state $^{32}PO_4$ activity in the sediment-overlying water.

Hayes & Phillips (1958) described results from various experiments in which they labelled sediment-overlying water with $^{32}PO_4$ and compared the transfer of ^{32}P to the sediment in presence and in absence of bacterial activity (bacterial activity in the overlying water was suppressed with the antibiotics terramycin or tetracycline). They found that inactivation of suspended bacteria increased the initial flux of ^{32}P from the water to the sediment drastically.

Jackson & Schindler (1975) conducted a similar experiment with water from lake 239 and sediment from Lake 227. They inactivated bacteria by addition of 0.48% formalin at the outset of the experiment and monitored the loss of radioactivity in the filtrate of the sediment-overlying water. In the system containing only lake water but no mud, losses of $^{32}PO_4$ were due to uptake by suspended matter (bacteria, planktonic organisms, organic and inorganic detritus) and to possible adsorption of ^{32}P to the walls of the container. In addition to the components mentioned above, two other systems contained sediment as an additional solid phase; one was poisoned with formaldehyde, the other was not. The three treatments can be ranked according to decreasing rates of radioactivity from solution, as follows: plankton with sediment > formalin-treated plankton with sediment = plankton with no sediment. At first, these results might seem to contradict those reported by Hayes & Phillips (1958). Whereas Hayes & Phillips (1958) concluded that suspended microorganisms retarded the transfer of ^{32}P to the sediment surface, Jackson & Schindler (1975) concluded that 'microorganisms accelerated the transfer of dissolved ^{32}P to the mud'. However, this conclusion is misleading, because Jackson & Schindler (1975) showed only the removal of dissolved ^{32}P and not of total ^{32}P from the water.

In summary, it can be concluded from the results presented by Hayes & Phillips (1958) and Jackson & Schindler (1975) that 'suspended bacteria' retard the transfer of $^{32}PO_4$ from the water to the sediment. This effect is due to scavenging of $^{32}PO_4$ in dissolved and particulate compounds, which move more slowly from the water to the sediment than does the $^{32}PO_4$. Bacteria were able to do so because of P content of the compartments BPP and DOP were large compared to the PO_4 concentration of the sediment-overlying water. However, it cannot be decided from such experiments if 'sediment bacteria' contain a significant amount of the total P content and if they significantly and directly control the flux of P across the sediment/water interface.

Hupfer & Uhlmann (1991) partly inactivated bacteria in sediment cores by the addition of antibiotics to the sediment-overlying water after it became anoxic. In the presence of active bacteria (no antibiotics added), the sediment took up 5.1 mg P m^{-2} day^{-1} and 9.5 mg P m^{-2} day^{-1} when NO_3^- or O_2 were available as the oxidant. When bacteria were inactivated, P uptake switched to net release of P (12.7 mg P m^{-2} day^{-1} and 9.8 mg P m^{-2} day^{-1} in the presence of NO_3^- or O_2 respectively). This observation suggests that antibiotics

(1) induced release of P from bacteria, or
(2) decreased the ability of the sediment to immobilize P either by chemical/physical adsorption or by biological uptake.

Even if these results do not allow discrimination between the two potential mechanisms, it seems unlikely that antibiotics decreased the abiotic sorption capacity of the sediment; O_2 is known to oxidize Fe^{2+} quickly at neutral pH, even in the absence of bacterial activity. Bacteria compete with Fe^{2+} for O_2 and, in addition, produce protons while decomposing organic material; hence, they lower the abiotic oxidation rate of Fe^{2+} in the sediment. Therefore, inactivation of bacteria is expected to increase, rather than decrease absorption of PO_4 by iron oxyhydroxides. Yet, the contrary was observed: In the presence of O_2 or NO_3^-, the control sediments took up P from the overlying water but the sediments released P when the bacterial activity was suppressed. This indicates that antibiotics either lowered the P uptake of bacteria or even induced P lysis from biota.

Sinke & Cappenberg (1988) gamma-irradiated sediment cores and compared initial short-term P release rates of irradiated (sterilized) and non-irradiated cores at temperatures ranging from 10 to 80 °C. Gamma irradiation killed sediment biota and likely induced cell lysis and release of P. P released from organisms equilibrates with abiotic surfaces in the sediment and diffuses to the overlying water. Increasing temperature enhances mineralization, favour desorption and increases diffusion.

As shown in Fig. 7, release rates of irradiated cores increased linearly with temperature and clearly exceeded those of the non-sterilized sediments, except for the highest temperatures (60 to 80 °C). In the lower temperature range, the differences in release rates between sterilized and non-sterilized sediment increased with increasing

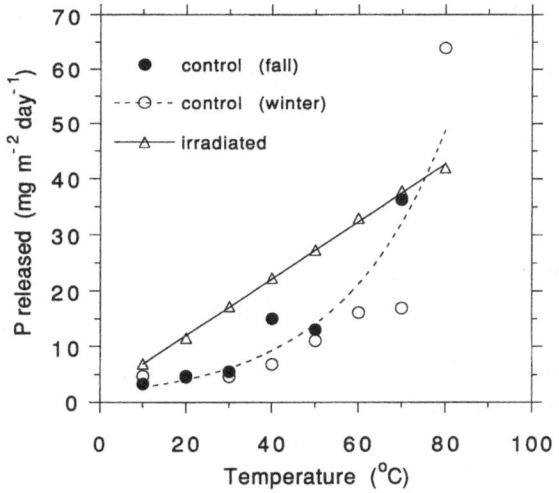

Fig. 7. Release of total dissolved P from gamma-irradiated and non-sterilized sediment cores; redrawn from Sinke & Cappenberg (1988).

temperature, in agreement with the expected increase of equilibrium concentration in the interstitial water and the increased molecular diffusion. If temperature exceeds a critical value, it gradually causes death of the most sensitive organisms and, hence, also induces lysis. Thus, release rates did not differ between irradiated and non-irradiated samples at 70 to 80 °C, probably it did not matter whether biota were killed by irradiation or by the high temperature.

Experiments conducted with sediments of Lac Tantaré, Quebec (Gächter *et al.*, 1991), support the findings of Sinke & Cappenberg (1988). Anoxic sediment sampled in the deepest part of the lake in October 1990 was sterilized with formaldehyde (0.16%), heat (20 minutes at 121 °C) or gamma irradiation (3 million Rads). Seventeen days after sterilization, total dissolved P concentration ([TDP]) was measured in the filtered (0.45 μm) interstitial water that was isolated by centrifugation. [TDP] in the unsterilized control (17.5 ± 7.5 μg liter^{-1}) was significantly lower than in the sterilized samples (heat: 388 ± 71 μg liter^{-1}; formaldehyde: 118 ± 42 μ liter^{-1}; radiation: 102 ± 7 μg liter^{-1}). This significant increase of P concentration of the interstitial water underestimates the amount of P incorporated in microor-

ganisms, probably because not all P was released and part of it was readsorbed to abiotic surfaces.

Such experimental results strongly support the hypothesis that bacteria incorporate P in their biomass that otherwise would contribute to increased interstitial water concentrations and, hence, to increased fluxes of P from the sediment to the overlying water. However, they do not allow us to quantify the bacterial contribution. In nature, sediments never become sterile. When bacteria die, e.g. due to changing redox conditions, they will be replaced by other species, better adapted to the new environment. As a consequence, part of the P released from dead bacteria will be taken up by their successors. Thus, under natural conditions, P released from biota always is partly masked by succession.

Interference of bacteria with the coupling of iron and phosphorus cycling in an aerobic sediment surface

It is well documented and widely accepted that aerobic sediment surfaces act as a P trap, which under reducing conditions release part or all of the previously trapped P. Since every sediment surface is eventually buried and becomes anoxic, it will lose the amount of P that exceeds its anoxic retention capacity. Thus, as illustrated in Fig. 5, the P retention by lake sediments depends on the settling flux of particulate P to the sediment ($P_{settled}$, expressed as g P m^{-2} yr^{-1}) and the P retention capacity of anoxic sediments (C_{anox}, expressed as g P m^{-2} yr^{-1}). As long as $P_{settled}$ does not exceed C_{anox}, the sediment will not release P under either oxic or anoxic conditions, and diagenesis of the sediment will not result in an accumulation of P at the sediment surface. If $P_{settled}$ exceeds C_{anox}, then the sediment surface is supplied with P from two sides: from above by settling particles, and from below by diffusion of dissolved P species originating from reductive dissolution of redox-dependent solid species. If the sediment surface is permanently oxic, it will accumulate P until its P content reaches its P binding capacity (C_{ox}) (Gächter, 1987). Then, the sediment starts to release P even under oxic conditions. If the sediment surface becomes periodically anoxic, it loses part of the P accumulated under previously oxic conditions. As a consequence of this 'anoxic regeneration', its P content may never reach C_{ox} and no P would be released under subsequent transient oxic conditions.

Thus, on a short-term basis, P fluxes across the sediment/water interface may be controlled by redox-dependent chemical or biological mechanisms, such as precipitation and dissolution of iron oxyhydroxides which adsorb PO_4, or redox-dependent uptake and release of P by microorganisms. Bacteria are known to be able to efficiently take up PO_4 from solution, if the substrate that they decompose does not fulfil their requirements (e.g., Barsdate et al., 1974; Planas, 1978; Fenchel & Blackburn, 1979; Fleischer, 1983; Lean & White, 1983; Lean, 1984; Currie & Kalff, 1984; Gächter et al., 1988; Stöckli, 1985). As a consequence, bacteria may have to complete with iron oxyhydroxides for PO_4. It was experimentally demonstrated by Fleischer (1986) that they are able to compete successfully. Their demand for P increases as the C:P ratio of the settled organic material increases. As discussed above, the C:P ratio of bacteria is approximately 20. Assuming a yield of 50% for aerobic bacteria, bacteria are only expected to release P if the C:P ratio of their substrate is smaller than 40. According to Gächter & Bloesch (1985), the C:P ratio of settling seston exceeds this value largely in oligotrophic and mesotrophic lakes. Thus, especially at the sediment surface of oligotrophic lakes, aerobic bacteria may have to complete with abiotic sorption sites for PO_4. In eutrophic aerobic lakes, the C:P ratio of the settled organic matter is closer to 40 and, hence, might satisfy the needs of bacteria for P. As a consequence, in eutrophic aerobic lakes, P diffusing to the sediment surface might to a lager extent be trapped by iron oxyhydroxides.

In most cases, retention of P in an oxic sediment layer is very likely due to adsorption to abiotic surfaces as well as to fixation by microorganisms. In eutrophic lakes, aerobic bacteria might store excess PO_4 as poly-P. If the sediment surface becomes anoxic, P is released due to reductive dissolution of iron oxyhydroxides as well as due to the hydrolysis of poly-P in bacteria.

In oligotrophic lakes, where $P_{settled}$ does not exceed C_{anox}, P does not accumulate in the oxic sediment surface. A large fraction of the P present is probably incorporated into not yet degraded organic material, into refractory compounds and into bacterial biomass. Since P supply of the bacteria is limited, they probably store little or no poly-P, and if they successfully complete with abiotic surface complexation, little PO_4 is adsorbed to iron oxyhydroxides. Hence, redox-dependent solid PO_4 is probably not an important fraction of the total P, and reduction of the sediment surface does not result in a sudden release of P. This could explain why the classical model of the coupled Fe and P cycling was derived in highly eutrophic lakes, whereas the view that bacteria might be important in controlling P flux across the sediment/water interface was first suggested from investigations of oligotrophic lakes. For examples, Levine (1975) and Levine & Schindler (1980) showed that hypolimnetic concentrations of PO_4 rarely exceeded 1 μg liter^{-1} in Lakes 227 and 302 S, even under anoxic conditions. Sorption of PO_4 to redox-independent surfaces such as hydrated aluminium oxides or humic complexes could, however, also explain why anoxic conditions did not result in a sudden release of PO_4 in these oligotrophic lakes.

Summary and conclusions

We have demonstrated that
(1) biomass of sediment bacteria is positively related to primary production or to the flux or organic matter to the sediment.
(2) most benthic bacteria are accumulated in a layer close to the sediment surface.
(3) the C:P ratio of bacteria is approximately 20; hence, it is significantly lower than that of phytoplankton, whose C:P ratio seems to vary between 200 and 40, depending on trophic state. Provided the P supply is sufficient and redox conditions are appropriate, the C:P

ratio in bacteria can decrease to values as low as 5, corresponding to a P content that is 10% of dry weight.

(4) in many lakes, a significant proportion of the permanently buried P is refractory organic P.

(5) comparing results obtained from experiments with sterilized and unsterilized sediments does not allow one to quantify directly the contribution of bacteria to P cycling across the sediment/water interface.

From this we conclude that

(1) sediment bacteria do not necessarily release P when they mineralize settled organic detritus. Depending on the C:P ratio of their organic substrate and their growth yield, they might transiently even take up dissolved P from their environment in order to fulfil their nutritional requirement.

(2) the ratio between the amount of P fixed in sediment bacteria and the amount of P being deposited per year as part of the organic material at the sediment surface increases with decreasing trophic state of a lake. In eutrophic lakes bacterial P in the sediment equals approximately 1 to 2 yrs worth of P deposition; whereas in oligotrophic lakes, it equals 5 to 9 yrs worth of deposition.

(3) seasonal changes of the bacterial P pool size (changing biomass and/or changing P content) may affect transport of P across the sediment/water interface.

(4) sediment bacteria may contribute to the production of organic P compounds; in oligotrophic lakes, they probably convert a larger fraction of the assimilated P into refractory organic P compounds than they do in eutrophic lakes. This mechanism stabilizes oligotrophic conditions.

In summary, we believe that, especially in shallow, oligotrophic lakes, bacteria may play a much more important role in P cycling across the sediment/water interface than has been generally assumed up to now. However, in order to test these conclusions unequivocally, techniques need to be developed to

(1) determine the P content of sediment bacteria as a function of varying environmental conditions,

(2) measure poly-P in sediments,

(3) differentiate between refractory and bioavailable P, and

(4) study production and chemical nature of refractory P generated by sediment bacteria.

Acknowledgements

We thank Th. Cappenberg, R. Carignan, M. Hupfer, D. Uhlmann and two anonymous referees for their valuable suggestions on the manuscript.

References

Barsdate, R. J., T. Fenchel & R. T. Prentki, 1974. Phosphorus cycle of model ecosystems: Significance for decomposer food chains and effect of bacterial grazers. Oikos 25: 239–251.

Borsheim, K. Y., G. Bratbak & M. Heldal, 1990. Enumeration and biomass estimation of planktonic bacteria and viruses by transmission electron microscopy. Appl. envir. Microbiol. 56: 352–356.

Boström, B., M. Jannson & C. Forsberg, 1982. Phosphorus release from lake sediments. Arch. Hydrobiol. Beih. Ergebn. Limnol. 18: 5–59.

Boström, B., I. Ahlgen & R. Bell, 1985. Internal nutrient loading in a eutrophic lake reflected in seasonal variations of some sediment parameters. Verh. int. Ver. Limnol. 22: 3335–3339.

Boström, B., 1985. The role of *Microcystis* colonies, its mucilage and associated bacteria, for nutrient fluxes from sediments to lake water. – A working hypothesis. In: M. Enell, W. Graneli & L.-A. Hansson (eds), Proc. 13th. Nordic Sump. on Sediments, Anneboda, Sweden: 6–8.

Boström, B., A.-K. Pettersson & I. Ahlgren, 1989. Seasonal dynamics of a cyanobacteria-dominated microbial community in surface sediments of a shallow, eutrophic lake. Aquat. Sci. 51: 153–178.

Boström, B. & E. Törnblom, 1990. Bacterial production, heat production and ATP-turnover in shallow marine sediments. Termochim. Acta 172: 147–156.

Brassard, P. & J. C. Auclair, 1984. Orthophosphate uptake rate constants are mediated by the 103–104 molecular weight fraction in Shield lakewater. Can. J. Fish. aquat. Sci. 41: 166–173.

Bratbak, G., 1985. Bacterial biovolume and biomass estimations. Appl. envir. Microbiol. 49: 1488–1493.

120

Bratback, H. & I. Dundas, 1984. Bacterial dry matter content and biomass estimations. Appl. envir. Microbiol. 48: 755–757.

Currie, D. J. & J. Kalff, 1984. The relative importance of bacterioplankton and phytoplankton in phosphorus uptake in freshwater. Limnol. Oceanogr. 29: 311–321.

Deinema, M. H., L. H. A. Habets, J. Scholten, E. Turkstra & H. A. A. M. Webers, 1980. The accumulation of polyphosphate in *Acinetobacter* spp. FEMS (Fed. Eur. Microbiol. Soc.) Microbiol. Lett. 9: 275–279.

Einsele, W., 1936. Über die Beziehungen des Eisenkreislaufs zum Phosphatkreislauf im eutrophen See. Arch. Hydrobiol. 29: 664–686.

Einsele, W. & H. Vetter, 1938. Untersuchungen über die Entwicklung der physikalischen und chemischen Verhältnisse im Jahreszyklus in einem mässig eutrophen See (Schleinsee bei Langenargen). Int. Revue ges. Hydrobiol. Hydrogr. 36: 285–324.

Fenchel, T. & T. H. Blackburn, 1979. Bacteria and mineral cycling. Academic Press, London.

Fleischer, S., 1983. Microbial phosphorus release during enhanced glycolysis. Naturwissenschaften 70: 415–416.

Fleischer, S., 1986. Aerobic uptake of Fe(III)-precipitated phosphorus by microorganisms. Arch. Hydrobiol. 107: 269–277.

Fricker, Hj., 1980. OECD Eutrophication program – Regional Project Alpine Lakes. Swiss Federal Board for Environmental Protection, Bern, Switzerland.

Gächter, R. & J. Bloesch, 1985. Seasonal and vertical in the C:P ratio of suspended and settling seston of lakes. Hydrobiologia 128: 193–200.

Gächter, R., 1987. Lake restoration. Why oxygenation and artificial mixing cannot substitute for a decrease in the external phosphorus loading. Schweiz. Z. Hydrol. 49: 170–185.

Gächter, R. & A. Mares, 1985. Does settling seston release soluble reactive phosphorus in the hypolimnion of lakes? Limnol. Oceanogr. 30: 364–371.

Gächter, R., J. S. Meyer & A. Mares, 1988. Contribution of bacteria to release and fixation of phosphorus in lake sediments. Limnol. Oceanogr. 33: 1542–1558.

Gächter, R., A. Tessier, E. Szabo & R. Carignan, 1991. Measurements of total dissolved phosphorus in small volumes of iron-rich interstitial water. Aquatic Sciences (in Press).

Hayes, F. R. & J. E. Phillips, 1958. Lake water and sediment. IV. Radiophosphorus equilibrium with mud, plants, and bacteria under oxidized and reduced conditions. Limnol. Oecanogr. 3: 459–475.

Hoffmeister, D., D. Weltin & W. Dott, 1990. Untersuchungen zur bakteriellen Phosphatelimninierung. II. Mitteilung: Physiologische Untersuchungen an Reinkulturen. Wasser, Abwasser, gwf. 131: 270–277.

Hupfer, M. & D. Uhlmann, 1990. Phosphate immobilization by microorganisms in lake sediments. Sediment/Water Interactions 5th International Symposium. August 6–9, 1990. Uppsala.

Hupfer, M. & D. Uhlmann, 1991. Microbially mediated phosphorus exchange across the mud-water interface. Verh. int. Ver. Limnol. 24: 2999–3003.

Jackson, T. A. & D. W. Schindler, 1975. The biochemistry of phosphorus in an experimental lake environment: evidence for the formation of humic-metal-phosphate complexes. Verh. int. Ver. Limnol. 19: 211–221.

Jewell, W. L. J. & P. L. McCarty, 1968. Aerobic decomposition of algae and nutrient regeneration. Stanford Univ. Techn. Rep. 91.

Jones, J. G., M. J. L. G. Orlandi & B. M. Simon, 1979. A microbiological study of sediments from the Cumbrian lakes. J. Gen. Microbiol. 115: 37–48.

Jordan, M. J., G. E. Likens & B. J. Peterson, 1985. Organic carbon budget. In: G. E. Likens (ed.). An Ecosystem Approach to Aquatic Ecology. Mirror Lake and its Environment. Springer-Verlag. ISBN 0–387–96106–2.

Kämpfer, P., A. Eisenträger, V. Hergt & W. Dott, 1990. Untersuchungen zur bakteriellen Phosphatelimninierung. I. Mitteilung: Bakterienflora und bakterielles Phosphatspeicherungsvermögen in Abwasserreinigungsanlagen. Wasser, Abwasser gwf. 131: 156–164.

Kulaev, I. S., 1979. The biochemistry of inorganic polyphosphates. Wiley.

Laczko, E., 1988. Abbau von planktischem Detritus in den Sedimenten voralpiner Seen: Dynamik der beteiligten Mikroorganismen und Kinetik des biokatalysierten Phosphoraustausches. Ph. D. Thesis Nr. 8371, Eidgenössische Tech. Hochschule (ETH), Zürich, Switzerland.

Lean, D. R. S., 1973a. Phosphorus dynamics in lake water. Science 179: 678–680.

Lean, D. R. S., 1973b. Movements of phosphorus between its biologically important forms in lake water. J. Fish. Res. Bd. Can. 30: 1525–1536.

Lean, D. R. S., 1984. Metabolic indicators for phosphorus limitation. Verh. int. Ver. Limnol. 22: 211–213.

Lean, D. R. S. & F. H. Rigler, 1974. A test of the hypothesis that abiotic phosphate complexing influences phosphorus kinetics in the epilimnetic lake water. Limnol. Oceanogr. 19: 787–788.

Lean, D. R. S. & E. White, 1983. Chemical and radiotracer measurements of phosphorus uptake by lake plankton. Can. J. Fish. aquat. Sci. 40: 147–155.

Levine, S. N., 1975. A preliminary investigation of orthophosphate concentration and the uptake of orthophosphate by seston in two Canadian Shield lakes. MS. Thesis, University of Manitoba, pp. 151.

Levine, S. N. & D. W. Schindler, 1980. Radiochemical analysis of orthophosphate concentrations and seasonal changes in the flux of orthophosphate to seston in two Canadian Shield lakes. Can. J. Fish. aquat. Sci. 37: 479–487.

Likens, G. E., 1985. The Lake-Ecosystem. In: G. E. Likens (ed.), An Ecosystem Approach to Aquatic Ecology. Mirror Lake and its Environment. Springer-Verlag. ISBN 0–387–96106–2.

Mortimer, C. H., 1941–1942. The exchange of dissolved substances between mud and water in lakes. 1 and 2. 3 and 4. J. Ecol. 29: 280–329; 30: 147–201.

Mortimer, C. H., 1971. Chemical exchanges between sediments and water in the Great Lakes – speculations and probable regulatory mechanisms. Limnol. Oceanogr. 16: 387–404.

Osgood, R. A., 1988. A hypothesis on the role of *Aphanizomenon* in translocating phosphorus. Hydrobiologia 169: 69–79.

Planas, D., 1978. Phosphorus uptake rates in planktonic communities related to light gradient. Verh. int. Ver. Limnol. 20: 2731–2736.

Psenner, R., R. Pucsko & M. Sager, 1984. Die Fraktionierung organischer und anorganischer Phosphorverbindungen von Sedimenten. Arch. Hydrobiol./Suppl. 70: 111–155.

Psenner, R. & R. Pucsko, 1988. Phosphorus fractionation: advantages and limits of the method for the study of sediment P origins and interactions. Arch. Hydrobiol. Beih. Ergebn. Limnol. 30: 43–59.

Psenner, R., B. Boström, M. Dinka, K. Pettersson, R. Puckso & M. Sager, 1988. Sediment phosphorus group: Working group summaries and proposals for future research. 4. Fractionation of phosphorus in suspended matter and sediment. Arch. Hydrobiol. Beih. Ergebn. Limnol. 30: 98–110.

Redfield, A. C., B. H. Ketchum & F. A. Richards, 1963. The influence of organisms on the composition of sea water. In: M. N. Hill (ed.), The Sea, v. 2. Interscience: 26–77.

Riemann, B. & R. T. Bell, 1990. Advances in estimating bacterial biomass and growth in aquatic systems; Arch. Hydrobiol. 118: 385–402.

Rigler, F. H., 1973. A dynamic view of the phosphorus cycle in lakes. In: E. J. Griffith, A. Beeton, J. M. Spencer & D. T. Mitchell (eds), Environmental Phosphorus Handbook. John Wiley & Sons: 539–572.

Shuter, B. J., 1978. Size dependence of phosphorus and nitrogen subsistence quotas in unicellular microorganisms. Limnol. Oceanogr. 23: 1248–1255.

Sinke, A. J. C. & T. E. Cappenberg, 1988. Influence of bacterial processes on the phosphorus release from sediments in the eutrophic Loosdrecht Lakes, The Netherlands. Arch. Hydrobiol. Beih. Ergebn. Limnol. 30: 5–13.

Stöckli, A. P., 1985. Die Rolle der Bakterien bei der Regeneration von Nährstoffen aus Algenexkreten und Autolyseprodukten. Experimente mit gekoppelten, kontinuierlichen Kulturen. Ph. D. Thesis Nr. 7850, Eidgenössische Technische Hochschule (ETH) Zürich. 183 pp.

Uhlmann, D. & H.-D. Bauer, 1988. A remark on microorganisms in lake sediments with emphasis on polyphosphate-accumulating bacteria. Int. Revue ges. Hydrobiol. 73: 703–708.

Uhlmann, D., I. Röske, M. Hupfer & G. Ohms, 1990. A simple method to distinguish between polyphosphate and other phosphate fractions of activated sludge. Wat. Res. 24: 1355–1360.

Vladstein, O., A. Jensen, Y. Olsen & H. Reinertsen, 1988. Growth and phosphorus status of limnetic phytoplankton and bacteria. Limnol. Oceanogr. 33: 489–503.

Van Groenestijn, J. W., 1988. Accumulation and degradation of polyphosphate in *Acinetobacter* sp. Ph. D. Thesis, Agricultural University, Wageningen, The Netherlands.

Vollenweider, R. A., 1974. A Manual on Methods for Measuring Primary Production in Aquatic Environments. IBP Handbook Nr. 12 ISBN 0–632–00531–9. 225 pp.

Wentzel, M. C., L. H. Lötter, R. E. Loewenthal & G. V. R. Marais, 1986. Metabolic behaviour of *Acinetobacter* spp. in enhanced biological phosphorus removal – a biochemical model. Water SA 12: 209–224.

Wetzel, R. G., 1983. Limnology. Second Edition. Saunders College Publishing. ISBN 0–03–057931–9. 767 pp.

Williams, J. D. H., 1973. Phosphorus in the sediments of Okanagan mainstem lakes. Supplement report to task 121 report by B. E. St. John entitled 'The limnogeology of the Okanagan mainstem lakes' (not published).

Williams, J. D. H., T. P. Murphy & T. Mayer, 1976. Rates of accumulation of phosphorus forms in Lake Erie sediments. J. Fish. Res. Bd. Can. 33: 430–439.

Wolf, G., 1986. Die Verteilung des partikulären Phosphors und sein Verhältnis zur Biomasse der Blaualge Oscillatoria limosa im Pelagial des Piburgersees. Diplomarbeit Universität Insbruck.

Hydrobiologia **253**: 123–129, 1993.
P.C.M. Boers, T.E. Cappenberg & W. van Raaphorst (eds),
Proceedings of the Third International Workshop on Phosphorus in Sediments.
© 1993 *Kluwer Academic Publishers.*

The role of *Gloeotrichia echinulata* in the transfer of phosphorus from sediments to water in Lake Erken

Kurt Pettersson, Eva Herlitz & Vera Istvánovics
Uppsala University, Institute of Limnology, Erken Laboratory, Norr Malma 4200, S-761 73 Norrtälje, Sweden

Key words: phosphorus, cyanobacteria, *Gloeotrichia echinulata*, sediment

Abstract

The abundance of *Gloeotrichia echinulata* colonies in the sediments of Lake Erken and their phosphorus content were investigated to determine the contribution of *Gloeotrichia* colonies to total sediment phosphorus. Moreover, the potential size of the algal inoculum and the migration to the water during summer were estimated.

The surplus phosphorus content of the resting colonies in the sediment was about 45% of total phosphorus, which maximized at 8.5 μg P (mg dw)$^{-1}$ or 81 ng P colony^{-1}. The C:P ratio (by weight) in the early colonies appearing in the lake water was 50:1, while the ratio stabilized at 150 during the major migration period. The internal supply of surplus phosphorus was used during the pelagic growth of the colonies.

The internal phosphorus loading to the epilimnion of Lake Erken due to *Gloeotrichia* migration could, from the measurements of the increase in particulate epilimnetic phosphorus, be estimated at 40 mg P m^{-2} or 2.5 mg P m^{-2} d^{-1} in late July and early August. Determination of the number of colonies in the sediment before and during the migration verified this value to be a conservative estimate of the internal phosphorus loading due to *Gloeotrichia* migration to the epilimnion in Lake Erken.

The sediment P content calculated from the P concentration in early epilimnion colonies resulted in a value of 35 μg P (g dw)$^{-1}$ as a maximum. This corresponds to only 3% of the total phosphorus content in Lake Erken sediment.

Introduction

Certain species of both spore- and non-spore forming cyanobacteria have long been suspected to spend an initial period of development in or on the sediment (Wesenberg-Lund, 1904; Goor, 1925). However, these early studies offered no quantitative evidence about the number of colonies in the sediment. More recent papers have shown large sediment populations of *Microcystis* (e.g. Boström *et al.*, 1989) and in the case of

Gloeotrichia echinulata the presence of akinetes in the sediment (Roelofs & Oglesby, 1970) indicates that a part of the life-cycle is benthic. The extent of migration of these algae into the epilimnion of the lake water remains largely unknown in the case of stratified lakes. Barbiero & Welch (1992) have presented results from a two-year study in the shallow, non-stratified Green Lake.

A result of algal recruitment from the sediments to the epilimnion would be a transport of phosphorus (P) from the sediments to the water col-

124

umn (internal P loading). Like other algal groups, cyanobacteria have the ability to store a luxury uptake of P (Fitzgerald & Nelson, 1966; Stewart et al., 1978). A benthic population has a good opportunity to take up P from the pore water, which is very rich in P, followed by transport to the water column where it can use this P for a number of cell divisions. Some shallow lakes have an internal loading of P that coincides with a development of cyanobacteria in the water column (e.g. Osgood 1988), indicating algal P transport from the sediments. This process might also take place in stratified lakes and the objective of this study was to quantify the role of *Gloeotrichia echinulata* in the internal transport of P to the epilimnion in the moderately eutrophic and during summer stratified Lake Erken.

Material and methods

Study site

Lake Erken ($A = 24$ km^2, ^{z}mean $= 9$ m) is a moderately eutrophic lake in southern Sweden (Pettersson, 1990). Summer stratification lasts usually from May until early September. During this period the external nutrient loading is less than 10% of the annual loading (Pettersson, 1985). In 1991 the development of stratification in Lake Erken was followed with a thermistor system connected to a data logger (Pierson, 1990).

Phosphorus and biomass in the lake water

The sampling of lake water was performed at five stations, where epilimnion, metalimnion and hypolimnion were determined from temperature profiles. In each two meter stratum lake water was sampled proportionally to the volume it represented in the lake. The water was pooled into a composite sample for the epilimnion, metalimnion and hypolimnion. The sampling was performed twice a week from early July to late August 1991. A 200 μm net was used to separate the *Gloeotrichia* colonies from the other algae. Particulate (oxidative digestion with persulfate) and

surplus phosphorus (with boiling water) was determined in the *Gloeotrichia* concentrate after filtration onto 0.2 μm membrane filters (after Pettersson 1980). Total phosphorus (persulfate digestion) and molybdate reactive phosphorus (Murphy & Riley, 1962) was determined in the composite whole water sample.

The biomass of *Gloeotrichia echinulata* was estimated from analyses of particulate carbon (CHN-analyses) in the > 200 μm fraction after filtration over glass-fibre filters (Whatman GF/F).

Sediment studies

The presence of *Gloeotrichia* in the sediments of Lake Erken was preliminarily investigated in the autumn of 1989 and documented in more detail 23 March 1991. On this day 37 samples of surface sediment (0–4 cm) were taken along two

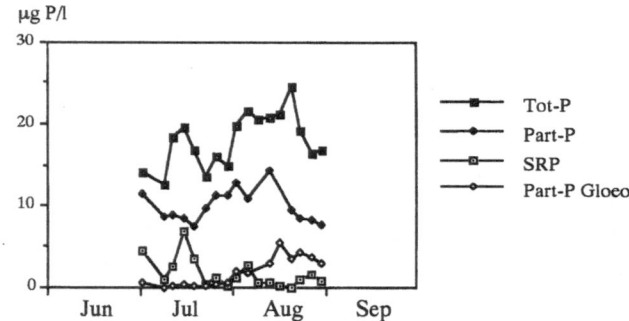

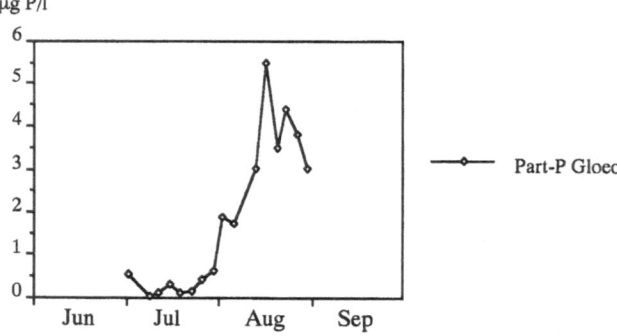

Fig. 1a. Total, particulate, soluble and particulate > 200 μm phosphorus in Lake Erken during the summer 1991. *1b* The development of the *Gloeotrichia* fraction (> 200 μm) in terms of phosphorus is enlarged.

transects. Two other sediment samplings were performed at five sites on 24 July and 3 August 1991. During the spring sampling the sediment was taken with an Ekman dredge and in the summer with a Willner core sampler. The algal colonies were washed out of the sediment and their phosphorus content and dry weight were determined. The dry weight of the sediment was also determined by drying at 105 °C overnight.

Results

Phosphorus in the epilimnion

The concentrations of total phosphorus, phosphate, particulate phosphorus as well as particulate phosphorus in the particle fraction > 200 μm (*Gloeotrichia*) during the summer of 1991 are shown in Fig. 1a and b. There was a gradual increase of total phosphorus in the epilimnion, which accelerated in early August, when the number of *Gloeotrichia* colonies in the epilimnion increased dramatically. The increase in epilimnion phosphorus due to particles > 200 μm was about 5 μg P l^{-1}, as is shown in Fig. 1b. At the same time the contribution of *Gloeotrichia* colonies to particulate phosphorus in the epilimnion increased from 1.1% to 53%. If the particle fraction 70–200 μm was included (mainly small *Gloeotrichia* colonies) the increase was somewhat larger. The phosphate concentration in the epilimnion was very low all summer in 1991, with the exception of a peak in mid July. A small increase (less than 2 μg P l^{-1}) occurred at the same time as the *Gloeotrichia* increase in early August.

Gloeotrichia in the sediments

The number of *Gloeotrichia* colonies found in the sediment along the major transect in March 1991 is shown in Fig. 2. The maximum number of colonies found was almost 800 000 m^{-2}, while the average was around 500 000 colonies m^{-2}. Over 10 m water depth the number of colonies in the sediments decreased to half of the average number, otherwise the concentration was fairly evenly distributed. A comparison of the number of colonies (g sediment dw)$^{-1}$ on the three sampling occasions is shown in Fig. 3. The average number of colonies decreased to one third from late

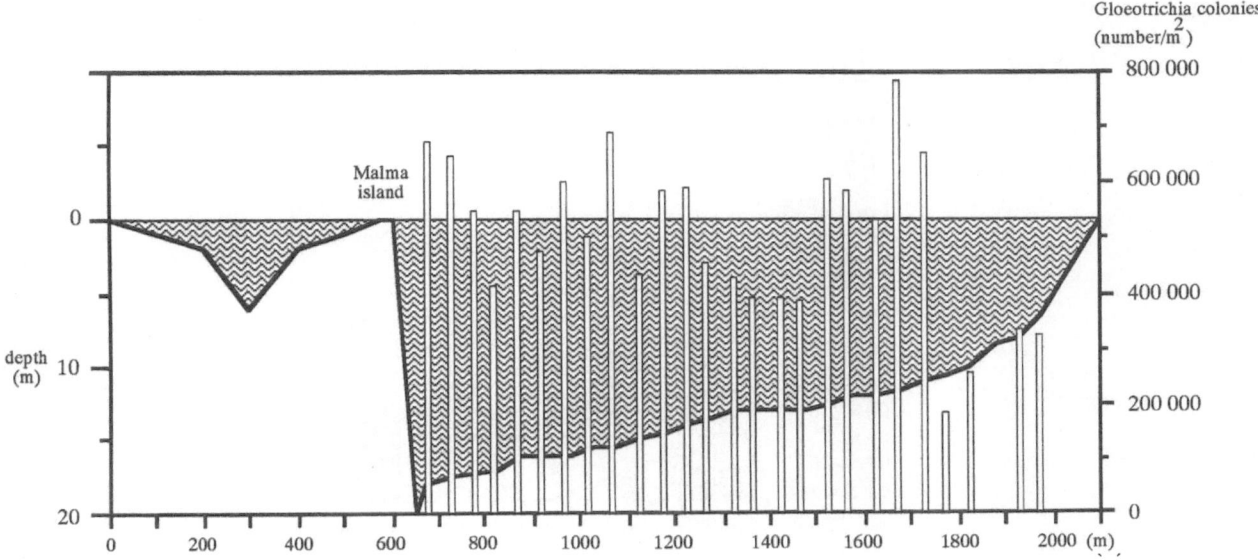

Fig. 2. The number of *Gloeotrichia* colonies along a transect (north-east) from the Malma island 700 m offshore in Lake Erken. The depth profile is also shown.

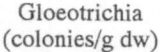

Gloeotrichia
(colonies/g dw)

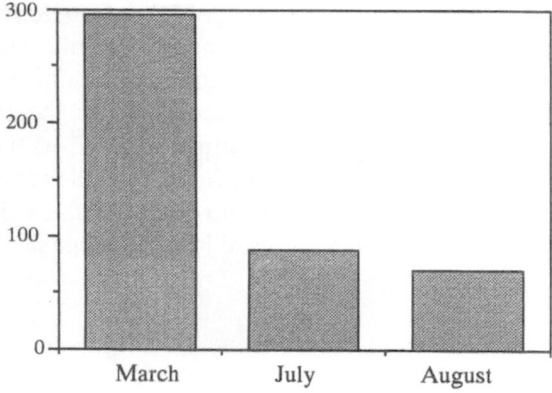

Fig. 3. The average number of *Gloeotrichia* colonies in the sediment on the three sampling occasions, 23 March (all samples), 24 July (five stations) and 3 August (five stations) 1991.

March to late July and the decrease continued until 3 August. A significant proportion of the benthic population left the sediment. As can be seen in Fig. 4, the summer samplings showed a decrease in colony numbers with decreasing water depth. The colonies in shallower areas were first affected by turbulence and warm water as the thermocline was depressed.

The average phosphorus content in the *Gloeotrichia* colonies was determined at 49 ng P colony^{-1} in the March sampling. In the earlier study during the autumn of 1989 the P content varied from 17 to 81 ng colony^{-1} with a tendency to lower values at the deepest sites. The phosphorus content on these locations could not be verified in March 1991. The surplus phosphorus content of the benthic colonies was 45% of total phosphorus on average with a range from 27% to 68% (autumn 1989). The average total phosphorus content was 5.2 μg P (mg dw)$^{-1}$ with a maximum value of 8.5 μg P (mg dw)$^{-1}$.

During the summer of 1991 the colonies appearing in the epilimnion were harvested and analyzed for phosphorus and carbon content. The C/P ratios (by weight) in July and August are shown in Fig. 5. There was a gradual increase from the middle of July until the end of the month and in early August the ratio stabilized around 150.

Discussion

The results of this study confirm our previous indications (Pettersson *et al.*, 1990; Istvánovics

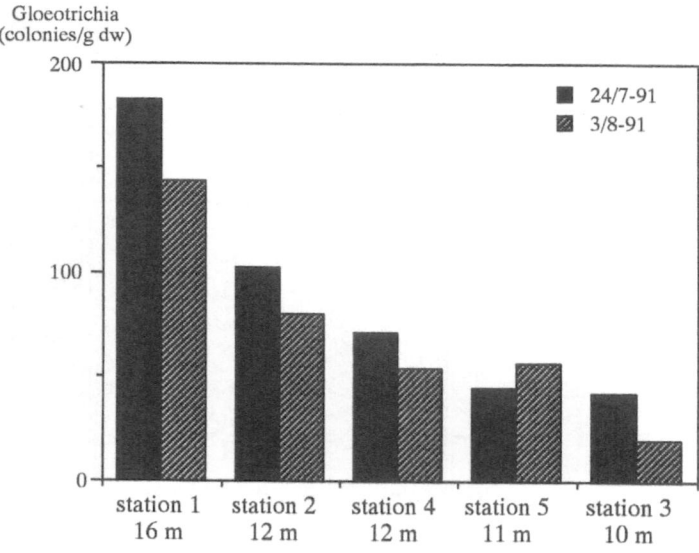

Fig. 4. The number of *Gloeotrichia* colonies at the five sampling stations used 24 July and 3 August 1991. The water depth at each station is given.

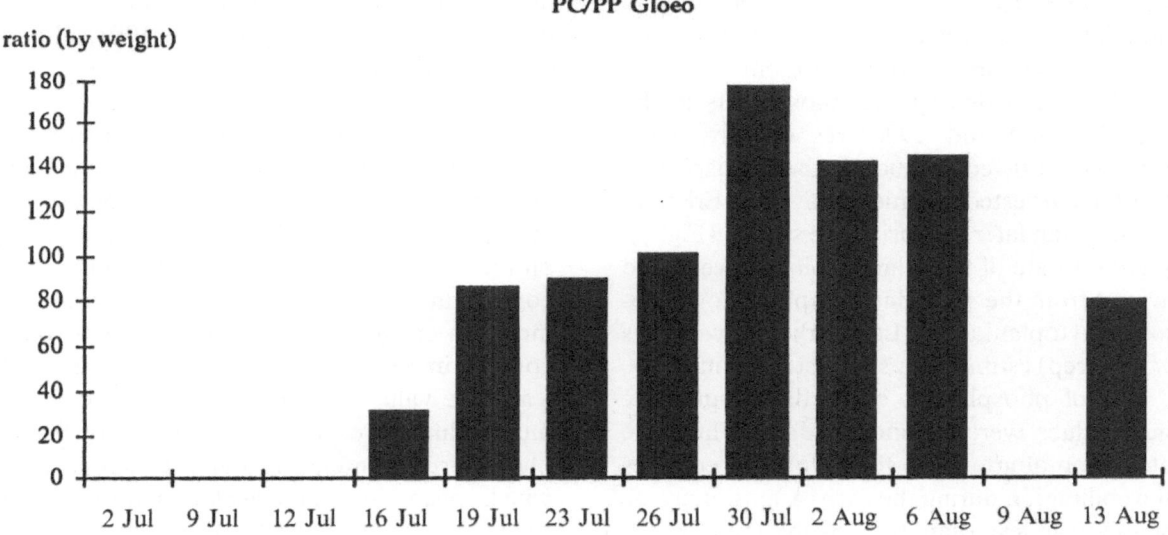

PC/PP Gloeo

Fig. 5. The C:P ratio in the *Gloeotrichia* fraction of the particulate material in the epilimnion of Lake Erken during July and August 1991.

et al., 1990; Istvanovics *et al.*, submitted; Pierson, 1990), that the planktonic population of *Gloeotrichia echinulata* is derived to a large extent from a benthic stock. From Fig. 6 (redrawn from Ulén, 1971) a peak P content of 12 μg P (mg dw)$^{-1}$ can be used to estimate the number of colonies responsible for the epilimnetic increase of particulate phosphorus in early August. An average dry weight of 9.5 μg colony^{-1} gives a P

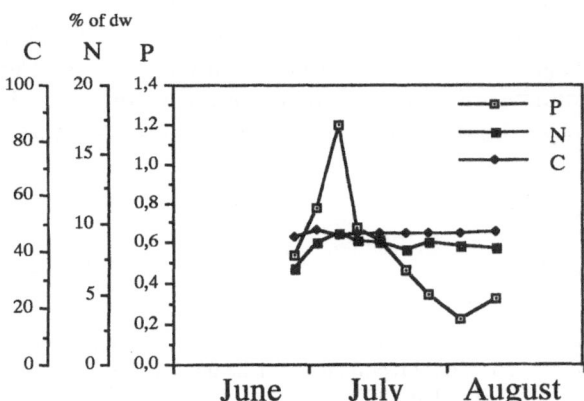

Fig. 6. The phosphorus, nitrogen and carbon content of *Gloeotrichia echinulata* in Lake Erken during the summer of 1969 (redrawn from Ulén, 1971).

content of 114 ng P colony^{-1}. Thus 5 μg P l^{-1} corresponds with 44 *Gloeotrichia* colonies l^{-1}. Thus, in the epilimnion (0–8 m) a number of 350 000 colonies m^{-2} were to be found in the middle of August 1991. A comparison to the decrease in number of colonies in the sediment is of course of interest. According to Fig. 3 about 70% of the colonies had left the sediments in August, which means 350 000 colonies m^{-2} (0–4 cm sediment depth). This is probably an underestimation, since the sampling technique (Ekman dredge) in March lead to dilution of the surface sediments with lake water. The sediment layer interacting with the overlying water could also have been deeper than 4 cm. Details about the number of colonies in the water, their size distribution and phosporus content will be given elsewhere (Pettersson *et al.*, in prep.). Anyhow, it can be stated that migrating colonies constitute a very significant part of the epilimnetic population. Istvánovics *et al.*(1990) showed that the share in P uptake of *Gloeotrichia echinulata* in Lake Erken was disproportionately low, if any, and concluded that growth was mainly or solely based on internal stores. Barbiero & Welch (1992) came to the same conclusion concerning *Gloeotrichia* in Green

Lake, Seattle USA. They could also show large accumulations of polyphosphates in migrating colonies in comparison with planktonic colonies and 1.8 and 2.8 times more phosphorus in the former in 1989 and 1990, respectively. Ulén (1971) also showed a much higher phosphorus content in harvested colonies from Lake Erken in early July than later on during the season (Fig. 6). The growth rate of these large colonies were also slow and from the 1989 data for phosphorus uptake by phytoplankton in Lake Erken Istvanovics *et al.* (in prep) estimated a shortest doubling time, in terms of phosphorus, of 12 days, but more typical values were around 30 days. Thus, the contribution of growth to the number of colonies in the epilimnion during the early August increase probably was less than 10%, if the initial phosphorus contribution is estimated at 0.5 μg P l^{-1} (Fig. 1). Barbiero and Welch (1992) also stated that the planktonic colonies were smaller in size and in Lake Erken this type of colonies appeared in August. They were harvested in the size interval 70–200 μm and thus did not contribute to the phosphorus increase in the fraction > 200 μm. In 1991 it was also demonstrated that phosphate uptake characteristics of *Gloeotrichia echinulata* were adapted to the phosphorus status of the sediment pore water (Istvánovics *et al.* submitted), which is a strong indication of a dominant input of colonies from the sediments to the epilimnion. From the thermistor recordings it was also possible to see that the major *Gloeotrichia* migration coincided with a depression of the thermocline from around 6 m to 10 m during the period 8 August to 12 August.

On the basis of the information above, an internal loading of 40 mg P m^{-2} could be estimated, which means 2.5 mg P m^{-2} d^{-1} for the main migration period (26 July to 11 August). In Green Lake, Barbiero & Welch (1992) estimated a rate of 2.25 mg P m^{-2} d^{-1} during August and early September 1989, while it was less, 0.24 mg P m^{-2} d^{-1}, in 1990. They also stated that a substantial part of the internal loading in Green Lake was probably explained by P translocation due to benthic recruitment of *Gloeotrichia* colonies. Our results indicate an even stronger significance of

the *Gloeotrichia* migration for the internal loading in the stratified Lake Erken in 1991, although earlier results (Pettersson *et al.*, 1990; Istvánovics *et al.*, accepted) showed summers with important inputs of phosphate from the hypolimnion.

The phosphorus pool that the *Gloeotrichia* colonies represented in the sediments could be estimated at 15 μg P (g dw)$^{-1}$. This estimate is based on the number of colonies and their phosphorus content in March 1991. The phosphorus accumulation calculated from the larger phosphorus content in the early epilimnion colonies would lead to a value of 35 μg P (g dw)$^{-1}$ as a maximum. This corresponds to only 3% of the total phosphorus content in Lake Erken sediment, which means that the translocation of phosphorus cannot be detected by sediment phosphorus analyses, but only by counting of colonies in the sediment in combination with registration of the migration and the phosphorus increase in the epilimnion. Gächter & Meyer (1993) presented the phosphorus content (mg P m^{-2}) of benthic bacteria in various lakes. A minimum value of 200 mg P m^{-2} was given for Mirror Lake (from Jordan *et al.*, 1985). The contribution of *Gloeotrichia* colonies to the phosphorus content in the sediments of Lake Erken was minor. From the March sampling an average of 25 mg P m^{-2} was calculated, but this was, as mentioned above an underestimation and a more probable figure would be around 100 mg P m^{-2} after phosphorus accumulation before migration. A more detailed evaluation of the information about planktonic colonies, migration and sedimentation gathered during the summer of 1991 will further improve our knowledge about the life strategy of *Gloeotrichia echinulata*.

References

Barbiero, R. P. & E. B. Welch, 1992. Contribution of benthic blue-green algal recruitment to lake populations and phosphorus. Freshwat. Biol. 27: 249–260.

Boström, B., A-K. Pettersson & I. Ahlgren, 1989. Seasonal dynamics of a cyanobacteria-dominated microbial community in surface sediments of a shallow, eutrophic lake. Aquat. Sci. 51: 153–178.

Fitzgerald, C. P. & T. Nelson, 1966. Extractive and enzymatic analyses for limiting or surplus phosphorus in algae. J. Phycol. 2: 32–37.

Goor, A. C. J.,1925. Contribution a la physiologie des Cyanophycees. Sur les pseudo-vacuoles rouges et leur signification. Revue Algologique 2: 19–38.

Gächter, R. & J. S. Meyer, 1993. The role of microorganisms in mobilization and fixation of phosphorus in sediment. In P. C. M. Boers, T. E. Cappenberg & W. van Raaphorst (eds), Proceedings of the Third International Workshop on Phosphorus in Sediments. Developments in Hydrobiology 84. Kluwer Academic Publisher, Dordrecht: 103–121. Reprinted from Hydrobiologia 253.

Istvánovics, V., K. Pettersson & D. Pierson, 1990. Partitioning of phosphate uptake between different size groups of planktonic microorganisms in Lake Erken. Verh. Int. Ver. Limnol. 24: 231–235.

Istvánovics, V., K. Pettersson, D. Pierson & R. Bell, 1992. An evaluation of phosphorus deficiency indicators for summer phytoplankton in Lake Erken. Limnol. Oceanogr.

Istvánovics, V., J. Padisak, K. Pettersson & D. Pierson. in prep. Growth and phosphorus uptake strategies of summer phytoplankton in Lake Erken (Sweden).

Istvánovics, V., K. Pettersson, M. A. Rodrigo, D. Pierson, J. Padisak & W. Colom. submitted. *Gloeotrichia echinulata*, a colonial cyanobacterium with a unique phosphorus uptake and life strategy. Plankton Research.

Jordan, M. J., G. E. Likens & B. J. Peterson, 1985. C. Organic carbon budget. In: G. E. Likens (ed.). An Ecosystem Approach to Aquatic Ecology. Mirror Lake and its Environment. Springer-Verlag. ISBN 0-387-96106-2.

Murphy, J. & J. P. Riley, 1962. A modified single solution method for the determination of phosphate in natural waters. Analyt. chim. Acta 27: 31–36.

Osgood, R, 1988. A hypothesis on the role of *Aphanizomenon* in translocating phosphorus. Hydrobiologia 169: 69–76.

Pettersson, K., 1980. Alkaline phosphatase activity and algal surplus phosphorus as phosphorus-deficiency indicators in Lake Erken. Arch. Hydrobiol. 89: 54–87.

Pettersson, K., 1985. Vattenöversikt Broströmmens vattensystem 1984. Report LIU 1985 B:1. Institute of Limnology, Uppsala university (in Swedish).

Pettersson, K., 1990. The spring development of phytoplankton in Lake Erken: species composition, biomass, primary production and nutrient conditions – a review. In P. Biró & J. F. Talling (eds), Trophic Relationships in Inland Waters. Developments in Hydrobiology 53. Kluwer Academic Publishers, Dordrecht: 9–14. Reprinted from Hydrobiologia 191.

Pettersson, K., V. Istvánovics & D. Pierson, 1990. Effects of vertical mixing on phytoplankton phosphorus supply during summer in Lake Erken. Verh. Int. Ver. Limnol. 24: 236–241.

Pettersson, K., D. Pierson & V. Istvánovics, in prep. The summer blooms of *Gloeotrichia echinulata* in Lake Erken.

Pierson, D., 1990. Effects of vertical mixing on phytoplankton photosynthesis and phosphorus deficiency. Thesis Uppsala university, Department of physical Geography, Division of Hydrology, 38 pp.

Roelofs, T. D. & R. T. Oglesby, 1970. Ecological observations on the planktonic cyanophyte *Gloeotrichia echinulata*. Limnol. Oceanogr. 15: 224–229.

Stewart, W. D. P., M. Pemble & L. Al-Ugaily, 1978. Nitrogen and phosphorus storage and utilization in blue-green algae. Mitt. int. Ver. Limnol. 21: 224–247.

Ulén, B., 1971. Elementarsammansättningen hos sötvattenplankton. Scripta Limnologica Upsaliensia 270, 12 pp. (in Swedish).

Wesenberg-Lund, C., 1904. Studier över de dansk söers plankton. I. Tekst. Copenhagen: Glydendanske Boghandel 223 pp.

Hydrobiologia **253**: 131–140, 1993.
P.C.M. Boers, T.E. Cappenberg & W. van Raaphorst (eds),
Proceedings of the Third International Workshop on Phosphorus in Sediments.
© 1993 *Kluwer Academic Publishers.*

Phosphorus composition and release in sediment bacteria of the genus *Pseudomonas* during aerobic and anaerobic conditions

Tomas Waara[1], Mats Jansson[2] & Kurt Pettersson[1]
[1] *Institute of Limnology, University of Uppsala, Box 557, S-751 22 Uppsala, Sweden;* [2] *Institute of Geography, University of Umeå, S-901 87, Sweden*

Key words: pseudomonas, sequential fractionation, bacterial P composition, bacterial P release

Abstract

A substantial amount of sediment phosphorus can be bound in bacterial biomass. In this study the fractional composition of phosphorus in the bacteria *Pseudomonas* was determined by sequential extraction with ammonium chloride, sodium hydroxide and hydrochloric acid according to the scheme of Hieltjes & Lijklema (1980). Both non-labelled and ^{32}P-labelled bacteria were used for fractionation. Up to 80% of the bacterial phosphorus was found in the NaOH-nRP fraction, which is in agreement with the results of Hupfer & Uhlman (1992) for *Acinetobacter* and activated sludge obtained with the sequential extraction scheme of Psenner *et al.* (1985). A significant correlation was found between bacterial biomass and the amount of phosphorus retained in the NaOH-nRP fraction when sediments were fractionated. Additional experiments with ^{32}P-labelled *Pseudomonas* in sediment–water systems were performed in order to follow bacterial release of phosphorus under aerobic and anaerobic conditions. These studies did not sustain the hypothesis that anaerobic conditions lead to rapid release of phosphorus from bacterial cells.

Introduction

Phosphorus leakage from the sediment-water interface might have a profound effect on the trophic state of a lake. The early work by Einsele (1938) and Mortimer (1941, 1942) laid the foundation for our present knowledge. They very clearly demonstrated that sediments retained phosphorus during oxic conditions but that a leakage occurred when the sediments turned anoxic. This was explained by chemical processes involving interactions between phosphorus and iron compounds. With decreased redox potential the reduction of ferric iron led to dissolution of iron-phosphorus complexes. Microbial activity was acknowledged to be important since decomposi-tion of organic material made the sediments reduced. Recent studies have suggested that microbial activity might have a more direct influence on anaerobic phosphorus release (see review of Boström *et al.*, 1988; Gächter *et al.*, 1988). Microorganisms have been shown to release phosphorus under anaerobic conditions (Fleischer, 1983; 1985; Fleischer *et al.*, 1988). This appears to be an adaptation to environments where fluctuations are common between aerobic and anaerobic conditions (Shapiro, 1967). A physiological explanation for this behaviour has been discussed by Wentzel *et al.* (1986). Under aerobic conditions bacteria seem to store phosphorus as polyphosphates as an energy reserve for later use. The release of phosphate under anaerobic conditions is caused by a shift in energy metabolism of the cells. Stored polyphosphate is transformed to ATP and the use of ATP leads to phosphate release. The subsequent increase of phosphate in the cell leads to a rapid diffusion out of the bacterial cell. The phosphorus mobilization process from bacteria under anaerobic conditions is fast and occurs at a redox potential where ferric iron is reduced (Gächter *et al.*, 1988). Therefore it could be difficult to distinguish between bacterial release and release caused by the dissolution of iron phosphorus complexes. It is not known if or to what extent phosphorus mobilization can occur from sediment living bacteria and thus the ecological significance of the process is unknown. From a theoretical stand-point it can be argued that a sudden phosphate release from bacterial cells can cause a substantial outflow of phosphorus from sediments.

Boström *et al.* (1988) estimated the phosphorus bound in microbial cells, including cyanobacteria, in Lake Vallentunasjön sediments to be 13% of the total sediment phosphorus. Internal phosphorus loading in Lake Vallentunasjön during summer was accompanied by a substantial depletion of the organic phosphorus pool in the sediment and changes in microbial cell-bound phosphorus could in theory explain the observed internal loading in Lake Vallentunasjön (Boström *et al.*, 1985). Fleischer (1986) estimated that the phosphorus content of bacteria in the uppermost 5 cm of the sediment in a eutrophic lake could cause an increase of the phosphorus concentration in an overlying 10 m water column with 30–150 μg l^{-1}. This study is an attempt to test the possibility that direct phosphate release from sediment bacteria can be a source of phosphorus release from reduced sediments. We isolated facultative anaerobic bacteria of the genus *Pseudomonas* from lake sediment and studied the phosphorus flux between bacteria, water and sediments by incubation with $^{32}PO_4$. Sequential extraction, previously successfully applied on bacteria by Hupfer & Uhlman (1992), was used to characterize and compare the phosphorus composition of bacteria and sediments during different incubation conditions.

Material and methods

Sediment from Lake Vallentunasjön was chosen for the experiment because of its well-known phosphorus composition (Petterson *et al.*, 1988) and the observed depletion of organic phosphorus in the sediment during summer internal loading (Boström *et al.*, 1985). Lake Vallentunasjön is a highly eutrophic lake situated 20 km north of Stockholm, Sweden. The lake received domestic sewage until 1970. Internal phosphorus loading dominates the annual phosphorus load. The sediment is organogenic and its total phosphorus content varies between 0.6 and 2.0 mg g^{-1} d.w.. Characteristics of the lake are given in Table 1.

Surface sediment (0–5 cm) from bottoms at about 2 m water depth was collected with an Ekman grab and stored at 4 °C before use. Lake water for the experiments was filtered (Whatman GF/C) and stored at 4 °C. Cultures of *Pseudomonas* were isolated from Lake Vallentunasjön sediment by incubation of sediment aliqouts with autoclaved, boiled rice and later on *Pseudomonas* selective agar plates (OXOID, Pseudomonas agar with C-F-C Supplement). For growing large cultures, nutrient broth (MERCK) was used and the bacteria were grown aerobically for 2 days at 26 °C. Labelled bacteria were obtained by adding ^{32}P (H$_3$ $^{32}PO_4$ in 0.02 N HCl, Amersham) to the nutrient broth. The sequential extraction scheme of Hieltjes & Lijklema (1980) was used to separate sediment as well as bacterial phosphorus into ammonium chloride, sodium hydroxide and hydrochloric acid soluble fractions. The nonreactive phosphorus in the sodium hydroxide fraction (NaOH-nRP) was calculated by sub-

Table 1. Morphometric and chemical data for Lake Vallentunasjön.

Lake surface area	6.2 km^2
Maximum depth	5.0 m
Mean depth	2.7 m
Volume	15.2 × 10^6 m^3
Alkalinity	2 meq l^{-1}
Total P	60–200 μg l^{-1}
Total N	1500–2500 μg l^{-1}

tracting molybdate reactive phosphorus from total phosphorus of that fraction. Residual phosphorus was calculated as a difference between total phosphorus content of sediment or bacteria and the sum of reactive phosphorus in the extracts plus the non-reactive phosphorus in the sodium hydroxide extract. In several experiments fractionation was made on ^{32}P-labelled bacteria and sediment. In those tests the method described by Orret & Karl (1987) was used to separate inorganic phosphorus from organic bound phosphorus by precipitation of inorganic phosphorus with tungstate. Samples were measured in a scintillation counter before and after precipitation. The residual-P fraction was measured as the activity remaining after the sequential extractions. By summing up all fractions we calculated the total-P content. The two methods yielded different residual-P fractions. Generally a higher residual-P fraction was obtained when labelled material was fractionated. Comparative tests between direct measurement of molybdate reactive phosphorus and calculated reactive phosphorus obtained by precipitation were made on bacteria. These showed a good correlation except for the sodium hydroxide fraction. The bacterial NaOH-RP fraction tends to be overestimated by the precipitation method. This could possibly be explained by coprecipitation of polyphosphates. Corrections for this error were made by analyzing the NaOH fraction for both non-radioactive molybdate reactive phosphorus and total phosphorus. The water content of the sediments was determined by drying at 105 °C for 24 hours. Total phosphorus content was analyzed as molybdate-reactive after oxidative acid digestion of dried sediment (Menzel & Cordwin, 1965).

The effect of aerobic and anaerobic conditions on bacterial phosphorus release was tested by incubating ^{32}P-labelled *Pseudomonas* together with sediment and water. The bacteria were grown in nutrient broth, containing ^{32}P (final conc. 200000 cpm ml^{-1}), for two days under aerobic conditions before use. The bacteria were then centrifuged and washed twice with sterile (autoclaved) sediment pore water. A suspension (1 ml) of bacteria and pore water was placed in a dia-

lysis bag after which an aliqout was removed for immediate phosphorus fractionation. Each bag was then sealed and the ^{32}P-activity was measured in a scintillation counter (Cherenkov radiation in water solution) as a measure of total phosphorus. The dialysis bags together with sediment (3.00 g ww) and water (10.0 ml Whatman GF/C filtered lake water) were added to injection vials (25 ml) which were closed and purged with either air or nitrogen for 30 minutes. Replicates of the aerobic and anaerobic samples were used in each experiment. The samples were incubated for 48 hours and were purged with air or nitrogen for 30 minutes each day. Before purging 0.15 mg glucose was added to each sample to avoid the risk of substrate depletion. After 48 hours the ^{32}P activity in the dialysis bag was measured again in a scintillation counter. Thereafter the sediment was phosphorus fractionated and the ^{32}P-activity was measured for each fraction. Dialysis bags, of the same quality as in this investigation, have earlier been used to study the ^{32}PO$_4$ release from labelled iron-phosphate complexes (Jansson, 1987). These studies showed that the appearance of ^{32}P in solution followed the same kinetics, whether the labelled complexes were contained behind a dialysis membrane barrier or not. Therefore, the dialysis membrane does not affect phosphate release from the bacteria inside the bag.

Another experiment was done to see if bacterial phosphorus composition changed at a shift from aerobic to anaerobic conditions. A suspension of *Pseudomonas* and sediment pore water was prepared as before, but without ^{32}P, and 10 ml subsamples were added to centrifugation tubes. Two series were made, one aerobic and

Table 2. Fractional composition of phosphorus in Lake Vallentunasjön test sediment (0–5 cm).

	mg g^{-1} d.w.	(%)
NH$_4$Cl	0.017	2.4
NaOH-RP	0.088	12.3
NaOH-nRP	0.131	18.3
HCl	0.353	49.2
Residual	0.128	17.9
Total P	0.717	

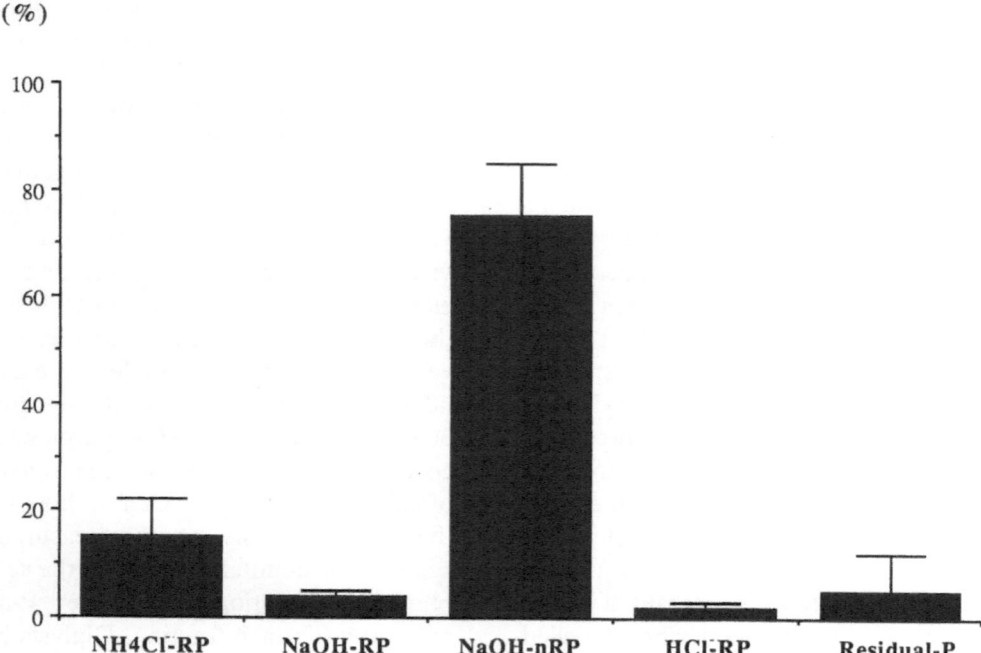

Fig. 1. Fractional composition of phosphorus in *Pseudomonas* (average of eight fractionations).

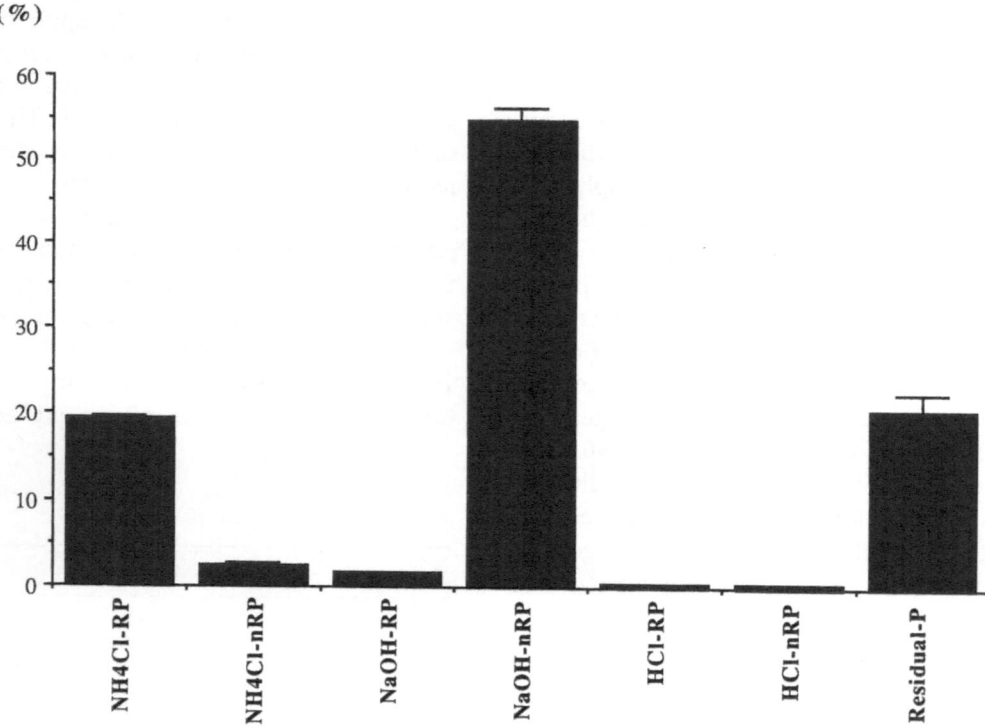

Fig. 2. The phosphorus (^{32}P) composition of *Pseudomonas* at the start of the aerobic/anaerobic dialysis experiment (average of duplicates).

one anaerobic with three parallels in each serie. The tubes were continuously purged with either air or nitrogen for 24 hours. Thereafter the content of each tube was phosphorus fractionated.

To test if the variation in bacterial biomass in the sediment could be detected by phosphorus fractionation of the sediment, a prolonged glucose incubation was performed with the aim of increasing bacterial biomass. Sediment (150 g w.w.) was incubated together with filtered lake water (250 ml, Whatman GF/C) in an Erlenmeyer flask. The flask was put on a shaking board, to maintain aerobic conditions, and incubated in darkness at 25 °C. Every second day 3.75 mg glucose-C and 0.375 mg ammoniumnitrate-N was added to the flask. Four parallel phosphorus fractionations were carried out at the start and after one and two weeks. For bacterial enumeration subsamples of 1 g wet sediment were diluted to 100 ml with a filter-sterilized mixture (1 + 1) of distilled and tap water containing 4% formaldehyde. Enumeration was done by an epifluorescence microscope (Zeiss filters KP 490, Rfl 510, LP 520) after staining with acridine orange (AO) to a final concentration of 0.005% AO and filtration onto 0.2 μm Nuclepore polycarbonate filters that where prestained with Sudan black. Biomass was calculated from size classification during enumeration by assuming that the density of bacteria is 1.00 g cm^{-3}.

Results

The fractional composition of phosphorus for the test sediment from Lake Vallentunasjön is shown in Table 2. Phosphorus fractionation of *Pseudomonas* done with the scheme of Hieltjes & Lijklema (1980) gave a 60 to 80% recovery of phosphorus in the NaOH-nRP fraction (Fig. 1).

The phosphorus composition of ^{32}P-labelled *Pseudomonas* at the start of the aerobic/anaerobic incubation is shown in Fig. 2, where corrections have been made for the error in the precipitation method (see above).

The release of ^{32}P from bacteria to sediment

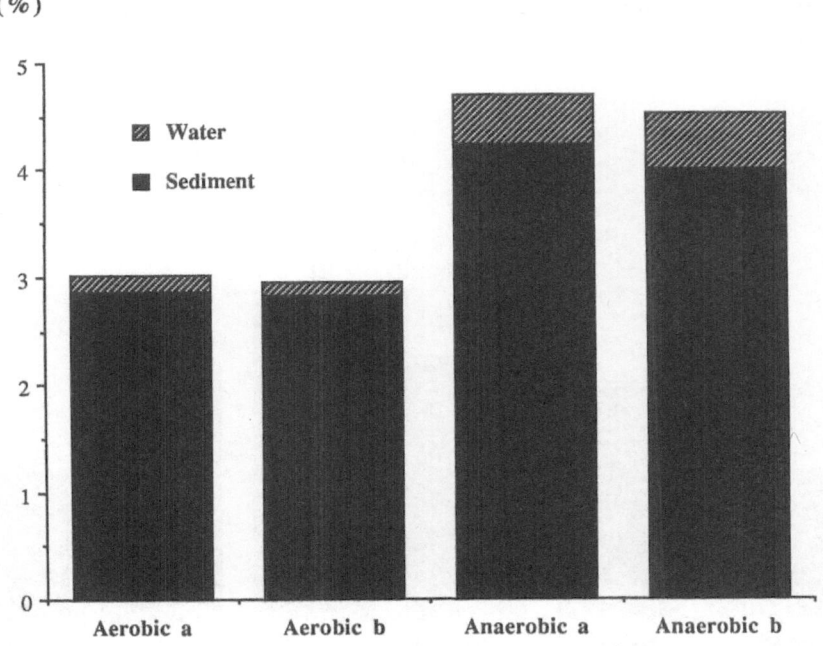

Fig. 3. Amount of phosphorus (in percent of total bacterial phosphorus) released through the dialysis membrane from *Pseudomonas* and its relative distribution between sediment and water under aerobic and anaerobic conditions with duplicates (a and b) in each series.

and water through the dialysis membranes after 48 hours of incubation is shown in Fig. 3. Anaerobic conditions gave a slightly higher bacterial release of ^{32}P. The distribution of ^{32}P into different fractions of the sediment outside the dialysis membrane is shown in Fig. 4. Compared to Table 2 it can be seen that 'new' phosphorus entering the sediment shows a different distribution

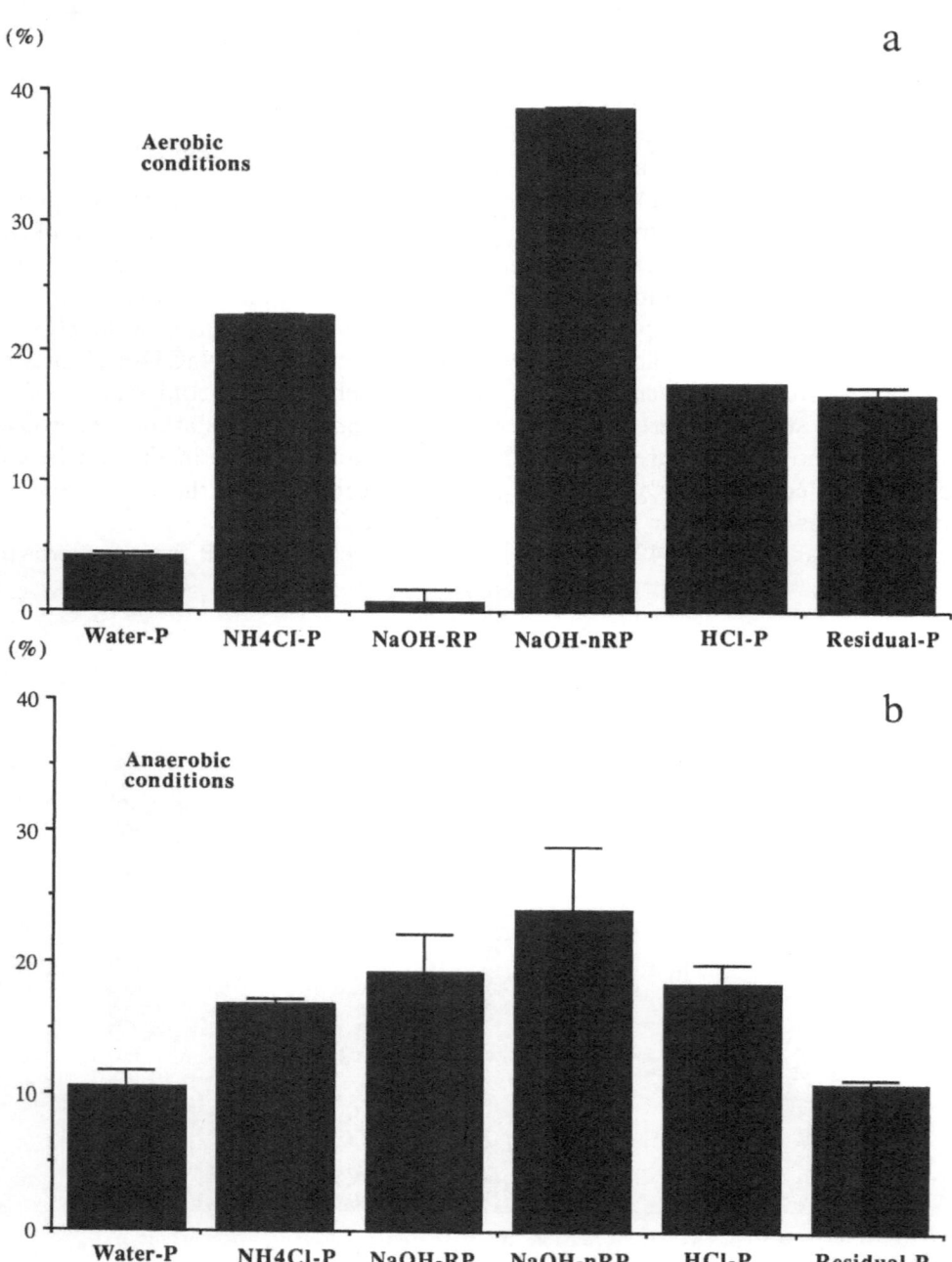

Fig. 4. Distribution of ^{32}P into different phosphorus fractions of the sediment outside the dialysis membrane under a) aerobic conditions and b) anaerobic conditions.

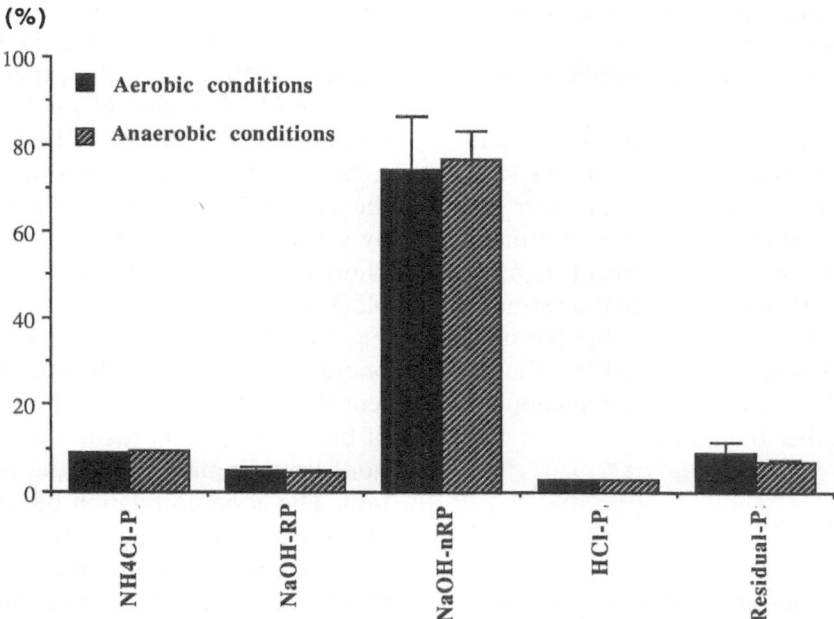

Fig. 5. Phosphorus composition of *Pseudomonas* (average of triplicates) after 24 h incubation in either aerobic (black column) or anaerobic (grey column) conditions.

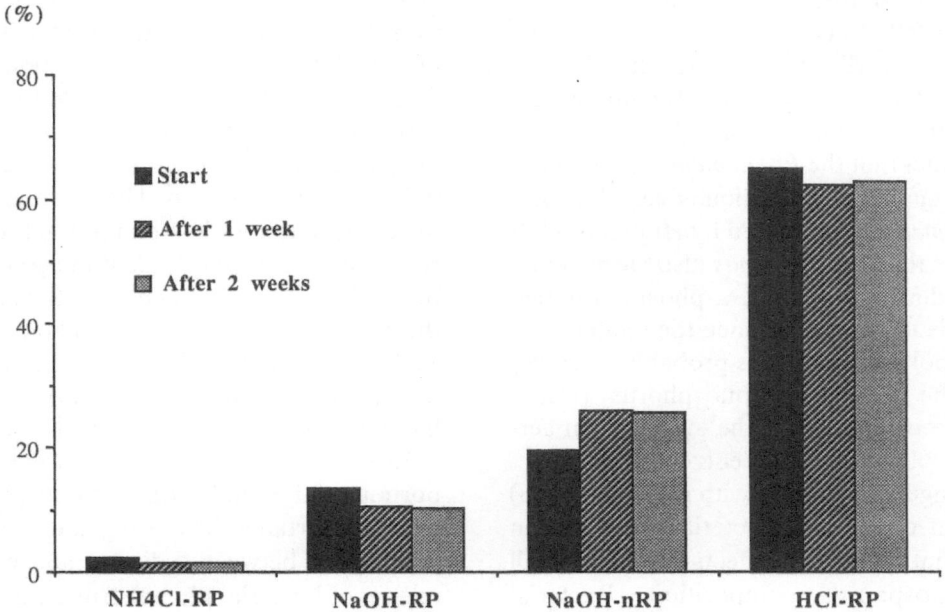

Fig. 6. Fractional composition of the extracted phosphorus from Lake Vallentunasjön sediment during the prolonged glucose experiment.

from the original sediment phosphorus pool. The largest proportion of the ^{32}P was found in the sodium hydroxide fraction. Furthermore, a different composition of released ^{32}P after aerobic and anaerobic incubation is evident.

The phosphorus composition of *Pseudomonas*

after 24 h incubation under either aerobic or anaerobic conditions is shown in Fig. 5. No apparent difference between the two series could be detected.

In the prolonged glucose incubation the bacterial biomass was 0.80 mg g^{-1} d.w. at the start, after one week 5.47 mg g^{-1} d.w. and after two weeks 5.52 mg g^{-1} d.w., i.e. the biomass multiplied seven times during the experiment. Figure 6 shows the fractional composition of the extracted phosphorus from the sediment during the incubation. At the end of the experiment the NaOH-nRP fraction had increased but a depletion occurred in all the other fractions.

Discussion

The pre-experimental treatment of the test bacteria in this investigation included growth in the presence of oxygen, energy and phosphorus. It is obvious that the dominating part of the bacterial phosphorus under these conditions is found in the NaOH-nRP fraction (Fig. 1). The same observation has earlier been made with *Acinetobacter* by Hupfer & Uhlman (1992) who also reported that this fraction consisted of polyphosphates. The fact that the *Pseudomonas* used in our study have a similar composition of cell phosphorus to the *Acinetobacter* studied by Hupfer & Uhlman suggests that *Pseudomonas* also stores polyphosphates during growth in a phosphorus-rich medium. This is important since the tendency to accumulate polyphosphates is probably a prerequisite for the substantial phosphorus release which has been reported at the switch from aerobic to anaerobic growth (Wentzel, 1986).

The prolonged incubation with glucose (Fig. 6) indicates that changes in bacterial biomass can have small but significant effects on the overall sediment phosphorus composition. Bacterial phosphorus can be calculated by using the biomass:C relation of $2.2 \cdot 10^{13}$ g C μm^{-3} (Bratbak & Dundas, 1984) and assuming a phosphorus to carbon ratio of 1:18 (Fenchel & Blackburn, 1979). With these assumptions the amount of phosphorus bound in bacteria at the start of the prolonged

glucose experiment was 9.6 μg P g^{-1} d.w. which corresponds to 1.3% of the total sediment phosphorus. At the end of the experiment the bacterial contribution was 67 μg P g^{-1} d.w. which accounts for 8% of the total phosphorus concentration of the sediment. Another way to estimate the contribution from bacterial phosphorus is by assuming that the increase in bacterial phosphorus is proportional to the increase in the NaOH-nRP fraction during the experiment (cf Fig. 6). The measured increase of bacterial phosphorus is then 6–8% of the total phosphorus content if calculated (assuming a 60–80% recovery of bacterial phosphorus in the NaOH-nRP fraction) from the increase in the NaOH-nRP fraction. The good correlation between the methods also suggests that the NaOH-nRP fraction can be a useful instrument for studies of bacterial phosphorus turnover in sediments. Furthermore, since two independent ways to estimate the changes in the bacterial phosphorus concentration gave almost identical results, it seems possible to conclude that the bacterial phosphorus increases from 1 to about 8% of the total sediment phosphorus during the glucose incubation experiment. The bacterial biomass at the end of the experiment (5.5 mg g^{-1} d.w.) was of the same order of magnitude as the annual mean bacterial biomass in Lake Vallentunasjön. Therefore, it is likely that the contribution of bacteria to the total phosphorus content in Lake Vallentunasjön is normally below 10%. Boström et al. (1989) showed that in the course of a year the bacterial biomass in Lake Vallentunasjön varied between 2 and 10 g m^{-2} in the upper 2 cm layer of the sediment. The lowest biomass values were found in the summer.

Even if bacterial phosphorus is a minor proportion of the total sediment phosphorus it may be of importance to the regulation of the phosphorus flux between sediment and water. A high release of bacterial phosphorus could have a pronounced effect since the sediment phosphorus pool is very large compared to the amount of phosphorus in the overlying water (Håkansson & Jansson, 1983). For example, if half of the bacterial phosphorus at a biomass of 10 g m^{-2} in the 0–2 cm surface sediment layer of Lake Vallentunasjön was released to the lake water, the mean phosphorus concentration in the lake would increase by about 25 μg l^{-1}. However, the experiments where labelled bacteria were enclosed in a dialysis bag (Fig. 3) indicate that shifts from aerobic to anaerobic conditions may not cause any dramatic increase in the flux of phosphorus from bacteria. Release of labelled phosphorus under aerobic conditions equalized 3% of the bacterial phosphorus content while the corresponding figure for anaerobic release was 4.5%. Since the membrane has a nominal cut-off of 10^4 Daltons this release has to include dissolved phosphorus. It is not possible to say if the bacterial release was different under anaerobic and aerobic conditions, but it is likely that the anaerobic release is different since anaerobic conditions should decrease the metabolism of the test bacteria. It is possible that the aerobic release reflects the phosphate excretion from growing bacteria (Jansson, 1987; Lean, 1973), while the anaerobic release to a large extent is caused by autolysis or the release of phosphate due to *e.g.* changed energy metabolism (Wentzel, 1986). In any case, the release of phosphorus from *Pseudomonas* at anaerobic conditions was not much higher than the release during aerobic conditions. The net release due to anaerobic conditions was that only about 1% of the bacterial phosphorus was transformed to dissolved phosphorus. This conclusion is also supported by the 24 h incubation experiment (Fig. 5) which demonstrates that the fractional composition of the bacterial phosphorus is more or less identical after incubation at aerobic or anaerobic conditions. The moderate release from *Pseudomonas* at the shift from aerobic to anaerobic conditions observed in this investigation can not therefore cause a substantial outflow of phosphorus from sediment to lake water. If sediment bacteria at *in situ* conditions behaved as *Pseudomonas* in our experiment a net release of approximately 1% of the bacterial phosphorus content (at a biomass of 10 g m^{-2}) at anaerobic conditions (Fig. 3) would increase the phosphorus concentration in a 2.7 m water column (mean depth of Lake Vallentunasjön) by only 0.4 μg l^{-1}. Our experiments were run for only two days and it is possible that longer incubation times could give higher values of bacterial phosphorus release. However, the rapid liberation of biomass phosphorus earlier reported from studies with active sludge (Shapiro, 1967; Wentzel *et al.*, 1986) and with the strict aerobic bacterium *Acinetobacter* (Fleischer *et al.*, 1988) was not found in this investigation. This study does not sustain the hypothesis that rapid release from bacterial cells causes a large mobilization of phosphate in lake sediments. On the other hand, our results are not enough to reject the hypothesis. We have studied facultative anaerobic bacteria of the genus *Pseudomonas* and it is possible that other sediment bacteria will behave differently. This investigation does show that considerable amounts of phosphorus can be bound and cycled in sediment living bacteria. Shifts in bacterial biomass in combination with the abiotic sorption characteristics of the sediments may therefore play an essential role in the phosphorus dynamics of lake sediments.

Acknowledgements

This study was supported by the National Swedish Environment Protection Board.

References

Boström, B., I. Ahlgren & R. Bell, 1985. Internal nutrient loading in a shallow eutrophic lake, reflected in seasonal variations of some sediment parameters. Verh. int. Ver. Limnol. 22: 3335–3339.

Boström, B., J. M. Andersen, S. Fleischer & M. Jansson, 1988. Exchange of phosphorus across the sediment–water interface. In G. Persson & M. Jansson (eds), Phosphorus in Freshwater Ecosystems. Developments in Hydrobiology 48. Kluwer Academic Publishers, Dordrecht: 229–244. Reprinted from Hydrobiologia 170.

Boström, B., A-K. Pettersson & I. Ahlgren, 1989. Seasonal dynamics of a cyanobacteria-dominated microbial community in surface sediments of a shallow, eutrophic lake. Aquat. Sci. 51: 153–178.

Bratbak, G. & I. Dundas, 1984. Bacterial dry matter content and biomass estimations. Appl. envir. Microbiol. 48: 755–757.

Einsele, W., 1938. Über chemische und kolloidchemische Vorgänge in Eisen-Phosphat-Systemen unter limnochemis-

140

chen und limnogeochemischen Gesichtspunkten. Arch. Hydrobiol. 33: 361 –387.

Fenchel, T. & T. H. Blackburn, 1979. Bacteria and Mineral Cycling. Academic Press, London.

Fleischer, S., 1983. Microbial phosphorus release during enhanced glycolysis. Naturwissenschaften 70: 415–416.

Fleischer, S., 1985. Microbial mediation of phosphorus exchange at the sediment–water interface. In M. Enell, W. Graneli & L. Hansson (eds), 13th Nordic Symposium on Sediments. pp. 9–16.

Fleischer, S., 1986. Aerobic uptake of Fe(III)-precipitated phosphorus by microorganisms. Arch. Hydrobiol. 107: 269–277.

Fleischer, S., M. Bengtsson & G. Johansson, 1988. Mechanism of the aerobic Fe(III)-P solubilization at the sediment water interface. Verh. int. Ver. Limnol. 23: 1825–1829.

Gächter, R., J. S. Meyer & A. Mares, 1988. Contribution of bacteria to release and fixation of phosphorus in lake sediments. Limnol. Oceanogr. 33: 1542–1558.

Hieltjes, A. H. M. & L. Lijklema, 1980. Fractionation of inorganic phosphates in calcareous sediments. J. envir. Qual. 9: 405–407.

Hupfer, M. & D. Uhlmann, 1992. Phosphate immobilization by microorganisms in lake sediments. Hydrobiologia (in press).

Håkansson, L. & M. Jansson, 1983. Principles of lake sedimentology. Springer, Berlin.

Jansson, M., 1987. Anaerobic dissolution of iron-phosphorus complexes in sediment due to the activity of nitrate-reducing bacteria. Microb. Ecol. 14: 81–89.

Lean, D. R. S., 1973. Phosphorus dynamics in lake water. Science 179: 678–680.

Menzel, D. H. & N. Cordwin, 1965. The measurement of total phosphorus in seawater based on the liberation of organically bound fractions by persulfate oxidation. Limnol. Oceanogr. 10: 280–283.

Mortimer, C. H., 1941. The exchange of dissolved substances between mud and water in lakes. I. J. Ecol. 29: 280–329.

Mortimer, C. H., 1942. The exchange of dissolved substances between mud and water in lakes. II. J. Ecol. 30: 147–201.

Orret, K. & D. M. Karl, 1987. Dissolved organic phosphorus production in surface seawaters. Limnol. Oceanogr. 32: 383–395.

Psenner, R., P. Pucsko & M. Sager, 1985. Die Fraktionierung organischer und anorganischer Phosphorverbindungen von Sedimenten. Arch. Hydrobiol. Suppl. 70: 111–155.

Pettersson, K., B. Boström & O-S. Jacobsen, 1988. Phosphorus in sediment – speciation and analysis. In G. Persson & M. Jansson (eds), Phosphorus in Freshwater Ecosystems. Developments in Hydrobiolgy 48. Kluwer Academic Publishers, Dordrecht: 91–101. Reprinted from Hydrobiologia 170.

Shapiro, J., 1967. Induced rapid release and uptake of phosphate by microorganisms. Science 155: 1269–1271.

Wentzel, M. C., L. H. Lötter, R. E. Loewenthal & G. v. R. Marais, 1986. Metabolic behaviour of *Acinetobacter* spp. in enhanced biological phosphorus removal – a biochemical model. Water S. A. 12: 209–223.

Hydrobiologia **253**: 141–150, 1993.
P.C.M. Boers, T.E. Cappenberg & W. van Raaphorst (eds),
Proceedings of the Third International Workshop on Phosphorus in Sediments.
© 1993 *Kluwer Academic Publishers.*

The relative importance of biological and chemical processes in the release of phosphorus from a highly organic sediment*

Chantal de Montigny & Yves T. Prairie
Université du Québec à Montréal, Département des Sciences Biologiques, C.P. 8888 Succ. A, Montréal (Québec), H3C 3P8, Canada

Key words: sediment, phosphorus, release, anaerobiosis, bacteria, pH, dystrophic lake

Abstract

Bacteria can play an important role in the process of anaerobic phosphorus release: they can act as a direct source of orthophosphates, or as a catalyst of iron hydroxyde reduction. We studied their influence on phosphorus release from highly organic sediments of a Canadian shield lake. Phosphorus and iron release were measured under aerobic and anaerobic conditions, with or without sterilization, and at different pH. We measured also the abundance and activity of bacteria in sediments. The increased P release after sterilization can be explained by cell lysis. Compared to sterilization, changing oxygen concentrations or acidification had little or no effect on P release. In these sediments, phosphorus and iron movements were independent. Most of the total dissolved iron seemed to be linked to humic acids, but not phosphorus.

Résumé

Les bactéries peuvent jouer un rôle considérable dans le relargage anaérobiques du phosphore, soit en catalysant la réduction des hydroxydes de fer, soit comme source directe d'orthophosphates. On a étudié leur importance dans la libération de phosphore à partir des sédiments d'un lac du bouclier canadien, très riches en matière organique. Le relargage de phosphore et de fer a été mesuré en condition aérobique et anaérobique, avec ou sans stérilisation, et à différents pH. On a également mesuré l'évolution du nombre et de l'activité des bactéries dans les sédiments. Le relargage de P après stérilisation peut être expliqué par la lyse cellulaire. Comparativement à la stérilisation, les changements dans la concentration en oxygène ou l'acidification ont peu ou pas d'effet sur le relargage du P. Les mouvements du fer et du phosphore se produisent indépendamment à partir de ces sédiments. La plus grande partie du fer total dissous semble lié aux acides humiques, ce qui n'est pas le cas du phosphore.

Introduction

The studies of Einsele (1936, 1938) and Mortimer (1941, 1942) demonstrated that high quantities of

phosphorus (P) can be released from sediments under anaerobic conditions. The classical Einsele-Mortimer model explains this fact by the reduction of ferric iron (Fe(III)) at the sediment surface when the redox potential drops below 200 mV. Thus ferrous iron (Fe(II)) and P return to solution by a purely chemical process. This

---

* A contribution to the GRIL (Groupe de Recherche Interuniversitaire de Limnologie)

model has been very successful in limnology because of its simplicity and elegance. Unfortunately, it created the general impression, for over forty years, that P release from sediments is dependent only on the behavior of iron (Fe) (Boström et al., 1988).

Evidence is mounting that the reduction of Fe(III) to Fe(II) is in fact of microbiological origin. Some anaerobic bacteria (in particular denitrifying bacteria) can use Fe(III) as electron acceptors when nitrates (NO_3) are absent (Sørensen, 1982; Jones et al., 1983; Jansson, 1987). Other bacteria can also solubilize and assimilate Fe-bound P in aerobic conditions, and then release it during anaerobic periods (Fleisher, 1986; Fleisher et al., 1988). Another possible mechanism is the chemical reduction of Fe(III) by sulfides (H_2S), after microbiological reduction of SO_4. Sulfate-reducing bacteria also generate alkalinity during this process, and hydroxyde ions can liberate bounded phosphorus by ligand exchange. However, SO_4 reduction requires a significantly lower redox potential (60–100 mV) than Fe(III) reduction (Boström, et al., 1982).

It is also possible that a large quantity of anaerobically released P originates directly from benthic microorganisms rather than Fe hydroxydes. It is now well known that different species of microorganisms (bacteria or cyanobacteria) can take up and release excess P under aerobic and anaerobic conditions, respectively (Shapiro, 1967; Fleisher, 1978; Fleisher, 1983; Ohtake et al., 1984; Gächter et al., 1988). Phosphorus is stored by bacteria either when their growth is limited by elements other than P or organic carbon (luxury uptake), or when P starved bacteria come in contact with this element (phosphate overplus phenomenon) (Fuhs & Chen, 1975). Phosphorus stored as polyphosphates (poly-P) in the presence of oxygen (O_2) is released in the cytoplasm as orthophosphates when conditions become anaerobic. These orthophosphates then diffuse out of the cells because of the increase in osmotic pressure (Malnou et al., 1983). This phenomenon occurs at the same redox potential as the chemical reduction of Fe (Gächter et al., 1988). Orthophosphates are also released during cell lysis of strictly aerobic bacteria killed by the absence of oxygen (Boström et al., 1988). Cell lysis occurs when cell metabolism is strongly perturbated. For example, the sudden elimination of O_2 in well-oxygenated cultures of Bacillus subtilis during the logarithmic phase caused cell lysis (Stolp & Starr, 1965).

It seems that a single mechanism is insufficient to explain the anaerobic release of P from sediments for all cases. More likely is that different mechanisms predominate in sediments of different composition.

In this study, we tested 1) whether P and Fe were released simultaneously during anaerobiosis, 2) whether sterilization and cell lysis can account for the observed release of P and 3) whether the stimulation of denitrifying bacteria through the addition of NO_3 can modulate P release. In addition, we examined the influence of pH on release of P from the sediment. Lastly, we investigated the possible role of SO_4-reducing bacteria in the process of P release.

Materials and methods

The release of Fe and P from sediments was studied in laboratory batch experiments. The sediments were obtained from the fully oxygenated littoral zone of the small dystrophic Lake Cromwell, a Canadian shield lake located about 80 km north of Montréal. More information on morphometric, physical and chemical features of this lake are provided in Lafond et al. (1990). The first 30 cm of sediments were sampled. Samples were taken at four different periods (28 May, 25 June, 09 July and 23 July). Laboratory flux chambers were constructed from plexiglass cylinders of 60 cm in diameter and 20 cm in height, for a total capacity of about 42 litres (Fig. 1). About 8 litres of sediments were added to each chamber and mixed (by the use of agitation system showed in Fig. 1) with about 32 litres of unfiltered lake water. Due to the low density of these sediments, agitation provided immediate and complete mixing of the whole system. The water contained 22 μg P $l^{-1} \pm 1$ of total phosphorus (TP), 10 μg P $l^{-1} \pm 1$

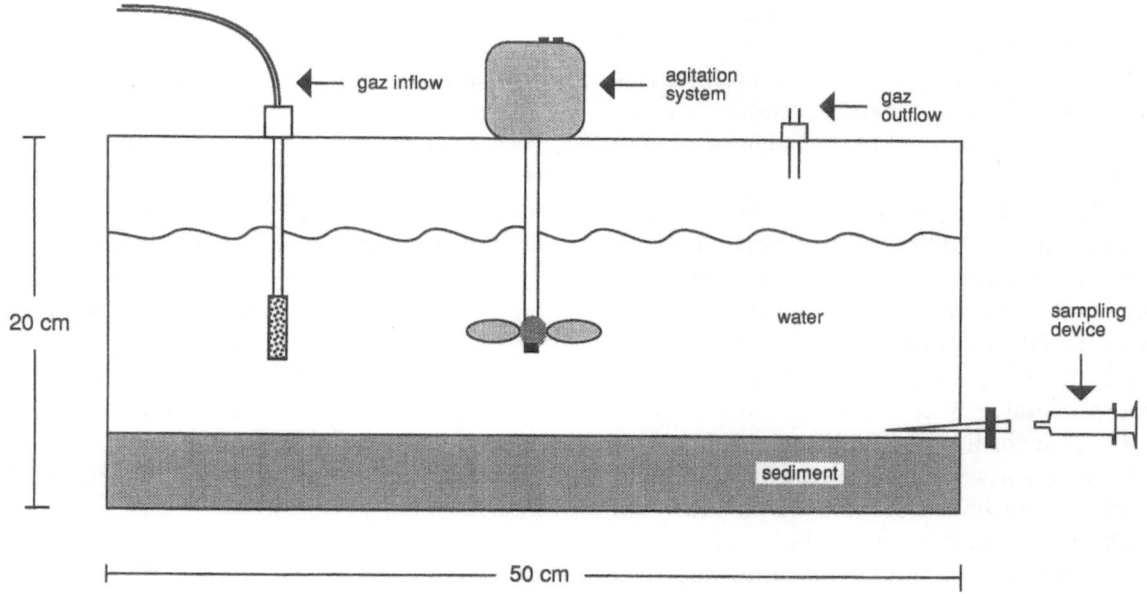

Fig. 1. Experimental flux chamber.

of total dissolved phosphorus (TDP), 4 μg N $l^{-1} \pm 2$ of NO_3, 79 μg $l^{-1} \pm 16$ of TDFe. The pH was 6.4 ± 0.1. Sediments and water were completely mixed twice daily during the experiments to maximize water-sediment exchange and to minimize the establishment of an O_2 concentration gradient between water and sediments.

The evolution of the water-sediment systems was studied under different conditions. Each batch of freshly sampled sediments was used in two different treatments. Each treatment was made in triplicate. We studied the effect of anaerobiosis by first bubbling air and then nitrogen. The NO_3 addition experiments were done with 1.0 g of KNO_3 (for a final concentration of 15 mg N l^{-1} of NO_3). Chemical sterilization was completed by the addition of 10 g of $HgCl_2$ per chamber in both aerobic and anaerobic conditions. The final concentration of Hg was approximately 176 mg l^{-1}. The effect of acidification was studied at pH 4.3 by the addition of about 3 ml of concentrated H_2SO_4. Alkalinization to pH 7.4 was reached with about 4 ml of 10 N NaOH.

Each day, 180 ml of water were sampled before agitation (12 hours after the last agitation), with

a disposable seringe attached to a spiget extracting water at the sediment–water interface (Fig. 1). We found that the chemical composition of this water was the same as that sampled in the middle of the chamber. An equal volume of lakewater (sampled the same day) was added to the chambers to maintain the volume constant. The dilution effect is virtually negligible (Sakata, 1985).

Immediately after sampling, the pH was measured electrometrically and dissolved O_2 was determined by the modified method of Winkler (azide) (APHA, 1985). Total dissolved phosphorus (TDP) and total dissolved iron (TDFe) were analyzed after filtration through a 0.45 μm Millipore filter, with an ALPKEM RFA300 autoanalyzer after persulfate digestion, and by the spectrophotometric 1,10-phenanthroline method after hot acidic digestion (APHA, 1985), respectively. Iron II was measured in anaerobic conditions by the same method but without filtration (0.45 μm) or digestion. Suspended particles were removed after color development by passage through a Whatman filter number 2. Sulfide were measured only at the end of the anaerobiosis experiments by the methylene blue method (APHA,

1985). Release rates of the various constituents were calculated as the least-squares slope of concentration as a function of time. Such release rate estimates were calculated before and after treatments and were restricted to the linear portion of the curves.

Sediments were analyzed at the beginning of each experiments. Organic matter was measured as loss on ignition at 550 °C. This method is recommended for organic sediments poor in carbonates (Byers *et al.*, 1978). Water content was estimated by evaporation at 105 °C after 24 hours. Total P was measured after ignition at 550 °C according to Andersen (1976). Dry ground sediments were ashed during one hour, then mixed and boiled (15 min.) with HCl 1 N. Phosphorus was measured in solution after filtration and dilution. Preliminary experiments using sediments from lake Cromwell showed that the recovery of TP was the same by this method as after digestion with perchloric acid. Total Fe was determined on the same extract. Bacterial counts were determined with DAPI (4′6-diamidino-2-phenylindole-dihydrochloride) under epifluorescence microscopy (Porter & Feig, 1980). The DAPI concentration used was 5 μg ml^{-1} (Schallenberg *et al.*, 1989). Bacterial activity in the sediments was estimated from the conversion of INT to INT-formazan, a measure of bacterial respiration (Trevors, 1984).

Two types of statistical analyses were performed. Differences in the concentration of the various constituents before and after treatments and among chambers were assessed using a two-way ANOVA. Differences in release rates (mg g^{-1} d^{-1}) were tested by comparing regression slopes of the concentration of TDP versus time before and after treatment with a *t*-test. Only the linear portions of the curves were used. To obtain the most precise estimate of the before treatment release rate, we used data from a preliminary set of experiments that ran for two full weeks untreated rather than the shorter 4 day periods preceding all treatments. All analyses were performed with SAS (Statistical Analysis System, version 6.03).

Results

The surface sediment of lake Cromwell is highly organic (57.4% ± 1.4) (mean ± 1 std. error) and very poor in both Fe (4.3 ± 0.4 mg Fe g^{-1} dry sediment) and P (1.1 ± 0.05 mg P g^{-1} dry sediment) (see Ostrofsky (1987) for a compilation of 66 North American lakes). The sediments of lake Cromwell contain an average of 95.5 ± 15.5 × 10^6 bacteria ml^{-1} of sediment, with a bacterial activity of 9.4 ± 0.9 mg INT-formazan g^{-1} dry sediment. The water content is 96.0% ± 0.4. These values represent the means of the four sampling over the summer. After mixing with lake water, the initial concentrations of TP, TDFe and TDP in the containers were about 16 μg l^{-1}, 100 μg l^{-1} and 10 μg l^{-1}, respectively. The pH varied between 6.3 and 6.6. The values are nearly identical.

The before treatment release rate obtained from the 2 week-long preliminary experiments was estimated as -0.06 ± 0.02 mg g^{-1} d^{-1} (Table 1). This value indicates a very slight but significant ($p < 0.05$) uptake of P by the sediments. This value is, however, not significantly different from the short-term release rather estimates obtained before the sterilizations, acidification and alkalinization experiments. Only in the non-sterile deoxygenation experiment was the before treatment release rate (0.08 ± 0.02 mg g^{-1} d^{-1}) significantly different ($p < 0.05$) from that obtained in the 2 week-long experiment. In that case, we used the

Table 1. Release rates of TDP before and after treatments.

Treatment	Phosphorus release rate (mg g^{-1} d^{-1})
Before treatment (2 weeks)	-0.06 ± 0.02
Sterilization	$3.26 \pm 0.04^{**}$
Sterilization + deoxygenation	2.43 ± 0.68
Alkalinization	0.19 ± 0.13
Acidification	0.05 ± 0.08
Before deoxygenation	0.08 ± 0.01
Deoxygenation	0.13 ± 0.03
Nitrates addition	-0.12 ± 0.07

The stars indicate the level of significance: $^*p < 0.05$; $^{**}p < 0.01$; $^{***}p < 0.001$.

latter value to compare before and after treatment release rates.

Deoxygenation

After bubbling nitrogen, O_2 level dropped below 1.0 mg l^{-1} (between 0.9 and 0.4 mg l^{-1}). At the same time a weak but significant ($P < 0.0001$) increase in the concentration of TDP (from 8 to 23 μg l^{-1}) and a strong increase in the concentration of TDFe ($P < 0.0001$) was observed (Fig. 2). However, the release rates of P were not significantly different before and after treatment (Table 1). The pH was not affected ($P > 0.30$) and no H_2S was detected. The sharp drop of pH on 7/28 was caused by the necessity to reseal the system with silicone (causing production of acetic acid) after the beginning of the experiment. The pH was restored naturally within two days. In a preliminary experiment where the same systems water-sediment were mixed during 2 weeks without any additional manipulation (sterilization, change of pH, etc), all water chemical concentrations remained stable. These observed changes are therefore not time effects. In one system, agitation failed for technical reasons. This failure did not affect the release of TDP, but the release of TDFe was prevented. After addition of 1.0 g of KNO_3, release of P and Fe ceased (Fig. 3). The concentration of NO_3 increased from 33 ± 2 μg l^{-1} to 2716 μg l^{-1}, and then decreased slowly for one week and stabilized at 1246 ± 52 μg l^{-1}. Bacterial activity dropped strongly after anaerobiosis from 11.2 ± 0.9 mg INT-formazan g^{-1} dry sediment to 3.2 ± 0.3 mg INT-formazan g^{-1} dry sediment, and bacterial counts in sediments dropped also from $95.5 \pm 15.5 \times 10^6$ bacteria ml^{-1} of sediment (2.34×10^9 bacteria g^{-1} dry weight) to $50.95 \pm 12.7 \times 10^6$ bacteria ml^{-1} sediment. Bacterial activity was not increased by the nitrate addition (the drop was also from 11.2 ± 0.9 mg INT-formazan g^{-1} dry sediment to 2.6 ± 0.2 mg INT-formazan g^{-1} dry sediment), suggesting a limitation by another nutrient, perhaps easily degradable organic C.

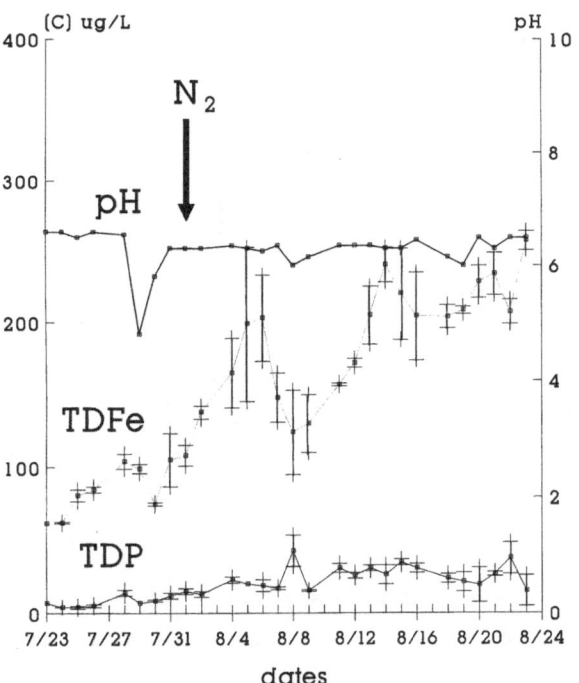

Fig. 2. Temporal evolution of TDP, TDFe, and pH in anaerobic non-sterile condition.

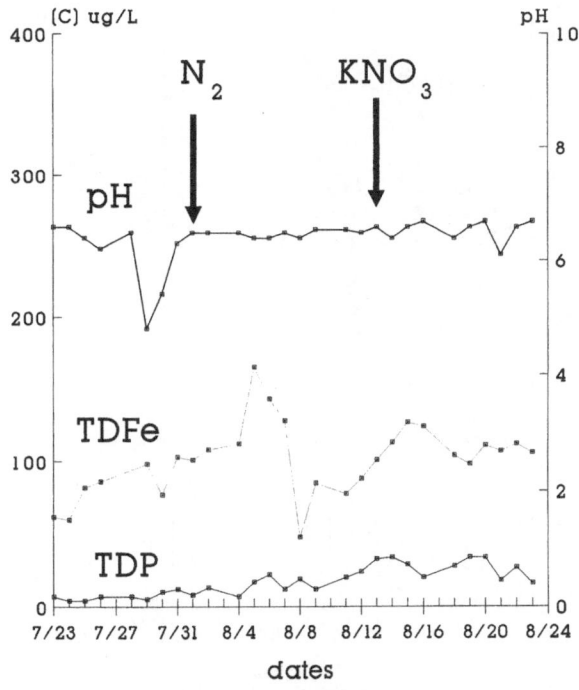

Fig. 3. Temporal evolution of TDP, TDFe, and pH in anaerobic non-sterile condition, without agitation.

146

Sterilization

In these experiments, the establishment of anaerobic conditions was preceded by chemical sterilization with $HgCl_2$ (Fig. 4). After addition of $HgCl_2$, the release rate of TDP increased from -0.06 mg l^{-1} d^{-1} to 2.43 mg l^{-1} d^{-1} (Table 1), albeit not significantly because of the high variability among chambers during release. After 3 days, the release rate returned to 0.09 mg g^{-1} d^{-1}. The concentration of TDP in water was stabilized at a level about five times higher than before sterilization. At the same time, the pH and TDFe dropped significantly ($P < 0.0001$), and water color changed from dark brown to nearly transparent. The subsequent anaerobiosis did not yield any additional effects. Similar results were obtained after chemical sterilization of the aerobic systems (Fig. 5). In this case, the increase of release rate of TDP (to 3.26 mg g^{-1} d^{-1}) is significant. The drop in TDFe concentration before sterilization seems related to the interruption of mixing during two days, inducing a clarification of

the water. Mercuric chloride suppressed bacterial activity completely in both aerobic and anaerobic systems and reduced bacterial counts from $95.5 \pm 15.5 \times 10^6$ bacteria ml^{-1} sediment to $27.7 \pm 11.6 \times 10^6$ bacteria ml^{-1} sediment.

In contrast with TDFe, ferrous Fe was measured in non-filtered water to avoid reoxygenation. This analysis was carried out only in anaerobic conditions and immediately before the bubbling of N_2. Total Fe(II) concentration increased strongly from 597.5 ± 100.6 to 1356.0 ± 151.5 $\mu g\,l^{-1}$ in the anaerobic non-sterile systems. However, when anoxia was preceded by sterilization with $HgCl_2$, no release occured. This clearly indicates that iron reduction is biotic in this system. Also, no release was observed in the system where mixing failed. The concentration of ferrous Fe was much higher in unfiltered water than total Fe in filtered water. A large fraction of Fe must therefore be linked to particles larger than 0.45 μm that were present in the unfiltered anoxic water or were formed after sampling but prior to filtration.

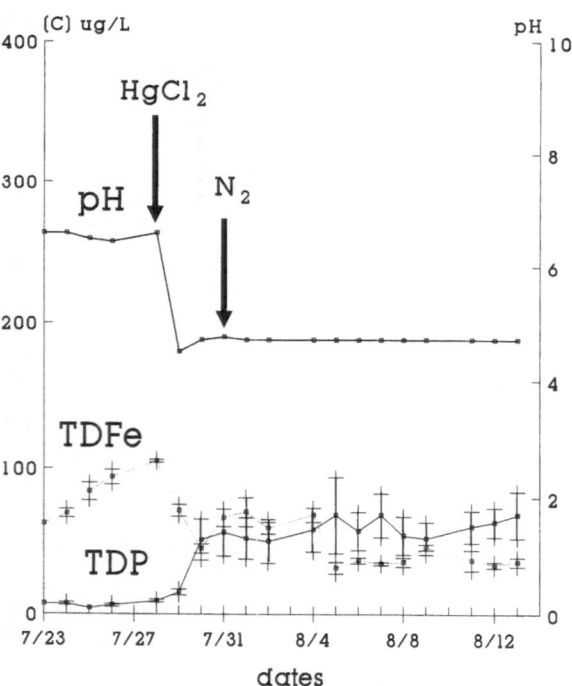

Fig. 4. Temporal evolution of TDP, TDFe, and pH in anaerobic sterile condition.

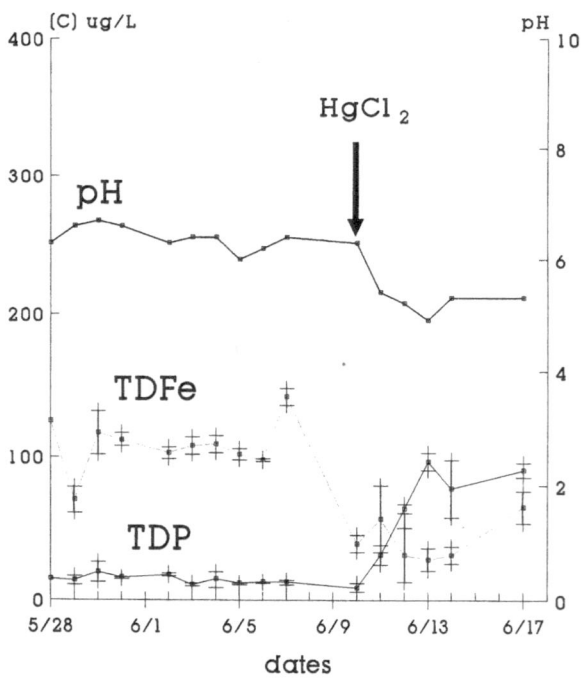

Fig. 5. Temporal evolution of TDP, TDFe, and pH in aerobic sterile condition.

Acidification and alkalinization

Acidification with concentrated H_2SO_4 had about the same effect on TDFe, pH and color as sterilization (compare Figs 5 and 6). However, in contrast to sterilization, acidification did not result in a massive release of TDP. The concentration of TDP in water increased almost instantly from 11 μg l^{-1} to 20 μg l^{-1}. After this initial increase, however, the release rate of P was not significantly different before and after acidification. Bacterial activity decreased from 3.1 ± 0.4 mg INT-formazan g^{-1} dry sediment to 0.8 ± 0.4 mg INT-formazan g^{-1} dry sediment. An addition of a few ml of NaOH, causing an increase of the pH by one unit, had the opposite effect: it resulted in a dramatic increase of the TDFe concentration accompanied by a darkening of the water. The concentration of TDP also increased suddenly the day following alkalinization (Fig. 7). The subsequent release rate was not significantly different than before treatment (Table 1). Bacterial activity slightly increased from

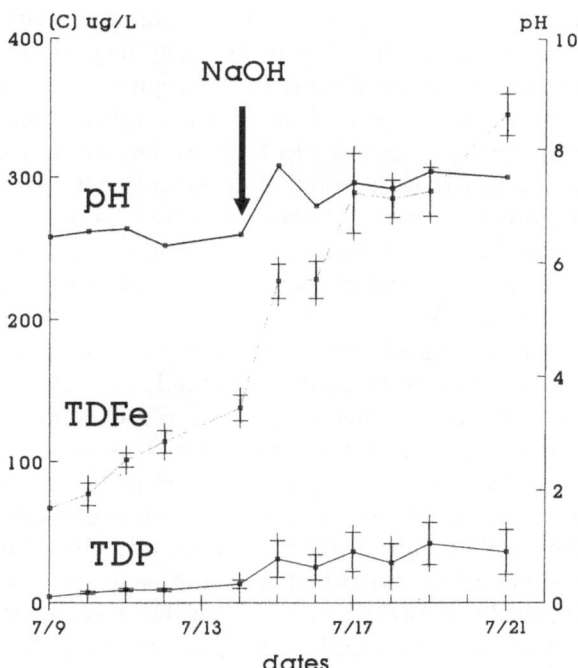

Fig. 7. Temporal evolution of TDP, TDFe, and pH after light alkalinization.

3.1 ± 0.4 mg INT-formazan g^{-1} dry sediment to 5.0 ± 0.4 mg INT-formazan g^{-1} dry sediment.

Discussion

From these experiments it can be concluded that in lake Cromwell the cycling of P and Fe are not tightly related as predicted by the model of Mortimer (1941, 1942). Sterilization as well as acidification caused a drop in TDFe concentration and a simultaneous increase in TDP concentration, albeit much more massive in the former.

These observations refute not only the Eisenle-Mortimer model for this sediment, but also the hypothesis of bacteria-catalyzed P release from Fe hydroxydes. Although bacterial reduction of Fe does appear to occur in our sediments, it did not mediate P release. In addition, if denitrifying bacteria were implicated, the addition of NO_3 should have stimulated these bacteria and increased the Fe release after NO_3 depletion. However, the NO_3 concentration was never depleted

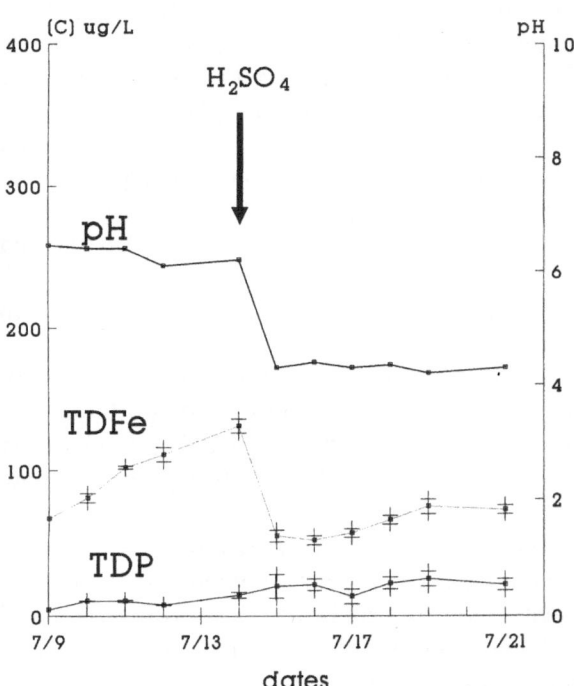

Fig. 6. Temporal evolution of TDP, TDFe, and pH in acidic condition.

in this case. It is possible that bacterial activity was limited by the supply of easily degradable organic substrate, if most of the organic C is refractory. The hypothesis involving SO_4 reduction must be also rejected. No H_2S production or increase in pH were noted after anaerobiosis, and dissolved O_2 level did not drop below 0.4 mg l^{-1} in the flux chambers as in the lake hypolimnion, maintaining a redox potential too high for SO_4 reduction.

The Fe movements between sediments and water appear to be controlled mainly by a chemical mechanism that is strongly pH-dependent and is reflected in the water color. This strongly suggests that TDFe is, for the most part, bound to humic acids. The interaction of metallic cations with humic acids is well-known. Goethite (α-FeOOH) is adsorbed to humic acids by ion exchange between acid carboxylic groups of humic molecules and surface oxydes (Tipping, 1981). For this reason, Fe concentrations in natural water are often several orders of magnitude greater than expected from equilibrium solubility of Fe hydroxydes, in particular in highly colored waters (Perdue et al., 1976). By definition, humic acids are precipited in acidic condition (Stumm & Morgan, 1981), probably with sequestred Fe. The pH-humic acids interactions depend strongly on the caracteristics on the humic acids, and can vary between lakes (Stewart & Wetzel, 1981).

The release of total Fe can be linked to the movement of dissolved humic acids, since acidified and non-mixed systems remain with a low water color. Another likely explanation for the release of Fe(II) is an active bacterial reduction of Fe, but another experiment in sterile pH-controled conditions must be carried out to test this hypothesis more rigourously.

Humic acids can also indirectly sorb P by binding with Fe(III) associated with humic complexes (Boström et al., 1982). Under low pH and low redox potential conditions, dissolved humic acid may associate with orthophosphate in the presence of Fe and render it inaccessible to phytoplankton (Francko & Heath, 1979; Stewart & Wetzel, 1981). However, in our experiments, no direct relationships between TDP and Fe could be established. On the contrary, when the concentration of TDFe decreased after acidification, the concentration of TDP continued to increase. After sterilization a decreasing TDFe concentration was even accompanied by a sudden increase in TDP. Thus, the often-invoked linkage of P with Fe-humic complexes appears to be of minor importance in lake Cromwell.

Our experiments suggest that almost all P released seems to be of direct bacterial origin. The maximum release occured after addition of $HgCl_2$, when bacterial activity and the number of bacteria dropped dramatically. Also, the fact that increased release ceased three days after sterilization suggests that the source of P (bacterial P) was exhausted. We observed the same phenomenon with or without anaerobiosis, and the two experiments were practically replicates. Thus, P release can probably be attributed to bacterial cell lysis. The increase of P mobilization after sterilization of sediments was also observed by Sinke & Cappenberg (1988) and Gächter & Meyer (1993). The fact that a lower release in P after acidification or anaerobiosis is linked to a lower decrease of bacterial activity (and bacterial abundance for the anaerobic condition) also supports the hypothesis that cell lysis is the key mechanism of P release. In a preliminary test, samples of 100 ml of unfiltered lake water were sterilized with $HgCl_2$ (at the same concentration used for the water-sediment systems), or bubbled 10 min. with N_2 or untreated. All samples were allowed to stand 24 hours before counting bacteria. The bacterial number was compared with those in freshly sampled lake water. No difference appeared between fresh water ($0.84 \pm 0.06 \times 10^6$ bacteria ml^{-1}) and control ($0.88 \pm 0.06 \times 10^6$ bacteria ml^{-1}), but chemical sterilization induced a 80% bacterial cell lysis ($0.18 \pm 0.01 \times 10^6$ bacteria ml^{-1}) while anoxia reduced bacterial abundance by only half as much ($0.51 \pm 0.03 \times 10^6$ bacteria ml^{-1}). In our sediment $HgCl_2$ experiments, approximately 67.8×10^6 bacteria ml^{-1} sediment were lysed by sterilization. This represent 5.424×10^{11} lysed bacteria for the whole system. Assuming an average individual bacterial biovolume of 0.18 μm^3 (Duarte et al., 1988), a density of 1, and a minimal 2% P content for bacteria (Heldal *et al.*, 1985; Schwartzbrod & Martin, 1985) we calculated that if all bacterial P was released, bacteriolysis can provide an amount of 500 μg l^{-1}. The observed release was about 100 μg l^{-1}, indicating that all P released can be easily accounted by the bacterial cellular P. It is also possible that a certain part of anaerobic P release originated from the dissolution of poly-P stored by facultative aerobic bacteria. However, 'luxury uptake' or 'phosphate overplus phenomenon' occurs only when P is in excess in the environment of bacteria and the conditions for these mechanisms are probably not favorable in the nutrient-poor sediments of Lake Cromwell, despite the possible limitation by easily degradable organic matter. Further studies are necessary to clarify this point.

Only the results of the alkalinization experiment are in agreement with the classical model of ligand exchange between hydroxides and P. The binding capacity of Fe and aluminium decrease markedly when the pH reaches 6.5, due to the substitution of phosphates by OH$^-$ (Swenson *et al.*, 1949). The effect of alkalinization of P release by sediments was more recently studied by Boström (1984). Boers (1991) noted that the addition of NaOH promoted the desorption of P from Fe(III) hydroxides more by increasing alkalinity than by increasing pH. In this experiment, release of P was not linked to a decreasing bacterial activity or biomass. To the contrary, bacterial activity increased after alkalinization. We suggest that because alkalinization increases bacterial activity, bacteriolysis is unlikely important and the classical mechanism is probably dominant. In fact, it is probable that bacteria took up part of the P released from hydroxides by ion exchange and hence masked a more important abiotic P release.

Conclusion

In summary, we conclude that, for the sediments of lake Cromwell, and probably of other oligodystrophic lakes as well, living micro-organisms act as an important P reservoir and that processes inducing cell lysis can result in a significant P release from the sediments.

Acknowledgements

We wish to thank J. Vaillancourt for field assistance and M. J. Carbonneau for TDP analyses. Counting of bacteria was executed by J. F. Greffe. We also thank D. Planas, R. Gächter and an anonymous reviewer for useful comments on the manuscript. This project was funded by NSERC in the form of a postgraduate scholarship to C. de Montigny and an operating grant to Y. T. Prairie.

References

Andersen, J. M., 1976. An ignition method for determination of total phosphorus in lake sediments. Wat. Res. 10: 329–331.

Andersen, J. M., 1982. Effect of nitrate concentration in lake water on phosphate release from the sediment. Wat. Res. 16: 1119–1126.

APHA, 1985. Standard Methods for the examination of water and wastewater, 16th edition. American Public Health Association, Washington, D.C., 1268 pp.

Boers, P. C. M., 1991. The influence of pH on phosphate release from lake sediments. Wat. Res. 25: 309–311.

Boström, B., 1984. Potential mobility of phosphorus in different types of lake sediment. Int. Revue ges. Hydrobiol. 69: 457–474.

Boström, B., M. Jansson & C. Forsberg, 1982. Phosphorus release from lake sediments. Arch. Hydrobiol. Beih. Ergebn. Limnol. 18: 5–59.

Boström, B., J. M. Andersen, S. Fleischer & M. Jansson, 1988. Exchange of phosphorus across the sediment–water interface. In G. Persson & M. Jansson (eds), Phosphorus in Freshwater Ecosystems. Developments in Hydrobiology 48. Kluwer Academic Publishers, Dordrecht: 229–244. Reprinted from Hydrobiologia 170.

Byers, C., E. L. Mills & P. L. Stewart, 1978. A comparison of methods of determining organic carbon in marine sediments, with suggestions for a standard method. Hydrobiologia 58: 43–47.

Duarte, C. M., D. F. Bird & J. Kalff, 1988. Submerged macrophytes and sediment bacteria in the littoral zone of Lake Memphremagog (Canada). Verh. int. Ver. Limnol. 23: 271–281.

Einsele, W., 1936. über die Beziehungen des Eisenkreislaufs zum Phosphatkreislauf im eutrophen See. Arch. Hydrobiol. 29: 664–686.

Einsele, W., 1938. Über chemische und kolloidchemische Vorgänge in Eisen-Phosphat-Systemen unter limnochemis-

150

chen und limnogeologischen Gesichtspunkten. Arch. Hydrobiol. 33: 361–387.

Fleischer, S., 1978. Evidence for the anaerobic release of phosphorus from lake sediments as a biological process. Naturwissenschaften 65: 109–110.

Fleischer, S., 1983. Microbial phosphorus release during enhanced glycolysis. Naturwissenschaften 70: 315.

Fleischer, S., 1986. Aerobic uptake of Fe(III)-precipitated phosphorus by microorganisms. Arch. Hydrobiol. 107: 269–277.

Fleischer, S., M. Bengtsson & G. Johansson, 1988. Mechanism of the aerobic Fe(III)-P solubilization at the sediment-water interface. Verh. int. Ver. Limnol. 23: 1825–1829.

Florentz, M., P. Granger & P. Hartemann, 1984. Use of ^{31}P nuclear magnetic resonance spectroscopy and electron microscopy to study phosphorus metabolism of microorganisms from wastewaters. Appl. envir. Microbiol. 47: 519–525.

Foy, R. H., 1986. Suppression of phosphorus release from lake sediments by the addition of nitrate. Wat. Res. 20: 1345–1351.

Francko, D. A. & R. T. Health, 1979. Functionally distinct classes of complex phosphorus compounds in lake water. Limnol. Oceanogr. 24: 463–473.

Fuhs, G. W. & M. Chen, 1975. Microbiological basis of phosphate removal in the activated sludge process for the treatment of wastewater. Microb. Ecol. 2: 119–138.

Gächter, R., J. S. Meyer & A. Mares, 1988. Contribution of bacteria to release and fixation of phosphorus in lake sediments. Limnol. Oceanogr. 33: 1542–1558.

Gächter R. & J. S. Meyer, 1993. The role of microorganisms in mobilization and fixation of phosphorus in sediments. In P. C. M. Boers, T. E. Cappenberg & W. van Raaphorst (eds), Proceedings of the Third International Workshop on Phosphorus in Sediments. Developments in Hydrobiology 84. Kluwer Academic Publishers, Dordrecht: 103–121. Reprinted from Hydrobiologia 253.

Heldal, M., S. Norland & O. Tumyr, 1985. X-ray microanalytic method for measurement of dry matter and elemental content of individual bacteria. Appl. envir. Microbiol. 50: 1251–1257.

Jansson, M., 1987. Anaerobic dissolution of iron-phosphorus complexes in sediment due to the activity of nitrate-reducing bacteria. Microb. Ecol. 14: 81–89.

Jones, J. G., S. Gardener & B. M. Simon, 1983. Bacterial reduction of ferric iron in a stratified eutrophic lake. J. Gen. Microbiol. 129: 131–139.

Lafond, M., B. Pinel-Alloul & P. Ross, 1990. Biomass and photosynthesis of size-fractionated phytoplankton in Canadian Shield lakes. Hydrobiologia 196: 25–38.

Malnou, D., P. Chopard & H. Andrearczyk, 1983. La déphosphatation biologique: peut-on y croire? T.S.M.-L'Eau 78: 63–71.

Mortimer, C. H., 1941. The exchange of dissolved substances between mud and water in lakes. I.J. Ecol. 29: 280–329.

Mortimer, C. H., 1942. The exchange of dissolved substances between mud and water in lakes. J. Ecol. 30: 147–201.

Ohtake, H., K. Takahashi, Y. Tsuzuki & K. Toda, 1984. Phosphorus release from a pure culture of Acinetobactre calcoaceticus under anaerobic conditions. Envir. Tech. Lett. 5: 417–424.

Ostrofsky, M. L., 1987. Phosphorus species in the surficial sediments of lakes of eastern North America. Can. J. Fish. aquat. Sci. 44: 960–966.

Perdue, E. M., K. C. Beck & J. H. Reuter, 1976. Organic complexes of iron and aluminium in natural waters. Nature 260: 418–420.

Porter, K. G. & Y. S. Feig, 1980. The use of DAPI for identifying and counting aquatic microflora. Limnol. Oceanogr. 25: 943–948.

Sakata, M., 1985. Diagenetic remobilisation of manganese, iron, copper and lead in anoxic sediment of a freshwater pond. Wat. Res. 19: 1033–1038.

Schallenberg, M., J. Kalff & J. B. Rasmussen, 1989. Solutions to problems in enumerating sediment bacteria by direct counts. Appl. envir. Microbiol. 55: 1214–1219.

Schwartzbrod, J. & G. Martin, 1985. Les micro-organismes, agents biogéochimiques. p. 1–45. In Point sur l'épuration et le traitement des effluents. 2.1 Bactériologie des milieux aquatiques. G. Martin (ed.), Technique et Documentation Lavoisier, Paris.

Shapiro, J., 1967. Induced release and uptake of phosphate by microorganisms. Science 155: 1269–1271.

Sinke, A. & T. E. Cappenberg, 1988. Influence of bacterial processes on the phosphorus release from sediments in the eutrophic Loosdrecht Lakes, The Netherlands. Arch. Hydrobiol. Beih. Ergebn. Limnol. 30: 5–13.

Sørensen, J., 1982. Reduction of ferric iron in anaerobic, marine sediment and interaction with reduction of nitrate and sulfate. Appl. envir. Microbiol. 43: 319–324.

Stewart, A. J. & R. G. Wetzel, 1981. Dissolved humic materials: photodegradation, sediment effects, and reactivity with phosphate and calcium carbonate precipitation. Arch. Hydrobiol. 92: 265–286.

Stolp, H. & M. P. Starr, 1965. Bacteriolysis. Ann. Rev. Microbiol. 19: 79–104.

Stumm, W. & J. J. Morgan, 1981. Aquatic Chemistry. An introduction emphasizing chemical equilibria in natural water. 2nd edition. John Wiley & Sons, Toronto. 780 pp.

Tipping, E., 1981. Adsorption by goethite (α-FeOOH) of humic substances from three different lakes. Chem. Geol. 33: 81–89.

Trevors, J. T., 1984. The measurement of electron transport system (ETS) activity in freshwater sediment. Wat. Res. 18: 581–584.

Hydrobiologia **253**: 151–163, 1993.
P.C.M. Boers, T.E. Cappenberg & W. van Raaphorst (eds),
Proceedings of the Third International Workshop on Phosphorus in Sediments.
© 1993 *Kluwer Academic Publishers.*

Bio-availability of phosphorus in sediments of the western Dutch Wadden Sea

Victor N. de Jonge, Menno M. Engelkes & Joop F. Bakker
Rijkswaterstaat, Tidal Waters Division, P.O. Box 207, 9750 AE Haren (GN), The Netherlands

Key words: Wadden Sea, sediments, phosphorus compounds, bioavailability, algae, iron, calcium redox potential, oxygen

Abstract

The purpose of this study was to make a prognosis of the effects of extended purification of terrestrial waste water, reaching the Wadden Sea by the River Rhine and Lake IJssel, on the phosphate concentration in the western Wadden Sea.

The quantities of different phosphorus fractions in intertidal and subtidal sediments of the Marsdiep tidal basin (western Dutch Wadden Sea) were measured. Different methods are applied to determine the amount of phosphorus that can be released from these sediments. The direct bioavailability is determined by inoculating sediment suspensions with a natural mixture of precultured micro-organisms from the sampling area. A second approach is the measurement of the phosphate release under different redox conditions. Sequential extraction of sediment samples with different solvents is also applied. Under the present conditions and compared to the nutrient loads from fresh water (Lake IJssel) and from the North Sea, the phosphorus stored in the sediments of the western Dutch Wadden Sea plays a minor role in the total supply to micro-algae and bacteria. The bulk of the biologically available phosphorus in the sediments originates from the metal-associated fraction. Releasable phosphate may contribute to the local annual primary production to an extent of *ca* 45 to *ca* 150 g C m^{-2} a^{-1}. The total amount of phosphorus in the sediment (mainly calcite associated) is twice to 6 times the biologically available amount.

Introduction

The pathways and fate of phosphorus in estuarine systems like the Wadden Sea (Fig. 1) are important, because this nutrient seems to be one of the major growth limiting factors for algae in such areas (De Jonge, 1990).

The input of phosphorus to the Wadden Sea comes from several minor sources and two major discharge points in the 'Afsluitdijk' (Fig. 1). Phosphorus is discharged in both dissolved and particulate form. In addition to fresh water sources, suspended matter (particulate organic matter and inorganic matter) and associated phosphorus from the coastal zone of the North Sea are transported into the Wadden Sea due to physical accumulation of suspended matter in the coastal direction (Postma, 1967).

The loading of suspended matter in the coastal zone with phosphorus is partly caused by the loads of phosphate and total phosphorus from the River Rhine. The concentration gradient of phosphate between the Wadden Sea and the North Sea and the residual current occurring from

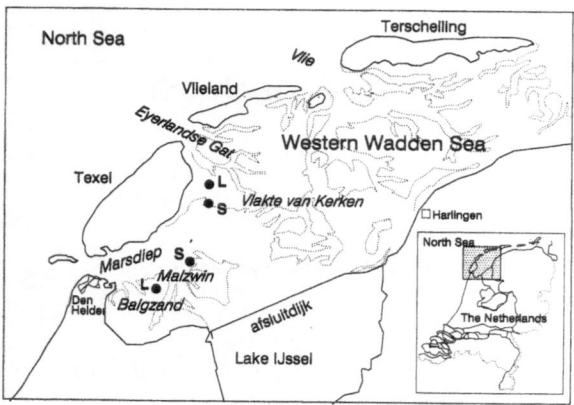

Fig. 1. Map of the western Dutch Wadden Sea with sampling locations.

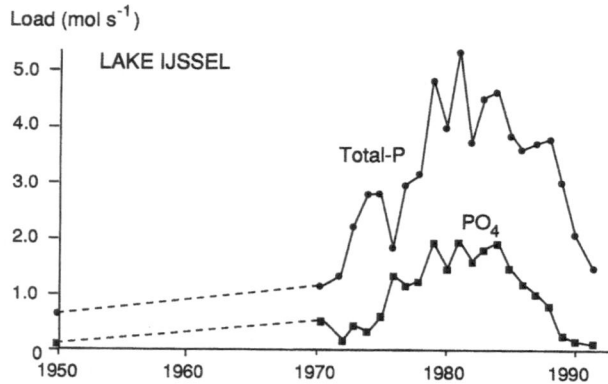

Fig. 2. Loads of phosphate and total phosphorus (this is the sum of dissolved and particulate fractions) from Lake IJssel (cf. Fig. 1).

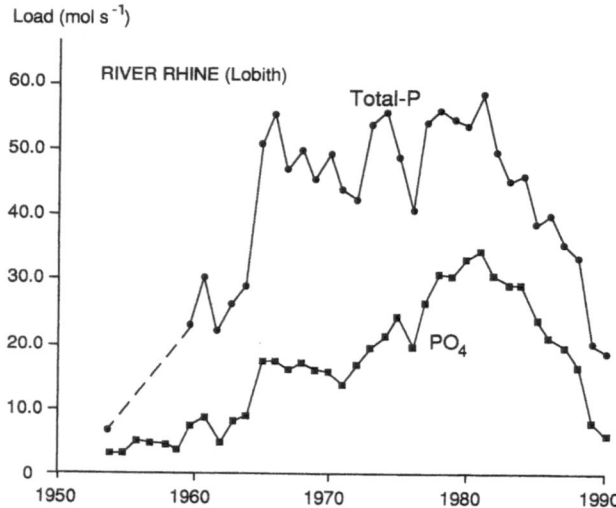

Fig. 3. Loads of phosphate and total phosphorus (this is the sum of dissolved and particulate fractions) from the River Rhine at station Lobith on the Dutch-German border.

the Vlie tidal inlet to the Marsdiep tidal inlet (Fig. 1) (van Raaphorst & van der Veer, 1990; Ridderinkhof & Zimmerman, 1990) are mainly responsible for the seaward transport of the dissolved fractions. Import and export of phosphorus are not balanced in the Wadden Sea. From data of Duursma (1961) and Cadée & Hegeman (1977) it can be calculated that ca 10% of the imported phosphorus or 0.5 g P m^{-2} a^{-1} may be buried in sediments of the western Dutch Wadden Sea.

The total input of phosphorus to the western Dutch Wadden Sea is not constant but varies as a function of the fresh water discharge and the phosphorus concentrations in this water (Fig. 2). For the mid 1970s the phosphorus budget presented by De Jonge & Postma (1974) indicated a total supply of ca 6 g P m^{-2} a^{-1} of which 0.5 g P m^{-2} a^{-1} (8%) would be buried in the sediment for a longer period (Duursma, 1961 and Cadée & Hegeman, 1977).

Since the early 1980s, the phosphorus loads of both Lake IJssel (Fig. 2) and the Rhine (Fig. 3) have been decreasing steadily. In 1990 the phosphate load from Lake IJssel (Fig. 2) reached the lowest values since 1970, and that of the Rhine (Fig. 3) reached its lowest values since the early 1960s. The effects of variations in phosphorus loads from both the North Sea and Lake IJssel on the annual budget were studied by van Raaphorst & van der Veer (1990). They concluded that

in the past the phosphorus import from the North Sea was less important as was thought previously and that since the mid 1970s the western Dutch Wadden Sea turned from a system importing particulate phosphorus to a system that exports it.

From the above-given information the question may arise whether the amount of buried phosphorus can support the local primary production. This is important because in a previous paper the fuelling role of phosphate from Lake IJssel on the primary production of the western Dutch Wadden Sea was demonstrated (De Jonge, 1990).

It is the aim of the present paper to determine the amounts of buried phosphorus in the western Dutch Wadden Sea and to find out which part of this pool of phosphorus is available to the primary production in that area.

Materials and methods

Study area

The western Dutch Wadden Sea is the most western part of the 600 km long Wadden Sea (Fig. 1) and consists of the tidal basins of the Marsdiep, Eyerlandse Gat and the Vlie.

The study area was restricted to the Marsdiep tidal basin. This area covers a surface of approximately 690×10^6 m^2, has a tidal prism of approximately 1×10^9 m^3 and a mean tidal range near Den Helder of 1.37 m. The residual current between the Vlie tidal inlet and the Marsdiep tidal inlet is approximately 900 m^3 s^{-1} (Ridderinkhof & Zimmerman, 1990). Four locations were selected for sampling of which two (one littoral and one sublittoral) were located on the 'Vlakte van Kerken' east of the island of Texel. Two other (one littoral and one sublittoral) were situated near the 'Malzwin', a channel that crosses the 'Balgzand', a tidal area that is connected with the mainland.

The sediment of the littoral Malzwin station is characterized by a silt content of ca 0.5% while the sublittoral sediment also shows a silt content of less than ca 0.25%. The littoral sediment of Vlakte van Kerken has a silt content of ca 0.5% similar to the sublittoral sediment.

The annual phytoplankton primary production in the Marsdiep area is not constant, but has varied from 150 to over 500 g C m^{-2} a^{-1} over the last two decades (Postma & Rommets, 1970; Cadée & Hegeman, 1974, 1979; Cadée, 1986; Veldhuis et al., 1988). Year to year variations are significantly correlated with annual changes in the supply of phosphate from Lake IJssel (De Jonge, 1990).

Field observations

Sampling of the littoral and the sublittoral was carried out from a ship by a hand operated piston corer. The cores were ca 20 cm long and had an inner diameter of 2.4 cm; 15 samples were taken per station.

On board the ship three cores were quickly sliced under normal air in layers of 0–0.5; 0.5–1; 1–2; 2–4; 4–6; 6–8; 8–10 and 10–12 cm. Per depth the samples were pooled, thoroughly mixed and stored at 4 °C. Chemical sequential extraction of phosphorus took place within 24 hours.

Pore water was collected by squeezing pooled subsamples of six cores, which had been sliced as indicated above, on board the ship under an atmosphere of nitrogen gas (ca 3 bar) and filtered through cellulose nitrate filters (Schleicher & Schuell, 0.2 μm). Pore water was collected in polyethylene bottles and immediately stored at −20 °C in a normal freezer.

The six remaining cores, reserved for experiments, were stored at −20 °C to stop all bacterial activity (mineralization) and changes in redox conditions.

Sample processing and chemical analyses

From each station and each depth a sample of 0.5 g of wet sediment (corresponding to ca 0.3 g dry weight) was transferred into a tube for a sequential extraction. Two nearly identical methods of sequential extraction were applied (Fig. 4). The first method was according to the extraction procedure of Hieltjes and Lijklema (1980). The second method differed from the first one in that an additional extraction step, directly after the NH$_4$Cl extraction, with 0.11 M NaHCO$_3$/ Na$_2$S$_2$O$_4$ was applied to obtain the redox sensitive phosphorus fraction. This fraction mainly consists of iron-associated phosphorus. These two different extractions were done to get insight in the kind of substrate that phosphorus is associated with and the possible effect of anoxia on the flux of phosphate from the sediment.

In a single-step extraction, 1 M HCl was used

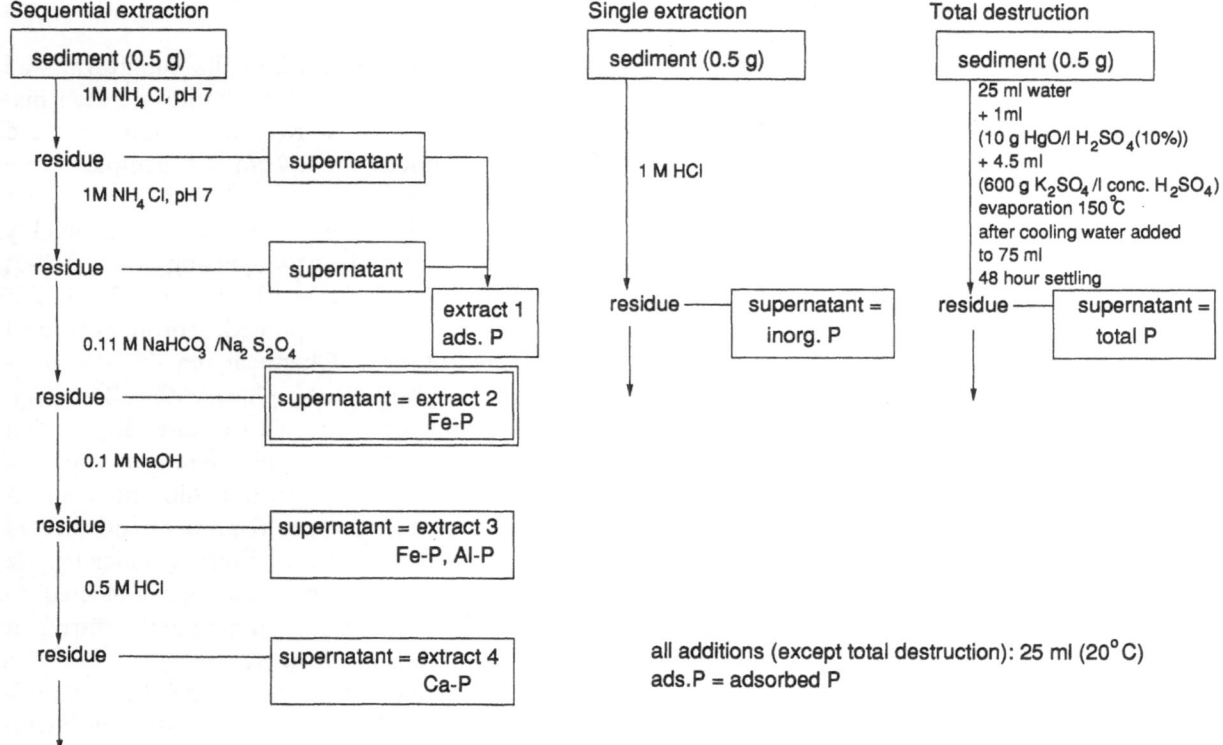

Fig. 4. Diagram of the extraction procedure followed to discriminate between different phosphorus fractions.

to obtain the total inorganic phosphorus (Krom & Berner, 1981). Total phosphorus was determined after total destruction with K_2SO_4/H_2SO_4 at 370 °C.

The organic phosphorus fraction was calculated by subtraction of the inorganic fraction (1 M HCl) from the total phosphorus obtained by total destruction (Fig. 4).

After extraction all samples were neutralized. Prior to analysis all samples were quickly thawed at 75 °C to minimize the total effect of freezing and thawing on the concentration of soluble reactive phosphate (SRP). SRP was determined with a modification of industrial method 113–73w (based on Murphy & Riley, 1962). Homogeneity of the pooled and mixed samples was controlled by a six-fold subsampling of the mixtures. The coefficient of variation varied from 10.5 to 11.2%.

Experiments

Diffusive phosphorus flux under anoxic conditions
To measure the total phosphorus flux under anoxic conditions, three deepfrozen cores for each location were thawed at room temperature. Deepfrozen cores were used because these experiments could not be carried out simultaneously with the other chemical analyses. Further, it is difficult to simulate natural fluxes of phosphate during experimental conditions. Therefore, we did not claim to determine the precise daily fluxes of phosphate but we were merely interested in the total possible flux of phosphate during the growing season. The water above the sediment was replaced by *ca* 30 ml of phosphate-free artificial sea water (Admiraal & Werner, 1983) which was poor in oxygen, and enriched with some glucose (5 mmol 1^{-1}) to maintain anoxic conditions by oxygen consumption of heterotrophic bacteria in

the sediment and the water. The cores were stoppered without leaving air bubbles in the water of the headspace on top of the sediment and the tubes were kept in the dark at 18 °C.

After the sediment surface had become anoxic, the headspace water was removed and replaced by a new water sample as described above.

Diffusive phosphorus flux under oxygenated conditions

To measure the phosphorus flux under oxygenated conditions, three deepfrozen cores for each location were thawed at room temperature. The water above the sediment was replaced by *ca* 30 ml of phosphate-free artificial sea water (Admiraal & Werner, 1983). These cores were stoppered with a cotton wool plug and placed in the dark at 18 °C. After six days the water of the headspace was removed and replaced by a new sample of water.

For the oxygenated series as well as the anoxic series, the removed water was immediately filtered through 0.45 μm cellulose nitrate filters (Sartorius) and stored at -20 °C. This procedure was repeated several times at intervals of six days.

Bio-available phosphorus

Approximately 0.5 g of homogenized wet sediment from the sliced and pooled sediment cores was transferred into tubes containing 25 ml of phosphorus-free artificial sea water (Admiraal & Werner, 1983). All other nutrients were present in excess. Samples were enriched with 15 mmol l^{-1} of sodium acetate and inoculated with a precultured batch of a natural mixture of heterotrophic bacteria from the sampling site. The tubes were closed with a cotton wool plug and vortexed frequently. Growth was followed by measuring the optical density. After 4 weeks, when growth had stopped, the total protein content in the samples was determined (Lowry *et al.*, 1951). Corrections were carried out for the initial protein content of the sediments. Protein values were converted into phosphorus equivalents, using a standard curve obtained from protein yields on known amounts of phosphorus added to the same medium and the

same bacteria cocktail as used during experiments on bio-availability of phosphorus (Fig. 5)

We have chosen for bacteria instead of algae because the culturing of bacteria requires less equipment than that of algae. Moreover, the natural bacteria population may contain many specialists so that the figures for bio-availibility of phosphorus may be maximum values.

Results

Field observations

A representative example of the results obtained by the sequential extraction procedure is given in Fig. 6. The results in this figure show that the fraction of adsorbed SRP (NH_4Cl fraction) decreased rapidly with depth. The metal-associated phosphorus (NaOH fraction) is highest near the sediment surface and does not decrease dramatically with depth. The calcium-associated phosphorus (HCl fraction) is rather constant with depth, although near the surface in Fig. 6a the concentrations are somewhat higher. The redox sensitive and mainly iron-associated phosphorus ($NaHCO_3/Na_2S_2O_4$ fraction) strongly decreases with depth. By comparison with total NaOH-P in first panel (Fig. 6a) this fraction also contains part

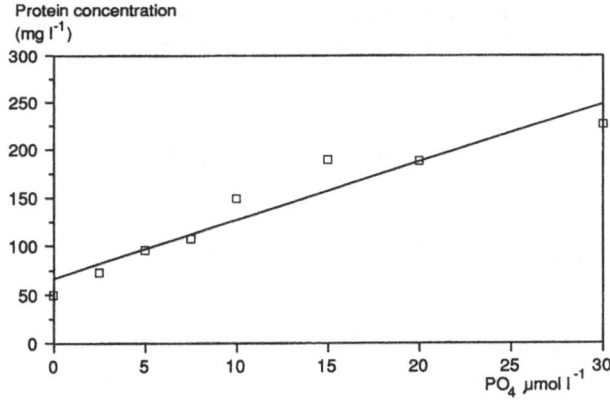

Fig. 5. Protein yield as a function of phosphate concentrations in a batch culture in which phosphorus was the growth limiting factor. The curve is a mean based on four independent series of experiments.

156

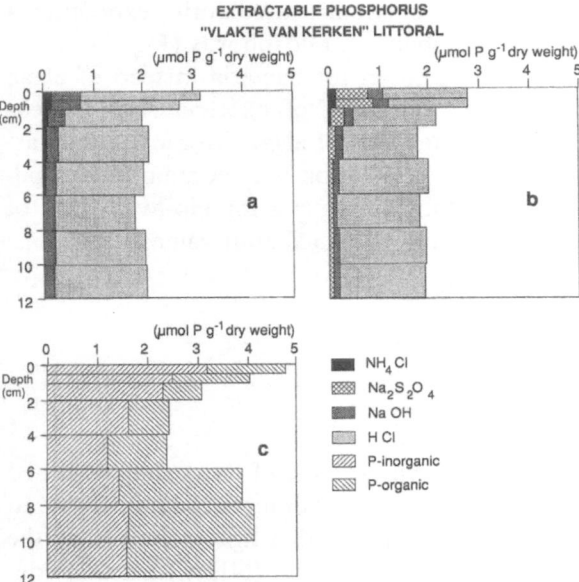

EXTRACTABLE PHOSPHORUS
"VLAKTE VAN KERKEN" LITTORAL

Fig. 6. Representative example of concentrations of different phosphorus fractions in sediment obtained for station 'Vlakte van Kerken' littoral, after application of the extraction procedure given in Fig. 4. a: extraction method after Hieltjes and Lijklema (1980); b: additional $Na_2S_2O_4$ extraction after NH_4Cl step in the procedure given under a; c: P-inorganic: 1 M HCl leach, P-organic: difference between total P (total destruction) and P-inorganic (1 M HCl leach).

of the metal-associated fraction. The total inorganic phosphorus is highest near the sediment surface (Fig. 6c). At a depth below 2 cm, the amount of inorganic phosphorus (Fig. 6c) is rather constant. The amount of organic phosphorus (Fig. 6c) shows highest concentrations at a depth of 6 to 8 cm. At the surface a second maximum in organic phosphorus occurs. Total phosphorus (Fig. 6c) concentrations are highest near the surface while a second concentration maximum is found at a depth of 6–10 cm.

We conclude from these profiles that the calcium-associated fraction is the largest. The second most important fraction is the organic fraction. The concentration profiles at the other stations are not identical but the relative proportions of the different fractions are approximately the same. Generally the calcite-associated and total phosphorus concentrations were considerably smaller in the sublittoral sediments.

The total amount of each of the phosphorus fractions was calculated by summation of the amounts present in the top 12 cm (Table 1). Table 1 shows that the smallest amount of phosphorus is present adsorbed to the sediment (NH_4Cl fraction) while the largest is the calcium-associated fraction (0.5 M HCl leach). The values in

Table 1. Amounts of different phosphorus fractions in the top 12 cm of the sediment at the 4 stations. Values are expressed in mmol P m^{-2}. The values in the columns 1, 3, and 4 are obtained by applying the method of Hieltjes and Lijklema (1980), the values in column 2 by applying sodium dithionite in between the NH_4Cl step and the NaOH step, the values in column 5 by applying the method of Krom and Berner (1981) and the values in column 7 by applying total destruction.

Location	Fraction				1M HCl			Assay (bio-available)
	NH_4Cl (adsorbed)	$NaHCO_3$/ $Na_2S_2O_4$ (redox sensitive)	NaOH (metal associated)	0.5 M HCl (calcium associated)	(inorganic)	(organic)	(total)	
Malzwin littoral	9	25	17	395	472	182	654	72
Malzwin sublittoral	8	23	30	142	242	11	253	58
Vlakte v. Kerken littoral	4	25	42	332	258	250	608	115
Vlakte v. Kerken sublittoral	7	13	24	140	171	172	343	71
Mean	7	21	28	252	311	154	464	79

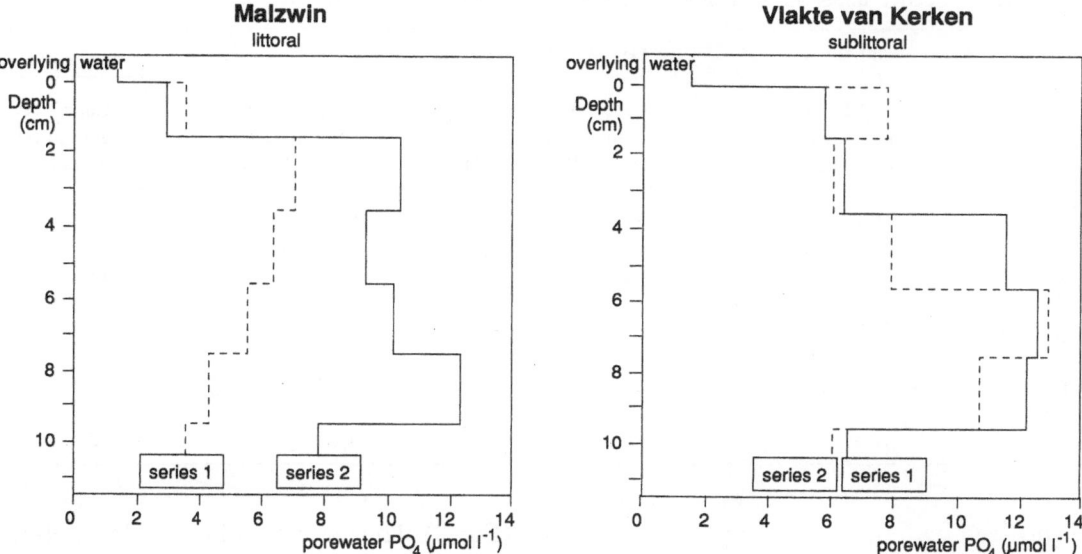

Fig. 7. Concentrations of pore water phosphate at two stations 'Malzwin' littoral and 'Vlakte van Kerken' sublittoral. Series 1 and 2 refer to two duplicate series of sediment cores.

Table 1 also allow a comparison of the sum of the mean inorganic phosphorus fractions (extracts 1 to 4 in Fig. 4) with the mean value of the single extraction method (1 M HCl in Fig. 4). The result shows that the values are very close to each other (308 mmol P m^{-2} and 311 mmol P m^{-2}).

The pore water concentrations are presented in Fig. 7. The pore water SRP concentrations in the top interval are significantly higher than those in the bottom water. At a depth of 6–10 cm the concentrations reach maximum values. Below this zone the concentrations decrease again.

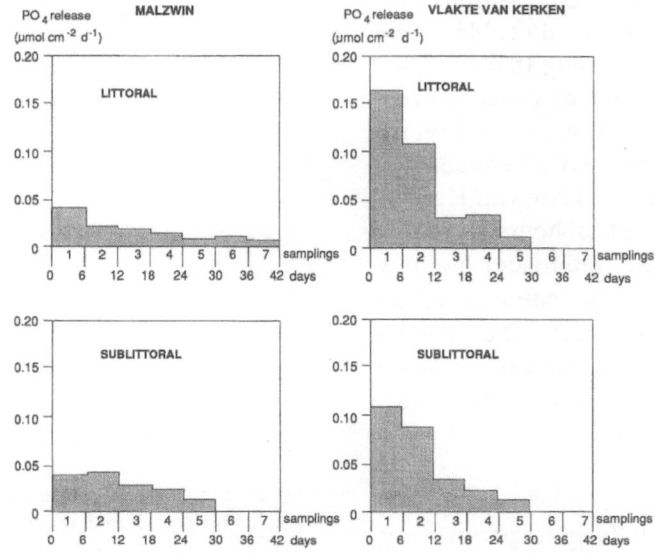

Fig. 8. Mean daily diffusive phosphate flux under anoxic conditions for the four stations ('Malzwin': littoral and sublittoral; 'Vlakte van Kerken': littoral and sublittoral).

Table 2. Total release of SRP under anoxic conditions and during oxygenated conditions for the 4 stations.

Location	Anoxic after first 5 samplings (mmol P m^{-2})	Oxygenated conditions		
		Calculated annual release (mmol P m^{-2} a^{-1})	After first 5 samplings (mmol P m^{-2})	Calculated annual release (mmol P m^{-2} a^{-1})
Malzwin littoral	6.1	34	0.6	7
Malzwin sublittoral	8.6	48	1.4	17
Vlakte v. Kerken littoral	20.8	57		
Vlakte v. Kerken sublittoral	15.8	52		
Mean	12.8	48	1.0	12

Experiments

Diffusive phosphate flux

The results of all the flux experiments are presented (Fig. 8). The initial diffusive flux under anoxic conditions differs significantly between the different sampling locations, ranging from 0.05 to 0.15 μmol cm^{-2} d^{-1}. The initial flux for station 'Vlakte van Kerken' was especially high. After one month (Fig. 8) the SRP release from sediments at each station reached a value of approximately 0.015 μmol cm^{-2} d^{-1}.

The total release under anoxic conditions during the first 30 days was calculated (Table 2). An annual estimate is given assuming that the SRP flux at the end of the experiment would be representative for an entire year. For comparison the diffusive SRP flux from oxygenated sediments of 'Balgzand' (the sediments of 'Vlakte van Kerken' became anoxic within a couple of hours and could therefore not be used as a reference) are presented. Anoxic conditions at the sediment surface of Balgzand do not usually occur. The diffusive SRP flux from oxygenated Balgzand sediments is *ca* 0.003 μmol P cm^{-2} d^{-1}, and more than a factor 7 less than the anoxic flux.

Bio-available phosphorus

The results on the bio-available phosphorus fraction show a remarkable difference both between stations and between littoral and sublittoral sites

(Fig. 9). The amount of bio-available phosphorus is generally larger in the fine grained sediments than in the coarse grained sediments, while the top 1 cm of the littoral site of station 'Vlakte van Kerken' contains more than twice as much as deeper layers. The same difference, but less pro-

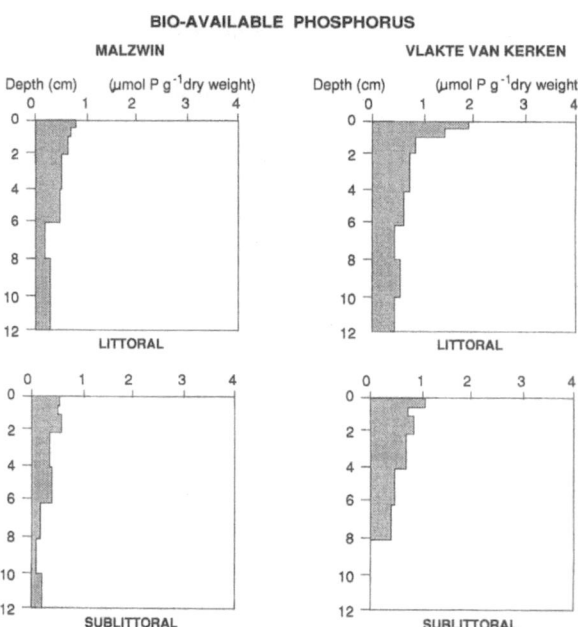

Fig. 9. Bio-available phosphorus as measured by phosphorus uptake by precultured natural bacteria added to sediment samples from different depth layers. Sampling stations were 'Malzwin' (littoral and sublittoral) and 'Vlakte van Kerken' (littoral and sublittoral).

nounced, is present at station 'Malzwin'. Overall the amounts vary from 0.1 to ca 2 μmol P g^{-1} dry sediment.

Discussion

Comments on the extraction method applied

Application of sequential extraction methods to sediment samples does not produce sharply defined phosphorus fractions. This can be deduced from the fact that both the NaOH and the HCl fraction are quantitatively influenced by a previous extraction step with $Na_2S_2O_4$ (Fig. 6). Apparently, part of the metal associated fraction and the calcium-associated phosphorus is mobilized by the sodium dithionite procedure. This conclusion is in agreement with the findings of Psenner and Pucsko (1988). Sherwood et al. (1987) have shown that the calcium associated phosphorus fraction may contain phosphorus associated with iron oxyhydroxide coated on shell material.

Concentrations in pore water

The SRP concentrations in the pore water (Fig. 7) follow a pattern with relatively low values at the sediment surface but still higher than in the overlying water. Deeper in the sediment the SRP concentrations increase over depth and at a maximal depth of 10 cm the values decrease again. The concentration levels are comparable with those given by Suess (1976), and Sundby et al. (1992), but substantially lower than the values reported by Bray et al. (1973), Boynton and Kemp (1985), and Balzer et al. (1987).

Mixing all the SRP from pore water in the 12 cm thick sediment layer with the overlying water column in the study area would increase the concentration of water column SRP by 0.1 μmol l^{-1}, assuming an average water depth of 3.9 m (Postma, 1954). This contribution of porewater phosphate to the water column is consequently low.

Phosphorus in sediments and anoxia

The magnitude of the redox sensitive (sodium dithionite) (Fig. 6) fraction is suggestive of what may happen if sediments of intertidal flats in the Wadden Sea become anoxic (Kolbe, 1991). Under conditions of low redox potential, the total sodium dithionite-sensitive phosphorus fraction might reach the water column within a short time. Based on the redox sensitive phosphorus pool in the 12 cm thick sediment layer (Table 1) this may lead to a considerable increase in SRP of the water column ranging from ca 3.5–6.5 μmol l^{-1}. The maximal measured increase in SRP over a period of 6 days (Fig. 8; Vlakte van Kerken, littoral) did correspond to ca 2.5 μmol l^{-1}. This value is close to the range given above.

The calculated iron/phosphorus ratio for the surface layers we sampled (not reported here) was 12 and very close to the ratio of 10 reported by Lijklema (1977) for phosphorus adsorbed to FeOOH. This gives us some information on the possible source of the Fe-P and the Al-P pool in Fig. 4. Iron oxyhydroxide is known to be a geochemically reactive compound, formed during the oxidation of Fe^{2+} by O_2 (Canfield, 1989). Manganese is also reported to be an important geochemical compound and strongly interacting with the iron geochemistry (Myers & Nealson, 1988; Sundby & Silverberg, 1985), but this compound was not examined in the present study. However, the similar chemical behaviour and the similar concentrations in pore water suggests that the role of manganese in the cycling of phosphorus may be comparable to that of iron. Iron (oxyhydr)oxides are known as scavengers of SRP by adsorption (Froelich et al., 1979; Golterman, 1976; Krom & Berner, 1981; Lijklema, 1977; 1980; Parfitt et al., 1975). But, redox-controlled dissolution of iron (oxyhydr)oxides presumably is a source of SRP under anoxic conditions (Krom & Berner, 1981; Sundby et al., 1992). The sediment redox cycles of Fe, Mn and O_2 result in (oxyhydr)oxide precipitation in the top layers of sediment. In Skagerrak sediments, maximum iron and manganese (oxyhydr)oxide concentrations coincided with the oxygen penetration depth.

Deeper in the sediments, both elements are present in reduced form (Bakker & Helder, in press; Rajendran *et al.*, 1992). The oxygen penetration at 'Vlakte van Kerken' location was limited to the upper 3–4 mm (pers. comm. van den Ende, 1991).

The redox sensitive phosphorus fraction can be compared with the results from the flux under anoxic conditions. The mean amount of redox sensitive phosphorus in the western Dutch Wadden Sea is given in Table 3 for a 12 cm thick sediment layer, while in the same table the phosphorus release under anoxic conditions is also given. Further details are given below. Interestingly the phosphorus flux under anoxic conditions is twice the redox sensitive amount. This means that despite the shortcomings in the flux experiments (addition of 15 mmol l^{-1} acetate which may have led to substantial incorporation of phosphorus into bacteria) the results certainly have not led to an underestimation of the phosphorus release.

Bio-available phosphorus in the sediments

When the results of the sediment extraction are compared with those of the bio-availability of phosphorus, it can be concluded that, despite the nearly absence of organic phosphorus at the Malzwin sublittoral station (Table 1), the growth of the micro-organisms was substantial.

The amount of bio-available phosphorus in the sediments is 10 to 20% of the total sedimentary phosphorus (Table 1). The bio-available fraction is usually assumed to consist of the adsorbed and metal-associated P fractions. However, the amount of bio-available phosphorus is much higher than the sum of the above-mentioned fractions. Therefore, a part of the organic phosphorus or the calcium associated fraction (cf. de Groot, 1990) must be available to micro-organisms in order to explain the currently observed growth of heterotrophic bacteria. The observed slow or even absent growth of bacteria when sediments from greater depths were exposed to our bacteria mixture (cf. Fig. 6 and 9; sublittoral) indicates that organic phosphorus and calcium phosphorus are relatively non-labile. The lack of growth with sediment from 8–12 cm depth at station 'Vlakte van Kerken' (Fig. 9) coincided with the total absence of metal-associated P fraction (data not shown).

Spearman's rank correlation test revealed that the highest correlation between the biologically available phosphorus and the different chemically determined phosphorus fractions was obtained for a combination of the adsorbed and the metal-associated fraction, and implicitly also with the redox sensitive fraction ($0.85 < r < 0.96$ and $P < 0.005$ for all stations).

The bio-available phosphorus fraction must be considered as an important source of phosphate

Table 3. Maximal possible annual release in mol a^{-1} of phosphorus calculated from the mean amount of the different phosphorus fractions present in a 12 cm deep sediment column in the Marsdiep tidal basin. Corresponding potential primary production values are given assuming that the release of phosphorus fractions occurs within one year. Values based on data presented in Tables 1 and 2.

	Phosphorus extraction fraction				1 M HCl				Anoxic flux	Oxygenated flux
	NH_4Cl (adsorbed)	$Na_2S_2O_4$ (redox sensitive)	NaOH (metal associated)	0.5 M HCl (Ca associated)	(inorganic)	(organic)	(total)	(bio-available)		
Phosphorus release ($\times 10^6$ mol a^{-1})	5	15	19	174	244	107	325	55	33	8
Potential primary production ($g\,C\,m^{-2}\,a^{-1}$)	9	28	36	320	400	200	600	100	60	15

in the Wadden Sea in addition to external sources. It can contribute between 15 and 25 μmol l^{-1}, assuming that all the bioavailable phosphorus mixes with the water column instantly (Table 1). The contribution of sediment phosphorus to the water column under anoxic conditions is maximally only 0.15 μmol P $cm^{-2} d^{-1}$ (Fig. 8; Vlakte van Kerken, littoral), a value that corresponds with *ca* 2.6 μmol P l^{-1}. However, the comparison can in fact not be made because we have no information about the rate with which the bioavailable phosphorus reaches the water column. The use of the values in Tables 1 and 2 gives the best possible comparison, as will be shown below.

Possible role of sediment phosphorus in the local primary production

Evaluation of the role of the release of sediment phosphorus in the local phytoplankton primary production can be done by converting the values in Tables 1 and 2 to amounts per annum per area. The realistic value to base these calculations on is the average of the four sampling stations from which a flux (load) is calculated under the assumption that the phosphorus in each fraction will reach the water column within a growing season.

For Table 1 it is assumed that the phosphorus amounts present in a certain fraction in the 12 cm sediment layer of the entire tidal basin becomes available in the course of one year. These values are presented in Table 3.

The calculated annual fluxes during oxygenated or anoxic conditions given in Table 2 are simply converted to the entire area and also presented in Table 3. The result of these calculations is a load for the total Marsdiep tidal basin expressed in mol a^{-1} (Table 3). These loads can be compared with the SRP loads of Lake IJssel and the River Rhine presented in Figs 2 and 3 (PO_4^{3-}-P) (cf. De Jonge, 1990). If for instance only the bioavailable P is considered, this fraction would amount to *ca* 80% of the PO_4^{3-}-P load from Lake IJssel (Fig. 2) and about 30% of the PO_4^{-3}-P load of the Rhine (Fig. 3) for the year 1990.

These values also offer a possibility for estimating the potential annual primary production resulting from this 'internal' phosphorus load. This was done by applying the atomic Redfield ratio (C:P = 106:1) and assuming no re-use of phosphorus in the tidal basin. These primary production values are given at the bottom part of Table 3 where the assumption was made that phosphorus is indeed the major growth limiting factor in the area (De Jonge, 1990). Potential primary production data are given to enable a comparison with the measured values in this tidal basin which for this area are listed in a previous paper (De Jonge, 1990).

The potential annual phytoplankton primary production that may result from the diffusive flux under oxygenated conditions is 15 g C $m^{-2} a^{-1}$ (Table 3). This is less than 10% of the mean primary production value (148 g C $m^{-2} a^{-1}$) for the Marsdiep tidal basin over the period 1964–1986 (cf. De Jonge, op. cit.).

The potential primary production that may result from the flux under anoxic conditions amounts to 60 g C $m^{-2} a^{-1}$. This value is substantially higher than the value under oxygenated conditions, but at present anoxia is not a common feature in the Dutch Wadden Sea. Therefore, it can be assumed that the diffusive phosphorus release is at an intermediate value between anoxic and oxygenated conditions in the sediment.

The bio-availability experiments indicate that organic phosphorus does not play an important role in fuelling local primary production, at least not in the short term.

The measured primary production values for the Marsdiep tidal inlet and the inner part of the Marsdiep tidal basin range from 50 in 1950 to over 500 g C $m^{-2} a^{-1}$ in 1981–82 in the tidal inlet and from 20 in 1950 to 210 g C $m^{-2} a^{-1}$ in 1986 for the inner area of the same tidal basin (see listing in De Jonge, 1990). Potentially (see above) the bio-available phosphorus can contribute substantially to the local primary production when this fraction reaches the water column during one growing season. This can result from resuspension (De Jonge, 1992) or due to bioturbation by macrozoobenthos such as the lugworm (*Arenicola*

162

marina) that lives in tubes up to a depth of 30 cm. Release may also be due to effects of bio-irrigation on redox conditions. However, the importance of such a release under natural conditions has not been quantified yet because of lack of data. Despite this, it can be assumed that the actual role of this phosphorus flux in fuelling the local primary production will be relatively small, because the bio-available phosphorus in the 30 cm sediment layer (notice that this is 2.5 times the layer which was investigated) that may become available to the water column due to bioturbating organisms may account for a one time mean annual primary production of *ca* $250 \text{ g C m}^{-2} \text{ a}^{-1}$ (cf. Table 3; column 'bio-available' multiplied by 2.5). Moreover, it may be questioned whether this bio-available fraction reaches the water column within a single growing season.

Conclusions

1. Most of the sedimentary phosphorus in the western Dutch Wadden Sea is calcium-associated phosphorus. This fraction is believed to be not available to primary producers on the short term.
2. Organic phosphorus does not play a significant role in fuelling the local primary production in the western Dutch Wadden Sea on the short term.
3. Bio-available phosphorus can substantially contribute to the local primary production when it becomes available within a single growing season. It is predominantly associated with redox sensitive metal (oxyhydr)oxides.
4. Phosphorus fluxes out of the sediment in the western Dutch Wadden Sea under both anoxic and oxygenated conditions are insignificant for enhancing the local primary production.

Acknowledgements

We are grateful to Dr. H. van Gemerden (University of Groningen) for his discussions and suggestions. This has significantly improved the results of this study.

References

Admiraal, W. & D. Werner, 1983. Utilization of limiting concentrations of orthophosphate and production of extracellular organic phosphates in cultures of marine diatoms. J. Plankton Res. 5: 495–513.

Bakker, J. F. & W. Helder, in press. Skagerrak (northeastern North Sea). Oxygen micro profiles and pore water chemistry in sediments. Marine Geology, special issue on the Skagerrak.

Balzer, W., H. Erlenkeuser, M. Hartmann, P. J. Müller & F. Pollehne, 1987. In: J. Rumohr, E. Walger & B. Zeitzschel (eds), Lecture notes on coastal and estuarine studies, Springer-Verlag, Berlin, 112–158.

Boynton, W. R., W. M. Kemp, 1985. Nutrient regeneration and oxygen consumption by sediments along an estuarine salinity gradient. Mar. Ecol. Progr. Ser. 23: 45–55.

Bray, J. T., O. P. Bricker & B. N. Troup, 1973. Phosphate in interstitial waters of anoxic sediments: oxidation effects during sampling procedure. Science 180: 1362–1364.

Cadée, G. C., 1986. Increased phytoplankton primary production in the Marsdiep area (western Dutch Wadden Sea). Neth. J. Sea Res. 20: 285–290.

Cadée, G. C. & J. Hegeman, 1974. Primary production of phytoplankton in the Dutch Wadden Sea. Neth. J. Sea Res. 8: 240–259.

Cadée, G. C. & J. Hegeman, 1977. Distribution of primary production of the benthic microflora and accumulation of organic matter on a tidal flat area, Balgzand, Dutch Wadden Sea. Neth. J. Sea Res. 11: 24–41.

Cadée, G. C. & J. Hegeman, 1979. Phytoplankton primary production, chlorophyll and composition in an inlet of the western Dutch Wadden Sea (Marsdiep). Neth. J. Sea Res. 13: 224–241.

Canfield, D. E., 1989. Reactive iron in marine sediments. Geochim. Cosmochim. Acta 53: 619–632.

De Jonge, V. N., 1990. Response of the Dutch Wadden Sea ecosystem to phosphorus discharges from the River Rhine. In D. S. McLusky, V. V. de Jonge & J. Pomfret (eds), North Sea–Estuaries Interaction. Developments in Hydrobiology 55. Kluwer Academic Publishers, Dordrecht: 49–62. Reprinted from Hydrobiologia 195.

De Jonge, V. N., 1992. Physical processes and dynamics of microphytobenthos in the Ems estuary (The Netherlands). Ph. D. Thesis, Groningen, 195 pp.

De Jonge, V. N. & H. Postma, 1974. Phosphorus compounds in the Dutch Wadden Sea. Neth. J. Sea Res. 8: 139–153.

Duursma, E. K., 1961. Dissolved organic carbon, nitrogen and phosphorus in the sea. Neth. J. Sea Res. 1: 3–147.

Froelich, P. N., G. P. Klinkhammer, M. L. Bender, N. A. Luedtke, G. R. Heath, D. Cullen, P. Dauphin, D. Hammond, B. Hartman & V. Maynard, 1979. Early oxidation of organic matter in pelagic sediments of the eastern equatorial Atlantic: suboxic diagenesis. Geochim. Cosmochim Acta 43: 1075–1090.

Golterman, H., 1977. The role of iron in the exchange of phosphate between water and sediments. In: H. L. Golterman (ed.), Interactions between sediments and fresh water. Dr W. Junk Publishers, The Hague: 286–293.

Groot, C. de, 1990. Some remarks on the presence of organic phosphates in sediments. In D. J. Bonin & H. L. Golterman (eds), Fluxes between Trophic Levels and through the Water–Sediment Interface. Developments in Hydrobiology 62. Kluwer Academic Publishers, Dordrecht: 303–309. Reprinted from Hydrobiologia 207.

Hieltjes, A. H. M. & L. Lijklema, 1980. Fractionation of inorganic phosphorus in calcareous sediments. J. Envir. Qual. 9: 405–407.

Kolbe, K., 1991. Zum Auftreten 'schwarzer Flecken', oberflächlich anstehender, reduzierter Sedimente, im ostfriesischen Wattenmeer. Report Niedersächsisches Landesamt für Wasser und Abfall, Forschungsstelle Küste Norderney, 24 pp.

Krom, M. D. & R. A. Berner, 1981. The diagenesis of phosphorus in a nearshore marine sediment. Geochim. Cosmochim. Acta 45: 207–216.

Lijklema, L., 1977. The role of iron in the exchange of phosphate between water and sediments. In: H. L. Golterman (ed.), Interactions between sediments and fresh water. Dr W. Junk Publishers, The Hague: 313–317.

Lijklema, L., 1980. Interaction of orthophosphate with iron(III) and aluminium hydroxides. Environ. Sci. Technol. 14: 537–541.

Lowry, O. H., N. J. Roseborough, A. L. Farr & R. J. Randall, 1951. Protein measurement with the folin phenol reagent. J. Biol. Chem. 193: 265–275.

Murphy, J. & J. P. Riley, 1962. A modified single solution method for determination of phosphate in natural waters. Anal. Chim. Acta 27: 31–36.

Myers, C. R. & K. H. Nealson, 1988. Microbial reduction of manganese oxides: interactions with iron and sulphur. Geochim. Cosmochim. Acta 52: 2727–2732.

Parfitt, R. L., R. J. Atkinson & R. S. C. Smart, 1975. The mechanism of phosphate fixation by iron oxides. Soil Sci. Soc. Amer. Proc. 39: 837–841.

Postma, H., 1954. Hydrography of the Dutch Wadden Sea. Arch. néerl. Zool. 10: 405–511.

Postma, H., 1967. Sediment transport and sedimentation in the estuarine environment. In: G. H. Lauff (ed.), Estuaries. Am. Assoc. Adv. Sci. Publ. 83: 158–184.

Postma, H. & J. W. Rommets, 1970. Primary production in the Wadden Sea. Neth. J. Sea Res. 4: 470–493.

Psenner, R. & R Pucsko, 1988. Phosphorus fractionation: advantages and limits of the method for the study of sediment P origins and interactions. Arch. Hydrobiol. Beih. Ergebn. Limnol. 30: 43–59.

Raaphorst, W. van & H. W. van der Veer, 1990. The phosphorus budget of the Marsdiep tidal basin (Dutch Wadden Sea) in the period 1950–1985: importance of the exchange with the North Sea. In D. S. McLusky, V. V. de Jonge & J. Pomfret (eds), North Sea–Estuaries Interaction. Developments in Hydrobiology 55. Kluwer Academic Publishers, Dordrecht: 21–38. Reprinted from Hydrobiologia 195.

Rajendran, A., S. Gupta & J. F. Bakker, 1992. Control of manganese and iron in Skagerrak sediments. Chem. Geology, 09: 111–129.

Ridderinkhof, H. & J. T. F. Zimmerman, 1990. Residual currents in the western Dutch Wadden Sea. In: R. T. Cheng (ed.) Residual currents and long-term transport. Coastal and Estuarine Studies 38. Springer-Verlag, New York: 93–104.

Sherwood, B. A., S. L. Sager & H. D. Holland, 1987. Phosphorus in foraminiferal sediments from North Atlantic Ridge cores and in pure limestones. Geochim. Cosmochim. Acta 51: 1861–1866.

Suess, E., 1976. Nutrients near the depositional interface. In: I. N. McCave (ed.), The benthic boundary layer. Plenum Press, New York: 57–79.

Sundby, B. & N. Silverberg, 1985. Manganese fluxes in the benthic boundary layer. Limnol. Oceanogr. 30: 372–381.

Sundby, B, C. Gobeil, N. Silverberg & A. Mucci, 1992. The phosphorus cycle in coastal marine sediments. Limnol Oceanogr. 37: 1129–1145.

Veldhuis, M. J. W., F. Colijn, L. A. H. Venekamp & L. A. Villerius, 1988. Phytoplankton primary production and biomass in the western Dutch Wadden Sea (The Netherlands); a comparison with an ecosystem model. Neth. J. Sea Res. 22: 37–49.

Hydrobiologia **253**: 165–177, 1993.
P.C.M. Boers, T.E. Cappenberg & W. van Raaphorst (eds),
Proceedings of the Third International Workshop on Phosphorus in Sediments.
© 1993 *Kluwer Academic Publishers.*

Pore water phosphorus and iron concentrations in a shallow, eutrophic lake – indications of bacterial regulation

Å. Eckerrot & K. Pettersson
Institute of Limnology, Uppsala University, Box 557, S-75122 Uppsala, Sweden

Key words: pore water, phosphorus, iron, bacteria

Abstract

Peak pore water SRP and iron(II) concentrations were found during summer in surface sediments in the shallow and eutrophic L. Finjasjön, Sweden, and the concentrations generally increased with water depth. The SRP variation in surface sediments (0–2 cm) was correlated with temperature ($R^2 = 0.82$–0.95) and iron(II) showed a correlation with sedimentary carbon on all sites ($R^2 = 0.42$–0.96). In addition, sedimentary Chl*a*, bacterial abundances and production rates in surface sediments (0–2 cm) varied seasonally, with peaks during spring and fall sedimentation. Bacterial production rates were correlated with phosphorus and carbon in the sediment ($R^2 = 0.90$–0.95 and $R^2 = 0.31$–0.95, respectively), indicating a coupling with algal sedimentation. A general increase in sediment Chl*a* and bacterial abundances towards sediments at greater water depth was found. Further, data from 1988–90 reveal that TP and TFe concentrations in the lake were significantly correlated during summer ($R^2 = 0.81$ and 0.76, in the hypolimnion and epilimnion, respectively). The results indicate that the increase in pore water SRP and Fe(II) in surface sediments during summer is regulated by bacterial activity and the input of organic matter. In addition, spatial and temporal variations in pore water composition are mainly influenced by temperature and water depth and the significant correlation between TP and TFe in the water suggests a coupled release from the sediment. These findings support the theory of anoxic microlayer formation at the sediment-water interface.

Introduction

Studies of internal P loading have a long tradition within limnological research. In the early works of Mortimer (1941, 1942), the redox-coupled release of phosphorus and iron in stratified lakes with an anoxic hypolimnia was described. However, later research has added new information to the original theories and several authors have shown that a release of phosphorus and iron from 'oxic' sediments in shallow and polymictic lakes occurs, and may even be an important potential source of P (Kamp-Nielsen, 1975; Holdren & Armstrong, 1980; Drake & Heaney, 1987; Löfgren, 1987; Jensen & Andersen, in press). Theoretically, the formation of anoxic microlayers at the sediment-water interface may cause a P and Fe release even at high oxygen levels in the water mass. Enhanced microbial activity in the sediment in summer during high water temperature is believed to contribute to the formation of these layers (Löfgren, 1987; Boström *et al.*, 1988).

Several papers discuss the effects of temperature on the P-release (e.g. Kamp-Nielsen, 1975; Holdren & Armstrong, 1980; Löfgren & Boström, 1989; Søndergaard *et al.*, 1990; Jensen & Andersen, in press) and various authors have focused on the importance of microbial mineralization to pore water increase and/or sediment release of phosphorus and iron (e.g. Gächter et al., 1988; Sinke & Cappenberg, 1988; Boström et al., 1989; Löfgren & Boström, 1989; Boers, in press). Phosphorus in the pore water is of great importance for the internal loading due to a high mobility of the dissolved ions and can be used as an indicator of internal P loading in lakes (Holdren & Armstrong, 1986). A simultaneous increase in soluble reactive phosphorus (SRP) and ferrous iron (Fe(II)) in the pore water is often observed in eutrophic lakes during periods of high water temperature, where peak concentrations often are found close to the sediment-water interface, indicating an impact of microbial activities (Theis & McCabe, 1978; Enell & Löfgren, 1988; Löfgren & Boström, 1989; Boers, in press). However, some studies suggest that the P and Fe release may be partially uncoupled and caused by separate processes in the sediment (Boström et al., 1988; Gächter et al., 1988).

Further, in sediments with large amounts of P adsorbed to iron(III) compounds, the diffusive flux of P across the sediment-water interface is known to be low at high redox potentials. The adsorption capacity is however reduced at high pH, as demonstrated in several studies (e.g. Boström, 1984; Boers, 1991; Jensen & Andersen, in press). In shallow lakes where intense photosynthesis by phytoplankton can induce a high pH, it may have a crucial effect on the P-release (Jacoby et al., 1982; Drake & Heaney, 1987).

The present investigation was performed in the eutrophic, shallow and polymictic L. Finjasjön, Sweden. In this lake, trophic recovery after a reduction in the external P loading has been delayed and to improve water quality, lake restoration by partial sediment removal was introduced in the mid 1980s. However, despite efforts to restore the lake it still demonstrates an internal P loading, dense algal blooms with elevated pH-levels (> 10) and a rapid increase in total iron during summer. The aim of this paper is to discuss the following theories: i) the peak in SRP and Fe(II) in the surface pore water during summer is mainly caused by the activities of sediment bacteria, ii) the formation of anoxic microlayers at the sediment surface during high sediment oxygen demand are promoting a P and Fe release, iii) year-to-year fluctuations in pore water concentrations of SRP and Fe(II) in surface sediments are regulating the internal loading in L. Finjasjön, and derive from differences in temperature, wind mixing and sediment bacterial activities.

Material & methods

Study site

The comparatively large and shallow L. Finjasjön (Fig. 1; lat. 56° 08′, long. 13° 42′) is a eutrophic and polymictic lake with a surface area of 11 km² and an average water residence time of 5 months. The mean depth is approximately 4 m and a maximum depth of ca. 12 m is found in the southern basin. Of the five inlets to the lake, two are recipients for chemically treated sewage (Löfgren, 1987). Chemical treatment of waste water was introduced in the late 1970s, which reduced the external P load to L. Finjasjön by approximately 90% (Forsberg & Ryding, 1985) and water from the sewage treatment plant in Hässleholm (chemical precipitation of P) is discharged into the southern basin of the lake (Fig. 1). In 1988–1990, the external P loading to the lake amounted to 6.2, 4.8 and 4.4 tons y^{-1}, respectively, whereas the net internal P loading reached 5.7, 4.1 and 6.9 tons y^{-1}, respectively (Löfgren, 1991).

Despite efforts to restore the lake by partial sediment removal, large algal biomasses, mainly Microcystis sp., reappear every summer, and chlorophyll a in the epilimnion commonly reaches 150 μg Chl $a\, l^{-1}$. Due to its shallowness and large surface area, the lake seldom stratifies for longer periods, and earlier studies (Löfgren, 1987) even found stratifications with a duration of a few hours. Hence, the hypolimnion is usually oxic. However, temperature data from 1988 (Hässleholm sampling program) reveal 2 distinct stratifications between June and September with a duration of 1 and 5 weeks. Oxygen levels in the

L. FINJASJÖN

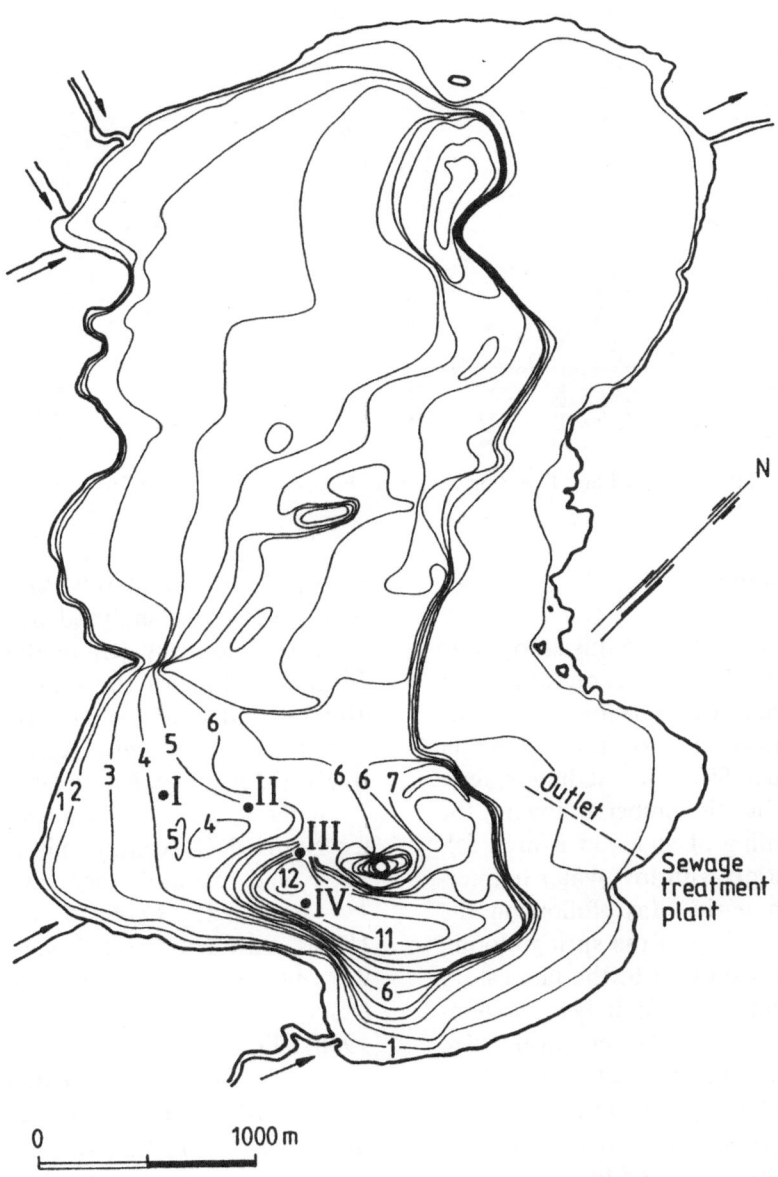

Fig. 1. Location of sampling sites I–IV in L. Finjasjön.

bottom water reached 0.1 mg O_2 l^{-1} during the longest period. At the breakdown of the stratification, total phosphorus and total iron concentrations in the epilimnion rapidly increased from *ca.* 0.07 to 0.34 mg TP l^{-1} and from *ca.* 0.28 to 0.90 mg TFe l^{-1}, respectively (Hässleholm data). In 1989, 4 stratifications lasting between 1 and 5 weeks were recorded between May and September. Consequently, TP and TFe concentrations varied in accordance with this (Fig. 2). Regarding 1990, hypolimnetic water data are not available for the whole summer, which is why the duration of stratifications is unknown.

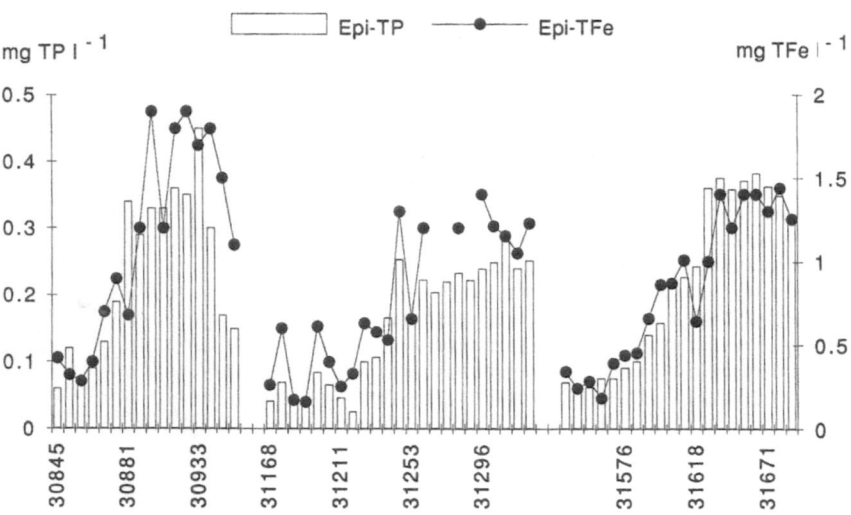

Fig. 2. Total phosphorus (TP) and total iron (TFe) (mg l^{-1}) in L. Finjasjön epilimnetic water, 1988–1990. Data supplied by Hässleholm municipality.

Field sampling and analyses

The four sampling sites chosen in this study were located at depths of 3–12 m in the southern basin (Fig. 1). Intact sediment was collected with a core sampler on 5 different occasions; in late summer (September '89/August '90) when sediment temperatures were high, in November following the fall overturn and settling of algae, in March following the winter period with low water temperature, and finally in early May, following the spring overturn and settling of the spring bloom. The sediment was transported to the laboratory and stored overnight at approximately *in situ* temperature after which the upper 0–2 cm layers were taken out and used for phosphorus fractionations and determination of bacterial abundance and production. Further, sediment samples were pre-frozen, freeze-dried and analysed for water content, chlorophyll *a*, C, P and Fe. For pore water analysis, duplicate sediment cores were incubated in the laboratory with segmented channels which were filled with distilled water and covered with a dialysis membrane, modified after Löfgren & Ryding (1985a). The cores were incubated in the dark for 5 days at approximately *in situ* temperature, after which the dialysates were taken out with a micropipette. 0.1 ml of the sample was

preserved in 4.9 ml 0.07 M HNO_3, whereafter total-Fe(II) was analysed by atomic absorption spectrophotometry (Shimadzu, AA-670; A_{248}). Further, 0.1 ml of the dialysate was diluted to 10 ml with distilled water and SRP was analyzed spectrophotometrically (A_{882}) as molybdate-reactive P (Murphy & Riley, 1968).

General sediment analyses were made on freeze-dried (Edwards, minifast 680) sediment. After acid oxidative digestion with a mixture of 1/4 conc. H_2SO_4 + 1/5 conc. HNO_3 + 1/20 $HClO_4$ and 1/2 distilled water, sedimentary TP were filtered and analysed spectrophotometrically (A_{882}) as molybdate-reactive P (Murphy & Riley, 1968). TFe was analysed by atomic absorption spectrophotometry (A_{248}) after oxidative acid digestion with 7 M HNO_3 according to Swedish Standard (SS 02 81 84). Sedimentary total carbon (TC) was analysed using a Carlo Erba Elemental Analyzer (Mod. 1106). The fractional composition of phosphorus (triplicate samples from one core) was determined by wet sediment sequential extraction, modified after Hieltjes & Lijklema (1980). 1 M NH_4Cl (pH 7), 0.1 M NaOH and 0.5 M HCl were used in the extraction procedure. Sedimentary chlorophyll *a* (triplicate samples from one core) was extracted from freeze-dried sediment with 96% ethanol and analysed spectrophotometrically at A_{750} and A_{665}. The method is modified after Hansson (1988). Bacterial abundances were determined from autofluorescence counting with epifluorescence microscopy, according to Boström et al. (1989). Formaldehyde-preserved (4%) sediment samples were diluted and filtered with 0.01% acridineorange on prestained (Sudanblack) 0.2 μm membrane filters. Bacterial production rates in surface sediments (triplicate samples from one core) were calculated from uptake of [^{3}H]-thymidine, after Bell & Ahlgren (1987). Incorporated [^{3}H]-thymidine was extracted in three steps; 1) hydrolysis with NaOH + sodium dodecyl sulphate (SDS) + EDTA, 2) precipitation with cold HCl + trichloroacetic acid (TCA), 3) hydrolysis with TCA. The samples were finally analysed on a liquid scintillation counter (LKB Wallace).

Data analyses

Relationships between pore water SRP, Fe(II) and various parameters were evaluated by single and multiple regression analyses on a StatView program (for MacIntosh). The number of observations (n) is given in Table 1.

Results

Water chemistry

Epilimnetic TP and TFe during 1988–1990 are shown in Figure 2 (data from Hässleholm sampling program). As expected, TP and TFe concentrations were correlated in the lake ($R^2 = 0.81$ and 0.76, in the hypolimnion and epilimnion, respectively). Phosphorus reached about the same level in 1988 and 1990, whereas in 1989, the concentrations were somewhat lower (Fig. 2). In addition, Fig. 3 illustrates SRP in the hypolimnion on sites I–IV. In accordance with epilimnetic data, much higher concentrations were found in the hypolimnion in late summer 1990, than in 1989.

Carbon, phosphorus and iron in surficial sediments

Generally, the sediment contained large amounts of organic material, which was reflected in the

Table 1. Characteristics of L. Finjasjön surficial sediment (0–2 cm), sites I–IV. Mean values and standard deviation (SD). P-fractionation, Chl$a$ and bacterial production are means from triplicate samples. Pore water data are means from duplicate cores.

	Site			
	I Mean ± SD	II[a] Mean ± SD	III Mean ± SD	IV Mean ± SD
Water depth (m)	3.5	4.5	7.0	11.5
Water content (%)	93.6 ± 2.0[b]	93.7 ± 1.9[c]	95.1 ± 1.4[b]	95.7 ± 1.9[b]
Total phosphorus (TP) (mg g^{-1} DW)	2.1 ± 0.3[c]	2.6 ± 0.4[d]	3.9 ± 1.0[c]	3.9 ± 0.8[c]
NH_4Cl-RP (% of TP)[d]	1.7	1.4	2.0	2.2
NaOH-RP (% of TP)[d]	51.6	52.4	55.8	48.6
HCl-RP (% of TP)[d]	26.8	21.9	20.2	13.9
Res-P (% of TP)[d]	16.9	24.1	21.8	35.1
Total iron (TFe) (mg g^{-1} DW)	69 ± 144[c]	59 ± 8[d]	61 ± 3[c]	76 ± 1[c]
Total carbon (TC) (mg g^{-1} DW)	115 ± 2[c]	102 ± 1[d]	125 ± 2[c]	132 ± 17[c]
SRP (mg l^{-1})	2.0 ± 1.8[e]	1.2 ± 0.7[f]	2.4 ± 1.8[e]	2.9 ± 2.9[e]
Fe(II) (mg l^{-1})	8 ± 7.3[e]	20.1 ± 4.1[f]	21.6 ± 7.7[e]	45.9 ± 28.3[e]
Chl$a$ (μg g^{-1} DW)	186 ± 32[g]	191 ± 44[g]	281 ± 20[h]	400 ± 104[h]
Bacterial abundance (10^{10} cells g^{-1} DW)	8.4 ± 1.3[c]	10.2 ± 4.5[d]	11.8 ± 4.7[c]	15.6 ± 4.7[c]
Bacterial production (μg C g^{-1} DW hr^{-1})	3.0 ± 3.5[g]	0.3 ± 0.2[i]	1.3 ± 0.4[h]	2.1 ± 2.4[h]

[a] Site II not included in August 1990, [b] $n = 5$, [c] $n = 4$, [d] $n = 3$, [e] $n = 32$, [f] $n = 24$, [g] $n = 12$, [h] $n = 15$, [i] $n = 9$.

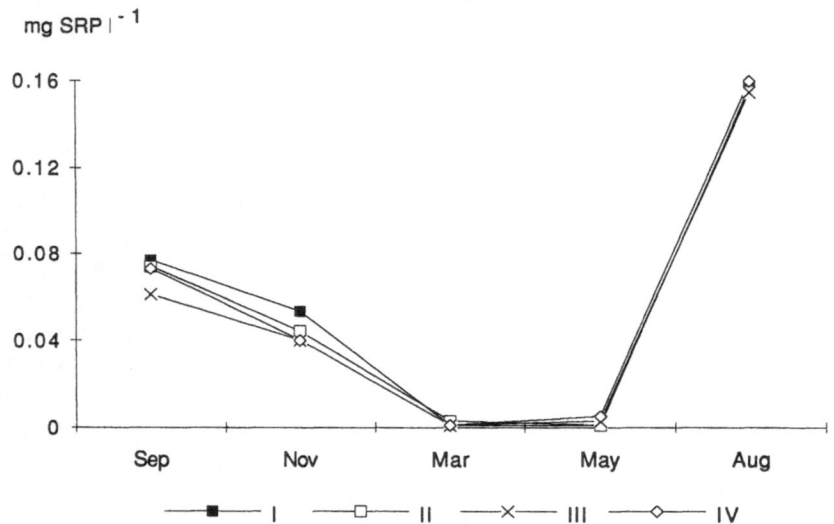

Fig. 3. SRP in hypolimnetic water (mg l^{-1}) on sites I–IV, September 1989–August 1990.

water content of the surficial layer (*ca.* 94–96%). The thickness of the organic, nutrient-rich sediment layer increased with increasing water depth and varied between *ca.* 0.3 and 1.5 m. TC varied between *ca.* 100 and 150 mg C g^{-1} DW (dry weight), as illustrated in Table 1. In addition, an increase with water depth was found during fall sedimentation 1989 ($R^2 = 0.85$, $p < 0.1$). Total phosphorus in the surficial sediment varied sea-

sonally and a general increase towards deeper areas was found ($R^2 = 0.76$, $p < 0.2$). The concentrations ranged from 2.0 to 4.2 mg TP g^{-1} DW (Table 1). The fractional composition of phosphorus in the top 2 cm sediment was analysed in November 1989. As expected, iron + aluminium-bound phosphorus (NaOH-RP) constituted the main sediment fraction, *ca.* 50–60% of TP (Table 1). The amount of calcium-bound phosphorus

Table 2. Results from single and multiple regression analyses[a], expressed as R^2.

Analysis	Site							
	I		II[b]		III		IV	
	R^2	p	R^2	p	R^2	p	R^2	p
P_w-SRP:Temp[c]	0.95	<0.05	0.86	<0.5	0.90	<0.06	0.82	<0.5
P_w-Fe(II):Temp[d]	0.37	<0.1	0.75	<0.1	0.13	<0.6	0.68	<0.1
P_w-SRP:Sed-TC[d]	0.12	<0.7	0.78	<0.4	0.15	<0.7	0.62	<0.3
P_w-Fe(II):Sed-TC[d]	0.62	<0.3	0.96	<0.2	0.42	<0.4	0.65	<0.2
Bact. prod:Sed-TP[c]	0.95	<0.2	–	–	0.90	<0.2	0.95	<0.2
Bact. prod:Sed-TC[d]	0.95	<0.03	0.31	<0.7	0.92	<0.05	0.62	<0.3

[a] *n*-values are given in Table 1.

[b] Site II not included in August 1990.

[c] Multiple regression analyses (polynomial) performed with pore water SRP vs temperature, and bacterial production vs sedimentary TP, respectively. Adjusted R^2-value.

[d] Single regression analyses performed with pore water iron(II), pore water SRP and bacterial production rates vs temperature and sedimentary TC, respectively.

(HCl-RP) varied between *ca.* 15 and 30% of TP and decreased with water depth. The loosely adsorbed phosphorus fraction (NH$_4$Cl-RP), reflecting the recently settled algal material, and the residual-P fraction (Res-P) both increased towards deeper sites (R^2 = 0.85 and 0.96, $p < 0.1$ and 0.05, respectively) and constituted between 1.4–2.2% and 17–35% of TP, respectively (Table 1). Further, L. Finjasjön sediment is characterized by its high iron content and concentra-

tions between 50 and 90 mg TFe g^{-1} DW were measured in the surficial sediment (Table 1). However, iron was not correlated with water depth and did not show a particular co-variation with the sedimentation events. Compared to the concentrations on other sites between May and August 1990, a drastic TFe increase on site I, from 66 to 90 mg TFe g^{-1} DW, is somewhat unexpected and the cause of this anomaly is unknown.

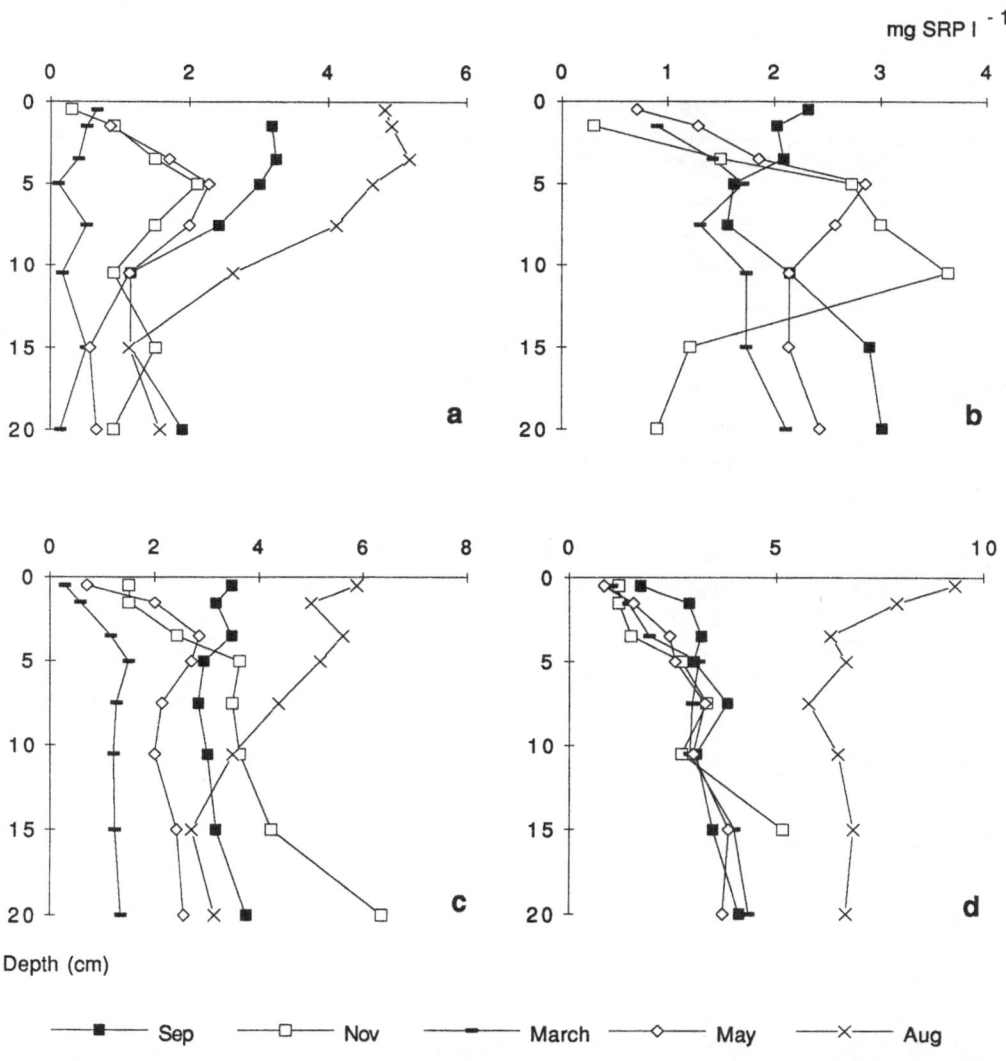

Fig. 4. 4a–d illustrate pore water SRP (mg l^{-1}) on sites I–IV, respectively. September 1989–August 1990. Mean values from duplicate cores, 0–20 cm sediment profile. Note different scales on X-axes.

Phosphate and iron (II) in pore water

Peak SRP concentrations were commonly found in surficial sediments, and varied between *ca.* 0.3 and 9.0 mg l^{-1} (means; 0–2 cm, Table 1). Generally, SRP increased with water depth (Fig. 4) and showed a temperature correlated variation on all sites ($R^2 = 0.82$–0.95, Table 2). A drastic change in the pore water was registered in 1990 when the concentrations in the 0–2 cm layer increased by 400–700% from May to August

(Fig. 4). Elevated SRP-concentrations were measured even at a sediment depth of 20–25 cm.

Further, pore water concentrations of iron(II) showed a seasonal variation, a general increase with water depth and were correlated with TC on all sites ($R^2 = 0.42$–0.96, Table 2). The concentrations (means; 0–2 cm) varied between *ca.* 20–300 mg Fe(II) l^{-1} and the peak value was found in sediments at site IV (Fig. 5). Further, iron(II) concentrations in the surface layer at site IV increased by nearly 375% from May to August

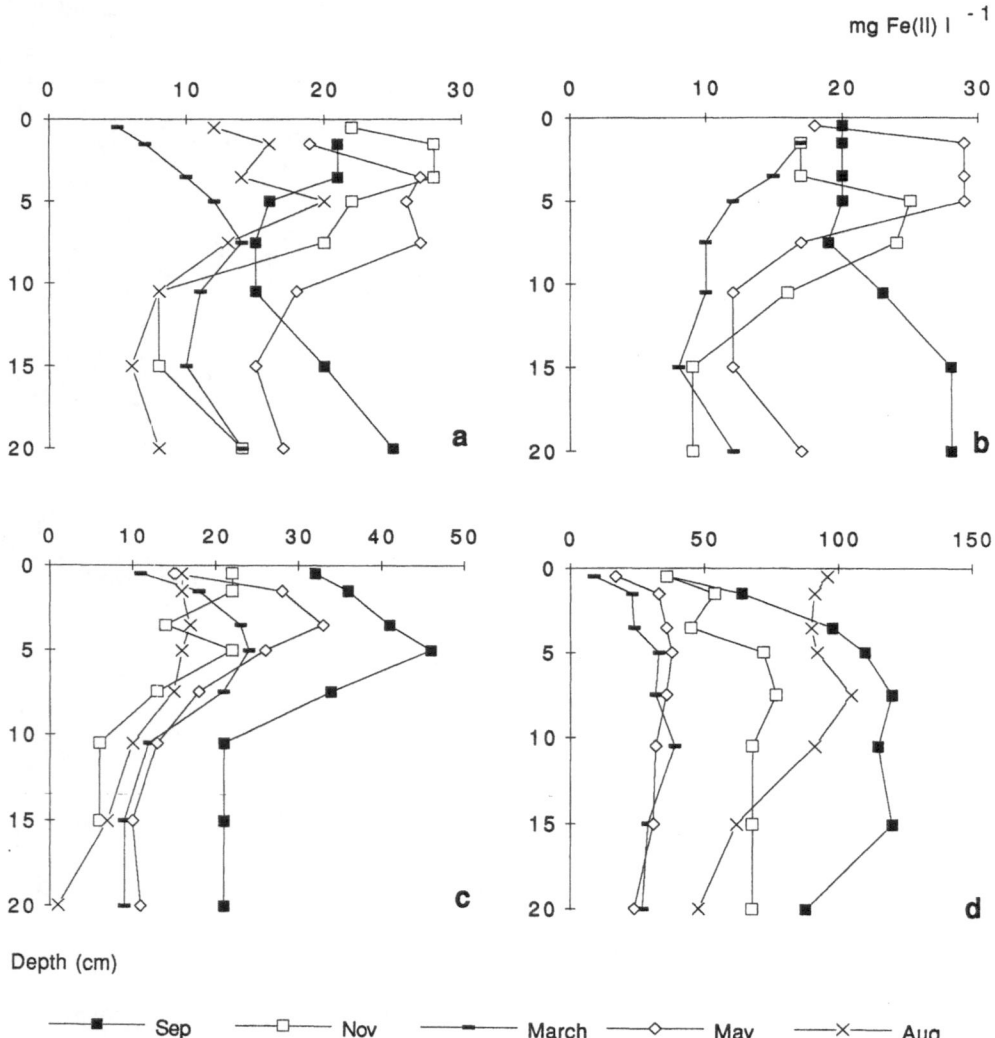

Fig. 5. 5a–d illustrate pore water Fe(II) (mg l^{-1}) on sites I–IV, respectively. September 1989–August 1990. Mean values from duplicate cores, 0–20 cm sediment profile. Note different scales on X-axes.

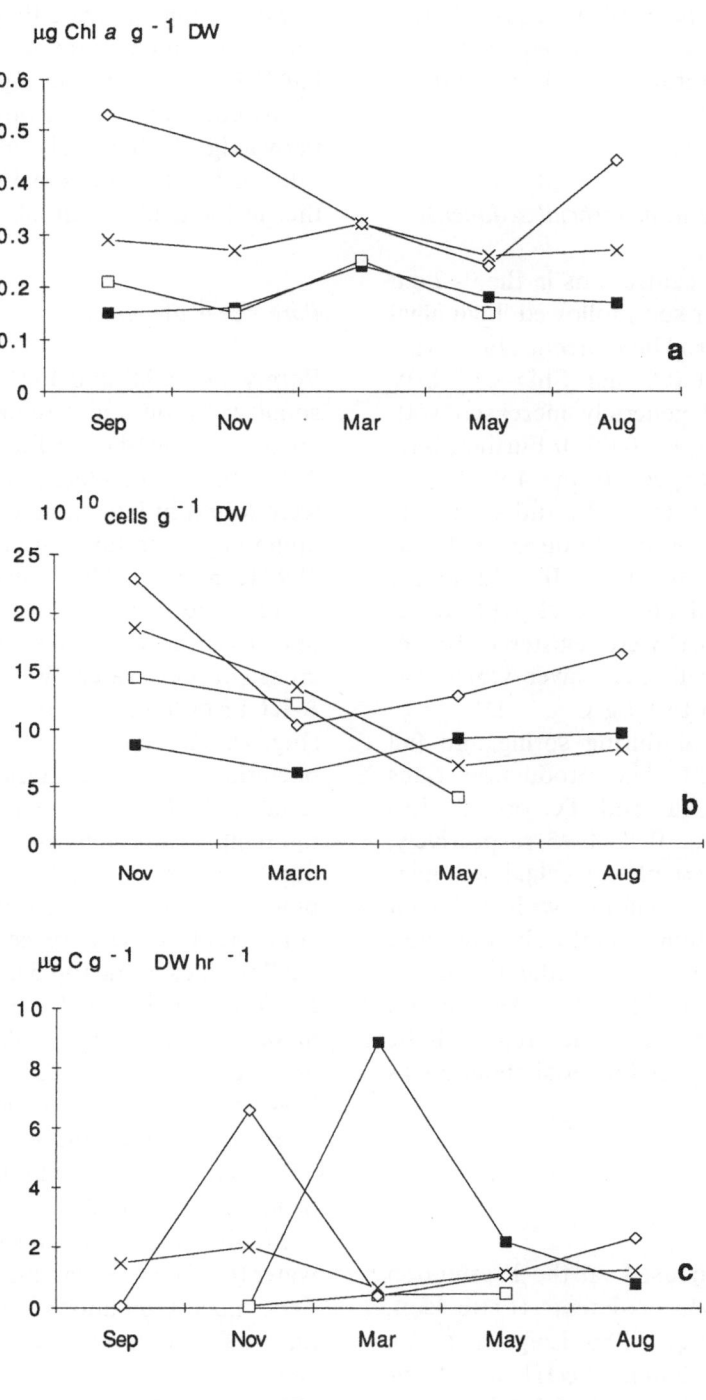

Fig. 6. Chlorophyll *a* (μg g^{-1} DW, 6a), bacterial abundances (10^{10} cells g^{-1} DW, 6b) and production rates (μg C g^{-1} DW y^{-1} hr, 6c) in L. Finjasjön surficial sediment (0–2 cm), sites I–IV.

1990 concurrent with the SRP increase. Hence, the iron(II) variation was correlated with pore water SRP and temperature ($R^2 = 0.85$ and 0.78, respectively).

Chlorophyll a and bacteria in surficial sediments

Peak chlorophyll a concentrations in the 0–2 cm layer coincided with, or soon followed, high algal biomasses in the water. The concentrations varied between 0.15 and 0.53 mg Chl$a$ g^{-1} DW (Fig. 6a, Table 1) and generally increased with water depth ($R^2 = 0.66$, $p < 0.001$). Further, bacterial abundances ranged from 4×10^{10} to 23×10^{10} cells g^{-1} DW (Fig. 6b) and were correlated to sediment chlorophyll$a$ during fall sedimentation ($R^2 = 0.98$, $p < 0.05$). In addition, a seasonal variation and an increase with water depth ($R^2 = 0.86$, $p < 0.1$) were registered. Bacterial production in the 0–2 cm layer varied between sites, from 0.05 to 9 μg C g^{-1} DW hr^{-1} (Table 1) and increased during spring and fall sedimentation (Fig. 6c). The production rates were correlated with TP and TC on all sites ($R^2 = 0.90$–0.95 and $R^2 = 0.31$–0.95, respectively, Table 2), indicating a coupling to algal sedimentation. A significant correlation with sediment Chl$a$ was accordingly found on the shallow site I ($R^2 = 0.997$, $p < 0.005$). Further, the maximum production rate (9 μg C g^{-1} DW hr^{-1}) was found on site IV in November 1989, concurrent with the fall sedimentation and peak bacterial abundances on this site (Fig. 6).

Discussion

Earlier studies have focused on the P retention capacity of L. Finjasjön sediments (Pettersson, 1984; Löfgren & Ryding, 1985b; Löfgren, 1987). Due to the high iron content, Fe(III) and P interactions have been known to dominate the sediment-water flux of P. In conclusion, previous investigations of the phosphorus retention capacity in L. Finjasjön sediments suggest a strong correlation to temperature, pH and redox conditions.

These findings indicate that bacterial activities in surface sediments may have a crucial impact on the P/Fe dynamics during summer and significantly contribute to the observed release. To our knowledge, bacterial abundances and production rates in L. Finjasjön have yet been analysed neither in the field nor in laboratory experiments.

Pore water dynamics

Pore water SRP and Fe(II) showed a clear seasonal variation and a temperature correlated increase was found regarding SRP ($R^2 = 0.82$–0.95, Table 2). As expected, maximum concentrations were found in late summer both years, and minimum concentrations after the winter, in March 1990 (Figs 4–5). These seasonal variations are in accordance with previous studies of other lakes (e.g. Holdren & Armstrong, 1986). A rapid increase in pore water SRP was measured on sites I–III prior to a rise on site IV in spring 1990 (Fig. 4). The high bacterial production rates and bacterial abundances at the time suggest that bacteria in shallow sediments rapidly responded to the temperature increase (6–12.5 °C) and to the input of algal material during the settling spring bloom (Fig. 6). This is further supported by the concurrent Fe(II) increase that co-varied with the SRP profiles in the top 5 cm of sediment (Figs 4–5). Although P and Fe levels in the pore water increased, the P/Fe molar ratios actually decreased. This indicates that some P may have been incorporated into bacterial biomass and/or released into the bottom water. The concurrent TP increase in the epilimnion indicates a P contribution from shallow areas.

Maximum P/Fe molar ratios in the surface pore water (0.18–0.63) were measured on sites I–III in September 1989 and August 1990. The sediment on site IV, however, differs considerably from the shallower locations. Thus, due to a much more reduced environment (the sediment is black, smells of H_2S and methane gas frequently generates bubbles in the cores in summer), iron(II) levels are substantially higher on this site and the P/Fe molar ratios are consequently lower. Maximum ratios were measured in March and August, 1990 (0.14 and 0.17, respectively). These results support the other data and illustrate the impact of water depth to the pore water composition. The elevated P and Fe concentrations in site IV indicate that phosphate mineral formation might occur, such as vivianite [Fe$_3$(PO$_4$)$_2$ ∗ 8H$_2$O], and that Fe(II) bound to e.g. carbonates might be dissolved (Holdren & Armstrong, 1986; Enell & Löfgren, 1988). Previous investigations (Löfgren & Ryding, 1985a) have suggested that apatite [Ca$_5$(PO$_4$)$_3$(OH)] may be important in regulating P concentrations in anaerobic sediment layers. According to Löfgren & Boström (1989), earlier studies have shown a P/Fe molar ratio of ca. 0.06–0.2 in the Fe(III)/hydroxide-P complex at pH = 7, which is the approximate sediment-pH in L. Finjasjön (measured in November and March). Hence, the shallow areas in the lake may be incapable of retaining P in the sediment in late summer due to the high P/Fe molar ratios (0.18–0.63), whereas the deepest sites appear to have low enough ratios all year round (0.05–0.17).

Further, the SRP increase by 400–700% (0–2 cm layer) over the summer of 1990 occurred during a temperature increase of 12.5° to 19.5 °C. The results can be compared to the results of Kamp-Nielsen (1975) who found that an increase in temperature by 5 °C would lead to an increase in the P release by 100% or more, at temperatures above 15 °C. Kamp-Nielsen (op. cit.) ascribed this to a breakdown of the oxidized microlayer on the sediment surface. Søndergaard et al. (1990) presented similar results in the temperature range of 14–18 °C for L. Søbygaard, Denmark and concluded that the temperature-dependent release might be explained by microbial and benthic activity. Thus, the coupled P and Fe increase in L. Finjasjön in summer (Fig. 2), indicates the formation of anoxic microlayers on the sediment surface. A pH dependent release is less probable, due to the simultaneous increase in TP and TFe. A high pH may induce a phosphorus release, but it does not enhance the solubility of iron (Löfgren, 1987). Further, the pH in the water mass may not affect the sediment pH, especially during periods

of high microbial activity, as demonstrated by Revsbech et al. (1983). Thus, elevated pH levels cannot explain the TFe increase in the lake during summer.

Implication of anoxic microlayers for P and Fe release

According to Jørgensen & Revsbech (1985), the potential for bacterial oxygen respiration may increase by 300–800% between 5–25 °C. Bacterial mineralization of organic matter increases the ion concentrations in the pore water and a simultaneous reduction of Fe(III) may occur (Lovley & Phillips, 1986). Lovley and Phillips (op. cit.) found the most intense Fe(III) reduction occurring in the top 0.5 cm sediment, and the reduction was limited to the top 4 cm or less. This is in accordance with the Fe(II) profiles in L. Finjasjön, where peak concentrations frequently were found in the top 5 cm of sediment (Fig. 5). In addition, the correlation between iron(II) and organic carbon (TC) on all sites (R^2 = 0.42–0.96, Table 2), strongly indicates that Fe(III) reduction was coupled to mineralizing bacteria, as proposed by Lovley & Phillips (op. cit.). However, iron can only be released to the overlying water when anoxic conditions develop at the sediment surface (Löfgren, 1987). The diffusion of Fe(II) through the diffusive boundary layer (DBL) is an oxygen consuming process which partially may explain the development of anoxic microlayers, and it is likely that the chemical oxidation of iron(II) to iron(III) can compete with the biological uptake of oxygen in this layer (Löfgren, 1987). The joint effect of biological respiration and chemical oxidation processes in summer lowers the likelihood of P adsorption at the sediment surface and favours a substantial P release. However, rapid shifts between oxidized and reduced conditions might occur in shallow lakes. Hence, low bacterial production rates were found in September 1989 and August 1990 at maximum SRP levels in the surface pore water (Figs 4, 6c). The profiles indicate a shift from oxidized to reduced conditions, as well as a collapse and sedimentation of a dense phytoplankton bloom (Enell & Löfgren, 1988). The high Chl$a$ values in the sediment at the time support this theory (Fig. 6a). Similar results were also documented by Löfgren & Boström (1989) in the eutrophic L. Vallentunasjön, where high SRP concentrations were found in the pore water despite a decrease in bacterial respiration. However, temporal and spatial variations in bacterial activity may be expected, and the limited number of observations during summer 1990 does not permit any firm conclusions.

Regulation of internal loading

The redox-coupled release of P and Fe in L. Finjasjön is apparent, as illustrated in Fig. 2. In 1990, P in the epilimnion increased from 0.07 to 0.36 mg TP l^{-1} over a 4-months' period, meaning that the P pool in the lake increased by *ca.* 12 tons. Porewater concentrations revealed a similar pattern. In shallow areas, the SRP increase in the 0–2 cm layer corresponds to *ca.* 0.7 mg SRP m^{-2} day^{-1}, whereas the increase in deeper areas amounted to *ca.* 1.2 mg SRP m^{-2} day^{-1}. The TP increase in the epilimnion, however, corresponds to an estimated flux of 104 mg P m^{-2} day^{-1} (mean; 4-months' period) meaning that also deeper sediment layers must be involved in the release. The results are comparable with the findings of Boström (1984) and Löfgren (1987) who estimated the net release rates in L. Finjasjön at 63 and 91 mg P m^{-2} day^{-1} from laboratory and field data, respectively. Further, the TFe increase in epilimnetic water amounted to 37 tons during the same period. The pore water Fe(II) increase on site IV (0–2 cm layer) was *ca.* 11.5 mg Fe(II) m^{-2} day^{-1}, as compared to the TFe increase in the epilimnion, corresponding to 3093 mg TFe m^{-2} day^{-1}. Thus, to explain the increasing TFe concentrations in the water, considerable amounts of Fe(II) must be released from deeper sediment layers.

The calculated net internal P loading was nearly 2.8 tons, or 70%, higher in 1990, than in 1989 (Löfgren, 1991) and may be related to pore water SRP concentrations which were *ca.* 60–80%

higher in the late summer of 1990 (Fig. 4). The results strongly indicate that yearly variations in pore water concentrations are regulating the internal loading in L. Finjasjön.

Conclusion

In conclusion, the results indicate that the phosphorus and iron increase in L. Finjasjön surficial pore water in summer is mainly governed by the activities of sediment bacteria. Further, spatial and seasonal variations in the lake are influenced by temperature, water depth and wind mixing. Thus, the aerobic P and Fe release will likely continue, although year-to-year shifts in the internal loading may be expected.

Acknowledgements

We thank S. Wallström, Hässleholm municipality, for his assistance with sediment sampling in 1989–90, and supplied data from Hässleholm municipality are acknowledged. Collegues at the Institute of Limnology, Uppsala, are recognized for their assistance, and Dr R. G. Titus and Prof. C. Forsberg are acknowledged for the fruitful discussions regarding data and manuscript preparation. Critical comments from two anonymous referees are especially recognized. This investigation was financially supported in part by the Hässleholm municipality, and in part by the J. Gust. Richert Foundation.

References

Bell, R. T. & I. Ahlgren, 1987. Thymidine incorporation and microbial respiration in the surface sediment of a hypereutrophic lake. Limnol. Oceanogr. 32: 476–482.

Boers, P. C. M., Ion concentrations in interstitial water as indicators for phosphorus release processes and reactions. Wat. Res. In press.

Boers, P. C. M., 1991. The influence of pH on phosphate release from lake sediments. Wat. Res. 25: 309–311.

Boström, B., 1984. Potential mobility of phosphorus in different types of lake sediment. Int. Revue ges. Hydrobiol. 69: 457–474.

Boström, B., A.-K. Pettersson & I. Ahlgren, 1989. Seasonal dynamics of a cyanobacteria-dominated microbial community in surface sediments of a shallow, eutrophic lake. Aquat. Sci. 51: 153–178.

Boström, B., J. M. Andersen, S. Fleischer & M. Jansson, 1988. Exchange of phosphorus across the sediment–water interface. In G. Persson & M. Jansson (eds), Phosphorus in Freshwater Ecosystems. Developments in Hydrobiology 48. Kluwer Academic Publishers, Dordrecht: 229–244. Reprinted from Hydrobiologia 170.

Drake, J. C. & S. I. Heaney, 1987. Occurence of phosphorus and its potential remobilization in the littoral sediments of a productive English lake. Freshwat. Biol. 17: 513–523.

Enell, M. & S. Löfgren, 1988. Phosphorus in interstitial water: methods and dynamics. In G. Persson & M. Jansson (eds), Phosphorus in Freshwater Ecosystems. Developments in Hydrobiology 48. Kluwer Academic Publishers, Dordrecht: 103–132. Reprinted from Hydrobiologia 170.

Forsberg, C. & S-O. Ryding, 1985. 'Hur tillfrisknade sjöarna?' Sammanfattning av Naturvårdsverkets RR-undersökning. Vatten: 3–19 (in Swedish).

Gächter, R., J. S. Meyer & A. Mares, 1988. Contribution of bacteria to release and fixation of phosphorus in lake sediments. Limnol. Oceanogr. 33: 1542–1558.

Hansson, L.-H., 1988. Chlorophyll *a* determination of periphyton on sediments: identification of problems and recommendation of method. Freshwat. Biol. 20: 347–352.

Hieltjes, A. H. M. & L. Lijklema, 1980. Fractionation of inorganic phosphates in calcareous sediments. J. envir. Qual. 9: 405–407.

Holdren, G. C. & D. E. Armstrong, 1980. Factors affecting phosphorus release from intact lake sediment cores. Envir. Sci. Tech. 14: 79–87.

Holdren, G. C. & D. E. Armstrong, 1986. Interstitial ion concentrations as an indicator of phosphorus release and mineral formation in lake sediments. In: P. G. Sly (ed.), Sediments and Water Interactions. Springer Verlag.

Jacoby, J. M., D. D. Lynch, E. B. Welch & M. A. Perkins., 1982. Internal phosphorus loading in a shallow eutrophic lake. Wat. Res. 23: 1357–1366.

Jensen, H. S. & F. O. Andersen. Importance of temperature, nitrate and pH for phosphate release from aerobic sediments of four shallow, eutrophic lakes. Limnol. Oceanogr. In press.

Jørgensen, B. B. & N. P. Revsbech, 1985. Diffusive boundary layers and the oxygen uptake of sediments and detritus. Limnol. Oceanogr. 30: 111–122.

Kamp-Nielsen, L., 1975. A kinetic approach to the aerobic sediment-water exchange of phosphorus in L. Esrom. Ecol. Model. 1: 153–160.

Lovley, D. R. & E. J. P. Phillips, 1986. Organic matter mineralization with reduction of ferric iron in anaerobic sediments. Appl. envir. Microbiol. 51: 683–689.

Löfgren, S., 1987. 'Aerobic' sediment iron and phosphorus release in L. Finjasjön, a shallow, eutrophic, Swedish lake. In: Phosphorus retention in sediments – implications for aerobic phosphorus release in shallow lakes. Ph.D. Thesis, Institute of Limnology, University of Uppsala. Sweden.

Löfgren, S., 1991. Finjasjön – miljösituation samt förslag till miljömål och åtgärder. Utredning åt Gatukontoret, Hässleholms kommun. 23 p. In Swedish.

Löfgren, S. & S.-O. Ryding, 1985a. Apatite ionic products in different eutrophic sediments. Verh. Int. Ver. Limnol. 22: 3323–3328.

Löfgren, S. & S.-O. Ryding, 1985b. Apatite solubility and microbial activities as regulators of internal loading in shallow, eutrophic lakes. Verh. int. Ver. Limnol. 22: 3329–3334.

Löfgren, S. & B. Boström, 1989. Interstitial water concentrations of phosphorus, iron and manganese in a shallow, eutrophic swedish lake – implications for phosphorus cycling. Wat. Res. 23: 1115–1125.

Mortimer, C. H., 1941. The exchange of dissolved substances between mud and water in lakes. I. J. Ecol. 29: 280–329.

Mortimer, C. H., 1942. The exchange of dissolved substances between mud and water in lakes. II. J. Ecol. 30: 147–201.

Murphy, J. & J. P. Riley, 1968. A single-solution method for the determination of phosphate in seawater. J. mar. biol. Ass. U.K. 37: 9–14.

Pettersson, K., 1984. Transfers between forms of sedimentary phosphorus induced by nitrate treatment. Verh. int. Ver. Limnol. 22: 233–238.

Revsbech, N. P., B. B. Jørgensen, T. H. Blackburn & Y. Cohen, 1983. Microelectrode studies of the photosynthesis and O_2, H_2S and pH profiles of a microbial mat. Limnol. Oceanogr. 28: 1062–1074.

Sinke, A. J. C. & T. E. Cappenberg, 1988. Influence of bacterial processes on the phosphorus release from sediments in the eutrophic Loosdrecht Lakes, The Netherlands. – Arch. Hydrobiol. Beih. 30: 5–13.

Swedish Standard, SS 02 8184. 1988. In Swedish.

Søndergaard, M., E. Jeppesen, P. Kristensen & O. Sortkjær, 1990. Interactions between sediment and water in a shallow and hypereutrophic lake: a study on phytoplankton collapses in lake Søbygård, Denmark. Hydrobiol. 191: 139–148.

Theis, T. L. & P. J. McCabe, 1978. Phosphorus dynamics in hypereutrophic lake sediments. Wat. Res. 12: 677–685.

Hydrobiologia **253**: 179–192, 1993.
P.C.M. Boers, T.E. Cappenberg & W. van Raaphorst (eds),
Proceedings of the Third International Workshop on Phosphorus in Sediments.
© 1993 *Kluwer Academic Publishers.*

Ecological significance of bacterial polyphosphate metabolism in sediments

Danielle Davelaar
Chemnitzerstr. 13, D-6702 Bad Dunkheim, Germany

Key words: Heterotrophic bacteria, polyphosphate metabolism, dissimilatory manganese and iron reduction, ecological niche, phosphorus cycle, eutrophication

Abstract

The purpose of this study was to find theoretical evidence that bacteria, in particular those capable of polyphosphate (polyP) metabolism, are directly implicated in sediment phosphorus (P) dynamics and control P metabolism of freshwater ecosystems. The specific attributes and functional role of such bacteria were investigated on successive levels of ecological organization: individual microorganism, microbial community, freshwater ecosystem. The results of this systematic approach have been formulated as a number of hypotheses.
(1) PolyP metabolism is the mechanism which enables individual polyP bacteria to survive and grow under the fluctuating redox conditions characteristic of their habitat at the sediment-water interface.
(2) PolyP metabolism together with anaerobic Mn and/or Fe respiration is the mechanism that confers upon polyP bacteria the advantage required to fill a unique ecological niche within the microbial community to which they belong.
(3) To the freshwater ecosystem as a whole bacterial polyP metabolism is a homeostatic mechanism which limits P availability and makes ecosystem productivity self-correcting as a function of oxygen availability. Bacterial polyP pools in the sediment are vital components of the P cycle. It was suggested that the impact of this bacterial mechanism should be tested with regard to the eutrophication issue.

Introduction

The internal phosphorus (P) loading of the sediment and the related periodical release of P into interstitial water and overlying water column are factors of major concern to water managers in charge of eutrophication control and restoration of freshwater ecosystems. It is important to understand which are the chief processes that control P dynamics at the sediment–water interface and to quantify them, to be able to act upon them. According to Boström *et al.* (1982), the essential environmental factor controlling P dynamics at the sediment–water interface is the oxidation-reduction potential (ORP). Because the two processes generally associated with P behaviour in sediments, chemical iron-phosphorus interaction and bacterial metabolism, are both redox driven reactions, it is difficult to obtain clear evidence as to what extent each mechanism is responsible. Classical theory (Einsele (1936, 1938) and Mortimer (1941, 1942) model) overemphasized the chemical side of the explanation while the role of microorganisms remained underestimated for a long time. However, researchers now regularly report on the importance of microbial processes in sediment P dynamics (Fleischer, 1978, 1983; Boström *et al.*, 1982, 1988; Gächter *et al.*, 1988).

Basically, sediment bacteria influence the dynamics of both (i) chemically bound P and (ii) biologically bound P. (i) Bacterial respiration alters ambient redox conditions, thereby indirectly favouring the chemical Fe/P solubilization process; (ii) is a direct action: P assimilated in biomass is again mobilized during mineralization. Normal bacterial P metabolism, however, is of minor quantitative importance in the overall P movement at the sediment–water interface, as the stoichiometric composition of living cell material ($C_{106}H_{263}O_{110}N_{16}P$) reflects a low P content: microorganisms normally do not fix large quantities of P. In the case of polyphosphate (polyP) metabolism this is different: some bacteria accumulate P intracellularly in luxury amounts, up to about 20% of the dry weight in *Acinetobacter* spp. (Deinema *et al.*, 1980). Could the impact of bacterial polyP metabolism therefore be a factor of importance in sediment P dynamics?

The biochemical and microbiological aspects of polyP metabolism in microorganisms have been studied extensively (Harold, 1966; Dawes & Senior, 1973; Kulaev, 1979). However, its physiological function (energy or phosphate reserve, metabolic regulator) could not be clarified. At the same time, bacterial polyP metabolism in the activated sludge process for wastewater treatment was discovered and extensively studied as well. Marais *et al.*(1983) and Toerien *et al.* (1990) give comprehensive reviews of three decades of research in this technological field. Although full-scale applications in many cases are successful, it is important to realize that the mechanism controlling polyP metabolism in the activated sludge process remains obscure from a microbiological point of view.

A priori there is no reason to believe that sediment bacteria would behave in a different way from sludge bacteria, when exposed to similar environmental conditions. This presumption challenged Fleischer (1983) and Gächter *et al.* (1988) to prove that polyP metabolism and the responsible microorganisms play a role of importance in sediment P dynamics. So far the process could only be demonstrated in laboratory experiments using bacteria isolated from sediments. It

has still to be proven that bacterial polyP metabolism is operative in natural environments and to what extent this process affects the exchange of P across the sediment–water interface. The difficult progress in this direction of research is due to a lack of adequate techniques to measure intracellular polyP concentrations and the activity of polyP organisms *in situ*. In our opinion, it is also due to a lack of theoretical support for the idea. The author's firm conviction is that from a sound appreciation of the fundamental, ecological meaning of this bacterial process we shall be in a better position to estimate its value, and to develop experimental research strategies aiming at verification of the hypotheses and quantification of the processes *in situ*.

Ecological approach

The main objective of this paper is to be able to recognize the ecological significance of bacterial polyP metabolism in relation to freshwater ecosystem functioning as a whole. We shall treat the subject on a theoretical basis, building our ideas and hypotheses on evidence from experimental work reported in literature.

Theory and data

Within the context of wastewater treatment technology, ecological studies on bacterial polyP metabolism have been very scarce and restricted to some general ecophysiological considerations. Important questions as to whether and for what purpose polyP bacteria could be operative under natural conditions, in what locations they would be encountered, can hardly be traced in the extensive literature on this subject. This is not surprising since environmental engineers and microbiologists generally abstract from the reality beyond the limits of the controlled systems they study. Therefore, technological know-how and biological knowledge have not developed in pace. Moreover, within the context of basic research, the biochemical approach has been disproportionately favoured above the ecological approach. As we proceed, the challenge will be to integrate existing ecological theory and available technological data. We have in hand a considerable number of experimental results on bacterial polyP metabolism and behaviour of polyP organisms in activated sludge systems (Marais *et al.*, 1983; Toerien *et al.*, 1990). These data constitute a useful piece of information indeed, to speculate upon natural processes in sediments on the basis of known reactions in controlled systems.

Procedure

An answer to the central issue in this paper is to be found within ecological theory at a high level of integration: the organization level of the freshwater ecosystem. This requires a synecological approach. To analyse and understand processes on the ecosystem level some knowledge is necessary of lower levels of organization. Consequently, we should first know more about the autecology of polyP bacteria and on the way these microorganisms fit into the microbial community to which they belong, before we try to investigate their functional role within the freshwater ecosystem as a whole. The 'ecological niche' concept, as broadly defined in Odum (1971), covers the three aspects of the ecology of polyP bacteria that are

of interest: (*i*) the spatial niche or habitat at the individual level, (*ii*) the dimensional niche or microhabitat within the microbial community and (*iii*) the trophic niche or functional position within the ecosystem. It provides the theoretical frame required for this study. The following sections of this paper are devoted to a systematic, ecological approach to polyP accumulating bacteria as indicated in Table 1. In the final section a brief summary and discussion of the results are given.

The individual microorganism

Microorganisms generally exhibit adaptive features that make them physiologically compatible with their physical and chemical environment. They must be capable of survival, growth and metabolic activity in that habitat (Odum, 1971). We shall examine how this ecological principle applies to polyP organisms.

Activated sludge environment

Microorganisms actively performing polyP metabolism were first encountered, by chance, in the engineered environment of activated sludge systems for biological wastewater treatment in the sixties. Researchers observed the overall result of

Table 1. Ecological study of polyphosphate (polyP) accumulating bacteria at successive levels of biological organization.

Subject studied Ecological approach	Level of ecological organization		
	Individual microorganism	Microbial community	Freshwater ecosystem
Abiotic component	Sediment layer at the mud-water interface	Sediment environment	Freshwater environment
Biotic component	Polyphosphate accumulating bacteria	Heterotrophic microbial populations	Whole biotic community
Relationship studied	Physiological adaptation of polyP bacteria to environment	Position of polyP bacteria in microbial community	Functional role of polyP bacteria in ecosystem
Approach	Autecological	Synecological	Synecological
Aspect of "ecological niche" concept	Spatial niche, or habitat	Dimensional niche, or microhabitat	Trophic niche, or functional status

this microbial activity as a removal of P from the bulk liquid, in excess of that indicated by normal metabolic requirements of the sludge microflora. During the following two decades numerous investigations were conducted to establish the biological nature of the phenomenon and to determine the factors controlling its application (Marais *et al.*, 1983). Today there is little doubt that activated sludge bacteria are indeed responsible for the observed biological excess P removal (BEPR). Of central importance is that in systems designed for BEPR an environment is created for the proliferation of polyP bacteria, by providing for successive stages in which the activated sludge is subjected to anaerobic and aerobic conditions alternatively (as shown in Fig. 1a). Anaerobic P release is a prerequisite to subsequent aerobic excess P uptake and a continual change in the redox potential is a necessary pre-condition for the continuous BEPR in the activated sludge process (Barnard, 1976).

Natural environment in sediments

Davelaar (1978) recognized that polyP bacteria probably are ubiquitous in nature and that polyP metabolism must be operative in soils and aquatic environments. Toerien *et al.* (1986, 1990) postulated that the mechanism of BEPR in activated sludge systems must depend on a group of microorganisms which in nature are favoured by fluctuating conditions of anaerobiosis-aerobiosis. Fleischer (1978, 1983) challenged the established theory on P fixation/solubilization by abiotic coupling of the phosphorus and iron cycles, by presenting evidence indicating that P release from sediments is a direct biological process. In subsequent papers Fleischer (1986, 1988), referring to BEPR in activated sludge studies, suggested that at least part of the anaerobically released P in reduced sediments has its origin in bacterial cells. Based on the known occurrence of polyP storage in some sedimentary microorganisms, Boström *et al.* (1982, 1988) also speculated that phosphate mobilization as a result of altered physiology in living bacterial cells could contrib-

ute to P release from anaerobic sediments. Gächter *et al.*(1988) found that in some deepwater lakes iron and phosphate are not released simultaneously into bottom water, suggesting that in these particular sediments microorganisms responding to variations in ORP may contribute considerably to uptake and release of P. They also found that oxic uptake and anoxic release of P by benthic bacteria occur at about the same ORP as precipitation and dissolution of iron oxyhydroxides (FeOOH). The interface layer between water and mud in the sediment of freshwater ecosystems, where conditions prevail that allow for spontaneous iron oxidation-reduction, seems to be a natural environment for polyP microorganisms.

Physiological characterization of polyP bacteria

Fuhs & Chen (1975) first isolated bacteria, identified as *Acinetobacter* spp., from a P removing activated sludge plant: these bacteria accumulated polyP. Thereafter the presence of *Acinetobacter* has repeatedly been shown in activated sludge systems, but it has also been recognized that a number of other bacterial species are implicated in BEPR either directly, because they store polyphosphates or indirectly, because they produce metabolites that serve as a substrate for polyP microorganisms (acidogenic bacteria). In current research on BEPR the *Acinetobacter* genus is generally accepted to represent the polyP group of bacteria. *Acinetobacter* spp are ubiquitous in nature and can be readily isolated from soil, water and sewage. Most strains of *Acinetobacter* can grow in a simple mineral medium containing a single carbon and energy source such as ethanol, acetate, lactate. The genus requires oxygen for catabolic metabolism. For this reason polyP bacteria are indisputably thought of as strict aerobes but we have serious doubts as to this classification (see next section of this paper). Bergey (1984) gives a complete taxonomic description of *Acinetobacter*. Although in this description nothing is mentioned on storage products, the genus is capable of storing phosphorus as polyP (Fuhs &

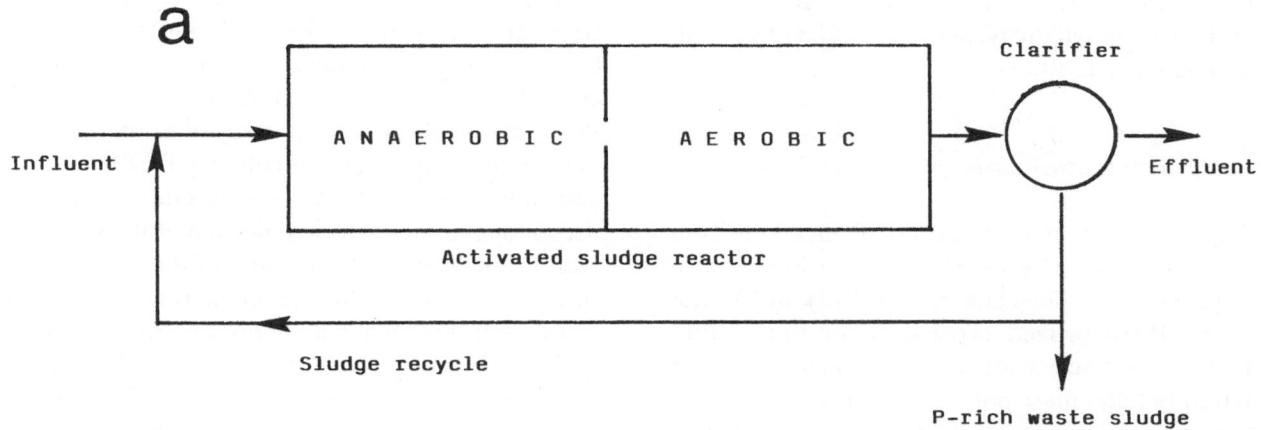

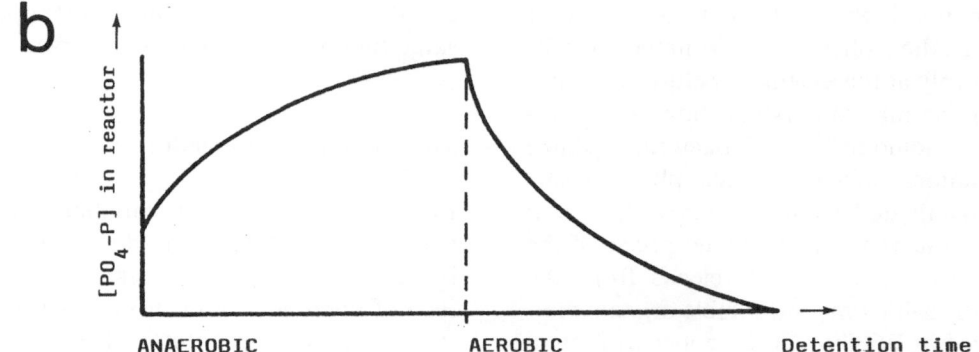

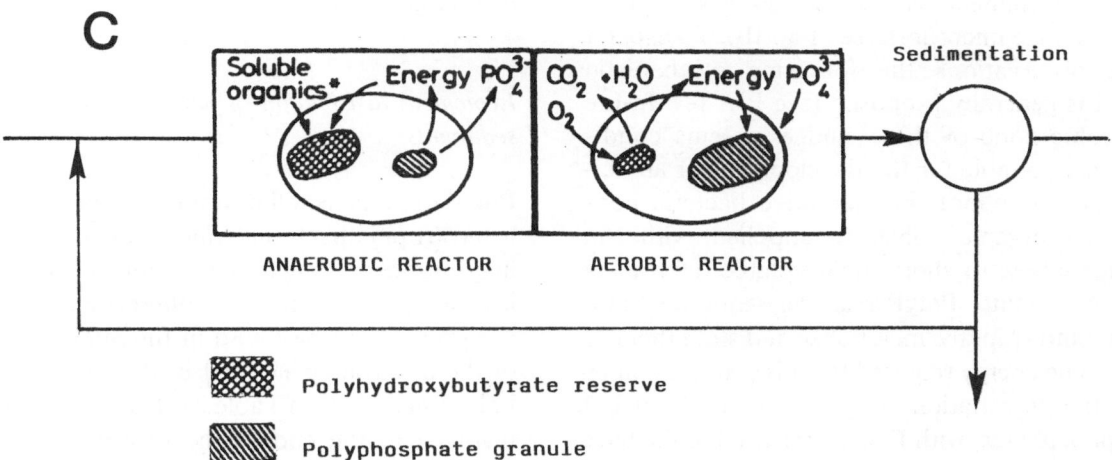

Fig. 1. Biological excess P removal in the activated sludge process; (a) process configuration; (b) PO$_4$-P concentration changes in the reactor; (c) cellular reactions as a function of the environmental conditions (Fig. 1c based on Arvin, 1985)

Chen, 1975; Deinema *et al.*, 1980) and organic carbon as polyhydroxybutyrate (PHB) (Lawson & Tonhazy, 1980).

PolyP metabolism: an ecophysiological adaptation

The activation of bacterial polyP metabolism in activated sludge systems requires a low oxygen concentration. Davelaar *et al.* (1978) and Lötter *et al.* (1986) proved experimentally that conditions do not select for a specific bacterial population but stimulate polyP metabolism in populations already present. Consequently, it would seem that bacteria proliferating under alternating conditions of anaerobiosis-aerobiosis are privileged with polyP metabolism as a means of adaptation to their physical environment. Looking more closely at the essential features of polyP metabolism, we may understand how this metabolic function could help polyP bacteria to utilize the opportunities offered by their physical surroundings. Activated sludge systems show a P concentration increase in the liquid phase of the anaerobic reactor, due to P release from the sludge. At the same time the soluble organic matter (measured as BOD or COD) concentration in this reactor decreases. In the next, aerobic reactor P concentration decreases to a level lower than in the influent, as a result of excess P uptake by the sludge microflora (see Fig. 1b). To explain these observations the following mechanistic model is generally proposed (see Fig. 1c). In the anaerobic zone of these sludge systems conditions are suitable for the development of an acidogenic microflora. Fermentative bacteria convert the organic substrate supplied with the sewage inflow to short-chain volatile fatty acids such as acetate. PolyP bacteria sequester these compounds rapidly, metabolize and store them as PHB. The energy required for this process comes from the degradation of previously accumulated polyphosphates, with P being released to the bulk solution. In the aerobic phase of the BEPR process polyP bacteria use the PHB store as a source of carbon and energy for growth and other cell functions, and as an energy source for polyP re-

synthesis which results in P uptake from the bulk solution. Detailed biochemical pathways have been set out in the models developed by Comeau *et al.* (1986) and Wentzel *et al.* (1986, 1991). PolyP metabolism is a key feature in the behaviour of microorganisms mediating BEPR in the activated sludge process. Today consensus has been reached as to the significance of this behaviour at the level of organization of the individual. Intracellular polyP storage serves the purpose of a source of biochemical energy to secure the life cycle of the organism, allowing it not only to survive but also to be physiologically active during the anaerobic period (Van Groenestijn & Deinema, 1987; Toerien *et al.*, 1990). From an autecological point of view, we postulate that polyP metabolism in sediment bacteria fulfils the same function of an energy reserve.

The microbial community

Based on their metabolic capabilities (carbon and energy sources, electron donor and acceptor), bacterial species can be arranged according to the type of microenvironment in which they flourish (Gaudy & Gaudy, 1981). This ordering often shows an ecological distribution of metabolically related microbial populations along an environmental gradient.

Ecological distribution of heterotrophic bacteria in sediments

PolyP organisms fall into the category of chemoheterotrophic bacteria, which can be subdivided in groups according to their mode of energy yielding metabolism. The only difference between the subgroups is to be found in the final disposition of the electrons generated in the oxidative catabolic reactions (see Table 2). The redox reactions involved at the end of the respiratory electron transport chain tend to occur in order of their thermodynamical possibility (Stumm & Morgan, 1981). The electron transfer being biologically mediated, the chemical reaction sequence (reduc-

Table 2. Classification of chemo-heterotrophic bacteria in metabolic groups (based on Gaudy & Gaudy, 1981).

First external electron acceptor		Metabolic groups		
Inorganic	Organic	Respiration		Fermentation
		Aerobic	Anaerobic	
O_2		Strict aerobes		
NO_3, NO_2			Denitrifiers	
Mn-, Fe-oxides			Mn-, Fe-reducers	
	Org. compounds			Fermentative bacteria
SO_4, S, S_2O_3			Sulfate reducers	
CO_2			Methane bacteia	

tion of O_2, NO_3^-, Mn^{4+}, Fe^{3+}, organic matter, SO_4^{2-}, CO_2) is paralleled by an ecological succession of microorganisms (strict aerobic bacteria, denitrifiers, manganese and iron reducers, fermentative bacteria, sulfate reducers, methane bacteria). In natural habitats bacteria capable of mediating the pertinent redox reactions are nearly always found present. The benthic microflora thus changes with the redox gradient, its local composition depends on microenvironmental conditions in the sediment.

Ecological niche of polyP bacteria

The ORP being indicative of the ecological milieu, we now want to position polyP bacteria more accurately in a well defined redox range that will define their place within the heterotrophic microbial community. In activated sludge systems appropriate redox conditions select for polyP metabolism and the responsible microorganisms. The primary selection factor for excess P removal is anaerobic-aerobic sequencing. But 'anaerobic' and 'aerobic' are vague qualifications, inadequate to describe the microhabitat where polyP bacteria are active. Besides oxygen other important physical and nutritional factors, susceptible to influence the ORP, were found to interact with polyP metabolism in activated sludge experiments. Such factors are: nitrates, acetates (and other fermentation products), hydrogen sulfites, carbon dioxide and the pH. The interference of

these oxidants and reductants with both the P release and P uptake is well documented in literature (see e.g. review by Toerien *et al.*, 1990). From this typical interaction pattern we reasoned (Davelaar, 1989) that polyP metabolism is operative in a narrow ORP interval delimitated by denitrification/true aerobic respiration at the upper end, and sulfate reduction at the lower end of the scale (see Fig. 2). It coincides with the beginning of fermentative metabolism. Enhanced P removal is indeed crucially dependent on the exposure of the activated sludge to short-chain compounds generated by acidogenesis in the anaerobic stage. Moreover, in the indicated ORP range, microbial mediation of Mn and Fe reduction is thermodynamically possible (Stumm & Morgan, 1981). Our theory seems to be in accordance with laboratory and field observations reported by Gächter *et al.* (1988), indicating that bacterial P uptake and release and oxidation-reduction reactions of FeOOH occur at about the same ORP range.

Mode of anaerobic respiration

The position along the redox gradient assigned to polyP bacteria should be related to their use of a specific first external electron acceptor. There is a general consensus on the obligate aerobic nature of the bacteria accumulating polyP in activated sludge systems (see previous section). However, Fleischer (1983, 1986) reported that

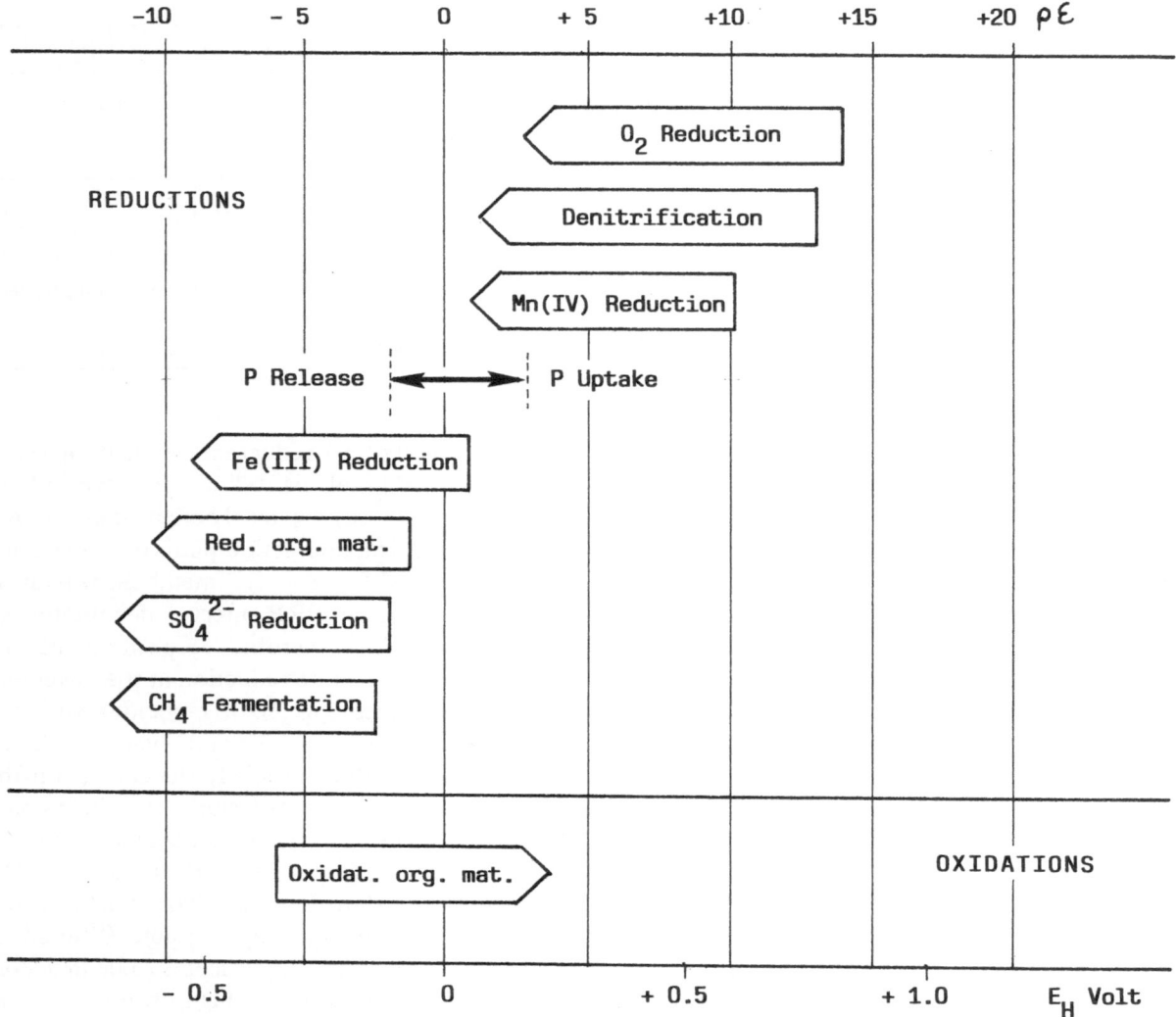

Fig. 2. Position of polyP metabolism in the sequence of redox processes mediated by heterotrophic bacteria (based on Stumm & Morgan, 1981 and Davelaar, 1989)

different types of facultative anaerobic microorganisms took part in rapid uptake and release of P in sediments. Toerien *et al.* (1986) raised the question whether one can still consider *Acinetobacter* an obligate aerobe taking into consideration that this bacterium apparently is physiologically active under anaerobic conditions. Davelaar (1989) postulated that P storing *Acinetobacter* spp and other polyP bacteria are per definition no strict aerobes, but a group of organisms

characterized by a facultative mode of respiration. The author furthermore advanced the hypothesis (Davelaar, 1989) that microorganisms adapted for life in the ecological niche defined above depend, for their energy metabolism, on fermentation products as electron donors and on manganese and/or iron oxides as final electron acceptors. In the virtual absence of free oxygen these organisms will shift to Mn and/or Fe respiration and invoke polyP metabolism. 'Anaerobic' conditions then refer to ORP values at which Mn and Fe occur mainly in the reduced form while sulfates are not yet subject to reduction. 'Aerobic' conditions indicate somewhat higher ORP values at which reduced Mn and Fe are easily oxidized and possibly nitrates reduced to nitrites. In case of an overall sufficient oxygen supply to sustain true aerobic respiration polyP bacteria will behave as strict aerobes and polyP metabolism will not occur. A direct implication of this hypothesis in connection with Fe-P interactions in sediment environments now emerges. PolyP bacteria, capable of mediating the oxidoreductive cycling of Mn and/or Fe, would achieve a *biotic* coupling of the Mn and Fe cycles with the phosphorus cycle.

PolyP metabolism coupled to Mn and/or Fe respiration: a competitive advantage?

It is now of interest to develop a hypothesis on the competitive advantage of polyP bacteria in the sediment environment, in terms of metabolic and respiratory mechanisms. In activated sludge studies, it has been observed that polyP bacteria are involved in a severe competition for substrate sequestration with associated microbial populations, the group of denitrifiers in particular. Marais *et al.*(1983) proposed that polyP accumulations serve as an energy reservoir during residence of polyP microorganisms in the anaerobic state to gain a positive advantage over non-P accumulating organisms, by partitioning off substrate for their exclusive use subsequently in the aerobic state. We believe that PHB/polyP metabolism is necessary but not sufficient in the struggle for fitness between neighbouring heterotrophic microbial populations. In the 'aerobic' phase of the life cycle of polyP bacteria, the polyP/PHB function by itself does not guarantee actual growth and replenishment of polyP reserves. Substrate is available in the form of PHB, but how about the supply of oxygen? In sediments oxygen generally is scarce; there should be a heavy competition for the available oxygen. Would strictly aerobic polyP bacteria eventually be able to outgrow facultative

and other microaerophilic populations, far better adapted to periodical low oxygen tensions as a result of their specialized modes of energy-yielding metabolism (e.g. nitrate respiration, fermentation)? If polyP bacteria indeed dispose of a mode of anaerobic respiration involving the reduction of Mn and/or Fe oxides, they certainly are advantageously equipped to make use of the restricted quantity of oxygen generally available in their natural environment. Mn and Fe compounds, relatively scarce but rapidly regenerated in their oxidized forms in the redox environment where polyP bacteria proliferate, are continuously available for reduction. The role of oxygen in the 'aerobic' phase becomes restricted to that of a second external electron acceptor. We come to the conclusion that PHB/polyP metabolism coupled to dissimilatory Mn and/or Fe reduction is a mechanism that confers upon polyP bacteria the competitive advantage, required to fill a unique ecological niche within the microbial community to which they naturally belong.

The freshwater ecosystem

The sediment is only one of the components of the abiotic freshwater environment and the biotic community of a freshwater ecosystem is much larger than the microbial community alone. It is an organized whole with characteristic trophic structures and metabolic patterns, that functions as a unit through coupled metabolic transformations (Odum, 1971).

Functional role of the microbial community in freshwater ecosystems

In freshwater ecosystems microorganisms are the principal producers as well as decomposers of organic matter. They also play a critical role in the transformations and cycling of numerous elements. The principal ecological functions of heterotrophic bacteria in sediment environments are: (1) to decompose dead organic matter, liberating mineral nutrients for primary production;

(2) to assimilate and reintroduce organic matter into the food web; (3) to perform mineral cycling activities (Atlas & Bartha, 1987). PolyP bacteria play their role in organic matter decomposition together with the other groups of aerobic and anaerobic heterotrophic benthic bacteria. An interesting question to speculate upon is whether polyP bacteria also fulfil such a vital function in mineral cycling as for instance denitrifiers or sulfate reducers are known to do in the N and the S cycles, respectively. Their most obvious function in this respect would be a mediatory role in the P cycle (see Fig. 3a).

Hypothesis on the functional role of polyP bacteria

Classically, most P transformations mediated by microorganisms are viewed as transfers of inorganic to organic phosphate (biosynthesis) or as transfers of phosphates from insoluble, immobilized forms to soluble or mobile compounds (mineralization) (Boström *et al.*, 1988). Bacterial polyP metabolism does not quite fit into this model. The role of polyP bacteria in sediment P dynamics is a dual one. Besides normal P assimilation and mineralization, it involves incorporation and intracellular polymerization of P alternated with polyP degradation and P release to the sediment environment (as shown in Fig. 3b). In chemical respect, the phosphate implicated in polyP metabolism remains in the inorganic form, outside as well as inside the bacterial cell. PolyP synthesis and degradation are P transformations that affect the exchange of P across the sediment-water interface in a direct way. Depending on the external redox conditions in the sediment polyP bacteria mobilize or immobilize P, in turn supplying it to or withholding it from the entire biotic community. Effectively and efficiently, they regulate P availability in their surroundings. They do so in a particularly critical range of the redox scale, namely the anaerobic/aerobic interface. From this, we reason that an ecologically meaningful feedback mechanism must exist between phosphorus and free oxygen availability in natural freshwater environments. This mechanism

seems to be governed by the metabolic activity of polyP bacteria autochthonous to the sediment. Transferring this hypothesis to an evolutionary context, it becomes clear that the 'productivity' and 'limiting factor' concepts of ecological theory explain the function of such a homeostatic mechanism at the ecosystem level of organization. Under the originally reducing atmospheric conditions, productivity remained low because of the virtual absence of free oxygen. As conditions became more and more oxidizing, the energy yield of biotic communities increased and so did productivity. Once oxygen was the limiting factor, now another essential factor started setting limits to ecosystem productivity: this factor was phosphorus. P is indeed vital to life but it should remain scarce in natural aerobic environments. It seems reasonable that free living organisms should have evolved means to self-regulate P availability so as to control productivity in a beneficial sense. From an evolutionary point of view, polyP bacteria must have had a particularly stabilizing influence at the shift of reducing to oxidizing atmospheric conditions by the first oxygen producing phototrophic organisms. Concluding, we advance the hypothesis that bacterial polyP metabolism is a homeostatic mechanism at the ecosystem level by which, P availability is limited and freshwater productivity controlled as a function of oxygen availability. PolyP accumulations in sediment bacteria are dynamic, highly recyclable biogenic P pools acting as temporary, redox dependent P sinks (see Fig. 3). The whole biotic community of freshwater environments is thus equipped to respond to slow seasonal and other, more frequent, redox fluctuations. Bacterial polyP pools in sediments are crucial components in the cycling of phosphorus, as P is retained within the biotic community but its availability remains restricted.

Concluding discussion

Bacteria that respond to fluctuating redox conditions by alternately consuming and renewing internal, energy-rich polyP reserves are natural in-

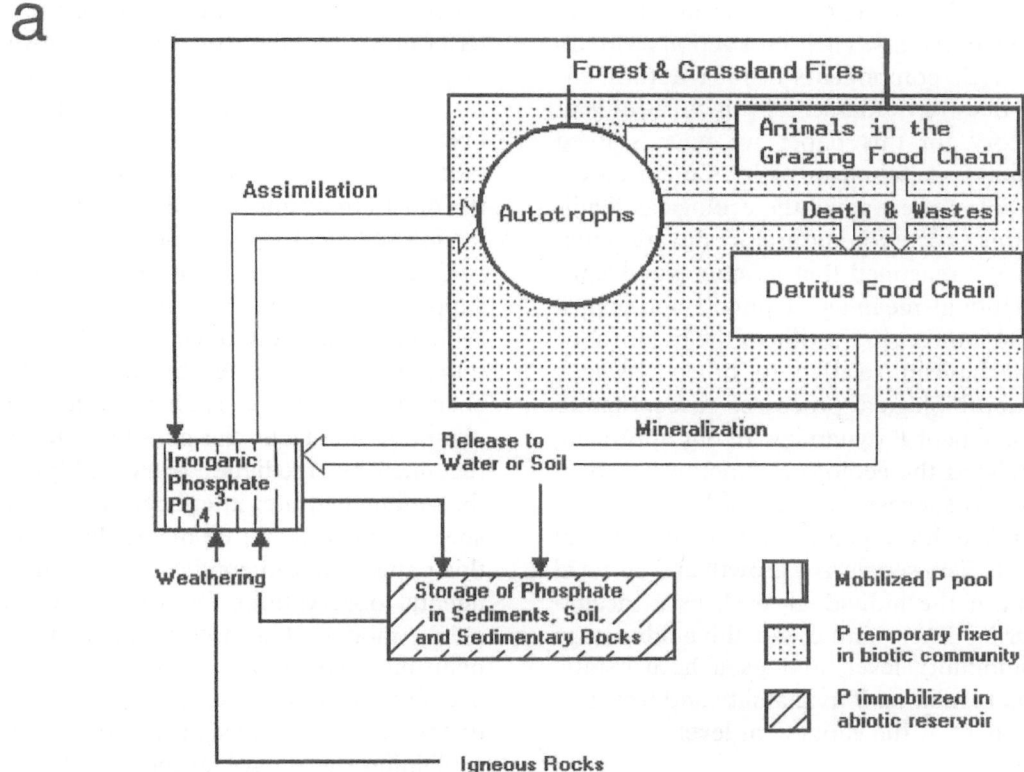

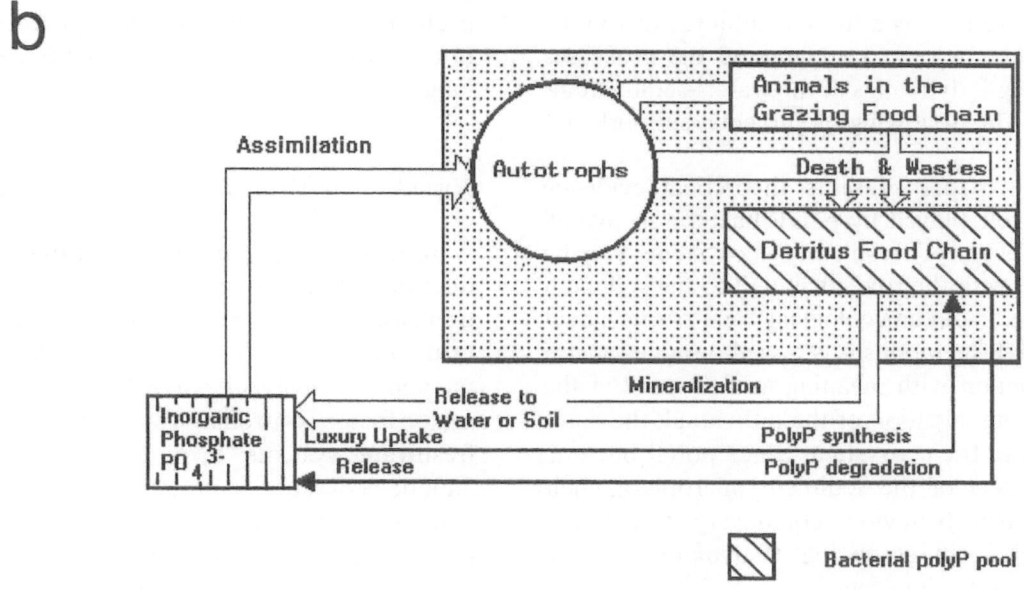

Fig. 3. The phosphorus cycle (redrawn from Clapham, 1973); (a) abiotic P reservoir and biotic P pool; (b) impact of bacterial polyP metabolism on P cycling in the biotic portion.

habitants of the sediment-water interface in freshwater ecosystems. Their interactions with biotic and abiotic components of their natural environment obey to fundamental principles of ecological theory. In this paper we have studied bacterial polyP metabolism with as main objective to be able to recognize the ecological significance of this process in natural freshwater environments. We reasoned that a better knowledge of the functional meaning of polyP metabolism would be an indirect indication of the importance of this mechanism relative to other physical, chemical and biological processes susceptible to influence sediment P dynamics. In our approach, we have related the ecological functions of bacterial polyP to successive levels of biological organization. We developed the view that bacterial polyP metabolism serves as a growth and survival mechanism at the individual level, as a mechanism ensuring fitness and competitive advantage at the community level, and as a homeostatic mechanism regulating P availability and freshwater productivity at the ecosystem level.

Impact of bacterial polyP metabolism in overall sediment P dynamics

From the present study we infer that the activity of polyP bacteria is a fundamental factor to take into consideration, to understand the processes controlling P dynamics at the water–sediment interface. The following arguments support our view.

(1) This paper presents theoretical evidence that the role played by polyP bacteria in overall P metabolism of freshwater ecosystems can be identified with the role attributed to the sediment in general. From an ecological point of view, bacterial polyP metabolism is a well designed metabolic function with meaning and purpose at the levels of organization of the individual, the community and the ecosystem. Since polyP bacteria are members of the sediment microbiota, their characteristic behaviour could very well be the key mechanism governing P dynamics at the sediment-water interface.

(2) From technological data there is evidence

for the potential quantitative importance of bacterial polyP metabolism. Experience with BEPR in the activated sludge process has learned that the amount of P stored within bacterial cells by luxury uptake can be considerable. From a quantitative point of view, polyP metabolism should be the most important microbial process directly affecting sediment P dynamics.

(3) Furthermore there is empirical evidence to support our view. The flux of P across the sediment–water interface was for a long time thought to be governed by the iron-phosphorus chemistry. Recent research has indicated that this abiotic model lacks the possibility to explain direct microbial mediation of Fe and P exchange at the sediment–water interface. The theory with respect to bacterial polyP metabolism presented in this paper is not in conflict with the many experimental observations on which the established, abiotic model is based. But it offers an alternative plausible explanation, biochemical instead of chemical, for well known phenomena. It responds to the call for a revision of earlier convictions, explaining the microbial mediation of P release and uptake phenomena and linking those to the oxidoreductive cycling of Mn and Fe. In many cases bacterial polyP metabolism and dissolution/precipitation of Fe-P complexes in the sediment probably are one and the same process, the nature of which could remain misunderstood for a long time.

Hypothesis verification and the eutrophication issue

A direct practical consequence of the hypotheses developed in this paper concerns the eutrophication issue of freshwater ecosystems. This issue in turn could be helpful as an indirect proof that the metabolic activity of polyP bacteria is indeed a major factor controlling overall P metabolism in freshwater systems. From the standpoint of the aquatic ecosystem, eutrophication is a metabolic disorder caused by local perturbations of the phosphorus cycle. Until today we are unable to control eutrophication in all situations. In fact, we lack knowledge of the key mechanism eventually governing P dynamics in the sediment environment as well as in the whole freshwater system. When we speculate that bacterial polyP metabolism is this key mechanism, it should be interesting to investigate the reasons for its failure in eutrophic systems, the failure being the very reason why the system became eutrophic. One could think of a deficiency of an element or nutrient essential to a good functioning of polyP metabolism, manganese or iron perhaps. These elements are relatively scarce in freshwater environments. The disproportionate enrichment with P due to eutrophication could result in a relative shortage of Mn and Fe. Moreover the treatment of sediments with Fe-compounds *in situ* has proven to be successful in some cases, irrespectively whether the mechanism actually influenced was a chemical or a biochemical one. One could also think of more complex reasons for the malfunctioning of polyP metabolism under eutrophic conditions. The study of bacterial polyP metabolism in sediments in relation to the trophic state of freshwater ecosystems could likely help resolve the eutrophication issue. On the same occasion it would produce clear evidence in favour or against the hypothesis that benthic polyP bacteria control the cycle of phosphorus in freshwater environments.

References

Arvin, E., 1985. Phosphatfällung durch biologische Phosphorentfernung. GWF-Wass./Abwass. 126: 250–256.

Atlas, R. M. & R. Bartha, 1987. Microbial ecology: fundamentals and applications. The Benjamin/Cummings Publishing Company, Inc., Menlo Park, CA, 533 pp.

Barnard, J. L., 1976. A review of biological phosphorus removal in the activated sludge process. Water SA 2: 136–144.

Bergey's Manual of Systematic Bacteriology, 1984. Williams & Wilkins, Baltimore, MD: 303–307.

Boström, B., M. Jansson & C. Forsberg, 1982. Phosphorus release from lake sediments. Arch. Hydrobiol. Beih. Ergebn. Limnol. 18: 5–59.

Boström, B., J. M. Andersen, S. Fleischer & M. Jansson, 1988. Exchange of phosphorus across the sediment–water interface. In G. Persson & M. Jansson (eds), Phosphorus in Freshwater Ecosystems. Developments in Hydrobiology 48. Kluwer Academic Publishers, Dordrecht: 229–244. Reprinted from Hydrobiologia 170.

Clapham, W. B., Jr, 1973. Natural Ecosystems. The Macmillan Company, New York, 248 pp.

Comeau, Y., K. J. Hall, R. E. W. Hancock & W. K. Oldham, 1976. Biochemical model for enhanced biological phosphorus removal. Wat. Res. 20: 1511–1521.

Davelaar, D., 1978. Biological removal of phosphorus from wastewater in a nitrifying/denitrifying activated sludge system. MSc Thesis, Agricultural University, Wageningen, The Netherlands.

Davelaar, D., T. R. Davies & S. G. Wiechers, 1978. The significance of an anaerobic zone for the biological removal of phosphate from wastewaters. Water SA 4: 54–60.

Davelaar, D., 1989. Manganese: a necessary micronutrient to enhance biological phosphorus removal? Wat. Sci. Tech. 21: 1711–1716.

Dawes, E. A. & P. J. Senior, 1973. The role and regulation of energy reserve polymers in micro–organisms. Adv. microb. Physiol. 10: 135–266.

Deinema, M. H., L. H. A. Habets, J. Scholten, E. Turkstra & H. A. A. M. Webers, 1980. The accumulation of polyphosphate in *Acinetobacter* spp. FEMS Microbiol. Lett. 9: 275–279.

Einsele, W., 1936. Über die Beziehungen des Eisenkreislaufs zum Phosphatkreislauf im eutrophen See. Arch. Hydrobiol. 29: 664–686.

Einsele, W., 1938. Über chemische und kolloidchemische Vorgänge in Eisen-Phosphat–Systemen unter limnochemischen und limnogeologischen Gesichtspunkten. Arch. Hydrobiol. 33: 361–387.

Fleischer, S., 1978. Evidence for the anaerobic release of phosphorus from lake sediments as a biological process. Naturwissenschaften 65: 109–110.

Fleischer, S., 1983. Microbial phosphorus release during enhanced glycolysis. Naturwissenschaften 70: 415.

Fleischer, S., 1986. Aerobic uptake of Fe(III)–precipitated phosphorus by microorganisms. Arch. Hydrobiol. 107: 269–277.

Fleischer, S., 1988. Mechanism of the aerobic Fe(III)-P solubilization at the sediment–water interface. Verh. int. Ver. Limnol. 23: 1825–1829.

Fuhs, G. W. & M. Chen, 1975. Microbiological basis of phosphate removal in the activated sludge process for the treatment of wastewater. Microb. Ecol. 2: 119–138.

Gächter, R., J. S. Meyer & A. Mares, 1988. Contribution of bacteria to release and fixation of phosphorus in lake sediments. Limnol. Oceanogr. 33: 1542–1558.

Gaudy, A. F. & E. T. Gaudy, 1981. Microbiology for environmental scientists and engineers. McGraw-Hill Kogakusha Ltd, Tokyo, 736 pp.

Harold, F. M., 1966. Inorganic polyphosphates in biology: structure, metabolism and function. Bact. Rev. 30: 772–794.

Kulaev, I. S., 1979. The biochemistry of inorganic polyphosphates. J. Wiley & Sons, New York, 253 pp.

192

Lawson, E. N. & N. E. Tonhazy, 1980. Change in morphology and phosphate uptake patterns of *Acinetobacter calcoaceticus* strains. Water SA 6: 105–112.

Lötter, L. H., M. C. Wentzel, R. E. Loewenthal, G. A. Ekama & G. v. R. Marais, 1986. A study of selected characteristics of *Acinetobacter* spp. isolated from activated sludge in anaerobic/anoxic/aerobic and aerobic systems. Water SA 12: 203–208.

Marais, G. v. R., R. E. Loewenthal & I. P. Siebritz, 1983. Observations supporting phosphate removal by biological excess uptake – a review. Wat. Sci. Tech. 15: 15–41.

Mortimer, C. H., 1941. The exchange of dissolved substances between mud and water in lakes I. J. Ecol. 29: 280–329.

Mortimer, C. H., 1942. The exchange of dissolved substances between mud and water in lakes II. J. Ecol. 30: 147–201.

Odum, E. P., 1971. Fundamentals of ecology. Saunders College Publish, Philadelphia, 574 pp.

Stumm, W. & J. J. Morgan, 1981. Aquatic chemistry. J. Wiley & Sons, New York: 448–462.

Toerien, D. F., A. Gerber & K. E. U. Brodisch, 1986. Biological phosphate removal in activated sludge systems. Proc. IV ISME: 66–73.

Toerien, D. F., A. Gerber, L. H. Lötter & T. E. Cloete, 1990. Enhanced biological phosphorus removal in activated sludge systems. In K. C. Marshall (ed.), Adv. microb. Ecol. 11: 173–230.

Van Groenestijn, J. W. & M. H. Deinema, 1987. The utilization of polyphosphate as an energy reserve in *Acinetobacter* sp. and activated sludge. In R. Ramadori (ed.), Biological phosphate removal from wastewaters, Adv. Wat. Pollut. Contr. 4: 1–6.

Wentzel, M. C., L. H. Lötter, R. E. Loewenthal & G. v. R. Marais, 1986. Metabolic behaviour of *Acinetobacter* spp. in enhanced biological phosphorus removal – a biochemical model. Water SA 12: 209–224.

Wentzel, M. C., L. H. Lötter, G. A. Ekama, R. E. Loewenthal & G. v. R. Marais, 1991. Evaluation of biochemical models for biological excess phosphorus removal. Wat. Sci. Tech. 23: 567–576.

Hydrobiologia **253**: 193–206, 1993.
P.C.M. Boers, T.E. Cappenberg & W. van Raaphorst (eds),
Proceedings of the Third International Workshop on Phosphorus in Sediments.
© 1993 *Kluwer Academic Publishers.*

Transformations between organic and inorganic sediment phosphorus in Lake Balaton

Vera Istvánovics
Balaton Limnological Research Institute of the Hungarian Academy of Sciences, H-8237 Tihany,
Hungary

Key words: separation of organic and inorganic ^{32}P radioactivity, microbial P content, biotic and abiotic P uptake by sediments

Abstract

In order to estimate microbial P content and biological P uptake in sediments, the tungstate precipitation method of Orrett & Karl (1987) was used in sediment extracts. This method allows a simple and rapid separation of organic and inorganic ^{32}P radioactivity. Either inorganic ^{32}P (as carrierfree $H_3{}^{32}PO_4$) or organic ^{32}P (as ^{32}P-labelled algal material) was added to surface sediment suspensions of shallow Lake Balaton. Inorganic ^{32}P was rapidly transformed into organic ^{32}P, and this process was completely inhibited by formaline. P content of living benthic microorganisms was estimated from steady state distribution of the radioactivity. Transformation of algal organic P into inorganic P could also be detected.

In extremely P limited Lake Balaton benthic microorganisms were shown to supplement their high P requirements by inorganic P uptake. The velocity of the inorganic into organic P transformation, i.e. the rate of microbial P uptake, was comparable to P uptake in the water column. Microbial P uptake contributed significantly to total P fixation by sediments, particularly at low (≤ 100 μg P l^{-1}) phosphate additions.

Introduction

The most important P cycling processes at the sediment-water interface have traditionally been considered to be abiotic with microorganisms only modifying those physico-chemical conditions which govern abiotic processes (e.g. Boström *et al.*, 1982). Recently an increasing amount of evidence from the most different experimental and theoretical approaches has been presented, which suggests that benthic bacteria represent a significant P pool in the sediments, and that they contribute directly to P fixation and release (Boström *et al.*, 1988; Gächter *et al.*, 1988; Gächter & Meyer, 1993). However convincing, these pieces of evidence are indirect, since no method is avail-

able to measure separately P content of benthic microorganisms.

$H_3{}^{32}PO_4$ could conveniently be used to estimate microbial P content, as well as P uptake and release in sediments, if a simple technique were available to separate radioactivities present in inorganic and organic forms. The earlier methods, such as extractions with organic solvents or gel filtration, were highly time-consuming, and prevented handling of a large number of samples. More recently, Orrett & Karl (1987) applied successfully a simple precipitation technique to separate ^{32}P labelled organic substances from orthophosphate in seawater.

The aim of this preliminary study was to test if the precipitation method of Orrett & Karl (1987) could be used to distinguish between organic and inorganic ^{32}P in sediment extracts, too. Firstly, transformation of inorganic ^{32}P into organic forms was followed in time in $H_3^{32}PO_4$-labelled sediments. Rate of transformation and the amount of benthic microbial P were estimated. These estimates were compared with literature data in order to see whether the new method provided reliable information. Secondly, ^{32}P–labelled dead algal material was added to sediment samples and regeneration of inorganic ^{32}P was followed in time. Finally, biotic P uptake was measured in P-enriched sediment suspensions, which allowed a direct comparison of the new method with chemically measured biotic P uptake.

Description of sites studied

Lake Balaton is a shallow, large water body, which can be divided into four successive basins

$(z_{mean} = 3.2$ m, $A = 594$ km^2; Fig. 1). Prior to the largescale eutrophication management measures taken during the mid-1980'ies, external P loading was 7.3 mg m^{-2} d^{-1} to the south-western Keszthely basin, and 1.0 mg m^{-2} d^{-1} to the north-eastern Tihany basin of the lake (Somlyódy & Jolánkai, 1986). This resulted in hypertrophic conditions in the southwestern areas with an annual primary production of about 600 g C m^{-2}. The north-eastern areas remained mesotrophic with an annual primary production of around 180 g C m^{-2} (Herodek, 1986). During the last 5 years biologically available P loading (BAP) has been reduced by about 40% (Jolánkai, pers. comm.). The water quality, however, remained unchanged.

The sediments are highly calcareous, CaCO$_3$ and high magnesian calcite make up 50% of dry weight. The mean particle size is 10–40 μm. The organic content is only 2–4%, since detritus is rapidly decomposed due to the high temperature and frequent sediment resuspension (Máté, 1987).

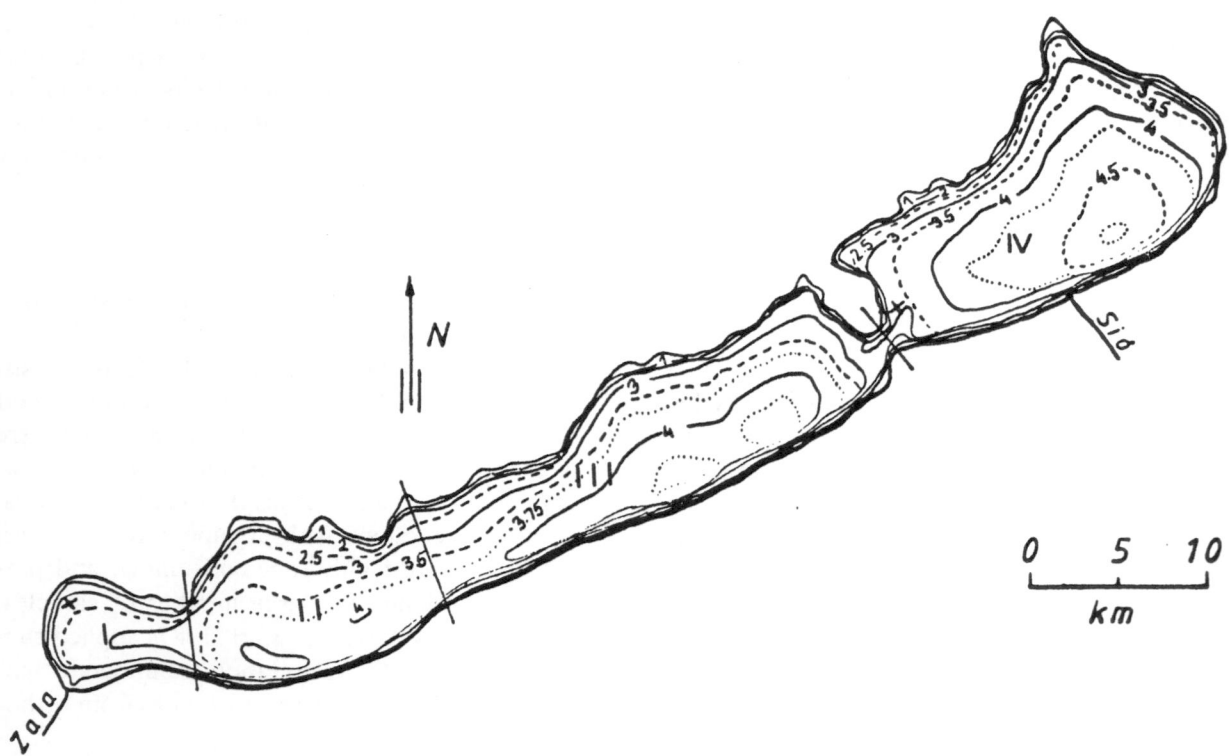

Fig. 1. Lake Balaton. X – sampling sites. I – Keszthely basin; IV– Tihany basin.

Anaerobic conditions were observed at the sediment-water interface only in the hypertrophic areas and for short periods of time (Oláh et al., 1983). The iron content is 1–2%. Concentrations of organic matter, iron and total phosphorus (500–700 μg P g^{-1} dw) are somewhat higher in the hypertrophic than in the mesotrophic sediments (Lijklema et al., 1986; Pettersson & Istvánovics, 1988).

Materials and methods

Sediment samples were taken with a Hargrave dredge on several occasions in the hypertrophic Keszthely and the mesotrophic Tihany areas of Lake Balaton (Fig. 1). Sediment suspensions were prepared from surface sediments (0–5 cm) with 0.45 μm membrane filtered lake water (Table 1). In order to compare living and sterile sediments, formaline was added in 5% final concentration to a set of samples. Since formaline decreases the pH, both living and poisoned samples were buffered with 5% 1 M borate buffer to pH = 7.9.

Either carrier-free $H_3^{32}PO_4$ or ^{32}P-labelled, dead algae were added to duplicate or triplicate sediment samples (Table 1). Sterile culture of Scenedesmus or lake water were preincubated with about 2 MBq $H_3^{32}PO_4$ ml^{-1} for at least one week. Algae were concentrated with centrifugation, killed with heat (60 °C, 5 minutes) and

washed several times before the experiments. Specific radioactivity of the algae was high enough to add sufficient radioactivity to the samples and keep P enrichment of the sediments negligibly low. Dissolved radioactivity (< 0.45 μm membrane filter) of the algal suspensions was less than 1% of the total radioactivity.

The sediments were incubated at two different temperatures (20, 22 or 24 °C and 4, 6 or 10 °C in various experiments) in the dark. Subsamples were taken immediately after the isotope addition and at roughly logarithmically increasing time intervals from several hours up to 18–22 days. The water was separated by centrifugation at 3000 g for 10 minutes and filtered through 0.45 μm membrane filters. Sediment subsamples (0.15 g wet weight) were fractionated with a sequential extraction procedure in 25 ml volume (Table 1). The rest of each sediment sample was used to determine water content as the difference in weight before and after drying at 105 °C for 24 hours.

In the first experiment we used the extraction method of Hieltjes & Lijklema (1980) (HL). This widely applied sediment fractionation scheme includes two extractions with 1 M NH_4Cl at pH = 7 for 2 hours, followed by an extraction with 0.1 N NaOH for 16 hours and with 0.5 N HCl for 24 hours. In order to measure rates of transformations, a relatively frequent subsampling was necessary. The complete HL scheme was too time-consuming, therefore we decided as a compromise

Table 1. Experimental conditions.

Date	Suspension (g dw l^{-1})	Type of experiment	Algae used	Fractionation
Tihany				
11.04.1988	85	$H_3^{32}PO_4$, ^{32}P-algae	Scenedesmus	LH
30.01.1089	cc. sed.	$H_3^{32}PO_4$, ^{32}P-algae	Scenedesmus	HL_m
28.02.1990	260	$H_3^{32}PO_4$	–	Krause
Keszthely				
11.04.1988	90	$H_3^{32}PO_4$, ^{32}P-algae	Scenedesmus	HL
08.11.1989	236	$H_3^{32}PO_4$, ^{32}P-algae	Natural phytopl.	HL_m
28.02.1990	240	$H_3^{32}PO_4$	–	Krause
26.11.1989	2.4	P fixation	–	HL_m

to leave out the NH_4Cl steps (HL_m procedure) instead of decreasing the number of samples. Since the HL procedure has been worked out to fractionate inorganic sediment P, and our aim was to separate organic and inorganic P forms, in one experiment we used a simplified extraction scheme of Krause (1964), which is used to fractionate microbial P. In this scheme inorganic P and low molecular weight polyphosphates, organic P esters, sugar phosphates, nucleotides, etc. are extracted in the first step with 10% ice-cold TCA for 30 min. Phospholipids appear in the 80% ethanol/diethylether fraction (1/1, 60 °C, 20 min). Finally the extraction with 1 M NaOH at 60 °C for 60 min yields nucleic acids, proteins, and long-chain polyphosphates (Lean & Cuhel, 1987).

Concentrations of soluble reactive phosphorus (SRP) and total dissolved phosphorus (TDP) were determined according to the FBA (1978) standards in the supernatants and in the sediment extracts after neutralization and dilution when necessary. Organic or non-reactive phosphorus (nRP) was calculated as the difference between TDP and SRP. Total P content of the sediments was measured after digestion with a mixture of oxidative acids.

Radioactivity of 5–10 ml of the supernatants and of different extracts was determined with an LKB RackBeta-2 liquid scintillation counter on the basis of the Cherenkov radiation. Counting efficiency was determined by internal standards, i.e. adding a known amount of ^{32}P to each sample and remeasuring the radioactivity.

An aliquot of the sediment extracts was neutralized with HCl or NaOH and chilled to 4 °C. An aliqout of the supernatants was also chilled. ^{32}P radioactivity present in inorganic and organic forms was separated by means of the precipitation method of Orrett & Karl (1987). According to this procedure, PO_4^{3-} precipitates as a complex with tungsten and procaine from the sample acidified with formic acid (pH ≈ 2). The precipitate can quantitatively be separated by low-speed centrifugation (3000 g; 5–10 minutes) and the supernatant contains organic P compounds. In order to increase efficiency of separation, we applied two precipitation and centrifugation steps, as described by Orrett & Karl (1987). Efficiency of separation was tested with $H_3^{32}PO_4$ standards added to the respective solvents. Separation efficiency was always $> 95\%$, when acidic or alkaline extracts were neutralized before the acidification with formic acid. Without a prior neutralization, however, the efficiency decreased to 60–80% in the HCl extract. Counting efficiency was determined in each sample with internal standards.

In the ethanol/ether step of the Krause (1964) scheme phospholipids were shown to appear when microorganisms were fractionated (Lean & Cuhel, 1987). During sediment extractions only $< 5\%$ of the total radioactivity was recovered in this step. The fraction was considered to contain organic ^{32}P in the sediments, too, without a further differentiation with the tungstate precipitation method.

P fixation by dilute aerobic sediment suspensions was measured as described earlier (Istvánovics et al., 1989), but ^{32}P radioactivity added to duplicate samples was separated into organic and inorganic forms. A set of samples was poisoned with 5% formaline. The phosphate addition was 0–1000 µg P l^{-1}, with roughly equal increases on a logarithmic scale.

Results and discussion

According to the HL procedure, fractional distribution of sedimentary P in the freshly collected sediments was similar to that described earlier (Pettersson & Istvánovics, 1988), i.e. the samples were representative for the lake. About 40% of total P was bound to calcium (NH_4Cl-RP and HCl-RP). This fractional distribution remained

Fig. 2. Fractional distribution of the radioactivity added as $H_3^{32}PO_4$(A) in the sediments of the hypertrophic Keszthely area according to the simplified version of the Hieltjes & Lijklema (1980) scheme (8 November 1989), and (B) in the sediments of the mesotrophic Tihany area according to the Krause (1964) scheme (28 February 1990).

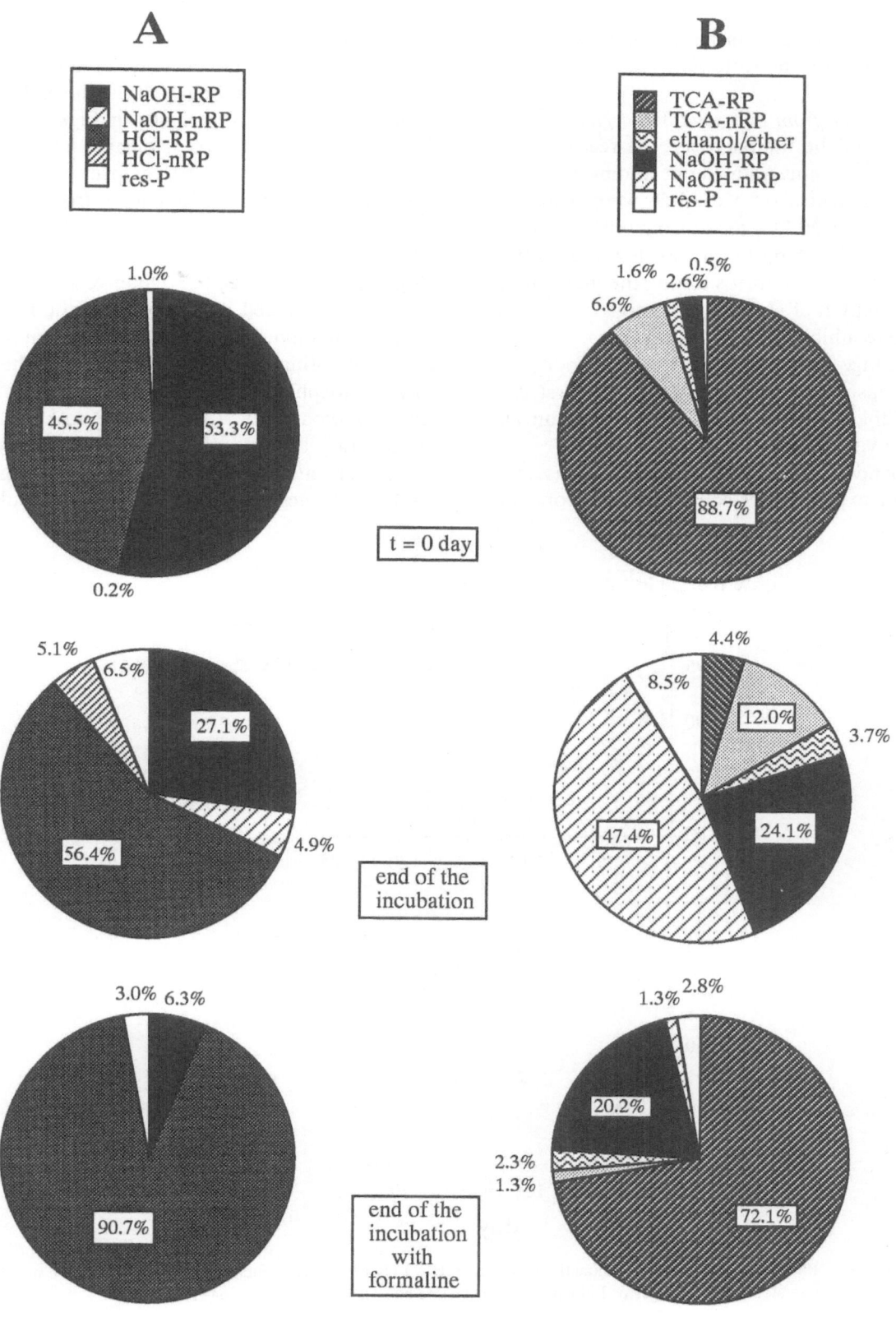

A

NaOH-RP
NaOH-nRP
HCl-RP
HCl-nRP
res-P

B

TCA-RP
TCA-nRP
ethanol/ether
NaOH-RP
NaOH-nRP
res-P

t = 0 day

end of the incubation

end of the incubation with formaline

essentially unchanged during the equilibration of the samples with the isotope.

Transformation from inorganic into organic ^{32}P
Almost all of the radioactivity was recovered as inorganic ^{32}P immediately after adding $H_3^{32}PO_4$ (Fig. 2; sediments from Keszthely were incubated for 18 days at 20 °C, those from Tihany for 21 days at 22 °C.). A part of it was, however, transformed into organic forms during the incubation. Transformation of $H_3^{32}PO_4$ into organic ^{32}P was completely inhibited by 5% formaline.

Percentage radioactivity present as inorganic ^{32}P decreased exponentially during the first 2–8 days of the incubation, and approached an asymptotic value thereafter (Fig. 3).

Independent of the type of the extractions, sediments or experimental conditions, the constant patterns of inorganic into organic ^{32}P transformation were that initial decrease of inorganic ^{32}P radioactivity was exponential and formaline inhibited the process (examples in Figs 2 and 3). This indicates that the transformation was mediated by sedimentary microorganisms.

In previous studies with carrier-free $H_3^{32}PO_4$ the pool of 'isotopically exchangeable P' (Li *et al.*, 1972), 'native surface P' (Ku *et al.*, 1978) or 'biologically available P' (Carignan & Kalff, 1979) has been calculated on the basis of the isotopic dilution equation. This equation states that in isotopic equilibrium specific radioactivity of the dissolved phosphate $(^{32}P_d/^{31}P_d)$ is equal to the specific radioactivity of the exchangeable sedimentary P pool $(^{32}P_s/^{31}P_s)$. In these studies, however, transformation of inorganic into organic ^{32}P has been neglected, and $^{31}P_s$ values calculated

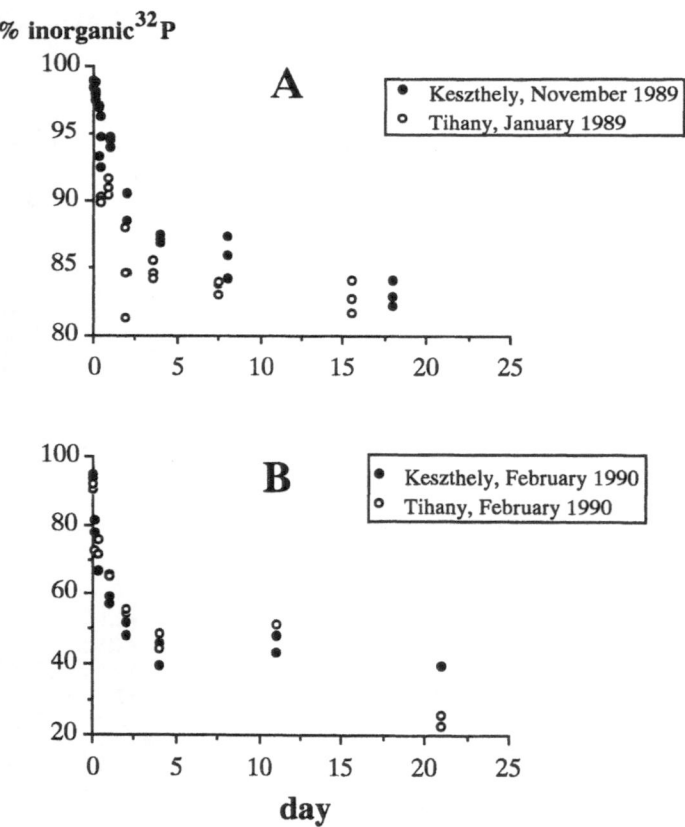

Fig. 3. Inorganic ^{32}P as percentage of total radioactivity added as $H_3^{32}PO_4$ to the sediments. Extractions were done according to (A) the simplified version of the Hieltjes & Lijklema (1980) scheme and (B) the Krause (1964) scheme.

from the above equation were implicitly assumed to be inorganic.

Our results show that only a part of $^{31}P_s$ is inorganic P ($^{31}P_{s,i}$), that part which is calculated from inorganic $^{32}P_s$ ($^{32}P_{s,i}$). The exchangeable inorganic P pool has a turnover within the sediments, since benthic microorganisms are able to assimilate inorganic P as indicated by the rapid transformation of inorganic ^{32}P radioactivity into organic. The rate constant (d^{-1}) of the initial decrease of the inorganic ^{32}P radioactivity was determined by linear regression fitted to *ln* transformed data (Fig. 3). Reciprocal of the rate constant is the turnover time (T, day) of exchangeable inorganic sedimentary P. T varied between 5 and 89 days (Table 2). In general, shorter values were obtained in the hypertrophic sediments and at higher temperature.

Pool sizes of inorganic and organic $^{31}P_s$ ($^{31}P_{s,i}$ and $^{31}P_{s,o}$) were calculated on the basis of the isotopic dilution equation. As a first approach, we assumed that $^{31}P_{s,o}$ calculated from $^{32}P_{s,o}$ gave the P content of metabolically active benthic microorganisms, although some experimental indications suggested that this assumption was not strictly valid. In the interstitial water or in the supernatants of the sediment suspensions a large part of the radioactivity was organic ^{32}P. This shows that uptake of inorganic P could be followed by a release of organic or colloidal P, which is a well documented process in planktonic microorganisms (Lean, 1973). The magnitude of the release, however, could not be estimated from our experiments, as total dissolved radioactivity was too low ($<0.05\%$ of the added radioactivity) to separate and measure exactly the amount of dissolved organic ^{32}P. If released organic ^{32}P was subsequently adsorbed to the solid phase, we overestimated microbial P content.

SRP concentration in the supernatants was

Table 2. Turnover times, uptake velocities, and poos sizes of exchangeable inorganic P and benthic microbial P in the sediments of Lake Balaton.

Date	Tihany		Keszthely	
	20–24 °C	4–10 °C	20–24 °C	4–10 °C
Turnover time of exchangeable inorganic P (day)				
11 Apr. 1988	20.4 ± 1.7	52.9 ± 5.5	4.6 ± 0.2	13.1 ± 1.6
30 Jan. 1989	59.5 ± 28.3	89.3 ± 24.0	–	–
8 Nov. 1989	–	–	18.3 ± 2.7	33.8 ± 2.3
28 Feb. 1990	9.9 ± 1.2	8.2 ± 0.7	9.4 ± 1.6	5.3 ± 0.3
Exchangeable inorganic P (μg P g^{-1} dw)				
11 Apr. 1988	84	82	49	116
30 Jan. 1989	57	44	–	–
8 Nov. 1989	–	–	125	131
28 Feb. 1990	48	72	153	91
Uptake rate of exchangeable inorganic P (μg P g^{-1} dw day^{-1})				
11 apr. 1988	4.1	1.6	10.7	8.9
30 Jan.1989	1.0	0.5	–	–
8 Nov. 1989	–	–	6.8	3.9
28 Feb. 1990	4.8	8.7	16.3	17.2
Benthic microbial P (μg P g^{-1} dw)				
11 Apr. 1988	42	19	87	84
30 Jan. 1989	12	11	–	–
8 Nov. 1989	–	–	27	17
28 Feb. 1990	118	91	180	153

often close to the detection limit ($\sim 1 \mu g$ P l^{-1}), especially when dilute sediment suspensions were incubated. Since the isotopic dilution equation is very sensitive to the SRP concentration, and dissolved organic ^{32}P could not be corrected properly, we assumed that the P content of the fraction with the highest specific radioactivity had the same specific radioactivity as dissolved phosphate. It was the NaOH-RP fraction that had the highest specific radioactivity according to the HL and HL_m schemes, and the NaOH-nRP fraction according to the Krause scheme. This assumption is also preliminary and needs experimental testing. Total $^{31}P_s$ values ($^{31}P_{s,i} + ^{31}P_{s,o}$) calculated in this way were of the same order of magnitude as $^{31}P_s$ values calculated on the basis of the specific radioactivity of the interstitial phosphate in our previous studies in which sediments were fractionated according to the Hieltjes & Lijklema (1980) scheme, but inorganic and organic ^{32}P forms have not been separated (Herodek & Istvánovics, 1986; Pettersson & Istvánovics, 1988). This suggests that the degree of overestimation due to the above assumption was probably not significant.

When the HL_m scheme was used, the amount of microbial P was estimated at 11–26 μg P g^{-1} dw in both the mesotrophic and hypertrophic sediments of Lake Balaton (Table 2). This corresponds to 160–390 mg P m^{-2} microbial P in the upper 5 cm sediment layer with an average water content of 75%. In the extractions performed according to the HL and particularly the Krause schemes, a much higher percentage of the radioactivity was recovered as organic ^{32}P by the end of the incubations than according to the HL_m scheme (e.g. Fig. 2). Consequently, estimated benthic microbial P was also much higher; 300–1,700 mg P m^{-2} in the mesotrophic Tihany sediments and 1,300–2,700 mg P m^{-2} in the hypertrophic Keszthely sediments (Table 2). The lower values were obtained from the results of the HL extractions, whereas the higher ones from those of the Krause extraction.

Benthic microbial P estimated in this way may include both algal and bacterial P. In January-April 1991 benthic algae were counted with an epifluorescence microscope in Lake Balaton sediments (G.-Tóth et al., in prep.). Algal biomass varied between 100 and 160 g ww m^{-2} in the upper 5 cm layer of the Keszthely area and between 40 and 130 g ww m^{-2} in the Tihany area with diatoms dominating the benthic community. As carbon content of diatoms is about 10% of their biovolume (Rocha & Duncan, 1985), and C:P ratio of P-rich planktonic diatoms is about 50 by weight (Pettersson, 1980), this corresponds to 200–320 mg P m^{-2} algal P content in the Keszthely area and 80–260 mg P m^{-2} in the Tihany area. In this way benthic algal P does not seem to exceed 40% of benthic microbial P in Lake Balaton (compared with the HL_m estimate), and its share is probably lower, 10–20% (compared with the HL and Krause estimates).

Only preliminary benthic bacterial numbers are available from Lake Balaton. Vörös (pers. com.) found 4.8–5.5×10^9 cells cm^{-3} with acridine orange staining and direct microscopic counting. The mean cell size was 0.5–1 μm^3. When we assume a bacterial carbon content of 220 μg C mm^{-3} (Bell, 1990) and a C:P weight ratio of 20 (Vadstein et al., 1988), the surface 5 cm sediment layer contains 1,300–3,000 mg P m^{-2} bacterial P. This value is similar to our microbial P estimate which was obtained from the results of the Krause extraction procedure.

Gächter & Meyer (1993) concluded from a comparison of a number of lakes that benthic bacterial carbon content of the surface sediments was equivalent to 10% of the annual primary production. Based on primary production in Lake Balaton (Herodek, 1986), and if the bacterial C:P ratio is assumed to be 20, the expected benthic bacterial P content would be 3,000 mg P m^{-2} in the sediments of the Keszthely area and 900 mg P m^{-2} in the sediments of the Tihany area.

Further experimental testing would be necessary to judge which sequential fractionation procedure provides a more reliable estimation of benthic microbial P content. Our results do not allow a strict comparison, since benthic microbial biomass and hence the share of organic ^{32}P radioactivity may vary in time, and different extraction procedures have not been used simultaneously. The great difference between values obtained in January 1989 with the HL_m scheme and those obtained in February 1990 with the Krause scheme (Table 2) may, however, indicate that the type of extraction strongly influenced our estimate of benthic microbial P content.

P contents both expected on the basis of the relationship given by Gächter & Meyer (1993) and estimated from benthic algal and bacterial cell counts (G.-Tóth et al., in prep., Vörös, pers. com.) were an order of magnitude higher than values estimated according to the HL_m scheme, but agreed with those estimated on the basis of the Krause scheme. Values obtained with the HL procedure were lower than the expected values by a factor of 2–3 (Table 2). These facts might suggest that a breakdown of benthic microbial P could occur during the HL_m extraction and to a lesser degree probably during the HL procedure, too. One must, however, also consider that a large fraction of benthic microorganisms is generally in a 'dormant' stage, as is indicated by the long average doubling times in the range of several years (Gächter & Meyer, 1993). In the upper 5 cm sediment layer of the Tihany area bacterial production measured with the ^3H-thymidine incorporation method varied between 100–1,300 mg C m^{-2} d^{-1} (recalculated from Zlinszky (pers. com.) using factors given by Bell, 1990). Doubling time would be 50–260 days, long enough to conclude that a fraction of benthic bacteria is not metabolically active in Lake Balaton sediments. Since only active microorganisms can be detected with the isotopic method presented in this paper, we cannot exclude that benthic microbial P content was overestimated when the Krause extraction scheme was used. If this is so, the possible reason is that the NaOH-nRP fraction which had the highest specific radioactivity during this procedure still had a lower specific radioactivity than the dissolved phosphate.

Benthic microbial P varied between 10–200% of the amount of exchangeable inorganic P (Table 2), and 3–28% of total sedimentary P. Gächter et al. (1988) estimated from bacterial cell counts, biovolumes, and C:P ratios determined in a flux chamber experiment, that in deep Lake Sempach 80% of total P was bacterial in the 0–0.5 cm sediment layer and it decreased to 22% in the 3–4 cm layer. Considering that here we studied the surface 5 cm sediment layer of a shallow lake, our estimates based on a totally different approach seem to agree with the findings of Gächter et al. (1988).

Transformation from organic into inorganic ^{32}P
When ^{32}P labelled algae were added to the sediments, organic ^{32}P was transformed into inorganic ^{32}P (Figs. 4 and 5; Keszthely sediments were incubated at 20 °C for 1 hour and 8 days, whereas Tihany sediments at 20 °C for 0.5 and 15 days.). In the heat-killed natural phytoplankton suspension, however, 46% of the total radioactivity was recovered as inorganic ^{32}P (Fig. 4A). We did not examine P status of the algae used in the experiments, but natural phytoplankton from Lake Balaton had to be extremely P deficient (Istvánovics & Herodek, 1985), and the initial share of inorganic ^{32}P was unexpectedly high.

Two sources of error could be suspected. Firstly, hydrolysis of organic P might occur when the inorganic P – tungsten – procaine complex was precipitated from the solution at low pH. Orrett & Karl (1987), however, tested the method with dissolved organic ^{32}P produced by a eutrophic freshwater plankton community, and found this source of error to be negligible. Secondly, the HL_m procedure might lead to a breakdown of organic P, at least during fractionation of microorganisms.

The share of inorganic ^{32}P radioactivity could not be neglected in the algal suspensions. Adding algal suspensions to the sediments, two opposite processes could occur simultaneously; inorganic ^{32}P could be transformed into organic ^{32}P and inorganic ^{32}P could be regenerated from organic ^{32}P due to e.g. autolysis, exoenzyme activity, or bacterial decomposition. Therefore, organic ^{32}P present during the incubation consists of the original algal ^{32}P which has not yet been mineralized, and organic ^{32}P which has been produced from inorganic ^{32}P. We assumed that the rate of inorganic into organic ^{32}P transformation was the

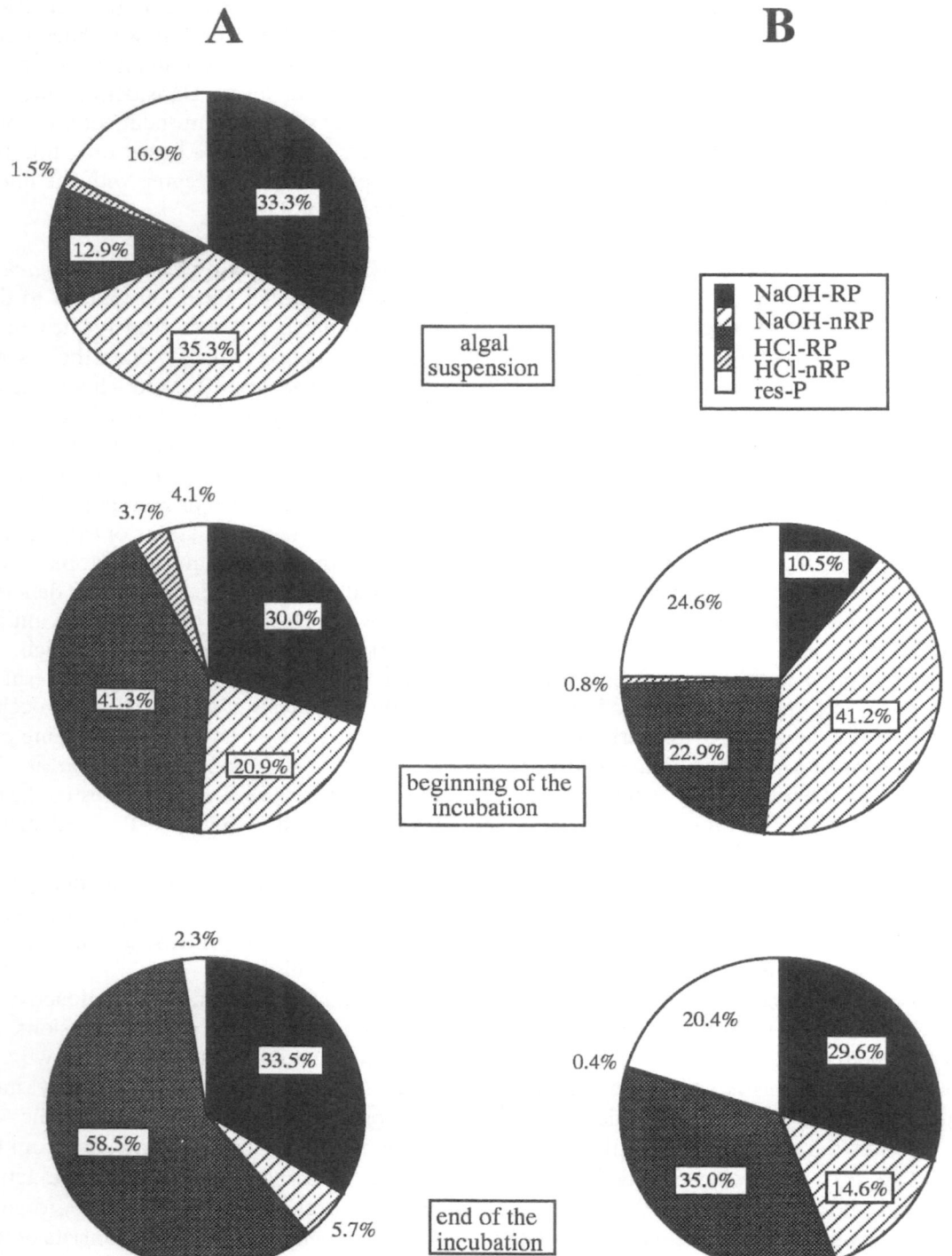

Fig. 4. Fractional distribution of radioactivity added as ^{32}P-labelled algae to the sediments of (A) the hypertrophic Keszthely area (8 November 1989) and (B) the mesotrophic Tihany area (30 January 1989) according to the simplified version of the Hieltjes & Lijklema (1980) scheme. The upper left-hand diagram shows fractional radioactivities in ^{32}P-labelled natural phytoplankton concentrate.

same when inorganic ^{32}P was added to the sediments either with the algal cells or as $H_3^{32}PO_4$. Therefore, organic ^{32}P values measured in the sediments incubated with ^{32}P-labelled algae were corrected with the respective values measured in the $H_3^{32}PO_4$-labelled samples (Fig. 5). The corrected value decreased exponentially at the beginning of the incubations. Turnover time of algal organic P was 1.8 ± 0.2 days in the sediments of the Keszthely area in November 1989, and 9.7 ± 2.9 days in the sediments of the Tihany area in January 1989. These turnover values seem to be realistic.

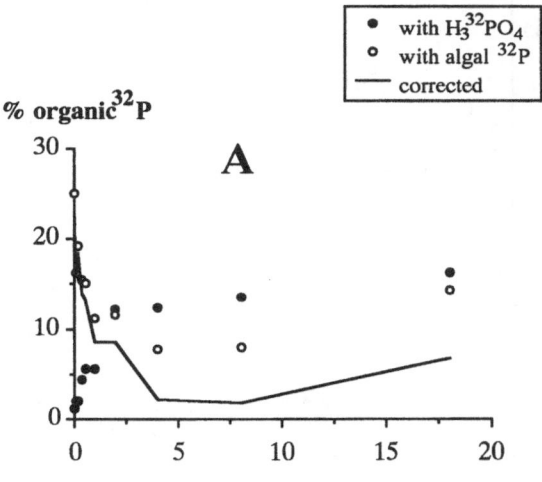

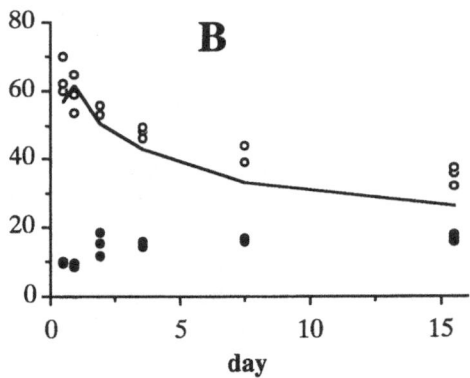

Fig. 5. Organic ^{32}P as percentage of total radioactivity in the sediments of (A) the hypertophic Keszthely area (8 November 1989) and (B) the mesotrophic Tihany area (30 January 1989) according to the simplified version of the Hieltjes & Lijklema (1980) scheme.

Aerobic P uptake of benthic microorganisms

Rapid transformation of inorganic ^{32}P into organic (Figs. 2 and 3) was a direct evidence of inorganic P uptake by benthic microorganisms. The rate of inorganic P uptake, i.e. transformation rate of inorganic P into organic, was calculated as the ratio of the pool size and turnover time of exchangeable inorganic sedimentary P. The uptake rates varied between $0.5-17$ μg P g^{-1} dw d^{-1} (Table 2), i.e. $7.5-255$ mg P m^{-2} d^{-1}. In general, higher values were obtained in the hypertrophic sediments and at a higher temperature than in the mesotrophic sediments and at a lower temperature. In comparison, the average gross P uptake velocity of planktonic microorganisms is $10-440$ mg P m^{-2} d^{-1} (Istvánovics & Herodek, 1985).

As discussed by Gächter & Meyer (1993), benthic bacteria are relatively richer in P than P deficient planktonic microorganisms, and the deposited organic P is not enough to cover their P requirements. Therefore, they must supplement their high P requirements with inorganic P uptake in sediments. In Lake Balaton planktonic microorganisms are extremely P deficient in both the hypertrophic and mesotrophic areas of Lake Balaton (Istvánovics & Herodek, 1985), and hence their C:P ratios are probably high. A comparison of the mass balance of biologically available P loading (Herodek, 1984) with benthic microbial P estimated in this study shows, that microbial P equals $0.6-4$ years organic P deposition in the Keszthely area, and $1.2-13$ years deposition in the Tihany area. The lower values are based on estimates obtained with the HL_m scheme, whereas the higher values on those obtained with the Krause scheme.

The average daily net organic P deposition can be estimated as < 0.5 mg P m^{-2} in Lake Balaton. Although the seasonal variation is certainly considerable, average organic P deposition rate is much lower than inorganic P uptake rates estimated in this study (Table 2). Even if benthic algae may contribute to inorganic P uptake in sediments, benthic bacteria seemed to assimilate more P from inorganic P uptake than from decomposition of the sedimented organic matter. In support of this conclusion, benthic bacterial production in the upper 5 cm sediment layer of the Tihany area varied between 100–1,300 mg C m^{-2} d^{-1} (Zlinszky, pers. com.), i.e. it was too high to assume that bacteria utilized only deposited organic P.

When benthic microorganisms take up inorganic P in sediments, they have to compete with inorganic sorbents. The potential role of benthic microorganisms in P fixation by sediments was studied in an aerobic P uptake experiment with 0–1000 μg P l^{-1} phosphate additions.

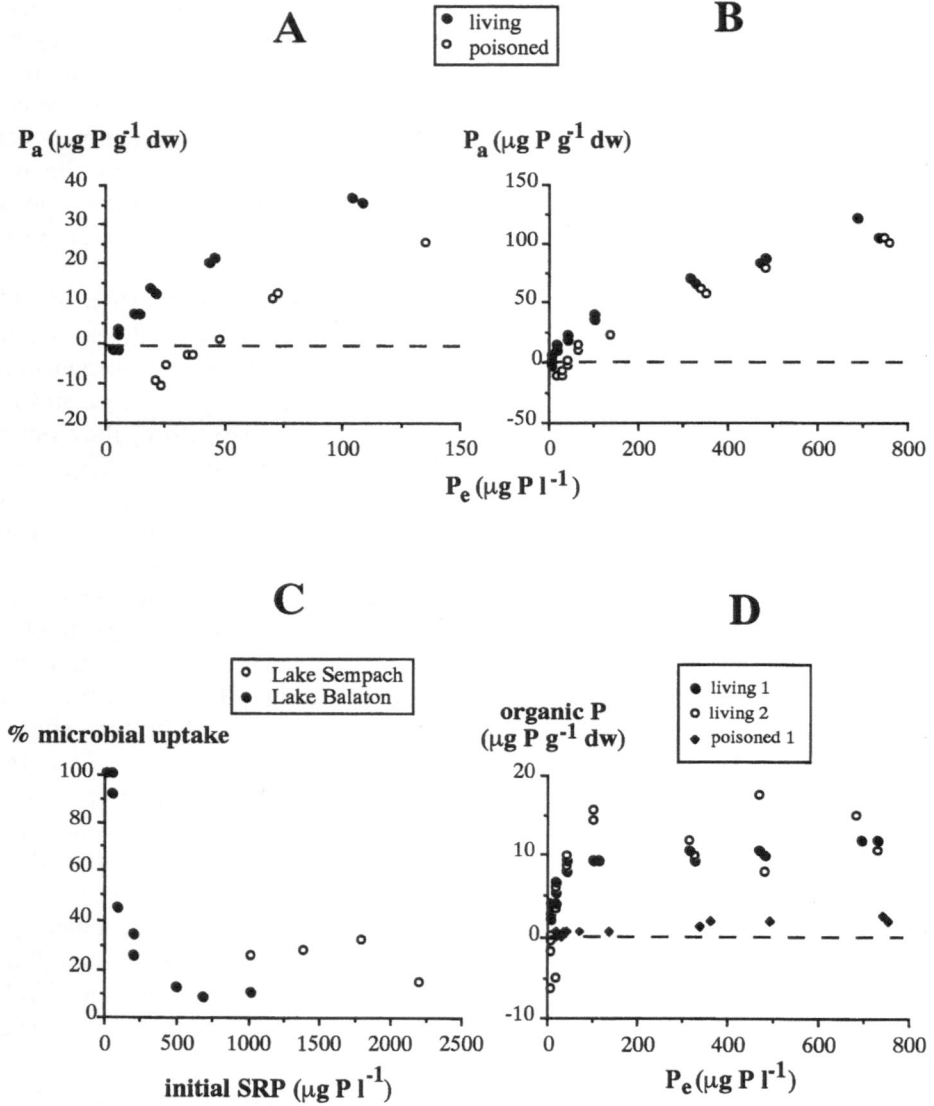

Fig. 6. The amount of P (P$_a$) taken up by formaline poisoned and living sediments of the hypertrophic Keszthely area (26 November 1989) as a function of the equilibrium phosphate concentration (P$_e$). (A) P addition 0–200 μg P l^{-1} and (B) P addition 0–1000 μg P l^{-1}. (C) Microbial P uptake as percentage of total P fixation by sediments as a function of the P addition. Data for Lake Sempach were taken from Gächter et al. (1988). (D) Benthic microbial P estimated from the tracer distribution in living and poisoned sediments (living 1 and poisoned 1), and increase of the organic P content measured chemically according to the simplified version of the Hieltjes & Lijklema (1980) scheme (living 2) as a function of P$_e$.

We calculated the amount of P taken up by living and formaline poisoned sediments as the difference between the initial and final SRP concentrations after 24 hours incubation (P_a, μg P g^{-1} dw, Fig. 6A and B). Poisoned sediments released P below ≤ 50 μg P l^{-1} addition. In contrast to this, living sediments released P only in unenriched samples, and the release was lower than in unenriched, poisoned sediments.

Gächter et al. (1988) presented a similar P uptake experiment with sediments of deep Lake Sempach, but their P addition ranged from 200 to 2200 μg P l^{-1}. These sediments took up much more P than Lake Balaton sediments, e.g. at 1000 μg P l^{-1} addition unsterilized sediments of Lake Sempach fixed about 420 μg P g^{-1} dw (recalculated from Fig. 3 of Gächter et al., 1988) in contrast to 120 μg P g^{-1} dw in Lake Balaton sediments (Fig. 6A). In Lake Sempach 14–31% of total P fixation by sediments could be attributed to benthic microorganisms. Similarly, in Lake Balaton benthic microbial uptake accounted for 9–36% of total sediment P fixation when 200–1000 μg P l^{-1} phosphate was added (Fig. 6C). It is, however, also evident from our results that the share of biotic uptake increased rapidly at lower concentrations, and only biotic uptake occurred when 10–50 μg P l^{-1} phosphate was added.

At the end of our P uptake experiment, sediments were fractionated according to the HL_m scheme. Separation of inorganic and organic ^{32}P in the extracts provided direct evidence, that the difference in P fixation by living and poisoned sediments was due to microbial P uptake. The amount of benthic microbial P was calculated on the basis of the isotopic dilution equation. Similarly to the experiments presented above (e.g. Fig. 2), formaline inhibited formation of organic ^{32}P, i.e. prevented biological P uptake (Fig. 6D). In living sediments benthic microbial P showed a saturation-type increase with the equilibrium phosphate concentration.

Increase of the organic P pool could also be calculated from the chemical measurements as the difference between the organic P contents in unenriched and enriched sediments. This differ-ence showed a good agreement with the amount of benthic microbial P estimated on the basis of the isotopic dilution equation (Fig. 6D).

Conclusions

The precipitation method of Orrett & Karl (1987) can be used to separate inorganic and organic ^{32}P in sediments, too. The facts supporting this conclusion were that inorganic ^{32}P was transformed into organic ^{32}P, the transformation was inhibited by formaline, benthic microbial P could realistically be estimated in Lake Balaton sediments, and ^{32}P-labelled dead algal material was transformed into inorganic ^{32}P. The procedure is simple and convenient, and its application can be recommended in more detailed studies of microbial mediated sedimentary P transformations.

A number of questions, however, remained unanswered in our preliminary study. Organic or colloidal P release from benthic microorganisms seemed to occur in the sediments, but further studies are needed to estimate its magnitude. The sequential extraction scheme applied may significantly influence the amount of inorganic ^{32}P that is recovered as organic ^{32}P. We could not decide with certainty which procedure is the 'best', although benthic microbial P content calculated from data obtained with the HL_m procedure seemed to underestimated. Alternative procedures should be tested simultaneously in the future.

Acknowledgements

This study was financially supported by the grant of the National Research Foundation (OTKA 88-1-596), 'The role of sediments in eutrophication of Lake Balaton'. I thank Dr. S. Herodek and two anonymous referees for their useful comments, Dr. H. Shafik for providing the algal cultures, and Mr. I. Báthori and his crew for field assistance.

206

References

Bell, R., 1990. An explanation for the variability in the conversion factor deriving bacterial cell production from incorporation of [^{3}H]thymidine. Limnol. Oceanogr. 35: 910–915.

Boström, B., M. Jansson & C. Forsberg, 1982. Phosphorus release from lake sediments. Arch. Hydrobiol. 18: 5–59.

Boström, B., M. Andersen, S. Fleischer & M. Jansson, 1988. Exchange of phosphorus across the sediment–water interface. Hydrobiologia 170: 229–244.

Carignan, R. & J. Kalff, 1979. Quantification of sediment phosphorus available to aquatic macrophytes. J. Fish. Res. Bd Can. 36: 1002–1005.

FBA, 1978. Water analysis. Mackereth, F. J., J. Heron & F. F. Talling (eds). FBA Sci. Publ. No 36.

Gächter, R. & J. S. Meyer, 1993. The role of microorganisms in mobilization and fixation of phosphorus in sediments. In P. C M. Boers, T. E. Cappenberg & W. van Raaphorst (eds), Proceedings of the Third International Workshop on Phosphorus in Sediments. Developments in Hydrobiology 84. Kluwer Academic Publishers, Dordrecht: 103–121. Reprinted from Hydrobiologia 253.

Gächter, R., J. S. Meyer & A. Mares, 1988. Contribution of bacteria to release and fixation of phosphorus in lake sediments. Limnol. Oceanogr. 33: 1542–1558.

Herodek, S., 1984. The eutrophication of Lake Balaton: Measurements, modeling and management. Verh. int. Ver. Limnol. 22: 1087–1091.

Herodek, S., 1986. Phytoplankton changes during eutrophication and P and N metabolism. In L. Somlyody & G. van Straten (eds), Modeling and Managing Shallow Lake Eutrophication. Springer, Berlin: 183–204.

Herodek, S. & V. Istvánovics, 1986. Mobility of phosphorus fractions in the sediments of Lake Balaton. Hydrobiologia 135: 149–154.

Hieltjes, A. H. M. & L. Lijklema, 1980. Fractionation of inorganic phosphates in calcareous sediments. J. envir. Qual. 9: 405–407.

Istvánovics, V. & S. Herodek, 1985. Orthophosphate uptake of planktonic microorganisms in Lake Balaton. Hydrobiologia 122: 159–166.

Istvánovics, V., S. Herodek & F. Szilágyi, 1989. Phosphate adsorption by different sediment fractions in Lake Balaton and its protecting reservoirs. Wat. Res. 23: 1357–1366.

Krause, H. R., 1964. Zur Methodik einer differenzierten Bestimmung von organischen Phosphor: Komponenten in Süsswasserplankton. Arch. Hydrobiol. Suppl. 28: 282–290.

Ku, W. C., F. A. DiGiano & T. H. Feng, 1978. Factors affecting phosphate adsorption equilibria in lake sediments. Wat. Res. 12: 1069–1074.

Lean, D. R. S., 1973. Movements of phosphorus between its biologically important forms in lake water. J. Fish Res. Bd Can. 30: 1525–1536.

Lean, D. R. S. & R. L. Cuhel, 1987. Subcellular phosphorus kinetics of Lake Ontario plankton. Can. J. Fish. aquat. Sci. 44: 2077–2086.

Li, W. C., D. E. Armstrong, J. D. Williams, R. F. Harris & J. K. Syers, 1972. Rate and extent of inorganic phosphate exchange in lake sediments. Soil. Soc. am. Proc. 36: 279–285.

Lijklema, L., P. Gelencsér, F. Szilágyi & L. Somlyódy, 1986. Sediment and its interaction with water. In L. Somlyódy & G. van Straten (eds), Modeling and Managing Shallow Lake Eutrophication. Springer, Berlin: 156–183.

Maté, F., 1987. A Balaton recens mederüledékének térképezése. M. Áll. Földtani Int. jelentése az 1985. évröl: 367–379.

Oláh, J., M. Abdelmoneim & L. Tóth, 1983: Nitrogen fixation in the sediment of shallow Lake Balaton, a reservoir and fishponds. Int. Revue ges. Hydrobiol. 68: 13–45.

Orrett, K. & D. M. Karl, 1987. Dissolved organic phosphorus production in surface seawaters. Limnol. Oceanogr. 32: 383–396.

Pettersson, K., 1980. Alkaline phosphatase activity and algal surplus phosphorus as phosphorus-deficiency indicators in Lake Erken. Arch. Hydrobiol. 89: 54–87.

Pettersson, K. & V. Istvánovics, 1988. Sediment phosphorus in Lake Balaton – forms and mobility. Arch. Hydrobiol. 30: 25–41.

Rocha, O. & A. Duncan, 1985. The relationship between cell carbon and cell volume in freshwater algal species used in zooplankton studies. J. Plankton Res. 7: 279–294.

Somlyódy, L. & G. Jolánkai, 1986. Nutrient loads. In L. Somlyódy & G. van Straten (eds), Modeling and Managing Shallow Lake Eutrophication. Springer, Berlin: 125–156.

Vadstein, O., A. Jensen, J. Olsen & H. Reinertsen, 1988. Growth and phosphorus status of limnetic phytoplankton and bacteria. Limnol. Oceanogr. 33: 489–503.

Hydrobiologia **253**: 207–216, 1993.
P.C.M. Boers, T.E. Cappenberg & W. van Raaphorst (eds),
Proceedings of the Third International Workshop on Phosphorus in Sediments.
© 1993 *Kluwer Academic Publishers.*

Extracellular phosphatase activity in sediments of the Breitenbach, a Central European mountain stream

Jürgen Marxsen & Hans-Heinrich Schmidt
Limnologische Flußstation des Max-Planck-Instituts für Limnologie, D-6407 Schlitz, Germany

Key words: stream, sandy sediment, phosphorus, extracellular enzymes, phosphatase

Abstract

Activity of extracellular phosphatases (phosphomonoesterases) was measured in sandy streambed sediments of the Breitenbach, a small unpolluted upland stream in Central Germany. Fluorigenic 4-methylumbelliferyl phosphate served as a model substrate. Experiments were conducted using sediment cores in a laboratory simulation of diffuse groundwater discharge through the stream bed, a natural process occurring in the Breitenbach as well as many other streams.

Streambed sediments contained high levels of particulate phosphorus, but concentrations of dissolved phosphorus in the interstitial water were 3 to 4 orders of magnitude lower. These interstitial concentrations were similar to those in the stream and groundwater. Extracellular phosphatase activity was high in the streambed sediments. These enzymes probably contribute significantly to the flux of phosphorus in sediment by hydrolyzing phosphomonoesters, making free phosphate available to the sediment microorganisms.

Factors influencing the kinetic parameters V_{max} (maximum activity) and apparent K_m (enzyme affinity) of phosphatase were discharge rates of water through the sediment, water quality (ground- or stream water), and substrate (phosphomonoesters) as well as dissolved ortho-phosphate concentrations. Enzymes are supposed to be effective at limiting substrate concentrations, where, in this study, changes in discharge rates had little influence on rates of hydrolysis. Higher V_{max} and lower K_m values were found during percolation of groundwater through the sediment cores, compared with stream water. This indicates that rates of hydrolysis were higher with groundwater, both at substrate limitation and at substrate saturation. This was probably a consequence of the lower levels of dissolved ortho-phosphate in the groundwater.

Introduction

Sandy streambed sediments in the Breitenbach, a small, unpolluted, upland stream in Central Germany, support high levels of bacterial biomass and productivity (Marxsen, 1988; Marxsen & Moaledj, 1988). Extracellular enzyme activity, including phosphatase, is also high (Marxsen & Witzel, 1990, 1991). Stream water concentrations

of soluble reactive phosphorus (SRP) were 30–40 μg P l^{-1}, total phosphorus was about 50 μg P l^{-1}. Concentrations in groundwater entering the stream by diffuse perfusion through the sediments were somewhat lower (about 20 and 40 μg P l^{-1}, respectively). Streambed sediments contained high levels of particulate (reactive and non-reactive) phosphorus, often exceeding 100 mg P (l sediment)$^{-1}$. Reactive phosphorus, experimentally soluble in stream or groundwater, was about 0.1 mg P 1^{-1}. Phosphorus concentrations in the sediment interstitial water were similar to those in groundwater. C:P ratios for total organic carbon and experimentally-soluble reactive P found in sediment cores (1987) were higher than 10000. However, if total phosphorus was considered, C:P ratios were between 14 and 62 (Marxsen & Witzel, 1991).

These data suggest that particulate phosphorus in sandy sediments, despite its unknown composition, could be a valuable source of P for microorganisms if it were made available for them. The high bacterial biomass and productivity occurring in the sandy sediments, along with the high level of phosphatase activity, suggest there is a considerable flux of phosphorus in this habitat, perhaps by rapid, closed-circuit, cycling. Phosphorus released by hydrolytic activity of extracellular phosphatases is probably not only important for heterotrophic microorganisms in the sediment, but also for autotrophic algae, which can develop high numbers on the surface of fine sediments (Cox, 1990).

In aquatic microbial ecology, phosphomonoesterases (phosphomonoesterhydrolases) are usually termed as phosphatases. These enzymes catalyze the hydrolysis of a variety of phosphomonoesters (Jansson *et al.*, 1988). Phosphatases in general (EC 3.1.3) also include phosphodiesterases, nucleosidases, and phosphoprotein phosphatases. In this paper, only phosphomonoesterases are considered. For the sake of simplicity, and as is usual for this group of compounds, we shall term these enzymes 'phosphatases'. We did not differentiate between alkaline (EC 3.1.3.1) and acid (EC 3.1.3.2) phosphatases. Acid phosphatases usually have pH optima between 4 and 6, whereas with alkaline phosphatases they are mostly well above 7, often between 9 and 10. pH of the Breitenbach water is circumneutral, so both types of phosphatases could be active, even if not at their optimum.

The term 'extracellular enzyme' refers to enzymes which function externally to cells (Wetzel, 1991; Marxsen & Witzel, 1991). They may be cell-bound, free in the water, or associated with inorganic or organic surfaces.

In this paper, experiments are described which investigate potential influences of the following factors on extracellular phosphatase activity: discharge rate through the sediments, water quality (ground- or stream water), and concentrations of competitive substrates and free phosphate. Experiments were conducted using an improved technique for measuring microbial activity in streambed sediments (Fiebig, in press), which resulted in less disturbance than with suspension techniques.

Study site

Investigations were conducted in the Breitenbach, a small, unpolluted, upland stream some 4 km in length in Central Germany. Stream pH is usually between 7.0 and 7.5, whereas the pH of groundwater entering the stream and sediment interstitial water is around 6.5 to 6.7. All waters are low in alkalinity. The stream is described in further detail by Marxsen, 1980a; Brehm & Meijering, 1982; Cox, 1990. Sediment cores were taken from a sandy part of the stream bed.

Methods

Phosphorus determination

Sediment cores were homogenized for 20 min in a centrifugal ball mill, and then fractionated according to Psenner *et al.* (1984). The sum of water-soluble P (extracted for 10 min in distilled water), reductant-soluble P (extracted for 30 min in sodium dithionite at 40 °C), and NaOH-soluble P (extracted for 16 h in 1 M NaOH) is considered to be the biologically-available fraction of phosphorus in sediments. Centrifuged extracts as well as stream, interstitial, and groundwater were analysed for orthophosphate (soluble reactive phosphorus, SRP) using the ammonia-molybdate method (Technicon Autoanalyser II). Total phosphorus was determined in the same way after UV digestion. Non-reactive phosphorus (NRP) was calculated as the difference between SRP and total P.

Measurement of enzyme activity

Extracellular phosphatase activity was determined from the liberation of fluorescent 4-methylumbelliferone (MUF) from non-fluorescent, 4-methylumbelliferyl phosphate disodium salt (MUF-P). This is a substrate for acid and alkaline phosphatases (Petterson & Jansson 1978, Jansson et al., 1981, 1988). MUF-P, as well as MUF for calibration, was obtained from Serva Feinbiochemica, Heidelberg, Germany. Detailed descriptions of the measurement techniques are provided by Marxsen & Witzel (1990, 1991).

Enzyme kinetic parameters were calculated using the non-linear regression analysis in the regression analysis program Enzfitter Version 1.05 (Leatherbarrow, 1987). This employs the algorithm of Marquardt (1963).

Experimental design

The homogenization (suspension) techniques which are frequently used in experimental investigations of microbial activity in sediments, and which we have also used previously, are prone to a number of disturbance artifacts. We improved our method for measuring extracellular enzymatic activity in streambed sediments by using a simulation of groundwater discharge through undisturbed cores of sandy sediment (Fiebig, in press). This simulates a natural process occurring in many streams, where such inputs of groundwater can be an important source of organic matter for microbial communities in stream beds.

At natural stream temperatures, sediment cores were perfused from below with stream or groundwater which, previously, had been filtered (Whatman GF/C) and boiled for 30 minutes, and to which appropriate concentrations of MUF-P had been added. Fluorescence of liberated MUF was measured in the water leaving the cores at the top.

Measurements were corrected for background fluorescence of uncleaved MUF-P.

Experiments performed

a. The time course of fluorescence was followed by perfusing stream water containing $100 \, \mu mol \, l^{-1}$ MUF-P at a rate of $11 \, ml \, cm^{-2} \, h^{-1}$ through seven cores (1 cm deep) taken from the sediment surface on 15.1.91 (temperature = 8.0 °C). For calibration purposes, one further column was perfused with stream water containing $10 \, \mu mol \, l^{-1}$ MUF. After 16 h of perfusion with pure stream water, the columns were perfused with water containing MUF-P and MUF for 24.5 h. The increase in fluorescence, the stability of MUF liberation rates and the decrease in fluorescence after switching back to MUF-P-free water were recorded.

b. The influence of different perfusion rates (1.9, 5.1, and $14.7 \, ml \, cm^{-2} \, h^{-1}$) on MUF-P cleavage was tested, in each case with five 1 cm cores from 28.1.91 (temperature = 8.0 °C). V_{max} and K_m values were calculated from hydrolysis rates measured at substrate concentrations of between 7 and $500 \, \mu mol \, l^{-1}$. Before perfusion with water containing MUF-P, unmanipulated stream water (filtered and boiled only) was percolated through the columns for about 15 h. During the last hour of this period, samples from the outflow of the cores were taken for phosphorus determinations. At the end of the experiment, after 41 h ($14.7 \, ml \, cm^{-2} \, h^{-1}$), 4 days ($5.1 \, ml \, cm^{-2} \, h^{-1}$), and 10 days ($1.9 \, ml \, cm^{-2} \, h^{-1}$), cores were perfused for a further 12 h with unmanipulated stream water and then frozen at −20 °C for subsequent analysis of the P fractions. Perfusion rates of 1.9 and $5.1 \, ml \, cm^{-2} \, h^{-1}$ corresponded to the natural rates of groundwater discharge into the Breitenbach during the greater part of the year, whereas the rate of $14.7 \, ml \, cm^{-2} \, h^{-1}$ is attained in the Breitenbach only during periods of heavy rainfall.

c. Differences between the cleavage of MUF-P in stream and groundwaters were investigated, in each case with 5 cores taken on 25.2.91 (cores

Table 1. Concentration of soluble reactive phosphorus (SRP) and non-reactive phosphorus (NRP) a. in stream and interstitial water from the site where sediment cores were extracted, b. in filtered and boiled stream and groundwater used for perfusion experiments, c. in stream water after percolation through sediment cores at different velocities (cores from 28.1.91).

a.

| Date | Interstitial water | | Stream water | |
	SRP $[\mu g\,P\,l^{-1}]$	NRP $[\mu g\,P\,l^{-1}]$	SRP $[\mu g\,P\,l^{-1}]$	NRP $[\mu g\,P\,l^{-1}]$
28.1.91	23	12	39	13
25.2.91	24	12	41	10

b.

| Date | Boiled stream water | | Boiled groundwater | |
	SRP $[\mu g\,P\,l^{-1}]$	NRP $[\mu g\,P\,l^{-1}]$	SRP $[\mu g\,P\,l^{-1}]$	NRP $[\mu g\,P\,l^{-1}]$
28.1.91	35	12	–	–
25.1.91	38	12	23	16

c.

Perfusion velocity $[ml\,cm^{-2}\,h^{-1}]$	SRP $[\mu g\,P\,l^{-1}]$	NRP $[\mu g\,P\,l^{-1}]$
1.9	28	15
5.1	31	21
14.7	34	18

were 1 cm deep, from the sediment surface, perfusion rate was 14.7 ml cm^{-2} h^{-1}, incubation temperature was 9.0 °C). V_{max} and K_m values were calculated from hydrolysis rates measured at substrate concentrations of between 8 and 500 μmol l^{-1}. Similarly, the influence of the competitive substrate glucose-1-phosphate (10 mmol l^{-1} dissolved in stream water) on MUF-P hydrolysis was investigated using a further five cores.

Results

Data for the different fractions of phosphorus in the surface sediments as well as the various types of water sampled (Table 1) were similar to previous values obtained from the Breitenbach (Brehm & Meijering, 1982; Schmidt, unpublished data).

At a perfusion rate of 11 ml cm^{-2} h^{-1}, the

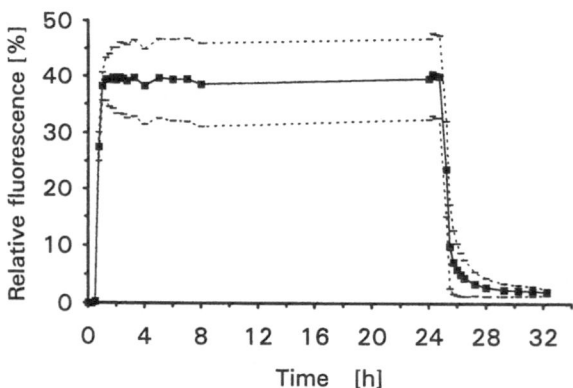

Fig. 1. Liberation of MUF from MUF-P over time. Data are reported as relative fluorescent units (%RFU). Mean values with 95% confidence limits are shown ($n = 7$). Perfusion with MUF-P began after 0.5 h and ceased after 25 h.

maximum level of free MUF concentrations in water discharging from the columns occurred after less than one hour of perfusion with MUF-P,

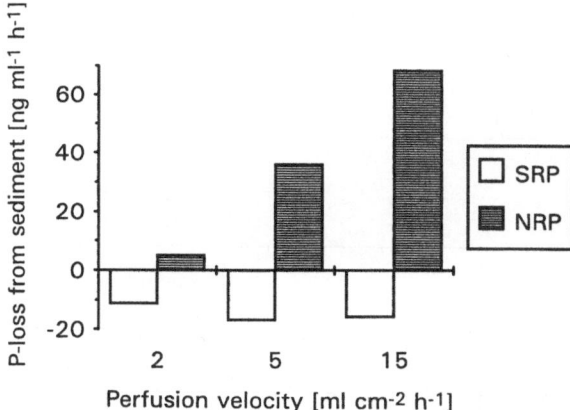

Fig. 2. Loss of phosphorus from sediment cores perfused with stream water at between 1.9 and 14.7 ml cm^{-2} h^{-1}, measured by determination of concentration differences between inflowing and outflowing water. Concentrations of NRP were higher in outflowing water, whereas SRP concentrations decreased slightly (see Table 1c).

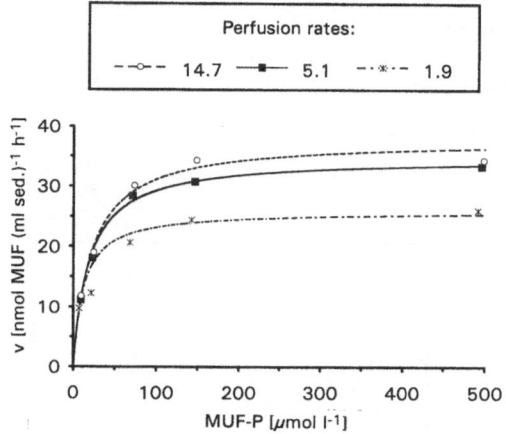

Fig. 3. Substrate saturation curves for the hydrolysis of MUF-P at perfusion rates of between 1.9 and 14.7 ml stream water cm^{-2} h^{-1} through sediment cores from the Breitenbach.

or after about 9 ml of water had perfused through 1 cm^2 of sediment core (Fig. 1). Similarly, fluorescence diminished when the cores were switched to MUF-P-free water. Rates of hydrolysis remained constant when the columns were perfused for an entire day (*ca.* 260 ml of water cm^{-2})

Concentrations of SRP decreased slightly during perfusion of stream water through the columns, whereas NRP increased (Table 1b, 1c). From these data, an increasing loss of NRP with increasing perfusion velocities was calculated, whereas there was a slight retention of SRP in the

sediment (Fig. 2). V_{max} increased with discharge up to the highest rate tested (14.7 ml cm^{-2} h^{-1}, Fig. 3, Table 3), which is rarely attained in the Breitenbach. K_m values also increased. At the end of the experiment testing different perfusion velocities, hardly any significant changes in the phosphorus fractions were measured in the columns with the two lower perfusion velocities (Table 2). At the highest velocity (14.7 ml cm^{-2} h^{-1}), however, concentrations of most of the phosphorus fractions, especially the water-soluble fractions, had increased.

The comparison of perfusion with stream water and groundwater revealed that V_{max} as well as the

Table 2. Concentrations of different phosphorus fractions in Breitenbach streambed sediment (0-1 cm). H_2O = water-soluble fraction; BD = bicarbonate-dithionite-soluble fraction; NaOH = NaOH-soluble fraction. Mean values and standard deviations (in parentheses) are given (*n* = 10). Sampling dates with figures in parentheses are for results from cores which were experimentally perfused with stream water at different velocities (1.9, 5.1, and 14.7 ml cm^{-2} h^{-1}; *n* = 5 in all cases). * = Data from experimentally-perfused cores which are significantly ($P < 0.05$) different from corresponding data for the original streambed cores from 28.1.91.

Sampling date	H_2O SRP	H_2O NRP	BD SRP [mg P l^{-1}]	BD NRP	NaOH SRP	NaOH NRP
28.1.91	6.9 (1.1)	2.3 (2.8)	31.6 (10.5)	5.2 (2.3)	32.9 (17.1)	56.7 (38.7)
28.1.91 (1.9)	9.0 (1.5)*	1.5 (0.9)	36.2 (11.3)	4.1 (3.3)	32.3 (11.0)	48.2 (40.6)
28.1.91 (5.1)	10.4 (1.0)*	1.7 (1.5)	40.4 (13.4)	7.1 (5.1)	36.5 (12.5)	50.2 (23.9)
28.1.91 (14.7)	24.5 (4.7)*	8.8 (2.6)*	53.9 (15.7)*	14.5 (3.1)*	38.2 (19.8)	94.4 (38.4)
25.2.91	7.5 (1.0)	0.3 (0.6)	55.7 (6.0)	12.2 (3.0)	41.4 (6.6)	109 (20)

Table 3. Kinetic parameters for extracellular phosphatase activity in Breitenbach streambed sediments subjected to different discharge rates.

Perolation rate [ml cm^{-2} h^{-1}]	V_{max} (S.E.) [nmol MUF (ml sed.)$^{-1}$ h^{-1}]	K_m (S.E.) [μmol MUF-P l^{-1}]
1.9	26.0 (1.7)	11.9 (1.9)
5.1	34.8 (1.0)	18.5 (1.3)
14.7	37.9 (1.6)	21.5 (2.3)

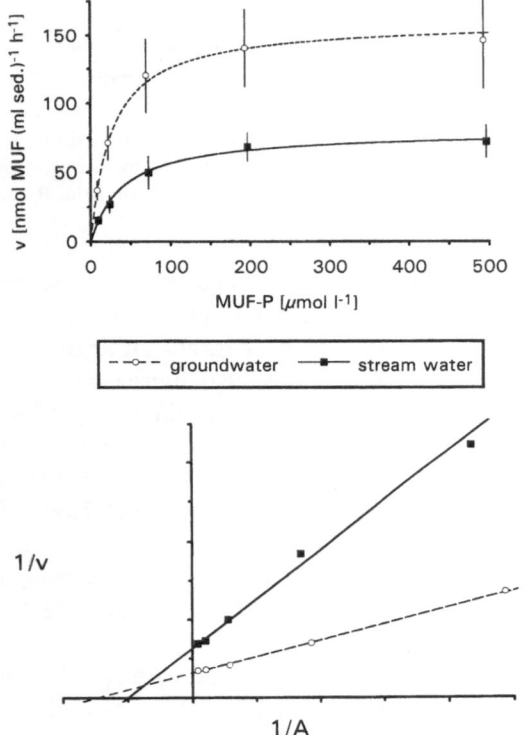

Fig. 4. Substrate saturation curves (a) and Lineweaver-Burk reciprocal diagrams (b) for the hydrolysis of MUF-P in Breitenbach sediment cores during percolation with ground- and stream water. Mean values (a, b) with 95% confidence limits (a) are shown (*n* = 5). The lines of he Lineweaver-Burk diagram are drawn by using V_{max} and K_m from a non-linear regression analysis of the untransformed data.

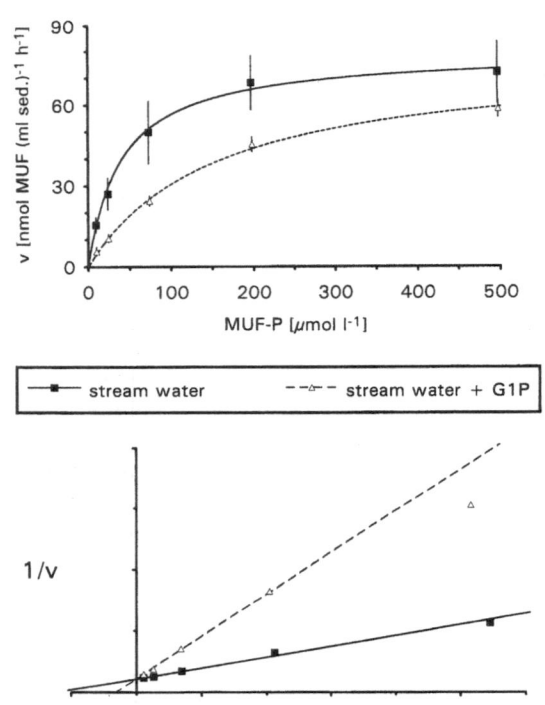

Fig. 5. Substrate saturation curves (a) and Lineweaver-Burk reciprocal diagrams (b) for the hydrolysis of MUF-P in Breitenbach sediment cores during percolation with pure stream water and with stream water containing 10 mmol l^{-1} of glucose-1-phosphate (G1P). Mean values (a, b) with 95% confidence limits (a) are shown (*n* = 5). The lines of the Lineweaver-Burk diagram are drawn by using V_{max} and K_m from a non-linear regression analysis of the untransformed data.

substrate affinity (a lower K_m) was higher with groundwater (Fig. 4, Table 4).

Addition of Glucose-1-phosphate (10 mmol l^{-1}) resulted in typical competitive inhibition of phosphatase activity: K_m increased and V_{max} remained constant (Fig. 5, Table 4).

Discussion

Exposure of the cores to the MUF-P substrate yielded a rapid response in fluorescence from liberated MUF, which remained constant for at least 24 hours (Fig. 1). This suggests that the majority

Table 4. Kinetic parameters for extracellular phosphatase activity in Breitenbach streambed sediments. comparison of discharge with groundwater, stream water, and stream water enriched with 10 mmol l^{-1} Glucose-1-phosphate (GIP). For concentrations of SRP see Table 1 (25.2.91).

	V_{max} (S.E.) [nmol MUF (ml sed.)$^{-1}$ h^{-1}]	K_m (S.E.) [μmol MUF-P l^{-1}]
Groundwater	159.4 (3.7)	26.9 (1.2)
Stream water	79.2 (3.3)	39.9 (3.5)
Stream water + GIP	77.4 (3.7)	152.4 (14.6)

of phosphatases are associated with surfaces. The implication that free dissolved phosphatases were quantitatively insignificant contrasts with the situation in lakes, where some 20 to 70% of phosphatases have been reported to be in the dissolved state (Jansson *et al.*, 1981; Chrost & Overbeck, 1987; Chrost *et al.*, 1989, Olsson, 1991; Wetzel,1991). In streambed sediments that are continuously perfused with water, similar high levels of dissolved phosphatases could only be maintained by a high production rate of new enzymes, which would be ecologically inefficient. However, definitive evidence for the prominence of surface-associated phosphatases in streambed sediments is still lacking.

The perfusion time (or water quantity) necessary for a maximum response can vary between core types. From other experiments (Marxsen, unpublished data), we know that phosphatase activity is stable for at least 5 days at natural temperatures and perfusion rates. This suggests that the method applied here did not induce significant changes in extracellular phosphatase activity of the sediment biota. This concurs with the results from Fiebig (unpublished data), where amino acid metabolism was not adversely affected during a six-month perfusion of groundwater through streambed cores.

The relevance of calculating kinetic parameters for heterogeneous microbial communities has received much discussion (e.g. Williams, 1973, Wright, 1973; Overbeck, 1974; Krambeck, 1979; Wright & Burnison, 1979; Marxsen, 1980; Azam & Hodson,1981). In general, it can be expected that a suite of enzymes with different V_{max} and K_m values act on a given substrate in the natural environment, as has been demonstrated by Azam

& Hodson (1981). Theoretically, the sum of these values should result in a deviation from Michaelis-Menten kinetics (Williams, 1973). Nevertheless, kinetic responses of natural microbial communities often fit the Michaelis-Menten equation rather well, especially if a narrow concentration range of about 1 order of magnitude is used (Wright & Burnison, 1979). This was the case in this investigation, where MUF-P concentrations of between 7 and 500 μmol l^{-1} were used (Figs 3–5). Possible explanations for this good fit, contrary to theory, could be the dominance of one single isozyme or a balancing out of the various opposing effects so that the net activity corresponds well with the Michaelis-Menten equation (Krambeck, 1979). Despite the gap between theory and practice, the Michaelis-Menten equation can still be regarded as a valuable 'workable model' (Wright, 1974) in aquatic microbial ecology.

Increases in V_{max} and K_m values with increasing velocity of water perfusion through the sediments indicate that hardly any change in rates of hydrolysis occurs at limiting substrate concentrations, with these rates only being influenced at, or near, substrate saturation levels (Fig. 3). Sediment concentrations of phosphomonoesters, the natural substrates comparable to MUF-P (Jansson *et al.*, 1988), are not known. If this fraction formed a significant proportion of the NRP content of sediments (total NRP on this occasion was 64 mg P (l sediment)$^{-1}$, Table 2), then substrate concentrations would be sufficiently high for rates of hydrolysis to increase under natural conditions.

However, there is evidence that the fraction of P that is hydrolyzable by extracellular phosphatases may be much lower. K_m values from all the experiments performed on phosphatase activity in this study varied between 12 and 40 μmol l^{-1}. These are only apparent K_m values, because they are the sum of the actual K_m and the natural substrate concentration (Marxsen, 1980b). Thus it can be concluded that natural levels of phosphorus compounds hydrolyzable by phosphomonoesterases were below these apparent values for K_m. If natural levels were 5 to 10 μmol l^{-1}, and assuming that these natural compounds (phosphomonoesters) contain only one atom of P, then concentrations of natural substrates can be estimated at around 150 to 300 μg P l^{-1}. These are much higher than levels of NRP in stream, ground-, and interstitial water of the Breitenbach (Table 1a, 1b), suggesting that the main source of substrates hydrolyzable by phosphomonoesterases might be the particulate P fraction of the streambed sediments. However, it cannot be concluded that these 150 to 300 μg P l^{-1} represent the entire amount of potentially hydrolyzable P in the sediment. Firstly, there are other P compounds which can be hydrolyzed only by enzymes other than phosphomonoesterases. Secondly, P compounds (including phosphomonoesters) in the particulate form may be only hydrolyzed by extracellular enzymes under specific circumstances, e.g. physical disruption of the particulate matter, lysis of cells, enhanced solubility of particulate P at higher perfusion rates. The last phenomenon may have been prominent during the experiment investigating the effect of different perfusion velocities. Concentrations of NRP in stream water increased during perfusion through cores (Table 1b, 1c), and loss of organic phosphorus from the cores (Fig. 2) as well as the apparent K_m values (Table 3) increased with increasing perfusion velocity. Thus, organic P compounds (including phosphomonoesters) were solubilized by the percolated water, especially at higher perfusion rates, contributing to the pool of dissolved substrates that could be acted upon by phosphatases.

At the end of the experiment investigating different perfusion velocities, SRP and NRP concentrations were highest at the highest perfusion velocity tested (Table 2). This concurs with changes in SRP levels in the perfused water at the beginning of the experiment, but not with changes in NRP levels, where decreases would have been expected. However, these long-term data would have been influenced by the additions of artificial MUF-P to the perfused stream water, resulting in an increased availability of phosphorus to the sediment, compared with the natural water. Thus, these data reveal the ability of sediment (along with its microbial community) to store temporarily-enhanced P loads supplied by the perfusing water, rather than the effect of naturally occurring increases in perfusion rates.

Competitive inhibition of phosphatase activity was observed on addition of Glucose-1-phosphate to the perfusing stream water (Fig. 5). However, the 10 mmol l^{-1} enrichment resulted in an apparent K_m of around 150 μmol l^{-1} only. This suggests that the natural phosphomonoesterases may operate with a different efficiency on the artificial model substrate (MUF-P) than on the naturally-occurring substrates. It is therefore questionable whether this substrate is acted upon by the entire suite of extracellular phosphomonoesterases occurring in the Breitenbach sediments.

Two different types of water can percolate through the Breitenbach sediments, namely stream water and groundwater. When these two types of water were perfused through the sediment cores, V_{max} as well as the substrate affinity was higher with groundwater (Fig. 4, Table 4). With extracellular β-glucosidases in the Breitenbach, ground- and stream water produced similar V_{max} values, but K_m was higher with groundwater (Marxsen, unpublished data). This was probably due to competitive inhibition from humic material in groundwater, an effect well known in limnic systems (Wetzel, 1990). The different trend with phosphatase is not surprising, because inorganic phosphorus concentrations were much lower in groundwater than in stream water (Table 1a, 1b). Thus, the P demand was higher in groundwater, which probably accounts for the differences between kinetic parameters in sediments exposed to these different types of water.

Perspectives

There are still many unanswered questions concerning the flux of phosphorus in streambed sediments. Nevertheless, there is evidence that phosphomonoesterases, which appear to be associated with both biotic and abiotic surfaces of sediments, play an important role in the supply of phosphorus to microorganisms in streambed sediments. The use of undisturbed cores for measuring microbially-mediated phosphorus flux, together with enhanced techniques for quantifying phosphorus, seems to be a promising approach for gaining further and more reliable insights into these important phenomena in streambed sediments.

Acknowledgements

The authors thank Beate Knöfel and Birgit Landvogt for technical assistance during this study, and Dr Douglas Fiebig for revising the English text.

References

Azam, F. & R. E. Hodson, 1981. Multiphasic kinetics for D-glucose uptake by assemblages of natural marine bacteria. Mar. Ecol. Prog. Ser. 6: 213–222.

Brehm, J. & M. P. D. Meijering, 1982. Fließgewässerkunde. Quelle & Meyer, Heidelberg, 311 pp.

Chrost, R. J. & J. Overbeck, 1987. Kinetics of alkaline phosphatase activity and phosphorus availability for phytoplankton and bacterioplankton in Lake Plußsee (North German eutrophic lake). Microb. Ecol. 13: 229–248.

Chrost, R. J., U. Münster, H. Rai, D. Albrecht, K. -P. Witzel & J. Overbeck, 1989. Photosynthetic production and exoenzymatic degradation of organic matter in the euphotic zone of a eutrophic lake. J. Plankton Res. 11: 223-242.

Cox, E., 1990. Studies on the algae of a small softwater stream. II. Algal standing crop (measured by chlorophyll-a) on soft and hard substrata. Arch. Hydrobiol./ Suppl. 83: 553–566.

Fiebig, D. M., 1992. Fates of dissolved free amino acids in groundwater discharged through stream-bed sediments. In B. T. Hart & P. G. Sly (eds), Sediment/Water Interactions. Developments in Hydrobiology 75. Kluwer Academic Publishers, Dordrecht: 311–319. Reprinted from Hydrobiologia 235/236.

Jansson, M., H. Olsson & O. Broberg, 1981. Characterization of acid phosphatases in the acidified Lake Gardsjön, Sweden. Arch. Hydrobiol. 92: 377–395.

Jansson, M., H. Olsson & K. Pettersson, 1988. Phosphatases; origin, characteristics and function in lakes. In G. Persson & M. Jansson (eds), Phosphorus in Freshwater Ecosystems. Developments in Hydrobiology 48. Kluwer Academic Publishers, Dordrecht: 157–175. Reprinted from Hydrobiologia 170.

Krambeck, C., 1979. Applicability and limitations of the Michaelis-Menten equation in microbial ecology. Arch. Hydrobiol. Beih. Ergebn. Limnol. 12: 64-76.

Leatherbarrow, R. J., 1987. Enzfitter. A non-linear regression data analysis program for the IBM PC/PS2. Biosoft, Cambridge, UK.

Marquardt, D. W., 1963. An algorithm for least-squares estimation of nonlinear parameters. J. Soc. ind. appl. Math. 11: 431–441.

Marxsen, J., 1980a. Untersuchungen zur Ökologie der Bakterien in der fließenden Welle von Bächen. I. Chemismus, Primärproduktion, CO_2-Dunkelfixierung und Eintrag von partikulärem organischen Material. Arch. Hydrobiol./ Suppl. 57: 461-533.

Marxsen, J., 1980b. Untersuchungen zur Ökologie der Bakterien in der fließenden Welle von Bächen. III. Aufnahme gelöster organischer Substanzen. Arch. Hydrobiol./Suppl. 58: 207–272.

Marxsen, J., 1988. Evaluation of the importance of bacteria in the carbon flow of a small open grassland stream, the Breitenbach. Arch. Hydrobiol. 111: 339–350.

Marxsen, J. & K. Moaledj, 1988. On the structure of bacterial communities in a Central European open grassland stream, the Breitenbach. Verh. int. Ver. Limnol. 23: 1855–1860.

Marxsen, J. & K. -P. Witzel, 1990. Measurement of exoenzymatic activity in streambed sediments using methylumbelliferyl-substrates. Arch. Hydrobiol. Beih. Ergebn. Limnol. 34: 21–28.

Marxsen, J. & K. -P. Witzel, 1991. Significance of extracellular enzymes for organic matter degradation and nutrient regeneration in small streams. In R. J. Chrost (ed.), Microbial Enzymes in Aquatic Environments. Brock/Springer Series in Contemporary Bioscience. Springer-Verlag, New York: 270–285.

Olsson, H., 1991. Phosphatase activity in an acid, limed Swedish lake. In R. J. Chrost (ed.), Microbial Enzymes in Aquatic Environments. Brock/Springer Series in Contemporary Bioscience. Springer-Verlag, New York: 206–219.

Overbeck, J., 1974. Microbiology and biochemistry. Mitt. int. Ver. Limnol. 20: 198–228.

Pettersson, K. & M. Jansson, 1978. Determinations of phosphatase activity in lake water – a study of methods. Verh. int Ver. Limnol. 20: 1226–1230.

Psenner, R., R. Pucsko & M. Sager, 1984. Die Fraktionierung organischer und anorganischer Phosphorverbindungen von Sedimenten. Versuch einer Definition ökologisch wichtiger Faktoren. Arch. Hydrobiol./Suppl. 70: 111–155.

Wetzel, R. G., 1990. Land–water interfaces: Metabolic and limnological regulators. Verh. int. Ver. Limnol. 24: 6–24.

Wetzel, R. G., 1991. Extracellular enzymatic interactions: Storage, redistribution, and interspecific communication. In R. J. Chrost (ed.), Microbial Enzymes in Aquatic Environments. Brock/Springer Series in Contemporary Bioscience. Springer-Verlag, New York: 6–28.

Williams, P. J. leB., 1973. The validity of the application of simple kinetic analysis to heterogeneous microbial populations. Limnol. Oceanogr. 18: 159–165.

Wright, R. T., 1973. Some difficulties in using ^{14}C-organic solutes to measure heterotrophic bacterial activity. In L. H. Stevenson & R. R. Colwell (eds), Estuarine Microbial Ecology. University of South Carolina Press, Columbia, SC: 199–217.

Wright, R. T., 1974. Mineralization of organic solutes by heterotrophic bacteria. In R. R. Colwell & R. Y. Morita (eds), Effect of the Ocean Environment on Microbial Activities. University Park Press, Baltimore: 546–565.

Wright, R. T. & B. K. Burnison, 1979. Heterotrophic activity measured with radiolabelled organic substrates. In J. W. Costerton & R. R. Colwell (eds), Native Aquatic Bacteria: Enumeration, Activity, and Ecology. ASTM Special Technical Publication 695. American Society for Testing and Materials, Philadelphia: 140–155.

Hydrobiologia **253**: 217–218, 1993.
P.C.M. Boers, T.E. Cappenberg & W. van Raaphorst (eds),
Proceedings of the Third International Workshop on Phosphorus in Sediments.
© 1993 *Kluwer Academic Publishers.*

Abstracts

P-flux regulation in sediments by coupled microbial processes

Stefan Heller
Gesellschaft für Gewässerbewirtschaftung, Berlin, Germany

Lake internal phosphorus-recycling from sediments is a significant process in most eutrophic shallow lakes.

The mechanisms for P-liberation in the sediments are coupled microbial processes depending on water transport at the sediment-water-interface. Thereby electron acceptors are supplied to sessile microassemblages of bacteria and inhibitory products, such as hydrogen sulphid, are removed by water transport.

Laboratory experiments with extractions of phosphorus by addition of hydrogen sulphid as reference (complete liberation of ironbound phosphorus) or by addition of glucose, iron and sulfate in different molar relation revealed the reaction mechanisms for various sediment types.

For example:

(1) With addition of hydrogen sulphid to sediments of 3 eutrophic lakes 14–39% of the total phosphorus are released to solution while this was only 2% with sediment of an oligotrophic lake.

(2) In Desulfurication-experiments addition of glucose and sulfate results in total phosphorus liberation of up to 50% within six weeks from sediments of eutrophic lakes. If also iron (as $Fe(HCO)_3$) was added in stöchometric relation to the experiment the desulfurication-rate increased significantly while only 1–7% of total phosphorus was liberated.

The sediment studies showed a monocausal hierachy of processes. The energyflow partly resulting in water transport and in chemical reactions control the sediment metabolism with respect to space and time distribution of microbial activity.

Suspended sediments as a source of P to summer biota of Lake Kinneret

A. Nishri
Kinneret Limnological Laboratory, Israel Oceanographic and Limnological Research, P.O. Box 345,
Tiberias, Israel 14102

Summer biota in the stratified Lake Kinneret are P limited. Lake external sources and/or internal cycling of P within the epilimnion cannot account for the P demand of summer biota. Another source of P must be the hypolimnion although calculated upward diffusional fluxes of SRP across the thermocline are negligible. Additional sources of hypolimnetic P are considered. Sediments collected in traps enabled us to measure downward fluxes of various species of P as well as other parameters. The role of resuspension in transportion of particulate P from near to coast zones to the center of the lake was evaluated. It is concluded that mixing of hypolimnic and epilimnic water, in near coast zones, due to internal seiche activity may be a mechanism by which either SRP or bioavailable particulate P (from anoxic bed sediments) is introduced to the epilimnin.

P-flux regulation in sediments by transport processes

Wilhelm Ripl

TU-Berlin, Institut für Ökologie, Fachgebiet Limnologie, Germany

Lake internal phosphorus-recycling from sediments is a significant process in most eutrophic shallow lakes.

The mechanisms for P-liberation in the sediments are coupled microbial processes depending on water transport at the sediment–water interface. Thereby electron acceptors are supplied to sessile microassemblages of bacteria and inhibitory products, such as hydrogen sulphid, are removed by water transport.

This relations where shown by timeseries in interstitial water strata and lake-budget calculations for several lakes in the Berlin-region, the Schlei-estuary and Swedish lakes.

In shallow eutrophic lakes up to $1 \, g/m^2/month$ phosphorus have been liberated by the process of internal fertilisation from sediments during short periods in summer and autumn. This was equal to the whole phosphorus content of the upper 2 mm of surface sediments and resulted in a large increase of phosphate concentration in water column. The periods of internal P-liberation where coupled with high rates of plankton sedimentation and depleted oxidized nitrogen by denitrification in sediment surface. In these periods desulfurication became the main respiration-process in sediment metabolism.

The sediment studies showed a monocausal hierachy of processes. The energyflow partly resulting in water transport and in chemical reactions control the sediment metabolism with respect to space and time distribution of microbial activity.

The role of microbial processes in the phosphorus flux regulation between sediments and water

Anja J. C. Sinke, Francis H. M. Cottaar, Kerst Buis & Peer Keizer

Limnological Institute, Rijsstraatweg 6, 3631 AC Nieuwersluis, The Netherlands

In aquatic sediments the mineralization of organic matter regenerates phosphates, resulting in an accumulation of phosphates in the interstitial water and the formation of a concentration gradient. The subsequent diffusive transport to the overlying water is affected by chemical and microbial processes. Especially the redox state of the sediment is known to be an important factor determining the actual phosphate flux. The presence of an oxidized microlayer at the sediment surface is considered to be an important trap for phosphates. The high adsorption capacity of the oxidized microzone is generally ascribed to the presence of iron(III)-hydroxides. Lately, also the role of microorganisms has come into focus. Microorganisms in the surface layer of the sediment might act as a redox dependent source/sink mechanism of phosphates.

This contribution is focused on the role of methane oxidizing bacteria in the sediment surface. In Lake Loosdrecht almost the entire methane flux diffusing upward is oxidized in the sediment surface layer. In periods with high methane production the major fraction of the oxygen consumption of the sediment can be ascribed to methanotrophic bacteria. An experimental setup will be described which enabled us to manipulate the methane flux toward the sediment surface. Microelectrode equipment was used to quantify the oxygen consumption and the oxygen penetration depth. In columns with high activity of methanotrophic bacteria a decrease in the oxygen penetration depth could be demonstrated. However, these columns appeared to have a higher phosphate adsorption capacity than the controls. Batch experiments were conducted to examine the role of methanotrophic bacteria in the uptake of phosphates.

Hydrobiologia **253**: 219–231, 1993.
P.C.M. Boers, T.E. Cappenberg & W. van Raaphorst (eds),
Proceedings of the Third International Workshop on Phosphorus in Sediments.
© 1993 *Kluwer Academic Publishers.*

Considerations in modeling the sediment–water exchange of phosphorus

Lambertus Lijklema
Nature Conservation Department, Agricultural University, P.O.Box 8080, 6700 DD Wageningen,
The Netherlands

Key words: sediment, water, phosphorus, models, temporal scales, spatial scales

Abstract

The potential to release accumulated phosphorus from sediments has been the major motive to study and to model the fate of this nutrient in sediments.

For the dynamics of the sediment–water interaction the sizes of the pools involved and the rates of conversion/transport from one pool to another are of primary interest. As the sediment pools for phosphate are generally much larger than the pools in the water column, a rather slow adjustment of the sediment to management measures will occur. For the analysis of management measures it is obvious that the gradual change in sediment composition must be taken into account. Only for rather short periods the sediment composition can be assumed to be constant; this may be appropriate for studies of e.g. the annual cycle.

The sediment-water interaction is a complex resultant of physical, chemical and biological processes, including:

– physical processes: advection due to seepage or consolidation, pore-water diffusion, transport and mixing of solids by resuspension, sedimentation and bioturbation.
– chemical processes: adsorption and desorption, dissolution and (co)precipitation, inclusion.
– biological processes: mineralization of a wide range of organic compounds by various (micro)organisms, each with their own nutrient requirements and electron acceptors.

Aspects which are discussed and need to be considered in application of a model in research or management are the level of aggregation and detail that is required and may still be practical, the spatial and temporal scales which are applicable for the processes mentioned and their influence upon the numerical dispersion and model stability, the availability of data for calibration/validation and the resolution of the analytical techniques. These aspects are not independent however.

Frequently models are not functional because they contain details which are either unnecessary or suggest a feigned accuracy which is not justified by analytical and experimental resolution of system characteristics.

Introduction

The potential to release accumulated nutrients or pollutants from sediments has been the major motive to study and to model the fate of these substances in sediments. Internal loading of

phosphate will sustain primary production and might delay the effects of restoration measures. Therefore, much of the research on eutrophication has been focused upon the role of sediments in the cycling of nutrients, see e.g. Baccini (1985).

The interaction between sediments and overlying waters is reciprocal: the loading of the sediments with phosphorus – generally in a particulate form – from the overlying water is controlled by its cycling and incorporation in different trophic levels in the water column, but this is in turn affected by the rate of release from the sediments which in a complex way is related to the loading of the sediment and the subsequent accumulation, redistribution and speciation. Especially in a transitional phase, when reduced external inputs start to restrict the productivity, the interaction between sediment and water is of great interest. Actually, managers want to know at what level this transition will occur and what the role of the sedimenta will be in the long and short term. They therefore generally want guidelines or rather simple predictive models relating readily accessible variables, some of which can be controlled, to trophic indicators. Such models may exist, but do no justice to the complexity of the system or only within certain limits, e.g. for the specific water body on which the model has been calibrated. In research, often much more complex models are used as a tool to collect one's thoughts, to design and evaluate experiments, to integrate results of different scientific disciplines, and eventually to generalize these for management purposes. One of the main dilemmas in research is the level of detail and complexity that should be introduced in the description of the system. Too simple models have very little predictive value outside the conditions for which they were calibrated, and prediction in the most essential function of models. On the other hand, complex models are hard to calibrate and validate, if possible at all, and have a high demand for parameters, input data, initial and boundary conditions, time and expertise. Although not addressing the role of sediments specifically, the empirical lake loading models belong to the first category; when error analysis is used the information from such models can be properly weighed and considered in the lake management (Reckhow, 1979). An example of a model that tends to be rather rich in variables and parameters can be found in Kamp Nielsen (1975).

In this contribution some aspects pertaining to the scope and complexity of models are discussed with a view to the function of the model.

Temporal scales

Whereas is the water column autotrophic processes and grazing food chains predominate, sediments are characterized by heterotrophic transformations and a detritus food chain. Thus in the overlying water phosphate is incorporated in algae and other organisms in a particulate form, but in the sediments the inorganic phosphate is (in part) regenerated and subsequently released, stored and/or buried. The proportions of release, fixation and burial are of interest for different objectives, in which from a methodological and modeling point of view the time scale is the most important.

Short term objectives, with a time horizon of about one year, may be to analyze the seasonal variation in internal loading as an important attribute of predicting chlorophyll levels, oxygen conditions, habitat quality etc. Within a year the sediment matrix will be fairly constant except for the pool of non-refractory organic material. Consequently the adsorptive capacity of the sediment for phosphate varies mainly due to changes in redox conditions related to the temperature cycle and the course of organic deposition. This means that a reasonable prediction of internal loading within a year does not require a complete sediment mass balance, but should take into account transient storage in pools related to the mineralization and the seasonal deposition of particulate organic phosphate from the water.

The buffer of phosphate in the microbial mass is one of these potentially important transient pools (Montigny & Prairie; Petterson; Gächter; 1993). Iron associated phosphate also can vary strongly due to changing redox conditions (Caraco et al.; 1993) and more general by Froelich (1988), Lijklema (1977, 1980).

Besides the effect of redox conditions also the change in the ambient concentration of phosphate will affect the fraction adsorbed on Fe (III)-(hydr)oxides, e.g. Breeuwsma (1973) and Al-

(hydr)oxides, e.g. Lijklema (1980), where further the adsorption and desorption have been shown to be time dependent, e.g. Hingston *et al.* (1974), Lijklema (1980), Sigg (1979).

For longer time scales, ranging over several years, the interest will be in the steady state conditions attained upon implementation of restoration measures such as reduced loading, flushing, dredging etc. In this case the sediment composition will gradually change over the years with still a superimposed annual fluctuation in organic and related pools as mentioned above. The gradual change in the matrix is particularly interesting with respect to the ratio of the phosphate loading to the loading of potential adsorption sites (iron, calcium etc.) and in relation with the overall sedimentation rate in mm y^{-1}. This ratio will determine the adsorptive capacity of the newly formed sediment. Via the pore water concentration this controls the release of phosphate. The ultimate phosphate adsorption capacity is dealt with in the studies by Danen; Forsgen & Jansson (1993), Brinkman & van Raaphorst (1986), Ryding (1985).

The net accumulation rate of the sediment (in mm y^{-1}) is important for the dilution of nutrients and adsorbents in the sediment and thus the capacity of the surface layer to store and release phosphate and for the rate at which an equilibrium will be attained in the composition of the top layer of the sediment. The time constant for the sediment dilution rate is the depth of the active sediment layer (the mixed layer, up to about 10 cm in fresh waters, which is grubbed up by macro-

fauna annually) divided by the accumulation rate, see Fig. 1. A more detailed semi-quantitative discussion of this sediment dilution rate in combination with sorption properties can be found in Lijklema (1986). A high dilution rate implies a speedy attainment of a new steady state in sediment composition – but generally still in the order of at least 10 years – and a lower buffer capacity for variable phosphate inputs.

In case the net accumulation is very small or zero, no effective burial of phosphate will take place and the adsorption capacity of the top layer will also change very little with virtually no net sedimentation. Hence the function of such sediment as a sink for phosphorus will be very limited.

In contrast to the short term description, modeling of the long term sediment properties requires a full sediment mass balance for phosphate, all potential adsorbents and for inert (with respect to P) material. Also the mineralization of poorly degradable organic components and desorption of non-surficially bound phosphate via solid phase diffusion should be considered. However the transient variations in bacterial pools etc. are less relevant and seasonal variations may be disregarded in favour of annual means.

The preceding discussion compares seasonal and long term aspects, but there is also frequently a diurnal variation in DO in the overlying water, that may be important for the dynamics of the sediment-water interaction. Particularly in highly productive lakes during the summer the daily DO variations may be high and affect the penetration

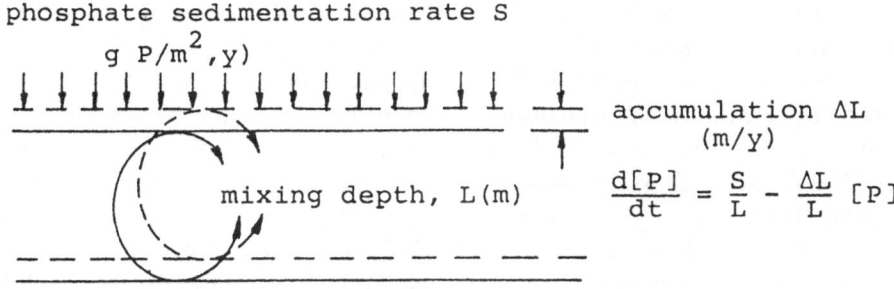

Fig. 1. Model for sediment-P dilution rate in an accumulating, bioturbated sediment. The analytical solution for the P concentration in the sediment (g m^3) is: $P = (S l^{-1})(1 - \exp.-\Delta L \, t \, l^{-1}) + P_0 \exp -\Delta L \, t \, l^{-1}$.

depth of oxygen. This depth is proportional to the square root of DO for zero order volumetric oxygen consumption rate in the sediment. Thus substantial variation in this penetration depth and associated secondary effects upon phosphate equilibrium concentration and flux can be expected and thus averaging over a day would introduce inaccuracies. The adjustment of the sediment to changing DO in the overlying water is more or less instantaneous in actively mineralizing systems. The partial differential equation

$$\frac{\delta C}{\delta t} = D \frac{\delta^2 C}{\delta z^2} - R,$$

in which C is the oxygen concentration (g m^{-3}); D the effective diffusion coefficient (m^2 s^{-1}); and R the volumetric oxygen consumption rate in the sediment (g m^{-3} s^{-1}); has a solution in which the time dependent term is controlled by $\exp - (D \pi^2 t / L^2)$ with L the penetration depth of oxygen in the sediment. Consequently the time constant Θ belonging to this process is $(L^2/D \pi^2)$. The steady state penetration depth is:

$$L = (2 c_0 D/R)^{0.5}$$

with c_0 the oxygen concentration in the overlying water.

With values for D, R and c_0 in the order of 3.10^{-10} m^2 s^{-1}, 5.10^{-3} g m^{-3} s^{-1} and 10 g m^{-3} respectively the time constant for approaching the equilibrium oxygen profile is in the order of 2 minutes. For all practical purposes this means instantaneous.

Frequently the sediment contains iron that may buffer the variation in oxygen penetration in the sense that an increase in depth is retarded due to oxidation of Fe^{2+}, whereas a decreasing DO in the overlying water will result in a delay in the lowering of the redox potential in the transition zone due to the necessary reduction of Fe^{3+}. Van Raaphorst et al. (1990) found in some marine sediments indications for a slow reduction and dissolution of iron (III) in anoxic zones and accordingly low phosphate concentrations.

For a sediment with a porosity of 0.8 in the top layer, a specific density of the sediment particles

of 2500 kg/m^3, an iron content of 0.2%, an O_2 penetration depth of 1 mm, an effective dispersion coefficient of $5 \cdot 10^{-9}$ m^2 s^{-1} and DO = 8 g/m^3, it takes about one hour to change the penetration depth 1 mm including the oxidation/reduction of all iron. This assumes immediate oxidation and reduction of the iron and neither of the two is necessarily true.

For most practical purposes the computation time may become excessive when too short time steps are needed; actually in more advanced models an implicit (or mixed implicit-explicit) numerical scheme is used to allow for long time steps (typically 1 hr to 1 day) without losing stability. With a time step of 1 hr it is still possible to account for diurnal variations in DO; time steps of one day exclude this. Hence, a functional model for systems with a high amplitude in the boundary DO condition should not use time steps exceeding 1 or 2 hours if sufficient accuracy in the daily variation in pore water P-concentration and release rates is desired.

The cause of the diurnal DO variation can be primary productivity in the water column, but also at the sediment water interface itself due to benthic algae. Here the picture is even more complex, because besides the DO effect on phosphorus release as discussed above also the local P-uptake during daytime and possible release at night by algae will affect the fluxes to the overlying water.

Spatial scales

Vertical stratification of sediment and pore water is commonly accounted for in studies and models dealing with P-release from sediments. In relatively simple approaches the sediment can be envisaged as an aerobic top layer overlying anaerobic sediment. If nitrate reduction is relevant an intermediate denitrifying anoxic layer can be defined (Snodgrass, 1985; Klapwijk & Snodgrass, 1985; Brinkman & van Raaphorst, 1986; van Eck & Smits, 1986; DiToro et al., 1990). Diffusive fluxes of electron acceptors and, if relevant, bioturbation are included. When one works with fixed depth, aerobic and anoxic layers may become unrealistic when transient conditions occur. However, calculation of the penetration depth of O_2 and nitrate and the redistribution of oxidized and reduced compounds over the layers after a change in depth leads into a jungle of computational problems related to the conservation of mass, electrons etc. So more advanced models capable to handle dynamic conditions mostly use multiple layers which in most cases increase in thickness with depth. This is important to obtain a proper representation of the usually steep gradients near the interface and at the same time to reduce the number of computational elements as much as possible. In these models mass balances for O_2 and other electron acceptors based on transport and reactions result in the assessment in which layer a substance will become exhausted and this controls in turn which redox reactions will occur. In order to prevent an excessive computational burden and instability, the fast processes are incorporated in an equilibrium computation and the system is driven by the slow processes, e.g. mineralization and diffusion.

Besides having effects on porosity and the vertical transport of interstitial water, bioturbation also moves particulate material in the vertical direction. In its most simple form this can be modeled by a 'box-model' approach, assuming immediate mixing of all material added at the surface over the mixed (bioturbated) top layer (Berner, 1980). This approach may be appropriate when slow adjustments to changes in external loading are considered, see e.g. Lijklema (1983) or very slow (e.g. radioactive decay) processes (Berner, 1980). However, more frequently the solid transport due to benthic organisms is modelled by a vertical dispersion term (Berner, 1980; Lijklema, 1986), sometimes assuming exponential attenuation of the dispersion coefficient with depth, but with the same values in the horizontal plane across the whole sediment. There are two problems associated with this. Firstly, the displacement of sediment particles by the organisms tends to be rather discrete: e.g. material is ingested at a certain depth and deposited at the interface. Hence the redistribution over the vertical is dif-ferent from a dispersive process (Davis, 1974a; 1974b).

Secondly, this process occurs only locally at discrete spots and therefore averaging in the modelling framework may be unrealistic. See e.g. the work of Aller (1991) on his hollow cylinder model. How serious the ensuing averaging error is, is difficult to predict and probably attempts to model this in a better way will fail due to computational complexity and lack of data to calibrate the process properly.

Horizontal gradients in sediment composition on a small scale are not only due to macrofauna, but also to local differences in bottom roughness through ripples, debris from animal and plant life, stones etc. These induce local variations in shear stress and consequently a preferential settling of fine and flocculent material in low shear stress areas. On a wider scale of the whole water body a similar segregation will be induced by variations in water depth and effective wind fetch (Håkanson & Jansson, 1983). Auer *et al.* (1993) report a coefficient of variation of the sediment release rate in a eutrophic lake of 24%. Hieltjes (1980) found in 7 cores sampled at one location within three days a coefficient of variation for the phosphate content in the top layer of 22%. These horizontal variations are real and beyond the experimental analytical error. They have consequences for models involving internal loading and for calibration studies of fluxes measured in situ or using intact sediment cores. Especially at high biological activity, when concomitant fluxes will be high, this heterogeneity will be pronounced. It will be impractical to account for the small-scale spatial variation and consequently averaging is common practise. However, calibrations frequently rely on just one single sampled core in which the composition and/or the release is measured. This thus may be a source of error in the order of at least 20%.

In systems with substantial internal transport inevitably a spatial distinction must be made. For short term simulations it may be acceptable to use constant sediment properties in time and spatially distributed values, normally as measured. All sediment computational elements thus have their own release rate as a function of overlying water concentrations of e.g. oxygen and phosphate. The water column can be modelled as one box if the mixing time is sufficiently short with respect to processes causing spatial gradients.

The combination of horizontal transport, resuspension and sedimentation will generate reallocation of materials and regional differences in accumulation rate and sediment dilution rate etc., which for a long term simulation make it necessary to set up a sediment mass balance that accounts for the resulting gradual change in sediment properties. The model becomes rather complex: a two dimensional horizontal transport model in the water for suspended solids distributed over a number of relevant fractions and sediment mass balances for all computational elements in the 2-D network must be integrated. To model the suspended solids in a functional way means that their settling properties and susceptibility for resuspension must be accounted for. In practice this means a fractional distribution on the basis of settling rates, not chemical composition (van Duin et al., 1992). Hence for transport of phosphate estimates of the phosphate content of the fractions of the suspended solids (organic, slowly settling fraction) or of the adsorbed phosphate in these fractions (especially the inorganic, fine fraction) must be made, which requires a lot of field work because this tends to be rather site specific. Such models become very large with a great demand on input data (initial sediment composition, critical shear stresses, wind-wave relationships, morphological data etc.), on computation time and on process information (e.g. sediment mixing rates). Including also a detailed description of the phosphate regeneration in the sediment, the chemical distribution and the release at the interface, makes this approach laborious and costly and thus it is only functional to account in some detail for spatial variations in properties and process rates in cases where this is of great importance.

The role of spatial heterogeneity and sometimes horizontal transport as well is discussed in this volume by Smits & Van der Molen and Auer et al.

Transport processes

The main transport processes in the sediment are advection due to infiltration, seepage or consolidation; dispersion and diffusion, bioturbation and bioirrigation and at the interface, resuspension and sedimentation. Of these bioturbation is the most difficult to assess due to the intrinsic problems of predicting densities of organisms, the variety of ways in which the macrofauna grubs up the sediment and induces water flow and the micro-environments created. The best that seems practical when not going into the field of very complex 'full-ecosystem models' is to assume a mixing rate in the sediments which may be a function of temperature or has an annual course based on field data on macrofauna densities or activities. Also an effect upon the porosity or the dispersion coefficient can be related to macrofauna (Berner, 1980). Van Raaphorst et al. (1990) present a (liquid) mass transfer coefficient which probably is a function of the abundance of macrofauna and is referred to as 'bio-irrigation'.

At the water-sediment interface small variations in pressure (wave action) will induce some dispersive transport into the pore waters that will be attenuated by dissipation of the energy with depth (Rutgers van der Loeff, 1981). Therefore the pore water molecular diffusion, corrected for tortuosity and porosity, should be increased with a dispersive term which decreases from the interface downwards. Portielje & Lijklema (1993) used an exponential decay of this term and obtained a good fit with field data. See also Booij (1989).

Consolidation and (net) settling of suspended solids cause a problem with respect to the vertical coordinates. For short simulation runs with limited compaction and/or sedimentation the shift of the interface will be small and the changes in mass of the different components in the (fixed) top layer of the sediment do not necessarily lead to a transport of solid material between sediment layers that has to be calculated. So, only diffusion and advection of the pore water cause vertical transport within the sediments. In cases where the porosity, which may be a function of depth, can be assumed not to change in time, the net deposition of solids at the interface can be compensated by the transport of an equivalent weight across the interfaces of all computational elements below the surface (Berner, 1980). In fact this represents the upward movement of the vertical coordinate. Burial of material below the active top layer of the sediment is then automatically included in the model. The downward flux of pore water due to the sedimentation can be described in the same way as the solids transport.

When consolidation is not a continuous process and thus the vertical porosity structure changes in space and time as well, a more complicated representation is needed. However, it is seldom that sufficient data are available and that a real improvement of the description of the system behaviour will be achieved by including this. It may be necessary for reservoirs in which sludge is dumped and e.g. deep oceans but for natural lakes the approaches outlined above will be practical and functional. A good background for this issue gives the classical publication by Berner (1980).

Seepage and infiltration can affect transport rates and concentration profiles if the rate is sufficiently high with respect to the effective diffusion coefficient; the dimensionless Peclet number $D_c/u.L$ should be smaller than 10 for advection to become important. L is the characteristic length considered (e.g. the active sediment layer depth or the oxygen penetration depth) and u is the advective vertical flow velocity (m s^{-1}). In systems where besides the physical transport also chemical reactions occur (e.g. phosphate adsorption/desorption) the criterion is more complex and involves also the reaction rate (van Raaphorst & Brinkman, 1985) and a higher flow rate u is needed for a substantial impact on the redistribution of the dissolved species than in the case of purely physical transport.

Physico-chemical processes

The most important physico-chemical processes affecting phosphate in sediments directly are sorption and precipitation/dissolution. Because pH and redox conditions are essential factors in these processes, indirectly also the oxygen and nitrate dynamics are involved and in sulfate rich waters also this ion. Thus in the marine and other high sulphate environments the precipitation of FeS reduces the availability of iron for phosphate adsorption. (Caraco et al., 1993).

The mineralization of organic material is the most important forcing factor for the inorganic system, but will be discussed in a separate section below.

Sorption of phosphate is mainly by iron- and aluminum-hydroxides. Iron is redox sensitive. Both hydroxides form polynuclear complexes with a charge that depends on the degree of oxolation (Lijklema, 1980) and the pH. Oxolation further is a rather slow process and the adjustment of the adsorption equilibrium to new conditions has been shown to contain a slow component (Bolan et al., 1985; Portielje & Lijklema, 1993). The assumption of instantaneous equilibration in several studies and models can only be justified when transient conditions are not important. However, inclusion of slow sorption in terms of solid phase diffusion (Portielje & Lijklema, 1993) or as a slow precipitation reaction (Smits & van der Molen, 1993) adds to the computational burden, parameters are difficult to assess and the transition from fast (instantaneous) adsorption to slow is rather arbitrary. However, in the more sophisticated models this transition point must be chosen because the fast reactions usually are included in an equilibrium calculation (e.g. by minimization of the Gibbs free energy, see de Rooy, 1991) and the slow reactions are calculated with comparatively large time steps together with transport and other 'slow' reactions. When the slow phosphate adsorption causes comparatively large changes in dissolved phosphate during a time step, this interferes in the next time step with the equilibrium part of the system and may induce instabilities.

Another complicating factor is the presence of organic matter that affects the polymer structure of the hydroxides in a complex way (Young & Comstock, 1984). Essentially these comments boil down to the fact that the phosphate – Fe(Al) chemistry is still poorly quantified and that attempts to detailed modelling soon approach a point where the results are not really improved by adding more equations or parameters to the model. Functional modelling then means to find a pragmatic description that covers most of the relevant interactions and can still be related to experimental data. This means measured adsorption isotherms rather than using an unknown Gibbs free energy for a continuum of intermediate Fe-phosphate-hydroxides.

Due to correlation of their content in sediments it is difficult to discern between the specific contribution of Al and Fe hydroxides in the phosphate adsorption, but there is experimental evidence that at pH = 7 Al has twice the affinity of Fe (Lijklema, 1980) and in principle it makes sense to correlate P with Fe + 2 Al (Danen et al.; Portielje & Lijklema, 1993). The overriding role of iron in the phosphate chemistry of sediments is evident from many studies, e.g. Boers (1991), Jensen et al. (1992); Forsgren & Jansson; Caraco et al.; Cooke et al.; Cooke et al.; 1993. It is frequently not clear whether the potential contribution of Al has been recognized in these studies.

Under reducing conditions Al maintains its adsorptive capacity but iron reduction often leads to enhanced Fe and phosphate concentrations. Equilibrium calculations are needed to decide whether or not in simulations certain compounds will precipitate, e.g. vivianite. The literature values of solubility products for pure compounds are not always adequate as a yardstick, and attempts to relate dissolved reactive phosphate concentration in natural systems to equilibrium solubility with some solid phase frequently have been unconvincing (Froelich, 1988; see also Fox, 1993). Impurities and irregularities in the lattice or rather the amorphous character of the solids in the sediments may affect the solubility to be much higher than for the corresponding pure mineral. Besides this some insoluble compounds are not formed at all. Kinetic barriers seem to be responsible for the absence of the highly insoluble apatites in most lake sediments. Precipitation of calcium bound phosphate apparently is restricted to co-precipitation with carbonates during episodes of algal bloom. The P:Ca ratio in the precipitate increases with the ambient phosphate concentration and the pH under which the carbonate was formed (Hieltjes, 1980). Thus again empiric relationships must be used if this process is relevant because a detailed description of the surface precipitation is impractical. After precipitation into the sediment it is not clear whether the partial dissolution by the generated carbondioxide is homogeneous or not. Even in hard water lakes like lake Balaton with a high carbonate content in the sediment, the Ca-related mechanism may dominate the cycling of phosphate (Lijklema et al., 1983), but a comparatively small amount of iron (and/or aluminum) tends already to take over the control of the phosphate behaviour.

Biological processes

Dominating is the mineralization of organic material from various sources. A wide range of compounds, with variable phosphate content and biodegradability is broken down or transformed by various micro-organisms, all with their own nutrient requirements and using specific electron acceptors. It is evident that in models only a rough approximation of the system can be attained, but some key features need to be preserved, depending on the modeling objectives.

For long term models the seasonal variations are less important and in its most simple form a sediment accumulation model could neglect the whole microbial sub-model. This means that it is assumed that virtually all the input of organic phosphates ultimately is transformed into inorganic compounds and burial below the active layer is also only in this form. In cases where a substantial, virtually inert organic-P fraction remains after mineralization, this can be modeled as a constant fraction of the input which is converted instantaneously into inert material. Gächter (1993) discusses the microbial production of inert P containing organic material in some detail. No complex kinetic formulations are thus required for long term steady state models.

When the seasonal variation is important much

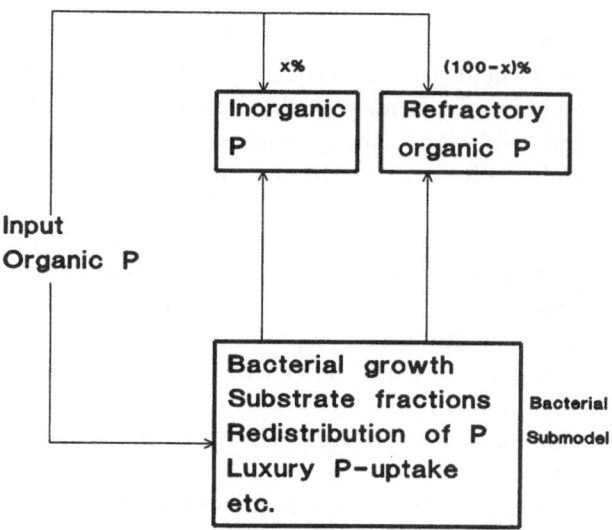

Fig. 2. Alternative model concepts for phosphorus regeneration from organic matter in sediments.

more detail is required because the mineralization drives the redox profiles, the regeneration of inorganic phosphate and the formation of other, labile organic phosphates due to the growth of micro-organisms. Under transient conditions, when the bacterial mass increases due to the influx of fresh organic matter, substantial amounts of phosphate can be stored in the newly formed cells. Because bacteria generally have a much higher phosphate content than algae, dead leaves and other detrital material, the degradation actually may lead to a decrease in the inorganic P-pool and result in desorption from iron etc., even when taking into account that only a part of the carbon and the associated P is available as an energy source for the bacteria. Petterson and Gächter & Meijer discuss several of these aspects (1993).

It is noteworthy that in biological phosphorus removal in activated sludge plants by alternating oxic and anaerobic conditions the growth of Acinetobacter species is favoured. This bacteria stores high quantities of phosphate as polyphosphate during aerobic conditions and uses the energy of the polyphosphates to assimilate organic matter under the anaerobic conditions induced by high organic loading. Essentially the same variable redox conditions exist in a small zone in the sediments due to diurnal variations in the oxygen concentration in the overlying water. Hence the biomass may contain a strongly variable phosphate content under varying redox conditions. However, when the biomass is small compared to the whole matrix and/or when little variation in oxygen profiles occurs the variation in this microbial pool can be neglected. If only the biomass is small, the variable redox conditions still may be an essential feature that must be modeled because the iron-phosphate system is affected.

The enhanced release of phosphate from sediments upon oxygen depletion in the top layer of the sediment thus may partly be due to microbial release and further to the well known reduction of iron (Einsele, 1936).

Another aspect that must be considered in modeling the mineralization is the gradually decreasing rate of decay because the more refractory materials remain and the readily degradable matter is metabolized fast. Berner (1980) has modeled this by starting with a multi-component Michaelis-Menten description of parallel reactions, which then is simplified depending on the conditions. Boudreau et al. (1991) propose a continuum representation. For a description of the regeneration of P and the oxygen consumption as processes forcing the sediment-water fluxes, a simple alternative way to account for the gradually decreasing overall rate is to assume a second order reaction (in organic material) or a two step sequential process according to

$$[\text{Organics I}] - K_1 \rightarrow [\text{Organic II}] - K_2 \rightarrow CO_2 + [\text{Refractory Org.}].$$

In either case calibration is not easy. The selection of rate constants and the fraction that ultimately will be converted into refractory material is arbitrary. In fact the latter will be dependent on the number of turnovers of the organic carbon through the bacterial pool, and this is not modeled. However, the main features can be represented and for a better representation in most cases the data are not available.

Connected to the variable degradability is also a variable stoichiometry. The refractory material usually has a lower P and N content. This stoechiometry is also difficult to infer from field data and mineralization experiments, especially when a variable pool in micro-organisms (e.g. as polyphosphates) and a redistribution over inorganic complexes occur. Again the more or less continuously decreasing P and N content can be represented roughly by attribution of an estimated P:N:C ratio to the arbitrary pools of (Organic I); (Organic II) and (Refractory Organic). In cases where the bacterial pool should be modeled in order to account for the dynamics associated with its variation, the stoechiometry becomes rather complex because the phosphorus content of the bacteria is a function of their physiological condition (e.g. the availability of substrate) and particularly because it is redox sensitive.

Calibration and validation

Two topics will be discussed under this heading: the need for data in relation with the complexity of the model and the adequacy of analytical techniques when variables and parameters are estimated experimentally.

It is obvious that with increasing complexity of models in terms of numbers of variables and processes modeled, more field and laboratory data are required for the input, the initial conditions, the boundary conditions and for the parameters (coefficients in the differential equations). It is a matter of experience to decide what the optimal complexity in a given situation is. By including more processes and interactions in principle a more realistic and complete representation of the sediment-water system can be achieved, but if the data base is incomplete, short, inaccurate or inhomogeneous, parts of the system will be described unsatisfactorily. Although more explanatory mechanisms are introduced, the net effect in terms of the quality of the simulations may be small or even negative. The propagation of errors in models is a difficult subject that requires the help of specialists for an accurate analysis, but awareness of potential problems can prevent already serious deceptions.

A small and/or poor data set can be reproduced in several ways by different combinations of rate constants and other parameters, several of which may be very unrealistic. Anyway the number of field data should be much higher than the number of parameters that must be calibrated and validation on independent data sets measured under different conditions should be standard practice. In the application of formal parameter estimation techniques the confidence contours in the parameter space can be assessed and this can contribute to an understanding of the interdependency of the errors in parameters (Beck & van Straten, 1983). The other way round, such information will help to direct experimental work to those processes from which most of the uncertainty arises.

It is evident that the predictive power of simple models is limited, especially when they are used outside the experimental range for which they were calibrated. By increasing the number of (essential) explanatory processes this predictive power can be enhanced but generally at the cost of an exponentially increasing requirement for data and up to a limit where the identification of the model becomes problematic. Identification means that from the data the underlying individual processes can be retrieved unambiguously.

The adequacy of analytical techniques used to measure processes or variables can be questioned in two ways: are they representative and are they homogeneous. A modeler should continuously be aware of the fact that the actual measured value, e.g. a concentration, is not necessarily equivalent to the variable modeled. A number of examples are:

– the chemical composition of a sediment as obtained by various extraction procedures is biassed; in fact the analytical methodology defines the (mixture of) compounds measured. The term 'Soluble Reactive Phosphate' illustrates this notion; it is not necessarily identical with ortho-phosphate. Numerous discussions and studies have been devoted to the question what the chemical nature of phosphate extracted with certain agents is (e.g. Jensen, 1992), but only approximate answers are possible.

- the pore water concentration as measured with peepers has a poor vertical resolution and under dynamic conditions also integrates over a certain time period.
- DO profiles measured with a micro-electrode include artefacts induced by the channel created by the insertion of the probe. The oxygen gradients are not only in the vertical direction, but also lateral across biofilms covering grains of different size and shape.
- the coring of sediments induces smearing, the separation in layers of a presupposed thickness also is not exact.

Estimates of confidence limits can help to direct sensitivity analysis.

Homogeneity of data means that all measurements are done with the same material, which is in practice impossible. Thus, the measurement of profiles of DO, SRP, pE, pH etc., of fluxes across the interface, of the chemical composition of slices of sediment, of reaction rates for nitrification, oxidation, sorption etc. are frequently done in intact cores and with sliced samples from different locations and collected at different times. Above it has been stated that a coefficient of variation of 20% for the composition of local samples is not unusual; hence a similar variation may be expected in measured rates, equilibria, DO penetration depth etc. In model simulations and calibration these inhomogeneous data are generally used to represent the conditions in one specific sediment or core with which release tests were done. This may be an important source of error. Multiple analyses and averaging may solve the problems partly at the cost of more experimental work. However, inhomogeneity is a fact of sediment-life.

Conclusion

Functional models are models which serve their purpose. For research generally the best approach is to work with self-made and tailor-made models. This is time consuming but the main objective of research: to gain insight, is very much promoted by organizing concepts and supposed underlying mechanisms in the framework of a mathematical model. Frequently only a submodel is needed of that part of the system that is under investigation. The use of an existing model may save time, but unless it is studied and analysed very carefully its contribution in terms of understanding system behaviour may be limited or even lead the researcher astray.

For management purposes and when limited experience and time is available, an existing 'universal' and generally commercial model may be useful. Still a substantial investment in time to understand the principles of the model and its power and its limitation are strongly recommended. A close interaction with the consultant in formulating the objectives, the features considered to be essential, the data requirements etc. will improve the quality of the predictions to be made.

References

Aller, R. C., 1982. The effect of macrobenthos on chemical properties of marine sediment and overlying water. In: P. L. McCall & M. J. S. Tavesz (eds), Animal-sediment relations. Plenum, New York: 53–102.

Atkinson, R. J., A. M. Posner & J. P. Quirk, 1972. J. Inorg. Nucl. Chem. 34: 2201–2211.

Auer, M. T., N. A. Johnson, M. R. Penn & S. W. Effler, 1993. Measurement and verification of rates of sediment phosphorus release for a hypereutrophic urban lake. In P. C. M. Boers, T. E. Cappenberg & W. van Raaphorst (eds), Proceedings of the Third International Workshop on Phosphorus in Sediments. Developments in Hydrobiology 84. Kluwer Academic Publishers, Dordrecht: 301–309. Reprinted from Hydrobiologia 253.

Baccini, P., 1985. Phosphate interactions at the sediment-water interface. In: W. Stumm (ed.), Chemical Processes in Lakes. Wiley, New York: 189–205.

Beck, M. B. & G. van Straten, 1983. Uncertainty and forecasting of water quality. Springer Verlag, Berlin.

Berner, R. A., 1980. Early diagenesis, a theoretical approach. Princeton Univ. Press, Princeton, USA.

Boers, P. C. M., 1991. The influence of pH on phosphate release from sediments. Wat. Res. 25: 309–311.

Bolan, N. S., N. J. Barrow, A. M. Posner, 1985. Describing the effect of time on sorption of phosphate by iron and aluminum hydroxides. J. Soil Sci. 36: 187–197.

Booy, K., 1989. Exchange of solutes between sediment and water. Thesis. Univ. of Groningen, The Netherlands.

Boudreau, P. & B. R. Ruddick, 1991. On a reactive continuum representation of organic matter diagenesis. Am. J. Sci. 191: 507–538.

Breeuwsma, A. & J. Lijklema, 1973. J. Colloid Interface Sci. 43: 437–438.

Brinkman, A. G. & W. van Raaphorst, 1986. De fosfaathuishouding van het Veluwemeer. PhD thesis. Twente University, Enschede, The Netherlands.

Caraco, N. F., J. J. Cole & G. E. Likens, 1993. Sulphate control of phosphorus availability in lakes: a test and re-evaluation of Hasler and Einsele's model. In: P. C. M. Boers, T. E. Cappenberg & W. van Raaphorst (eds), Proceedings of the Third International Workshop on Phosphorus in Sediments. Developments in Hydrobiology 84. Kluwer Academic Publishers, Dordrecht: 275–280. Reprinted from Hydrobiologia 253.

Cooke, G. D., E. B. Welch, A. B. Martin, D. G. Fulmer, J. B. Hyde & G. D. Schrieve, 1993. Effectiveness of Al, Ca and Fe salts for control of internal phosphorus loading in shallow and deep lakes. In: P. C. M. Boers, T. E. Cappenberg & W. van Raaphorst (eds), Proceedings of the Third International Workshop on Phosphorus in Sediments. Developments in Hydrobiology 84. Kluwer Academic Publishers, Dordrecht: 323–335. Reprinted from Hydrobiologia 253.

Danen-Louwerse, H., L. Lijklema & M. Coenraats, 1993. Iron content of sediment and phosphate adsorption properties. In: P. C. M. Boers, T. E. Cappenberg & W. van Raaphorst (eds), Proceedings of the Third International Workshop on Phosphorus in Sediments. Developments in Hydrobiology 84. Kluwer Academic Publishers, Dordrecht: 311–317. Reprinted from Hydrobiologia 253.

Davis, R. B., 1974a. Tubificids alter profiles of redox potential and pH in profundal lake sediments. Limnol. Oceanogr. 19: 342–346.

Davis, R. B., 1974b. Stratigraphic effects of tubificids in profundal lake sediments. Limnol. Oceanogr. 19: 466–488.

DiToro, D. M., P. R. Paquin, K. Subburamuu & D. A. Gruber, 1990. Sediment oxygen demand model: methane and ammonia oxidation. J. Env. Eng. Div. ASCE 116: 945–986.

Duin, E. H. S. van, G. Blom, L. Lijklema & M. J. M. Scholten, 1992. Aspects of modeling sediment transport and light conditions in lake Marken. In B. T. Hart & P. G. Sly (eds), Sediment/Water Interactions. Developments in Hydrobiology 75. Kluwer Academic Publishers, Dordrecht: 167–176. Reprinted from Hydrobiologia 235/236.

Eck, G. Th. M. van & J. G. C. Smits, 1986. Calculation of nutrient fluxes across the sediment water interface in shallow lakes. In: P. G. Sly (ed.), Sediments and Water Interactions, Springer Verlag, New York: 289–301.

Einsele, W., 1936. Ueber die Beziehungen des Eisenkreislaufes zum Phosphatkreislauf im Eutrophen See. Arch. Hydrobiol. 29: 664–686.

Forsgren, G. & M. Jansson, 1993. Sedimentation of phosphorus in limnetic and estuarine environments in the river Öre system northern Sweden. In: P. C. M. Boers, T. E. Cappenberg & W. van Raaphorst (eds), Proceedings of the Third International Workshop on Phosphorus in Sediments. Developments in Hydrobiology 84. Kluwer Academic Publishers, Dordrecht: 233–248. Reprinted from Hydrobiologia 253.

Fox, L. E., 1993. The chemistry of aquatic phophate: inorganic processes in rivers. In: P. C. M. Boers, T. E. Cappenberg, & W. van Raaphorst (eds), Proceedings of the Third International Workshop on Phosphorus in Sediments. Developments in Hydrobiology 84. Kluwer Academic Publishers, Dordrecht: 1–16. Reprinted from Hydrobiologia 253.

Froelich, P. N., 1988. Kinetic control of dissolved phosphate in natural rivers and estuaries: a primer on the phosphate buffer mechanism. Limnol. Oceanogr. 33: 649–668.

Gächter, R., J. S. Meyer & A. Mares, 1988. Contribution of bacteria to release and fixation of phosphorus in lake sediments. Limnol. Oceanogr. 33: 1542–1558.

Gächter, R. & J. S. Meyer, 1993. The role of microorganisms in mobilization and fixation of phosphorus in sediments. In: P. C. M. Boers, T. E. Cappenberg & W. van Raaphorst (eds), Proceedings of the Third International Workshop on Phosphorus in Sediments. Developments in Hydrobiology 84. Kluwer Academic Publishers, Dordrecht: 103–121. Reprinted from Hydrobiologia 253.

Hakanson, L. & M. Jansson, 1983. Principles of lake sedimentology. Springer Verlag. Berlin.

Hieltjes, A. H. M., 1980. Eigenschappen en gedrag van fosfaat in sedimenten. PhD thesis. Twente University, Enschede, The Netherlands.

Hingston, F. J., A. M. Posner & J. P. Quirk, 1971. Discuss. Faraday Soc. 52: 334–340.

Jensen, H. S., P. Kristensen, E. Jeppesen & A. Skytthe: Iron-phosphorus ratio in surface sediments as an indicator of phosphate release from aerobic sediments in shallow lakes. Hydrobiologia (1992). In press.

Kamp Nielsen, L., 1975. A kinetic approach to the aerobic sediment-water exchange of phosphorus in Lake Esrom. Ecol. Modeling 1: 153–160.

Klapwijk, A. & W. J. Snodgrass, 1986. Concepts for calculating the stabilization rate of sediments. In: K. J. Hatcher (ed.), Sediment Oxygen Demand; processes, modeling and measurement. Inst. of Natural Resources; Univ. of Georgia, Athens: 75–98.

Lijklema, L., 1977. The role of iron in the exchange of phosphate between water and sediments. In: H. L. Golterman (ed.), Interactions between sediments and freshwater. Dr W. Junk Publishers, The Hague: 313–317.

Lijklema, L., 1980. Interactions of ortho-phosphate with iron (III) and aluminum hydroxides; Envir. Sci. Technol. 14: 537–542.

Lijklema, L., P. Gelencser & F. Szilagyi, 1983. Sediments and sediment-water interaction. In: L. Somlyody (ed.), Eutrophication of shallow lakes: modeling and management. The lake Balaton case study. IIASA Collab. Proceed. Series CP-83-S3, Laxenburg, Austria: 81–100.

Lijklema, L., 1983. Internal loading. Wat. Supply 1: 35–42.

Lijklema, L., 1986. Phosphorus accumulation in sediments and internal loading. Hydrobiol. Bull. 20: 213–224.

Montigny, Ch. de & Y. T. Prairie, 1993. The relative importance of biological and chemical processes in the release of phosphorus from a highly organic sediment. In: P. C. M. Boers, T. E. Cappenberg & W. van Raaphorst (eds), Proceedings of the Third International Workshop on Phosphorus in Sediments. Developments in Hydrobiology 84. Kluwer Academic Publishers, Dordrecht: 141–150. Reprinted from Hydrobiologia 253.

Pettersson, K., E. Herlitz & V. Istvánovics, 1993. The role of Gloeotrichia echinulata in the transfer of phosphorus from sediments to water in Lake Erken. In: P. C. M. Boers, T. E. Cappenberg & W. van Raaphorst (eds), Proceedings of the Third International Workshop on Phosphorus in Sediments. Developments in Hydrobiology 84. Kluwer Academic Publishers, Dordrecht: 123–129. Reprinted from Hydrobiologia 253.

Portielje, R. & Lijklema, L., 1993. Sorption of phosphate by sediments as a result of enhanced external loading. In: P. C. M. Boers, T. E. Cappenberg & W. van Raaphorst (eds), Proceedings of the Third International Workshop on Phosphorus in Sediments. Developments in Hydrobiology 84. Kluwer Academic Publishers, Dordrecht: 249–261. Reprinted from Hydrobiologia 253.

Raaphorst, W. van & A. G. Brinkman, 1985. The calculation of transport coefficients of phosphate and calcium fluxes across the sediment water interface from experiments with undisturbed sediment cores. Wat. Sci. Techn. 17: 941–951.

Raaphorst, W. van, H. T. Kloosterhuis, A. Cramer & K. J. M. Bakker, 1990. Nutrient early diagenesis in the sandy sediments of the Doggerbank area, North Sea: Pore water results. Neth. J. Sea Res. 26: 25–52.

Reckhow, K. H., 1979. Empirical lake models for Phospho-rus: Development, Applications, Limitations and Uncertainty. In: D. Scavia & A. Robertson (eds), Perspectives on Lake Ecosystem Modeling. Science Publ. Inc., Ann Arbor: 193–221.

Rooy, N. M. de, 1991. Mathematical simulation of biochemical processes in natural waters by the model CHARON. Documentation Report, R1310-10, Delft Hydraulics, The Netherlands.

Rutgers van der Loeff, M. M., 1981. Wave effects on sediment water exchanges in a submerged sandbed. Neth. J. Sea Res. 15: 100–112.

Ryding, S. O., 1985. Chemical and microbiological processes as regulators of the exchange of substances between sediments and water in shallow eutrophic lakes. Int. Revue Ges. Hydrobiol. 70: 657–702.

Sigg, L., 1979. Die Wechselwirkung von Anionen und schwachen Säuren mit α-FeOOH (Goethit) in wässeriger Lösung. Dissertation, ETH, Zurich.

Smits, J. G. C., D. T. van der Molen, 1993. Application of SWITCH, a model for sediment–water exchange of nutrients, to Lake Veluwe in The Netherlands. In: P. C. M. Boers, T. E. Cappenberg & W. van Raaphorst (eds), Proceedings of the Third International Workshop on Phosphorus in Sediments. Developments in Hydrobiology 84. Kluwer Academic Publishers, Dordrecht: 281–300. Reprinted from Hydrobiologia 253.

Snodgrass, W. J., 1986. Comparison of kinetic formulations of three sediment oxygen demand models. In: K. J. Hatcher (ed.), Sediment Oxygen Demand, processes, modeling and measurement. Inst. of Natural Resources; Univ. of Georgia, Athens, USA: 139–170.

Young, T. C. & W. G. Comstock, 1986. Direct effects and interactions involving iron and humic acid during formation of colloidal phosphorus. In: P. G. Sly (ed.), Sediments and Water Interactions; Proc. of the third Intern. Symposium on Interactions between Sediments and Water, 1984, Geneva: 461–470.

Hydrobiologia **253**: 233–248, 1993.
P.C.M. Boers, T.E. Cappenberg & W. van Raaphorst (eds),
Proceedings of the Third International Workshop on Phosphorus in Sediments.
© 1993 *Kluwer Academic Publishers.*

Sedimentation of phosphorus in limnetic and estuarine environments in the River Öre system, northern Sweden

Gunilla Forsgren & Mats Jansson
Department of Physical geography, University of Umeå, S-90187 Umeå, Sweden

Key words: phosphorus, iron, organic carbon, sedimentation, estuary, lake

Abstract

Sedimentation of phosphorus was studied during the spring flood in April and May 1991 in Lake Örträsket and the Öre Estuary, northern Sweden. Lake Örträsket has an area of 7.3 km^2 and a mean depth of 22 m and is located 100 km from the coast halfways along the course of the River Öre. The river ends in a semi-closed low salinity estuary with an area of *ca.* 50 km^2 and a mean depth of 16 m. Sedimentation of phosphorus, iron and organic carbon were measured with sediment traps in Lake Örträsket and in the Öre Estuary. Characterization of particulate phosphorus in river water, sediment trap material and sediments were performed by the sequential extraction procedure proposed by Hieltjes & Lijklema (1980). Apart from being an efficient trap for suspended particles including particulate phosphorus, Lake Örträsket was shown to serve as a source for particulate material during spring 1991. The Öre Estuary, on the other hand, constitutes an efficient trap for the total supply of river-borne phosphorus during the spring flood period. Phosphorus was shown to be closely related to iron in particulate material in both the lake and the estuary. Adsorption of phosphorus on settling inorganic particles seems to be an important process, which is particularly pronounced in the estuary.

Introduction

Lakes and estuaries often constitute efficient traps for phosphorus introduced by riverine input (Larsen & Mercier, 1976; Ahl, 1988; Froehlich, 1988). The sedimentation of phosphorus depends on a combined effect of abiotic factors such as the relationship between lake morphometry and hydraulic loading (Vollenweider, 1976), chemical differences between river water and lake water (Stabel & Geiger, 1985; Stafford-Glase & Barlow, 1986), the mixing in saline gradients and related physicochemical reactions in estuaries (Bale & Morris, 1981; Froelich, 1989) and on the biological incorporation of phosphorus in organic particles (Persson & Broberg, 1988).

Resuspension of sediments can cause considerable redistribution of sedimented material in lakes (Håkansson & Jansson, 1983) and estuaries (Brydsten, 1992; Malmgren & Brydsten, 1992). The microbial activity in organogenic sediments may have a great impact on the phosphorus retention capacity of sediments and the microbially mediated anaerobic phosphate release from lake sediments (Mortimer, 1941; Mortimer, 1942) can in hypertrophics lakes balance or exceed the sedimentation (Boström *et al.*, 1982).

Most studies on phosphate sedimentation and phosphate sediment dynamics have been carried out in eutrophic lakes (Håkansson & Jansson, 1983). Phosphorus sedimentation in estuaries is less studied than in lakes and most investigations have been made in high salinity estuaries with considerable tidal influence (Bale & Morris, 1981; van Eck, 1982; Froehlich, 1988). Few, if any, studies are reported where the sedimentation characteristics of phosphorus are compared in lacustrine and estuarine environments of the same river system.

Since 1986 an extensive sediment dynamics research program has been carried out in the Öre Estuary in northern Sweden, which is a low salinity estuary with no tides receiving water from the soft water, humic rich River Öre. It has been demonstrated that the estuary is an efficient trap for river borne particles and that resuspension may cause dramatic redistribution of sediment particles (Brydsten & Jansson, 1989; Malmgren & Brydsten, 1992). In a previous study on the turnover of iron, phosphorus and organic carbon in the low salinity Öre Estuary (Forsgren & Jansson, 1992) it was shown that phosphorus and iron supplied by the River Öre during the spring flood was completely removed from the water column by sedimentation. This sedimentation was rapid – the average retention time for suspended riverine particles was only about one day – and comprised not only river-borne particulate species. A considerable amount of dissolved phosphorus was converted to particulate form during the mixing of fresh and saline water in the estuary. Adsorption of phosphorus to settling inorganic particles and aggregation and sedimentation of colloidal iron-phosphate complexes due to increasing salinity and pH were suggested to be the primary causes of this estuarine phosphorus sedimentation.

In order to further elucidate the phosphorus dynamics of the Öre Estuary and the River Öre system a study was carried out during the spring 1991. The purpose was to compare the sedimentation of phosphorus and the fractional composition of sedimenting phosphorus in the Öre Estuary and in Lake Örträsket. Both the lake and the estuary receive water from the River Öre, which offers an opportunity to evaluate the sedimentation processes in saline water in comparison with those in fresh water. The River Öre is a non-polluted soft water river with low concentra-

tions of nutrients and high concentrations of iron and DOC. Large transport of inorganic particles of the silt and clay fraction takes place during high flow episodes. The River Öre system is therefore particularly suited for studies of physical and chemical phosphorus sedimentation processes.

The spring flood period was chosen for this study for several reasons. Approximately 70–80% of the annual river transport of suspended solids take place during the snow melt period in April and May (Brydsten & Jansson, 1989). This period is therefore of utmost importance for the annual sediment balance and it supplies an extremely suitable situation for sedimentation studies. Non-biological processes should be dominant during this period particularly in Lake Örträsket, which normally is ice covered until the end of the spring flood.

Study area

The River Öre flows through the county of Västerbotten, northern Sweden (Fig. 1). The river is non-regulated and has a boreal regime. The drainage area is 2940 km^2. The length of the river is 225 km. Some chemical characteristics of the River Öre system are shown in Table 1.

Lake Örträsket, the largest lake in the River Öre, is situated in the main furrow halfways along the river course (Fig. 2a). The lake is elongated in a north-south direction and has an area of 7.3 km^2, a water volume of ca. $1.6 \cdot 10^8$ m^3 and a maximum and a mean depth of 60 m and 22 m, respectively. The lake has two major tributaries, River Örån and River Vargån. The lake is covered with ice from the beginning of December to the middle of May. Treated sewage water from ca. 300 persons is drained to the lake. The sewage input of phosphorus is negligable compared to the natural sources.

The river mouth, the Öre Estuary, is situated in the northern part of the Bothnian Sea (Fig. 2b). The Öre Estuary is a wind-dominated, semi-closed water-body with an area of approximately 50 km^2, a water volume of ca. $1.0 \cdot 10^9$ m^3 and a mean depth of 16 m. The estuary is well mixed

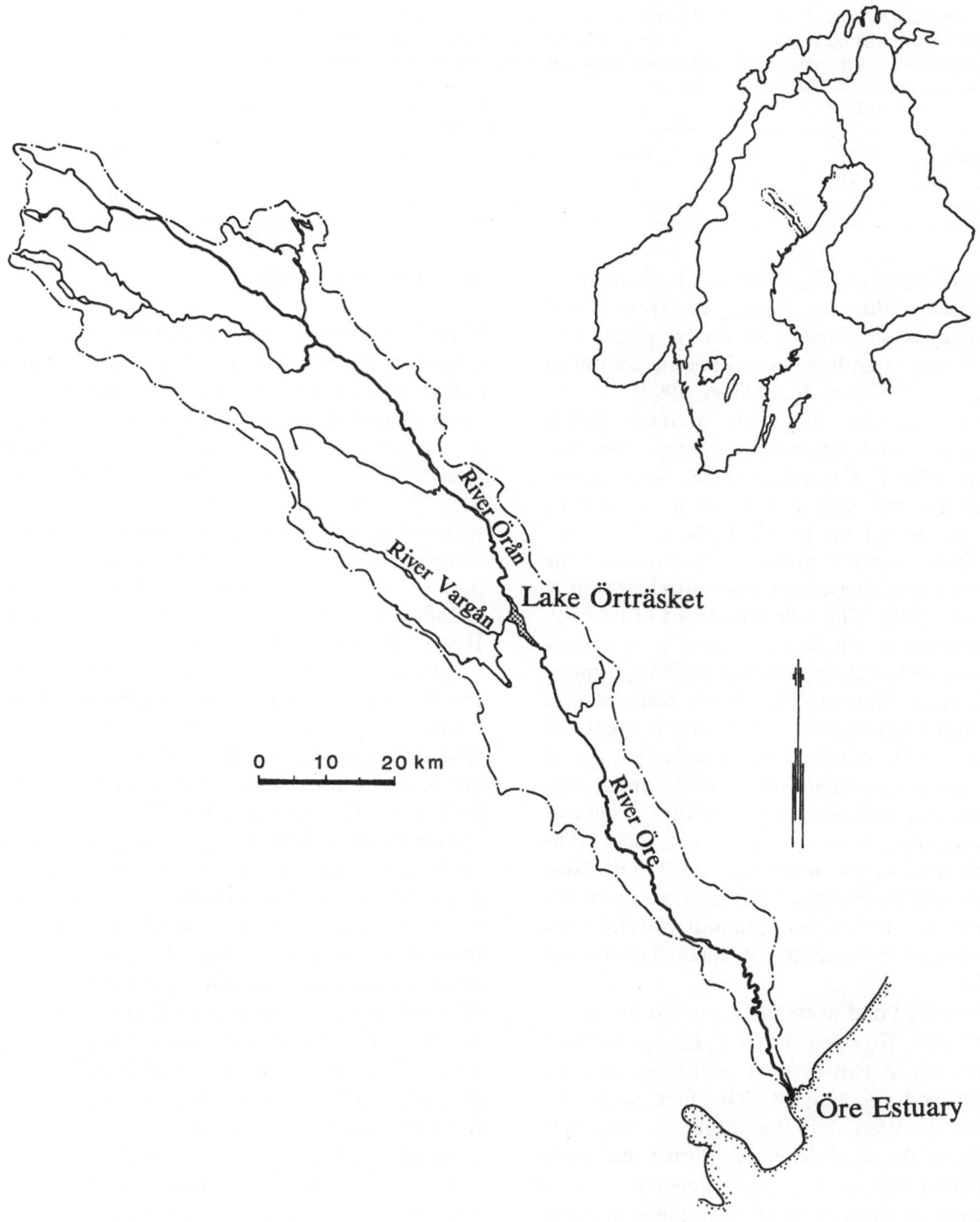

Fig. 1. Location of the River Öre catchment.

Table 1. Chemical characteristics of the River Öre system. The conductivity is given in $\mu S \cdot m^{-1}$ and the concentrations of the elements are given in $mg \cdot l^{-1}$. The chemistry from L. Örträsket (surface waters) and the R. Öre (Håknäs, 5 km upstreams the river mouth) represents mean values during the period March–November 1991 (8 samples) and the chemistry from the Öre Estuary (samples from 10 m water depth) represents mean values during the period April–May 1990.

	pH	Cond.	Tot-P	Tot-N	Tot-Fe	Tot-Al	Tot-Mn	TOC
L. Örträsket	6.55	0.24	0.010	0.32	0.83	0.12	0.04	10.1
R. Öre	6.71	0.27	0.021	0.34	0.77	0.11	0.03	9.6
Öre Estuary	7.7	50	0.008	0.24	0.11	0.03	<0.005	3.7

with a salinity of *ca.* 5‰ and there is virtually no tidal action in the area. During the spring flood the estuary is stratified, with a river plume (*ca.* 3–4 m deep) extending several kilometers out in the estuary (Forsgren & Jansson, 1992).

The highest shore line in the drainage area is 220 m above sea level and located approximately 5 km upstream L. Örträsket. The R. Öre is deeply eroded into the silty marine delta sediments, which are found up to the highest shore-line. These easily eroded marine sediments are the main source of suspended particles (Ivarsson & Forsgren, 1988). The following types of bedrock are dominant in the drainage area (Ivarsson & Karlsson, 1992); Revsunds granite (58%), gneiss (23%), metasediment (9%), basic metavulcanic (2%) and granodiorite (8%). The contents of phosphate-rich minerals are relatively high in basic- and intermediate rocks, basic metavulcanics and granodiorites, while apatite occurs as accessory minerals (<1%) in acid rocks. The bedrock around and upstreams the L. Örträsket is dominated by granites while gneisses dominate downstreams the lake. The granodiorite and basic metavulcanic rocks are mainly located upstreams the lake.

The spring flood normally occurs in the beginning of May. However, in 1991, the spring flood was divided in two distinct peaks occuring on April 16 and on May 18. The first peak was of short duration and the discharge was only *ca.* 20% of the total discharge during the whole spring flood period. Since the sediment traps were placed out on April 24 and thus not influenced by the first short peak, we focus this study on the fate of river borne material during the second peak of the spring flood.

Material and methods

Water samples were taken with a Ruttner sampler in the two inlets (R. Örån & R. Vargån) and in the outlet of L. Örträsket on six dates between April 24 and May 23 (Fig. 2a). Water samples were also taken in the R. Öre 5 km upstreams the river mouth, on five dates between April 24 and May 23. Water was collected at defined sections with turbulent water in order to get representative samples. Acid-washed PE-bottles were used and part of the water sample (metal analysis) was acidified down to pH < 1 with HNO_3 (20 μl conc. HNO_3 ml sample^{-1}) directly after sampling.

Sediment traps were placed at 4–8 m depth, 15–20 m depth and 1 m above the bottom at four stations (stations LA-LD) in L. Örträsket (Fig. 2a) and at 4 m depth, 8–10 m depth and 1 m above the bottom at three stations (stations EA-EC) in the Öre Estuary (Fig. 2b). The mooring system of the sediment traps were based on two anchors (connected to each other with a rope at a distance of 10–20 m apart) with the free-floating vessels fixed at the desired depth at a rope from the first anchor and a surface buoy connected to the second anchor. At each depth parallel vessels of cylindrical type with a inner diameter of 5.4 cm and a hight of 30 cm were used. The traps were set out from the lake ice on April 24 in L. Örträsket and emptied when the trap stations were free from ice. The traps at stations LA and LB were emptied on May 14 and on 23 and those at stations LC and LD were emptied on May 23. The traps in the Öre Estuary were set out on April 18 and emptied on May 6 and on May 22.

Intact cores were taken with a gravity corer from the surface sediments (0–10 cm) at 11 sta-

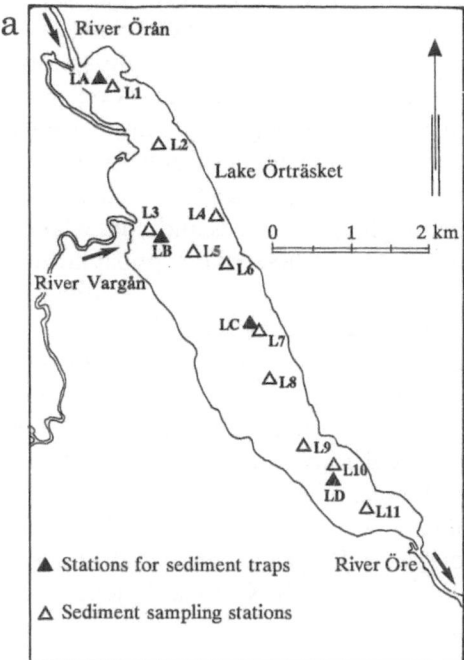

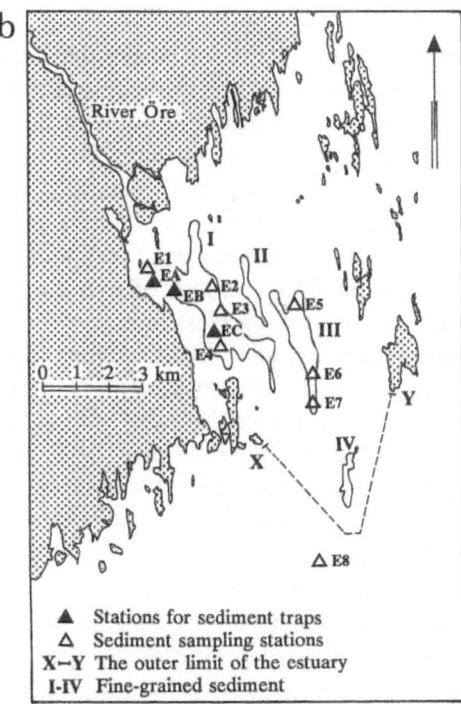

Fig. 2. Locations of stations for sediment traps and sediment sampling in a) the Lake Örträsket b) the Öre Estuary.

tions in L. Örträsket between April 10–12 (Fig. 2a). Six sediment cores from L. Örträsket (stations L1, L2, L3, L7, L8 and L11) were sliced in one cm thick layers and eight cores (stations L1, L3, L4, L5, L6, L9, L10 and L11) were sliced in five cm thick layers.

Composite samples (0–5 cm) (based on sediment cores from 3 stations) were prepared from each of two well defined fine-grained sediment bottoms in the Öre Estuary (bottoms I and III in Fig. 2b). Two of these cores (station E4 and E6) were sliced in one cm thick layers (0–1 cm and 4–5 cm). In addition, sediment samples (0–5 cm) were collected from bottoms just outside the river mouth (station E1) and just outside the estuary (station E8, Fig. 2b). All cores from the estuary were collected on May 6.

Data from continuous discharge registration at Torrböle (25 km upstreams the river mouth) during April and May 1991 were obtained from the Swedish Meteorological and Hydrological Institute. The discharge at the outlet and at the two inlets of L. Örträsket were calculated as proportional to the areal distributions of the drainage areas. Discrete determinations of discharge with a current meter in the inlet and outlet of L. Örträsket made at different discharge during 1990 and 1991 verified the proportionality between discharge and catchment area. The discharge in the R. Örån, the R. Vargån, diffuse inlets and the outlet of the L. Örträsket were estimated as 59, 12, 3 and 74%, respectively, of the total water discharge at Torrböle. The discharge at the river mouth was estimated to 102% of the discharge at Torrböle.

Chemical analysis

Temperature, pH and conductivity of water samples were measured in the field immediately after sampling. Samples stored in a refrigerator were analyzed for particulate and dissolved phosphorus, iron, manganese, aluminium and organic carbon. Particulate and dissolved material were separated by filtration of 50 ml samples through Sartorius membrane filters (0.45 μm cellulose acetate). Samples for organic carbon analysis were filtrated (50–100 ml) through ignited (400 °C during 5 hours) glassfiber filters (Whatman GF/F).

The concentration of suspended solids was measured by weighing of material collected on filters (Whatman GF/F) after drying at room temperature in a desiccator during 3 days. The suspended material from the main inlet to L. Örträsket on May 2 and at the river mouth on April 28 and May 5 were concentrated by centrifugation (for 20 minutes at 4500 rotations min^{-1}) of approximately 40 l of water and analyzed for the fractional composition of phosphorus.

The material from sediment traps and sediments were frozen and freeze-dried before homogenization and analysis for phosphorus, iron, manganese, aluminium and organic carbon. Six samples from the sediment traps in L. Örträsket, three samples from the sediment traps in the Öre Estuary and eight samples of sediments from L. Örträsket (0–5 cm from stations L1, L3, L9 and L11 and 0–1 cm and 4–5 cm from stations L2 and L8) and the Öre Estuary (0–5 cm composite samples from stations E2, E3, E4 and E5, E6, E7, 0–5 cm from stations E1 and E8 and 0–1 cm and 45 cm from stations E4 and E6), respectively, were analyzed for the fractional composition of phosphorus. These samples were stored wet in a refrigerator for *ca.* one month before analyses.

Phosphorus, iron, manganese and aluminium in sediment trap material and sediment was analyzed with an ICP-AES (inductively coupled plasma-atomic emission spectrometry) instrument (ARLA-ICP, model 3410, preparation: fusion with metaborate and subsequent dissolution in 10% nitric acid and the sample to flux ratio was 1:3 and the final volume 25 ml). The method has been described and discussed in detail by Burman (1987). Total phosphorus in water samples were analyzed with Autoanalyzer, Technicon Traacs 800 (the samples were prepared with potassium-peroxo-disulfate in acid solution). Iron, manganese and aluminium in water samples were analyzed directly with the ICP-AES instrument. Particulate organic carbon was analyzed on filter with an elemental analyzer (Carlo Erba,

model 1106) and dissolved organic carbon was analyzed with a Shimadzu TOC 500.

The fractional composition of phosphorus was analyzed by the sequential extraction procedure of Hieltjes & Lijklema (1980). This procedure has been demonstrated to be useful for the characterization of a large variety of lake sediments including highly calcareous and humic lakes (Pettersson *et al.*, 1988). The phosphorus fractionation was performed on wet samples. Triplicates were made on each sample. The standard deviation for each fraction was ± 10% (Pettersson, pers. comm.).

All analyses were made at the Water Chemistry Analytical Laboratory at the Norrbyn Laboratory, Umeå, except for the fractionation of phosphorus, which was made by the Institute of Limnology in Uppsala. Financial support was given by the Swedish National Science Research Council.

Calculations

The transported amounts of suspended particles and phosphorus, iron, manganese and organic carbon in total and particulate form in the tributaries to the lake, the lake outlet and at the inlet to the estuary were determined by multiplying the discharge by particle and element concentrations. The calculation was made for every day of the sampling period. Daily values of particle and element concentration were estimated by linear interpolation between each of the values obtained from analyses during the investigation period.

The sedimentation in the lake was estimated from the trap yield. Since the trap yield with one exception was approximately the same in surface and bottom traps, mean values of surface and bottom traps at each station and time period were used. The lake was divided into four subareas, where each area was represented by one set of traps. The subareas represented by the stations LA and LB were further divided into four zones. This was done because of the sharp gradient in the amounts of settling material between the traps located directly outside the inlets and the traps located further out in the lake. The amount and composition of the material settled in each of the additional areas were determined by linear interpolation between the traps located outside the inlets (LA and LB) and the traps located further out in the lake (LC and LD).

No calculations on the total transport and sedimentation of suspended particles and particulate phosphorus in the estuary were made in 1991 since sampling to obtain data for such calculations is extremely time consuming in the estuary and could not be performed simultaneously with the study in L. Örträsket. Comparisons of the total sedimentation in L. Örträsket and the Öre Estuary are therefore made with data from the extensive study made in the estuary in the spring of 1989 (Forsgren & Jansson, 1992). Here the transport and sedimentation of particles was calculated in the same way as in the lake. The areas that were represented by each of the sediment traps was determined by constantly measuring the extension of the fresh water plume in the estuary. Detailed information on the calculations in the Öre Estuary are given in Forsgren & Jansson (1992).

Data were statistically analysed using simple and multiple regression routines (on 'SYSTAT'). The data were close to normal distribution and regressions with a probability level of < 0.05 are excepted as significant.

Results

Water and sediment chemistry

The discharge maximum during the observation period occurred on May 18. The estimated values in R. Örån, R. Vargån and in the outlet of L. Örträsket were 100 $m^3 \cdot s^{-1}$, 20 $m^3 \cdot s^{-1}$ and 120 $m^3 \cdot s^{-1}$, respectively. The whole lake was free of ice around May 20 but small areas outside the inlets of the lake and near the outlet were free of ice already in the beginning of May.

The weighted (with consideration taken to the water discharge) mean concentration of total phosphorus during the period April 24 to May 23 was 8.9 $\mu g \cdot l^{-1}$ in the R. Örån, 11.4 $\mu g \cdot l^{-1}$ in the R. Vargån and 15.1 $\mu g \cdot l^{-1}$ in the outlet of L. Örträsket. The mean share of particulate phosphorus in R. Örån, R. Vargån and in the outlet of the lake were 41, 42 and 36%, respectively. The weighted mean concentration of total phosphorus in the R. Öre during the same period was 23.8 $\mu g \cdot l^{-1}$ of which 57% was in particulate form.

The amount of material collected in the sediment traps (expressed as the mean amount from the surface to the bottom traps) decreased with the distance from the inlets of L. Örträsket from 250 $\pm$ 19 $g \cdot m^{-2}$ at location LA and 440 $\pm$ 15 $g \cdot m^{-2}$ at location LB to 30 $\pm$ 10 $g \cdot m^{-2}$ at location LC and 20 $\pm$ 3 $g \cdot m^{-2}$ at location LD. There was a clear increase in the content of phosphorus in sedimenting material with the distance from the two inlets of the lake. The mean content (from surface to bottom traps) increased 1.5–1.7 times from location LA and LB (1.28 $\pm$ 0.02 $mg \cdot g^{-1}$ dry weight and 1.10 $\pm$ 0.04 $mg \cdot g^{-1}$ dry weight respectively) to location LC and LD (1.90 $\pm$ 0.04 $mg \cdot g^{-1}$ dry weight and 1.80 $\pm$ 0.09 $mg \cdot g^{-1}$ dry weight respectively). A similar decrease of the amount of sedimenting material was observed in the Öre Estuary, going seawards from the river mouth. The proportions between the amounts of material in traps from stations EA, EB and EC were approximately 7:2:1. The content of phosphorus in sedimenting material was lower in the estuary than in the lake but, similar to the behaviour of phosphorus in the lake, the mean content (from surface to bottom traps) increased 1.2–1.3 times from station EA and EB (1.02 $\pm$ 0.02 $mg \cdot g^{-1}$ dry weight and 1.06 $\pm$ 0.04 $mg \cdot g^{-1}$ dry weight, respectively) to station EC (1.29 $\pm$ 0.06 $mg \cdot g^{-1}$ dry weight). The contents of phosphorus in sediments in L. Örträsket and in the Öre Estuary are shown in Table 3. The sediments near the inlets (L1 and L3) have phosphorus contents less than 1 $mg \cdot g^{-1}$ dry weight, while the sediments in the deeper parts of the lake have phosphorus contents between 1–1.6 $mg \cdot g^{-1}$ dry weight. The situation is similar in the estuary with phosphorus contents about 1 $mg \cdot g^{-1}$ dry weight near the river mouth (E1)

Table 2. The mass balance of suspended particles (Susp) and total phosphorus (T-P) and particulate phosphorus (P-P) in Lake Örträsket during the period April 24 to May 23 1991 and in the Öre Estuary during the period May 3 to May 18 1989. *) 'surplus' is defined as (output + sedimentation)-input. All figures in tonnes.

| | L. Örträsket 1991 | | | | | | | | | Öre Estuary 1989 | |
| | Input | | | | Output | Sedimentation | Retention | 'Surplus' | | Input | Sedimentation |
	River Örån	River Vargån	Diffuse inlets	Total	River Öre	(traps)	(input-ouput)	*)		River Öre	(traps)
Susp	770	160	40	980	720	830	260	560		14 900	12 600
T-P	1.30	0.40	0.09	1.80	2.70		− 0.90			6	
P-P	0.54	0.15	0.03	0.70	1.00	0.97	− 0.30	1.27		3.3	16

and with seawardly increasing phosphorus contents up to 1.8 mg·g^{-1} dry weight at location E8.

Mass balance calculations

Table 2 shows the mass balance for suspended particles and phosphorus in total and particulate form transported in Lake Örträsket for the period April 24 – May 23. No net retention of phosphorus was found in the lake. However, considerable amounts of phosphorus were found in the sediment traps. The calculated phosphorus sedimentation was *ca.* 1.4 times the amount of particulate phosphorus supplied from tributaries. The sedimentation of particulate material on the other hand was approximately equal to the input of particles in the lake. No mass balance was calculated for the Öre Estuary in 1991. However, the investigation made in 1989 (Forsgren & Jansson, 1992) demonstrated that the sedimentation of phosphorus in the estuary during two weeks in the spring was *ca.* 4–5 times the input of particulate phosphorus while sedimentation of total suspended material approximately equaled the river input in the area affected by the river plume. It can therefore be concluded that the settling material in both the lake and the estuary contained more phosphorus than the incoming particles. The mean content of phosphorus in the 'surplus' in L. Örträsket calculated from data in Table 2 was 2.3 mg·g^{-1} which is considerably higher than the phosphorus content in trap material and sediment (Table 3).

Table 3. The content of total phosphorus (T-P) in sediments (0-5 cm) at different water depth in Lake Örträsket and in the Öre Estuary during the spring period 1991.

	Station	T-P (mg·g^{-1}·dw^{-1})	Depth (m)
L. Öträsket	L1	0.89	10
	L3	0.90	22
	L4	1.31	5
	L5	1.21	38
	L6	1.31	42
	L9	1.55	42
	L10	1.65	42
	L1	1.47	35
Öre Estuary	E1	0.94	7
	E2, E3, E4	1.23	20-25
	E5, E6, E7	1.75	30-35
	E8	1.83	45

Phosphorus fractionation

Figure 3 illustrates the total content of phosphorus and the relative contribution of the three dominating phosphorus fractions in suspended particles in the river, in sedimenting material and sediments. The NH_4Cl-P was always *ca.* 1% and is not included in the figure. The largest part of the input of phosphorus to L. Örträsket consists of Res.-P (44%). The material collected in sediment traps in the lake was richer in total phosphorus than the suspended material transported in the inlets. The mean relative content (weighted with consideration taken to the total amount of mate-

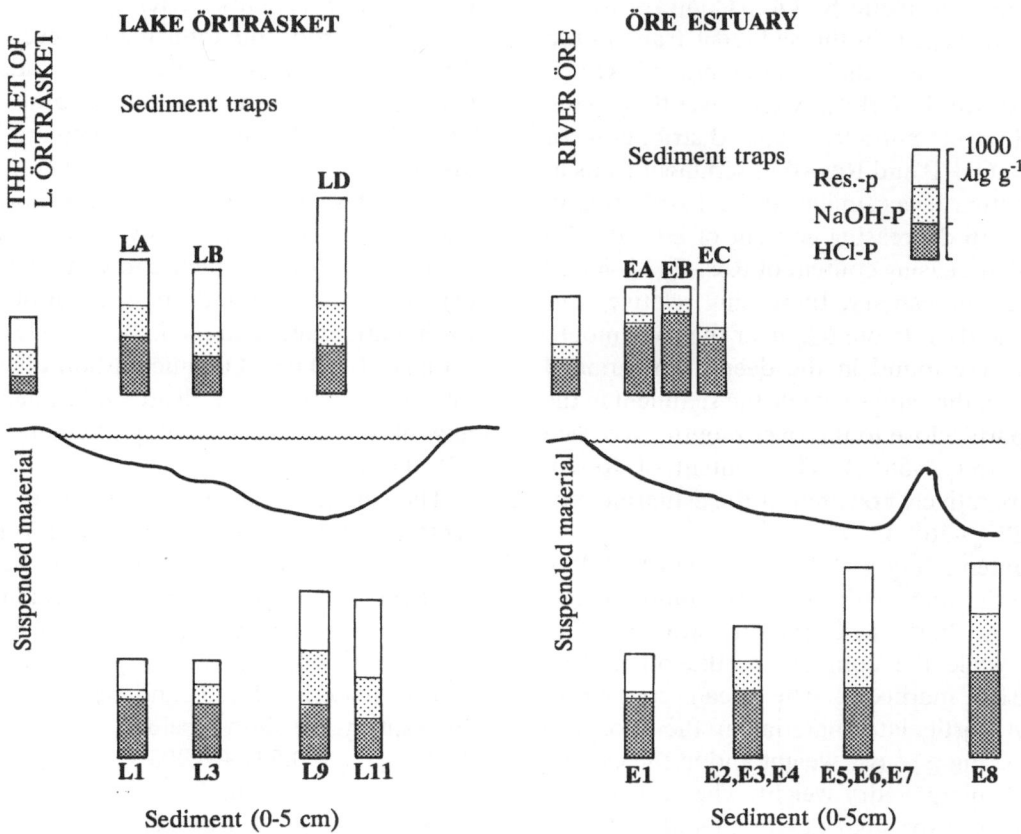

Fig. 3. The concentration of total phosphorus, HCl-P, NaOH-P and Res.-P in river borne particles, sedimenting material and in sediment in L. Örträsket and the Öre Estuary during the spring 1991.

rial collected in the sediment traps) of HCl-P in the lake particles was 32% which is 1.6 times higher than in the suspended material in the river (20%). The mean relative content of Res.-P (49%) was only slightly higher than in the riverborne particles (44%), while the mean relative content of NaOH-P (19%) was lower than in the riverborne particles (34%). The relative content of HCl-P decreased downstream in the lake from 42% (station LA) to 24% (station LD), while the relative content of Res.-P increased from 36% (station LA) to 54% (station LD). The relative content of NaOH-P was almost constant (17–21%) in all sediment traps. The same trend was even more pronounced in the sediment samples, except for the content of NaOH-P. The sediments in the shallow northern parts of the lake were extremely rich in HCl-P (57% at station L1 and

62% at station L3) while the deep parts of the lake were rich in NaOH-P (32% at station L9 and 26% at station L11) and Res.-P (35% at station L9 and 48% at station L11).

The suspended material in R. Öre has a similar relative content of Res.-P (49%) as in R. Örån but a higher relative content of HCl-P (33%) and a much lower relative content of NaOH-P (17% compared to 34% in the R. Örån). The composition of the suspended material in the R. Öre is almost identical to the mean composition of the trap material in the lake. Similar to the situation in the lake the material collected in sediment traps in the estuary was richer in total phosphorus than the suspended material transported in the R. Öre. The mean relative content of HCl-P in the sediment traps was 66%, which is 2 times higher than in the sediment traps in the lake and in the sus-

242

pended material in the R. Öre. Contrary to the behaviour of HCl-P in the sediment traps in the estuary, the mean relative contents of Res.-P (24%) and NaOH-P (10%) were lower than in the lake and river particles. The distribution of HCl-P, NaOH-P and Res.-P in sediment traps in the Öre Estuary was similar to the distribution in the lake, with decreasing content of HCl-P (73–45%) and increasing content of Res.-P (16–44%) seaward in the estuary. Increasing relative contents of NaOH-P (from 6% near the river mouth to 30%) were found in the deeper fine-grained sediments in the estuary, while the sediment in the shallow parts, close to the river mouth, were rich in HCl-P (50%–58%). The content of Res.-P was comparatively constant in the estuarine sediments (27%–36%).

The contents (mg·g^{-1} dry weight) of HCl-P in trap material and sediment were comparatively constant in both L. Örträsket and the Öre Estuary, while the contents of the other fractions varied markedly. The mean content of HCl-P in particulate material in the lake was 0.41 ± 0.02 mg·g^{-1} dry weight and in the estuary 0.46 ± 0.02 mg·g^{-1} dry weight. The corresponding mean value for suspended material in the rivers was 0.30 ± 0.03 mg·g^{-1} dry weight.

Relationships between phosphorus and iron, manganese, aluminium and organic carbon

The retention was positive for total iron, total carbon and particulate manganese and carbon in L. Örträsket. Negative retention was found for particulate iron and total manganese (Table 4). This indicates that, similar to the situation for both total and particulate phosphorus, there exists an internal source for particulate iron and dissolved manganese in L. Örträsket.

The interactions between phosphorous, iron and carbon are elucidated by multiple regression analysis (Table 5), which shows that iron alone explains the variation in the content of phosphorus in particulate material in both the lake and the estuary ($P < 0.05$). Organic carbon does not significantly increase the degree of explanation for any of the four types of particulate materials ($P > 0.05$).

The simple linear regression between the contents of total phosphorus and total iron in sediment trap material and sediments in both L. Örträsket and the Öre Estuary is illustrated in Fig. 4. The equations for the regression lines in sediment (Eq. 1) and trap material (Eq. 2) in the lake and in trap material (Eq. 3) and sediment (Eq. 4) in the estuary are shown below;

1) $P_{sediment} = 0.597 + 0.008 \times Fe_{sediment}$
   ($r^2 = 0.946$; $P < 0.001$; $n = 36$)
2) $P_{trap} = 0.406 + 0.015 \times Fe_{trap}$
   ($r^2 = 0.969$; $P < 0.001$; $n = 14$)
3) $P_{trap} = -0.160 + 0.035 \times Fe_{trap}$
   ($r^2 = 0.922$; $P < 0.001$; $n = 50$)
4) $P_{sediment} = 0.187 + 0.027 \times Fe_{sediment}$
   ($r^2 = 0.723$; $P < 0.001$; $n = 22$)

Three different relationships can be distinguished. The content of phosphorus per unit of iron is higher in material collected from sediment

Table 4. The mass balance for total and particulate iron, manganese and carbon (in tonnes) in the Lake Örträsket during the period April 24 to May 23 1991.

Element	Input				Output	Sedimentation	Retention
	R. Örån	R. Vargån	Diffuse inlets	Total	R. Öre	(trap yield)	(input-output)
Tot-Fe	130	30	7.1	160	160		1
Part-Fe	40	6.2	1.6	50	55	40	– 6
Tot-Mn	6.7	1.3	0.3	8.3	9.6		– 1.3
Part-Mn	2.5	0.2	0.05	2.8	2.6	4.4	0.2
Tot-C	1740	400	110	2250	2180		70
Part-C	110	20	5.3	140	130	30	3

Table 5. Results from the multiple regression routine with phosphorus as the dependent variable and iron and carbon as the independent variables for sediment trap material and sediments in Lake Örträsket during the spring 1991 and in the Öre Estuary during the spring periods 1990-1991. Probability level (P), number of observations (*n*) and degree of explanation (r^2).

	Lake Örträsket		Öre Estuary	
	Trap	Sediment	Trap	sediment
Fe (P)	0.000	0.000	0.000	0.038
C (P)	0.645	0.096	0.112	0.098
n	14	36	50	22
r^2	0.969	0.950	0.926	0.759

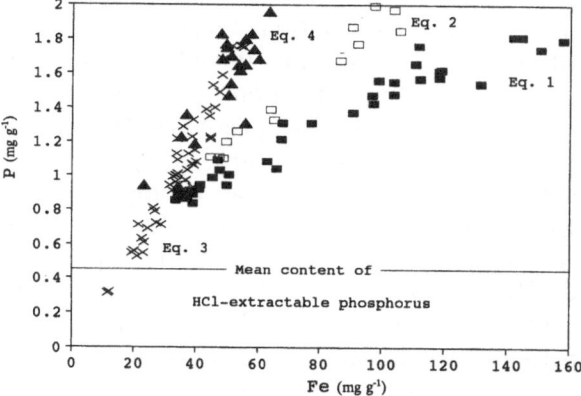

Fig. 4. The relations between concentrations of phosphorus and iron in particulate material in L. Örträsket in the spring 1991 and in the Öre Estuary in the spring 1990-1991. Trap material (□) and sediment (■) in L. Örträsket and trap material (×) and sediment (▲) in the Öre Estuary. Eq. 1–4 are defined in the text.

traps than in the sediments in L. Örträsket and still higher in both trap material and sediments in the Öre Estuary.

Aluminium showed no correlations with phosphorus in sediment traps or sediments in L. Örträsket or the Öre Estuary. This, plus the fact that aluminium contents are considerably lower than the iron concentrations in the R. Öre system suggest that aluminium-phosphorus interactions are quantitatively less important.

Discussion

Phosphorus balance

The sedimentation of river supplied particles and particulate phosphorus in the Öre Estuary have earlier been shown to be rapid and complete during the spring (Forsgren & Jansson, 1992). It was therefore assumed that L. Örträsket, which is a large water body receiving the same type of particulate matter as the estuary, should be an efficient trap for river-borne particulate phosphorus during the spring flood. However, the turnover of phosphorus in L. Örträsket appears to be more complex than what can be predicted by assuming that retention equals sedimentation.

The mass balance calculations for particles and particulate phosphorus (Table 2) show that the calculated sedimentation in the lake is of the same magnitude as the input from tributaries and diffuse inlets. The total sedimentation as calculated from the yield of a limited number of sediment traps should be considered as a rough estimate. However, the use of traps is the only practical way to measure the magnitude of the sedimentation and traps of the same type as those used in this study have been proven useful for calculations of sedimentation *in situ* (Håkansson & Jansson, 1983; Bloesch & Burns, 1984; Forsgren & Jansson, 1992). The representativity of the trap yield for whole lake sedimentation will always depend on the conditions specific for the actual lake. L. Örträsket is a comparatively simple environment in this context since it is a typical elongated through with the inputs in one end and the output in the other end. Moreover, most of the lake surface was covered with ice during the observation period which prevented the influence of wind and wave action on sedimentation.

The results in Table 2 agree well with the assumption that most of the riverine particle input should sedimentate in the lake. The fact that 70–80% of the total sedimentation took place in the areas close to the inlets demonstrates that the major share of the total sedimentation was of riverine origin. Phosphorus sedimentation was greater than the input of particulate phosphorus which means that the average concentration of phosphorus in settling particles was higher than in the suspended material of the river. This in turn indicates an internal lake source of particulate phosphorus. Such a source is also stressed by the totally unexpected result that the retention (input minus output) for both phosphorus and suspended material was substantially different from the calculated sedimentation. In principle, sedimentation and retention should be of the same magnitude provided that sedimentation is the dominant retention process, which is to expect in a large oligotrophic lake during a period with high load of suspended solids (Håkansson & Jansson, 1983).

In spite of the high sedimentation the retention is close to zero or negative i.e. the output of both total and particulate phosphorus equal or exceed the input. This situation is not uncommon in highly eutrophic lakes with substantial internal loading (Boström *et al.*, 1982; Boström *et al.*, 1988) but has not been reported for oligotrophic lakes. The average phosphorus retention for temperate forest lakes is more than 50% (Larsen & Mercier, 1976). The retention of L. Örträsket during 1991 was less than 15% (calculation from unpublished data of Ivarsson & Jansson), which low retention is explained by the obviously extreme conditions during the spring flood period which supplied 60% of the phosphorus loading during 1991.

In principle, the low retention during the period studied in this investigation can be explained if some component of the mass balance was false. In this case in means that either is the input underestimated or the output or the sedimentation overestimated. The input and output calculations should not be inaccurate to the extent needed to explain the discrepancies of the mass balance. The main inlets and the outlet were sampled six times each during the observation period. The concentration variations between the sampling events were small. The discharge was estimated from very accurate continous registrations downstreams the lake. Moreover, if the input-output budget was wrong it should be due to a system-

atic error and not result in an underestimated input and an overestimated output.

In theory, the sedimentation could be overestimated due to resuspension. However, this is very unlikely in the present investigation. The lake was ice-covered during almost the entire investigation period which excludes, the possibility of resuspension caused by wind and wave action. Another possible resuspension mechanism is erosion of shallow sediments close to the river inlets caused by inflowing water. This type of resuspension should be reflected in considerably higher amounts of material in the bottom traps compared to the surface traps near the inlets (cf Håkansson, 1989). However, the vertical distribution of trap material was the same throughout the water column at all stations except at station LC which is not influenced by the inlets. A third argument against resuspension is the composition of the particulate surplus (Table 2). The surplus is (in terms of weight) 50–60% of input for total particulate material while it is approximately 180% for phosphorus. The theoretical average content of phosphorus in the material supplied by the lake is thus 2.3 $mg \cdot g^{-1}$ dry weight which high content does not exist in any surface sediment particularly not near the inlets where it is less than 1 $mg \cdot g^{-1}$ dry weight (Table 3). The phosphorus budget could also be influenced by material from the first short discharge peak before the investigation period. However, this is not the case. We made weekly sampling of the inlet and outlet since late March and the retention of phosphorus in the lake is slightly negative also when the first peak is included in the budget calculations. It therefore appears as if another internal source exist of particulate phosphorus in L. Örträsket. This result was not expected and our investigation was not designed to collect data which could help to fully elucidate this phenomenon.

Few comparable data on phosphorus from similar type of lakes have been found in the literature. However, extremely increased concentrations (up to ten times) of particulate iron and dissolved manganese in the outlets of forest lakes in central and northern Sweden have been reported during spring periods (Andersson, 1991;

Pontér *et al.*, 1992). This was suggested to be caused by release of these elements from reduced sediments during winter and subsequent oxidation and export (in particulate form for iron and mainly dissolved form for manganese) in connection with the spring flood. Even if phosphorus was not discussed in the cited studies, it is likely that phosphorus should have a behaviour similar to iron according to the wellknown iron-phosphate relationships (Mortimer, 1941; Mortimer, 1942; Boström *et al.*, 1982). It is therefore interesting to compare the phosphorus budgets in L. Örträsket with that of iron, manganese and also carbon (Table 4). Both iron and manganese have a net turnover similar to that of phosphorus. Carbon, on the other hand, which is not affected by redox conditions and to a minor extent bound in particulate form shows a different balance, where the retention of total carbon is of the same magnitude as the sedimentation of particulate carbon.

No systematic sampling for water chemistry was performed in connection with this investigation. However, three vertical profiles sampled on April 16 have been analyzed and demonstrated that concentrations of total iron and total manganese in waters below 30–40 m (*ca.* 5–10 mg·l^{-1}) were 5–10 times higher for iron and 50–100 times higher for manganese than in the upper water column (*ca.* 1 mg-Fe·l^{-1} and 0.1 mg-Mn·l^{-1}). In principle, the amounts of total iron and total manganese below 30 m depth should be sufficient to cover the total export of total iron and total manganese from L. Örträsket during the whole investigation period.

Therefore, a hypothesis to explain the phosphorus budget of L. Örträsket during the spring is that a release of iron and phosphorus occurs from anoxic sediments in winter during which period large internal stores of these elements can be build up in the stagnant water body. Part of the released iron could be transformed to particulate material by formation of iron(III)oxyhydroxides in connection with the spring flood when oxygenated water is supplied to the lake and the whole water mass is circulating. Sorption of phosphorus to these oxides should be highly significant since the pH of 6.5 in the lake is favourable for this process (Stumm & Morgan, 1981). However, an intensive study of the hydrology and turnover of oxygen and redox sensitive elements during late winter and spring is necessary to test this hypothesis.

Composition of settling material and sediment

The large sedimentation of phosphorus in L. Örträsket and in the Öre Estuary takes place near the river inlets. A comparatively large part of this phosphorus is in the form of HCl-P. This is logical since the percentage of HCl-P in the particles which flowing into the lake and the estuary is 20–30% of total phosphorus. It is also logical that HCl-P dominates close to the inlets since HCl-P should represent apatite-P which is bound in minerogenic particles with comparatively rapid sedimentation rates.

However, it appears as if the settled material in both the lake and the estuary are richer in HCl-P than in the riverborne particulate material. The fractional composition of river material and trap material are not entirely comparable however, since the fractionation of river particles represents one occasion during maximum discharge in the R. Örån and two occasions during maximum discharge in the R. Öre, while the trap material fractionation represents a mean for the whole observation period. HCl-P is generally considered to represent apatite-phosphorus (Psenner *et al.*, 1984; Pettersson *et al.*, 1988) and it is difficult to see how the content of apatite should increase during sedimentation in this soft water lake at pH 6.5. If the sedimentation of HCl-P exceeds the input some alternative explanation should exist. Riverine particles introduced in Cayuga Lake in New York sorbed considerable amounts of phosphorus from the lake water at pH between 6.5 and 7.5 (Stafford-Glase & Barlow, 1986). Unfortunately no characterization of the particles was carried out. However, particles of the silt and clay fractions from a tropical river released substantial amounts of phosphate after extraction with sulfuric acid (Viner, 1982). Moreover, these particles readsorbed approximately the same amount of phosphorus after being transfered to neutral medium. It is therefore possible that minerogenic material sorbs phosphorus from lake water and estuarine water during sedimentation and that this phosphorus is extractable by strong acid. If so, the common interpretation of HCl-P has to be revised. Further studies in the River Öre system should be carried out to check for this explanation.

The contents of NaOH-P and Res.-P in the trap material increases with increasing distance from the inlets both in L. Örträsket and in the Öre Estuary suggesting that the sedimentation of phosphorus in water less influenced by riverine input depends on interactions between iron, organic material and phosphorus. This result was not unexpected because of the high concentrations of iron and dissolved organic carbon in the R. Öre system. A substantial part of the total phosphorus in the studied system is therefore potentially associated with complexes of iron and humic material (Ohle, 1937; Tipping, 1981).

The clear correlations between iron and phosphorus in trap material and sediment (Fig. 4, Table 5) indicate that iron-phosphorus aggregates play an essential role for the sedimentation of phosphorus in both the lake and the estuary. The empirical relationship between iron and phosphorus in Fig. 4 is evident for phosphorus contents in the particulate material higher than approximately $0.5-0.8$ mg·g^{-1} dry weight. This can be interpreted so that the particulate material in both the lake and the estuary have a fairly constant content of HCl-P of $0.4-0.5$ mg·g^{-1} dry weight plus a highly variable portion of phosphorus, the amount of which is controlled by iron-phosphorus relationships.

The trap material and the sediment show identical correlations between iron and phosphorus in the Öre Estuary, which is consistent with the conclusion that the sedimentary P:Fe ratio is largely determined by processes occurring during sedimentation (Forsgren & Jansson, 1992). The P:Fe ratio is generally higher in the estuarine particles than in the lake. Since iron controls all

phosphorus sedimentation except that of HCl-P in the investigated systems the different P:Fe ratios should depend on varying conditions for sorption of phosphorus on the iron oxyhydroxides. The higher P:Fe ratio in the estuary compared to the lake can be an effect of the higher pH in the estuary (7.5 compared to 6.5 in the lake) and the slightly higher concentrations of phosphorus in estuarine water. Furthermore, the disruption of the iron-organic material association which take place in the Öre Estuary (Forsgren & Jansson, 1992) can contribute to higher P:Fe ratios in the estuary. Organic material is complexed by iron hydroxides in fresh water (Tipping, 1981) and in principle the organic material and phosphorus thereby compete for binding sites on the iron hydroxides (Syers *et al.*, 1973). When the iron-carbon complexes break up, sites should theoretically be available for sorption of phosphorus.

Conclusions

1) Lake Örträsket is both a sink and a source for phosphorus during the spring. River borne particulate phosphorus sedimentates in the lake. In addition, particulate phosphorus is formed in the lake. The net retention of phosphorus during the spring is therefore close to zero.

2) The Öre Estuary constitutes an efficient trap for the total supply of river borne phosphorus during the spring.

3) Sorption of phosphorus to settling particles seems to be an important process, which is particularly pronounced in the estuary. The possibility that this sorbed phosphorus is extractable by strong acid must be further investigated.

4) The sedimenting material and the surface sediment in both the L. Örträsket and the Öre Estuary have a content of HCl-P around $0.4-0.5$ mg·g^{-1} dry weight plus a highly variable portion of phosphorus, which is controlled by iron-phosphorus relationships.

5) The P:Fe ratio is higher in the particulate ma-

terial in the estuary than in the lake. The different P:Fe ratios should depend on varying conditions for sorption of phosphorus on the iron oxyhydroxides.

Acknowledgements

This study was carried out with financial help from the Swedish National Science Research Council. We thank F.K. Louise Malmgren and F.K. Rolf Zale for stimulating cooperation during field work and Dr Erik Lundberg, Carl-Henrik Stangenberg and Dr Kurt Pettersson who made the chemical analyses.

References

Ahl, T., 1988. Background yield of phosphorus from drainage area and atmosphere: an empirical approach. In G. Persson & M. Jansson (eds), Phosphorus in Freshwater Ecosystems. Developments in Hydrobiology 48. Kluwer Academic Publishers, Dordrecht: 35–44. Reprinted from Hydrobiologia 170.

Andersson, P., 1991. Annual variations of dissolved and particulate Fe, Mn, major components in stream- and lakewaters in central Sweden. Thesis Institute of geology and geochemistry, Stockholm University, Stockholm, Sweden. 284: 1–66.

Bale, A. J. & A. W. Morris, 1981. Laboratory simulation of chemical processes induced by estuarine mixing: the behaviour of iron and phosphate in estuaries. Estuar. coast shelf Sci. 13: 1–10.

Bloesch, J. & N. M. Burns, 1980. A critical review of sedimentation trap technique. Schweiz. Z. Hydrol. 42: 15–55.

Boström, B., M. Jansson & C. Forsberg, 1982. Phosphorus release from lake sediment. Arch. Hydrobiologia 18:5–59.

Boström, B., J. M. Andersen, S. Fleischer & M. Jansson, 1988. Exchange of phosphorus across the sediment–water interface. (Ibid) pp. 229–244.

Boyle, E. A., J. M. Edmond & E. R. Sholkovitz, 1977. The mechanism of iron removal in estuaries. Geochim. Cosmochim. Acta. 41: 1313–1324.

Burman, J. O., 1987. Inductively coupled plasma emission spectroscopy: geological applications. In P. J. Elving & J. D. Winefordner (ed.), Chemical analysis 90: 27–47.

Broberg, O. & G. Persson, 1988. Particulate and dissolved phosphorus forms in fresh water: composition and analysis. In G. Persson & M. Jansson (eds), Phosphorus in Freshwater Ecosystems. Developments in Hydrobiology 48. Kluwer Academic Publishers, Dordrecht: 61–90. Reprinted from Hydrobiologia 170.

Brydsten, L., 1992. Wave-induced sediment resuspension in the Öre estuary, northern Sweden. In B. T. Hart & P. G. Sly (eds), Sediment/Water Interactions. Developments in Hydrobiology 75. Kluwer Academic Publishers, Dordrecht: 71–83. Reprinted from Hydrobiologia 235/236.

Brydsten, L. & M. Jansson, 1989. Studies of estuarine sediment dynamics using ^{137}Cs from the Tjernobyl accident as a tracer. Estuar. coast. shelf Sci. 28: 249–259.

Forsgren, G. & M. Jansson, 1992. The turnover of river-transported iron, phosphorus and organic carbon in the Öre Estuary, northern Sweden. In B. T. Hart & P. G. Sly (eds), Sediment/Water Interactions. Developments in Hydrobiology 75. Kluwer Academic Publishers, Dordrecht: 585–596. Reprinted from Hydrobiologia 235/236.

Froelich, P. N., 1988. Kinetic control of dissolved phosphate in natural rivers and estuaries: A primer on the phosphate buffer mechanism. Limnol. Oceanogr. 33: 649–668.

Hieltjes, A. H. M. & L. Lijklema, 1980. Fractionation of inorganic phosphates in calcareous sediments. J. envir. Qual. 9: 405–407.

Håkansson, L. & M. Jansson, 1983. Principles of lake sedimentology. Springer-Verlag Berlin Heidelberg, Germany. 316 pp.

Håkansson, L., S. Floderus & M. Wallin, 1989. Sediment trap assemblages – a methodological description. In P. G. Sly & B. T. Hart (eds), Sediment/Water Interactions IV. Developments in Hydrobiology 50. Kluwer Academic Publishers, Dordrecht: 481–490. Reprinted from Hydrobiologia 176/177.

Ivarsson, H. & G. Forsgren, 1988. Erosion and human land-use in the valley of River Öre, northern Sweden. SGÅ 64. Lund, Sweden.

Ivarsson, H. & L-I. Karlsson, 1991. Geological and geochemical conditions in the River Öre drainage area, northern Sweden. GERUM Nr 15. Umeå, Sweden.

Larsen, D. P. & H. T. Mercier, 1976. Phosphorus retention capacity of lakes. J. Fish Res. Bd Can. 33: 1742–1750.

Malmgren, L. & L. Brydsten, 1992. Sedimentation of river-transported particles in the Öre Estuary, northern Sweden. In B. T. Hart & P. G. Sly (eds), Sediment/Water Interactions. Developments in Hydrobiology 75. Kluwer Academic Publishers, Dordrecht: 59–69. Reprinted from Hydrobiologia 235/236.

Mortimer, C. H., 1941. The exchange of dissolved substances between mud and water in lakes I. J. Ecol. 29: 280–329.

Mortimer, C. H., 1942. The exchange of dissolved substances between mud and water in lakes II. J. Ecol. 30: 147–201.

Ohle, W., 1937. Kolloidgele als Nährstoffsregulanten der Gewässer. Naturwissenschaften. 25: 471–474.

Pettersson, K., B. Boström & O-S. Jacobsen, 1988. Phosphorus in sediments – speciation and analysis. In G. Persson & M. Jansson (eds), Phosphorus in Freshwater Ecosystems. Developments in Hydrobiology 48. Kluwer Academic Publishers, Dordrecht: 91–101. Reprinted from Hydrobiologia 170.

Pontér, C., J. Ingri & K. Boström, 1992. Geochemistry of

manganese in the Kalix River, northern Sweden. Geochim. Cosmochim. Acta. 56: 1485–1494.

Psenner, V. R., R. Pucsko & M. Sager, 1984. Die Fractionierung organisher und anorganisher Phosphorverbindungen von Sedimenten. Versuch einer definition ökologisch wichtiger Fraktionen. Arch. Hydrobiol. 1: 111–155.

Stabel, H. H. & M. Geiger, 1985. Phosphorus adsorption to riverine suspended matter. Implications for the P-budget of Lake Constance. Wat. Res. 19: 1347–1352.

Stafford-Glase, M. & J. P. Barlow, 1986. The role of lake water in enhancing sorption of phosphorus by stream particulates. In P. G. Sly (ed.), Sediments and water interactions. Springer-Verlag, New York, USA. pp. 451–460.

Stumm, W. & J. J. Morgan, 1981. Aquatic Chemistry. John Wiley & Sons, Inc., New York. 780 pp.

Syers, J. K., R. F. Harris & D. E. Armstrong, 1973. Phosphate chemistry in lake sediments. J. envir. Qual. 2: 1–14.

Tipping, E., 1981. The adsorption of aquatic humic substances by iron oxides. Geochim. Cosmochim. Acta. 45: 191–199.

van Eck, G. T. M., 1982. Forms of phosphorus in particulate matter from the Hollands Diep/Haringvliet, Hydrobiologia 92: 665–681.

Viner, A. B., 1982. A quantitative assessment of the nutrient phosphate transported by particles in a tropical river. Revue Hydrobiol. trop. 15: 3–8.

Vollenweider, R. A., 1976. Advances in defining critical loading levels for phosphorus in lake eutrophication. Mem. Inst. ital. Idrobiol. 33: 53-83.

Hydrobiologia **253**: 249–261, 1993.
P.C.M. Boers, T.E. Cappenberg & W. van Raaphorst (eds),
Proceedings of the Third International Workshop on Phosphorus in Sediments.
© 1993 *Kluwer Academic Publishers.*

Sorption of phosphate by sediments as a result of enhanced external loading

Robert Portielje & Lambertus Lijklema
Nature Conservation Department, Agricultural University, Ritzema Bosweg 32A, P.O. box 8080, 6700 DD Wageningen, The Netherlands

Key words: phosphorus, sediments, adsorption, dispersion, solid-phase diffusion, Al-hydroxide

Abstract

In artificial test ditches, originally poor in nutrients, the effects of enhanced external loading with phosphorus were studied. An important term in the mass balance of phosphorus is retention by sediment. Parameters concerning the uptake of phosphorus by the sandy sediment of a ditch have been measured or were obtained from curve-fitting and were used in a mathematical model to describe diffusion into the sediment and subsequent sorption by soil particles.

On a time scale of hours uptake of phosphorus from the overlying water by intact sediment cores could be simulated well with a simple diffusion-adsorption model. Mixing of the overlying water resulted in an enhanced uptake rate caused by an increased effective diffusion coefficient in the top layer of the sediment.

Laboratory experiments revealed that after a fast initial adsorption, a slow uptake process followed that continued for a period of at least several months. This slow sorption can immobilize a substantial part of the phosphorus added. It may physically be described as an intraparticular diffusion process, in which the adsorbed phosphate penetrates into metaloxides, probably present as sand grain coating, and thereby reaches sorption sites not immediately accessible otherwise.

The total sorption capacity of the soil particles is *ca.* 3.3 times the maximum instantaneous surficial adsorption capacity.

Introduction

Over the last decades eutrophication in surface waters has received ample attention in water research, resulting in numerous publications. The role of the sediments as a buffer for phosphorus has been a central issue. During periods of enhanced external loading the sediments can readily take up large amounts of phosphorus, thereby keeping bio-availability relatively low (Golterman, 1977). On the other hand, after a reduction of the external loading the sediments can release the sorbed phosphorus, thus counteracting beneficial effects (Lijklema, 1986; Ahlgren, 1977).

An important requirement for a realistic description and prediction of phosphorus fluxes across the sediment-water interface is a good apprehension of the underlying mechanisms. These include physical processes that affect transport (diffusion, dispersion) and the factors controlling them and the kinetics and time scales of the physical/chemical processes involved (adsorption, desorption, precipitation; Lijklema, 1991).

In this paper the accumulation of phosphorus

in the sediment of a test ditch receiving a high external loading with nutrients is discussed. Following other authors (e.g. Kamp-Nielsen, 1982) a multi-layer sediment model has been used to simulate uptake and subsequent sorption by the sediment particles. The description is in terms of diffusive transport mechanisms, both in the interstitial water and in a solid phase consisting mainly of aluminum-hydroxides, where partial conversion of the hydroxides to Al-phosphates follows (Van Riemsdijk, 1984; Bolan *et al.*, 1985; Van der Zee, 1988). The diffusive transport parameters are derived by fitting the model to observations.

Sites studied

The research is done in test ditches at Renkum in the Netherlands. The ditches have a length of 40 m and the water depth is maintained at 0.5 m. The bottom material is a 0.25 m layer of sand with a width of 1.6 m. On both lateral sides is a bank consisting of gravel with a slope of 30°. The volume of the ditches is 47 m³. The ditches are fed by precipitation, and, during dry periods, by the inlet of groundwater from a local well. The mean residence time of the water is 0.26 year.

A water-impermeable foil under the sediment and banks excludes infiltration or seepage. There are four ditches with a sandy sediment and four with a clay bottom, receiving four different levels of external nitrogen and phosphorus input. The atmospheric input of P and the input resulting from the inlet of groundwater is *ca.* 0.1 g P m^{-2} yr^{-1}.

Most of the time the water can be assumed to be mixed well horizontally and vertically.

In this study bottom material from the reference sand ditch, receiving no additional P-input, has been used for uptake experiments. The results are used to simulate the uptake of P by the sediment in the ditch receiving the highest load, 12 g P m^{-2} yr^{-1}, supplied in monthly dosages as K_2HPO_4 dissolved in tap water and spread evenly in the overlying water. Each dosage corresponds

to an instantaneous increase in the dissolved phosphorus concentration of 2.63 mg P l^{-1}.

Model concept

Following increased dissolved phosphorus concentrations in the water, phosphorus will diffuse into the sediment as a result of the imposed vertical concentration gradient. In a situation with no vertical advective flow the resulting concentration change $\delta C_w/\delta t$ in the overlying water is

$$\partial C_w/\partial t = D_{\text{eff}, z = 0} \cdot \varepsilon_{z = 0} \partial C/\partial z \big|_{z = 0} A/V . \quad (1)$$

For explanation of symbols see Table 1.

The change of the concentration in the pore water C_z (Berner, 1980) is

$$\partial C_z/\partial t = \partial (D_{\text{eff}, z = 0} \cdot \varepsilon_z \cdot \partial z)/\varepsilon_z \partial z - R_z . \quad (2)$$

The effective diffusion coefficient $D_{\text{eff}, z}$ is the sum of all diffusion processes involved in the vertical transport of solutes in sediments (Booij, 1989). It can be approximated by

$$D_{\text{eff}, z} = (D_{\text{mol}} + D_{\text{dis}, z} + D_{\text{bio}})/\theta^2 . \quad (3)$$

Bioirrigation depends on the number and species of benthic fauna. Solid mixing is neglected, based on the observation that detritus on top of the sediment does not mix with the sand to an appreciable extent.

The tortuosity $1/\theta^2$ accounts for the slow-down of vertical transport caused by an extended pathway due to the travel around particles (Berner, 1980). The tortuosity is related to the porosity by a formation factor that accounts for the ratio of the electrical resistivity of the sediment to that of the pore fluid (McDuff & Ellis, 1979), but the theoretical value for θ^{-2} of 0.7 is generally acceptable (Brinkman *et al.*, 1987). The molecular diffusion coefficient is ion-specific and depends on temperature. For $H_2PO_4^-$ it is $7.34 \cdot 10^{-5}$ $m^2 \cdot day^{-1}$ at 20 °C (Brinkman & Raaphorst, 1986).

The effective vertical diffusion coefficient in the

Table 1. Summary of symbols used.

Symbol	Explanation	Dimension
A	Area of sediment	L^2
C	Concentration in overlying water	$M \cdot L^{-3}$
C_z	Conc. in porewater at depth z	$M \cdot L^{-3}$
C_s	Conc. in solid	$M \cdot L^{-3}$
C_{eq}	Equilibrium conc. (fast step) in porewater	$M \cdot L^{-3}$
D	Water depth	L
$D_{eff, z}$	Effective diffusion coeff.	$L^2 \cdot T^{-1}$
D_{mol}	Molecular diffusion coeff.	$L^2 \cdot T^{-1}$
D_s	Solid-phase diffusion coeff.	$L^2 \cdot T^{-1}$
$D_{dis, z}$	Dispersion coefficient	$L^2 \cdot T^{-1}$
D_{bio}	Bioirrigation coefficient	$L^2 \cdot T^{-1}$
H	Wave height	L
K	Langmuir adsorption constant	$L^3 \cdot M^{-1}$
O	Specific surface area	$L^2 \cdot M^{-1}$ solid
R	Source and sinks	$M \cdot L^{-3} \cdot T^{-1}$
S	Sum of squares of deviations	–
T	Wave period	T
V	Volume of overlying water	L^3
VM	Concentration of solids	$M \cdot L^{-3}$
X_z	Amount of P adsorbed (fast step) at depth z	$M \cdot M^{-1}$
X_m	Max. adsorption capacity (fast)	$M \cdot M^{-1}$
h	Wave number	L^{-1}
k	Attenuation of D_{tur} over depth	L^{-1}
k_p	Permeability coefficient	$L \cdot T^{-1}$
k_s	Rate of adsorption	$L^3 \cdot M^{-1} \cdot T^{-1}$
$k_{s, act}$	Actual adsorption rate	T^{-1}
l	Thickness of outer layer involved in surficial adsorption	L
r	Depth in Me-oxide coating	L
z	Depth in sediment	L
α_z	Mechanical dispersion	$L^2 \cdot T^{-1}$
ε	Porosity	$L^3 \cdot L^{-3}$
λ	Wave length	L
θ	Tortuosity	–

sediment can be enhanced by horizontal pressure gradients, caused by wind induced wave action or seiches. Rutgers van der Loeff (1981) gives an overview of the percolation of shallow sand beds caused by wave action and the resulting vertical dispersion in the sediments. In our column experiments pressure gradients were caused by aeration of the overlying water and in the field-experiments in the ditches by wind. The resulting dispersion, $D_{dis, z}$ can be assumed to decrease exponentially with depth, in accordance with the analysis by Rutgers van der Loeff, 1981):

$$D_{dis, z} = D_{dis, z = 0} \cdot \exp(-kz), \qquad (4)$$

in which the attenuation coefficient k depends on sediment characteristics like porosity and particle size distribution.

The equilibrium of short-term sorption onto sediment particles is calculated using the Langmuir adsorption isotherm:

$$X_z = X_m K C_{eq, z}/(1 + K C_{eq, z}). \qquad (5)$$

X_m and K have to be determined experimentally for each type of sediment. X_z and $C_{eq, z}$ are time and depth dependent. A mass balance for the change in the amount of sorbed P has to be implemented in the model.

Short-term sorption equilibrium is not reached instantaneously, so in a short-term experiment adsorption processes are not at steady state. The sorption rate depends on the difference between actual porewater concentration C_z and the $C_{eq, z}$ calculated from (5). Thus, considering only adsorption, R_z can be described as (Brinkman *et al.*, 1987):

$$R_z = k_{s, act} (C_z - C_{eq, z}). \qquad (6)$$

The actual rate constant $k_{s, act}$ depends on the surface available for adsorption per volume pore water and, therefore, is a rate constant k_s, corrected for the concentration of adsorbents in the solid phase and the porosity:

$$k_{s, act} = k_s VM/\varepsilon. \qquad (7)$$

An explicit numerical calculation of the time-porewater-concentration course is used. Discretization with respect to depth yield

$$\frac{dC_i}{dt} = D_{eff}(i-1) * \frac{(\varepsilon_i + \varepsilon_{i-1})}{2} * \frac{(C_{i-1} - C_i)}{\varepsilon_i (\Delta z)^2}$$

$$- D_{eff}(i) * \frac{(\varepsilon_i + \varepsilon_{i+1})}{2} * \frac{(C_i - C_{i-1})}{\varepsilon_i (\Delta z)^2}$$

$$- R_i \qquad (8)$$

Index i refers to a horizontal layer and increases with depth. The layers are chosen equidistant.

True equilibrium of sorption usually is not reached within a few days, and after a fast initial adsorption a slow step often follows (e.g. Barrow & Shaw, 1975). The slow step is assumed to result from diffusion of adsorbed P into the interior solid phase consisting of Fe- and Al-(hydr)oxides, and this assumption has been shown to explain experimental observations satisfactorily (Van Riemsdijk *et al.*, 1984; Van der Zee, 1988).

Solid diffusion of adsorbed P can occur when the thickness of a metal-oxide particle or coating around the soil particles exceeds a treshold value at which not all the adsorbent is immediately accessible to adsorbing phosphate ions. Surficial adsorption induces a gradient towards the interior of the particles (Barrow, 1983). Although the exact chemical or physical mechanism by which ion-groups diffuse through a solid is not considered (solid phase or micropore diffusion), it is usually described as being analogous to diffusion in a liquid medium, hence

$$\partial C_s / \partial t = D_s \partial^2 C_s / \partial r^2 . \tag{9}$$

Since the thickness of the coating is very small compared to the radius of the sand particles spherical geometry has not been employed. To be able to calculate the concentration gradient in the coating, conversion of surficially adsorbed P, X_z, measured as $g\,P\,g^{-1}$ dry matter, to an internal concentration $C_{s,z}$ ($g\,P\,m^{-3}$) in the outer layer of the coating involved in surficial adsorption, is needed. The conversion is executed by dividing X_z by the volume of the outer layer per unit of dry matter:

$$C_{s,z} = X_z / O * l . \tag{10}$$

The time-concentration course is calculated numerically after dividing the inner part of the coating, not involved in surficial adsorption, in equidistant sublayers with thickness Δr:

$$dC_{s,j}/dt = D_s (C_{s,j-1} - C_{s,j})/(\Delta r)^2$$
$$- D_s (C_{s,j} - C_{s,j+1})/(\Delta r)^2 . \tag{11}$$

Index j refers to the number of the layer in the coating. Combining (6), (9) and (10) yields a mass balance for the internal concentration in the outer layer of the coating involved in surficial

adsorption ($j = 1$):

$$\frac{dC_{s,j=1}}{dt} = \frac{- D_s (C_{s,j=1} - C_{s,j=2})}{^{1}/_{2}\Delta r * l}$$
$$+ \frac{k_{s,\,act}(C_z - C_{eq,\,z}) * (\varepsilon/VM)}{O * l} \tag{12}$$

in which the term ε/VM is again introduced to convert a porewater concentration change to a change per amount of dry matter. $^{1}/_{2}\Delta r$ is the distance between the boundary between the layers $j = 1$ and $j = 2$ and the middle of the layer $j = 2$.

Figure 1 presents the processes implemented in the model for a horizontal layer in the sediment.

Materials and methods

Chemical analyses

Dissolved phosphorus was determined with the modified molybdate-blue method according to Murphey & Riley (1962) on a Skalar SA-40 Autoanalyser or on a Vitatron Photometer.

Total dissolved iron was measured after reduction with hydroxylammoniumchloride to Fe^{2+} with tripyridiltriazine as colour reagent on a Skalar SA-40 Autoanalyser.

Aluminum was determined on a Spectro Analytical Instruments ICP type Spectroflame with inductively coupled plasma atomic emission spectrometry (ICP-AES) (Novozamsky *et al.*, 1986).

Short-term adsorption characteristics

The maximal short-term adsorption capacity X_m and the adsorption constant K of the Langmuir equation (5) were determined by adding different amounts of dissolved phosphate to suspensions of the ditch sediment in filtered (0.45 μm) ditch water. After 48 hrs shaking at 20 °C and pH 7 it was assumed that the fast equilibrium had been reached. The amount of adsorbed P (X) and dis-

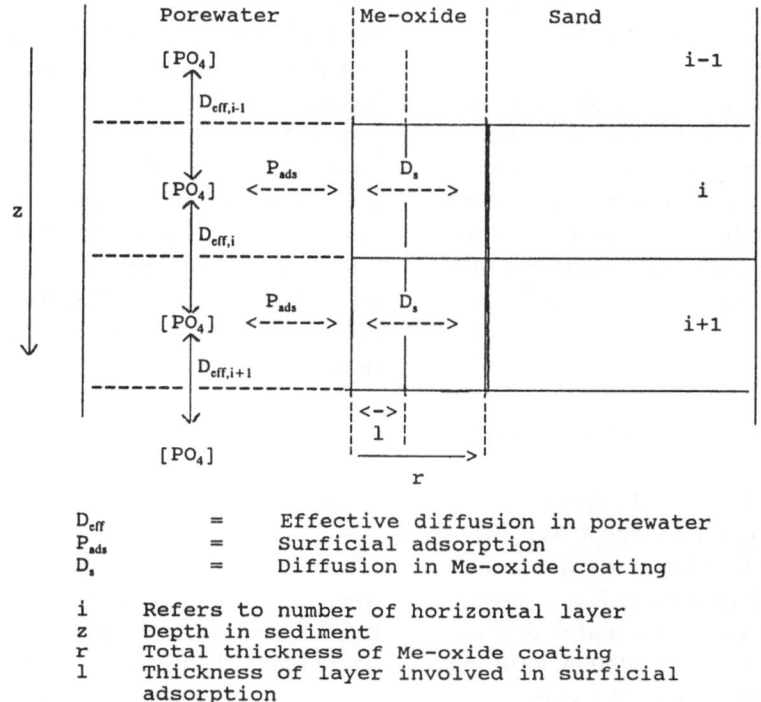

D_{eff}	=	Effective diffusion in porewater
P_{ads}	=	Surficial adsorption
D_s	=	Diffusion in Me-oxide coating

i	Refers to number of horizontal layer
z	Depth in sediment
r	Total thickness of Me-oxide coating
1	Thickness of layer involved in surficial adsorption

The volume of the porewater (porosity) decreases with depth.
The thickness of the coating is constant.

Fig. 1. Schematization of the processes implemented in the model.

solved P (C_{eq}) were determined. The native adsorbed P was estimated from extrapolation to $C_{eq} = 0$ and added to the measured additional sorption.

K and X_m were estimated by least-squares fitting of eq. (5) to the data. 90% confidence intervals for K and X_m were calculated from (Draper & Smith, 1966):

$$S = S_{min} \{1 + (p/n - p) \, F(p, n - p, 90\%)\} , \quad (13)$$

with S the sum of the squares at the 90% confidence contour, S_{min} the minimum sum of the squares, n the number of samples, p the number of parameters and F(p, n–p, 90%) the F-distribution according to Fisher.

Short-term uptake experiments

To study the uptake of phosphorus by sand and to gain qualitative insight into the effect of mixing of the overlying water, intact sediment cores were collected in a polythene cylinder with a diameter of 5.3 cm and immediately transported to the laboratory. The overlying water and a watery benthic layer consisting of algae and detritus with a thickness of 1.5–2 cm were removed carefully without resuspending the sand and replaced by filtered (0.45 μm) phosphate-enriched (added as K_2HPO_4) ditch water. The pH was adjusted to 7 with HCl and/or NaOH. The columns were incubated for two or three days. In one experiment the overlying water was not mixed, in another experiment it was mixed by moderate aeration.

The initial PO_4-P concentrations in the overlying water of the non-mixed column and the aerated column were 1.71 mg l^{-1} and 2.33 mg l^{-1} respectively, somewhat lower than the highest loading in the field experiment in the ditches. The initial heights of the overlying water columns were 9.9 cm and 12.0 cm respectively.

The change in PO_4-P concentration in the water was measured in 5 ml samples collected at regular time intervals.

The vertical profile of porosity and solid concentration in the sediment of the ditch was known from an inventory made after the construction of the ditches, in which intact sediment cores were used. The reference ditch, from which the cores were collected, has a very low primary production and virtually no macrophytes, so almost no changes had occurred.

Experiments on long-term sorption by sand particles

To study the long-term sorption of phosphorus by the sand, 5 g dried (40 °C) sediment was incubated in 250 ml at 18 °C. Phosphate was added as K_2HPO_4 up to 1, 2 and 5 mg P l^{-1} respectively. Samples of 5 ml were collected over a period of 150 days, with a higher frequency during the first weeks after the start of the experiment than later, and analysed for dissolved phosphorus.

Determination of total sorption capacity

As total equilibration of the coated sand grains with the solution takes a long time due to the very slow second phase uptake, it is difficult to estimate the equilibrium solid phase concentration $C_{s, eq}$ corresponding to the ambient P concentration and pH. Hence, a direct estimation of the total sorption capacity of the sand without the inconvenience of being dependent on the time scale of the solid-phase diffusion process has been applied. The method is based on the principle that dissolution of the Al- and Fe-(hydr)oxides in acid medium and subsequent homogeneous precipitation in the presence of phosphate at the original pH enables immediate occupation of all sorption sites present.

The metal-oxide coating was dissolved by shaking 1 g dried sediment during 48 hrs in 0.1 M HCl. After centrifugation (10 mins at 3500 rpm) Al was measured in the supernatant. Three series were prepared: in one the supernatant and the sand were not separated; from the second the supernatant was removed and replaced by demineralized water; the third consisted of the removed supernatant of series 2 only. After phosphate addition to all series (up to 0, 1, 2 and 5 mg P l^{-1} resp.) the pH was adjusted to 6.8 (the original value), so that formation of the Al-OH-phosphate complexes could occur. After two and three days Al and dissolved phosphorus were determined in the supernatant after centrifugation.

Simulation of vertical concentration profiles in the porewater

Since the start of the programme of monthly nutrient dosages in May 1989, the dissolved phosphorus concentration has been measured in the water regularly, usually at 1, 3, 6, 10 and 20 days after each monthly dosage and on the day before the next dosage. The concentrations on intermediate days were obtained by linear interpolation. These concentrations have been used as a boundary condition: $c_{z=0} = c_w$. This might slightly over- or underestimate $c_{z=0}$ due to the presence of a benthic algae layer or a diffusive sublayer on top of the sand, but the very loose structure of this layer (porosity $> 99\%$) and the relatively small rate at which P can be released from or taken up by the layer with respect to the high phosphate concentration and the turbulence in the overlying water most of the time makes this deviation negligible.

The amount of phosphorus adsorbed to the original sediment matter has been measured and used as an initial condition in the simulation of [P] in the porewater. Since aluminum, which is insensitive to changing redox conditions, is the main P-sorbing element in the ditches (see next section), redox conditions have not been incorporated into the model.

Results

Adsorption characteristics

The short-term Langmuir adsorption isotherm is given in figure 2. The values for X_m and K, obtained from least-squares fitting are

$X_m = 101 \ \mu g \ g^{-1}$ dry matter;
$K = 0.57 \ m^3 \ g^{-1}$.

The native adsorbed P was estimated at $3 \ \mu \ g^{-1}$ of dry matter. The boundaries of the 90% confidence intervals for both parameters are

X_m:83–124 $\mu g \ g^{-1}$ dry matter;
K:0.32–1.03 $m^3 \ g^{-1}$.

Short-term column uptake experiments

Figures 3 and 4 present observed and simulated [P] in the overlying water for the short-term column uptake experiments. Simulations include the role of dispersion induced by aeration. In case of no mixing of the overlying water, a good agreement is obtained if, next to molecular diffusion, $D_{dis, z=0}$ is set equal to the low value of $4.7.10^{-5}$ $m^2 \ d^{-1}$, which is about 65% of D_{mol}. The boundaries of the 90% confidence interval for $D_{dis, z=0}$ are $2.6.10^{-5}$–$7.0.10^{-5}$ $m^2 \ d^{-1}$.

At the aeration rate used in the second experiment a dispersion coefficient at the water-

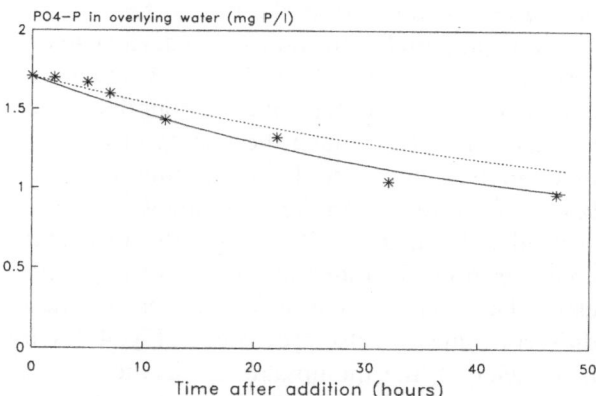

Fig. 3. Dissolved phosphorus concentrations in overlying water of sediment column without mixing. ($\blacksquare$ = measured, - - - = simulated, $D_{dis, z=0} = 0$, —— = simulated, $D_{dis, z=0} = 4.7.10^{-5} \ m^2 \ d^{-1}$)

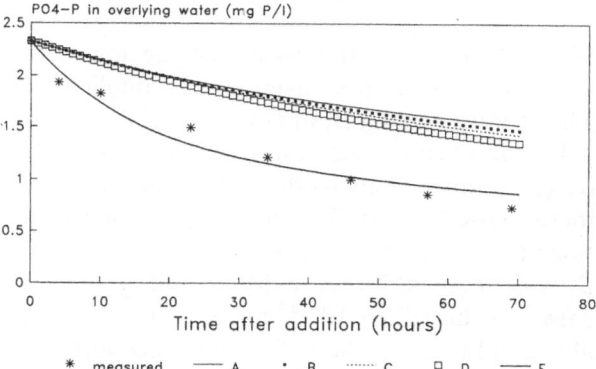

Fig. 4. Dissolved phosphorus concentrations in overlying water with mixing by aeration. (for values of parameters in A–E see Table 2)

sediment interface $D_{dis, z=0}$ of $4.6.10^{-4}$ $m^2 \ d^{-1}$ yields good results. The attenuation constant k in (4) has been set at $75 \ m^{-1}$. The simulation results are relatively insensitive to k. $D_{dis, z=0}$ and k are positively correlated. The boundaries of the 90% confidence interval for $D_{dis, z=0}$ are $3.7.10^{-4}$–$5.9.10^{-4}$ $m^2 \ d^{-1}$.

The presence of a diffusive boundary layer may have reduced the phosphate fluxes in the non-aerated column. This would lead to an underestimation of $D_{dis, z=0}$ in the non-aerated column. Erosion of the diffusive boundary layer by turbulence in the overlying water cannot explain the

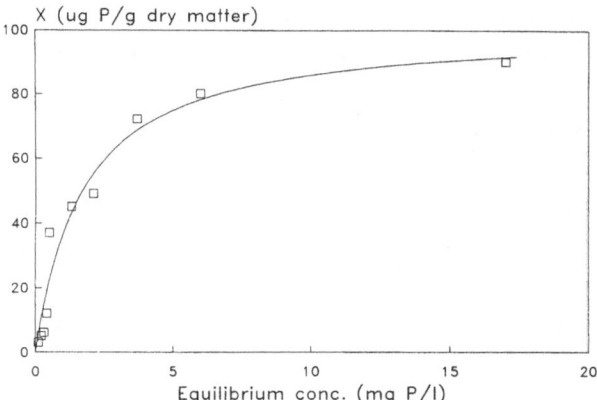

Fig. 2. Surficial Langmuir adsorption isotherm.

measured increase in uptake rate in the aerated column compared to the non-aerated, since in the simulation of the non-aerated column already a complete absence of this diffusive boundary layer was assumed. In Table 2 the results of a sensitivity analysis are given, by comparing the simulated concentrations in the overlying water of the aerated column at $t = 69$ hr with the measured concentrations for different values of the parameters. The time-concentration courses for these different values are also presented in Fig. 4. It can be concluded that the uptake rate in the aerated column can only be explained with an enhanced effective diffusion coefficient caused by the aeration.

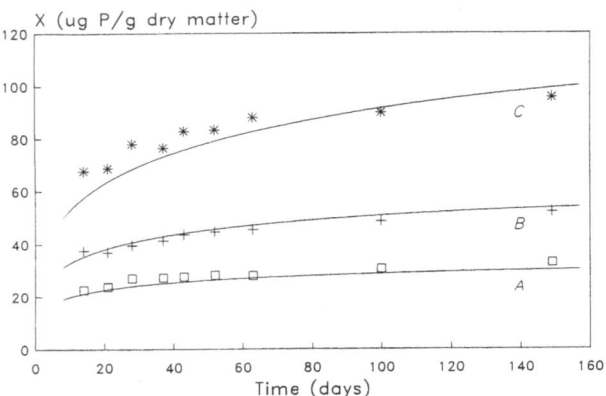

Fig. 5. Long-term sorption experiments, as measured in suspensions and simulated. (At $t = 0$: A: 1 mg P/l; B: 2 mg P/l; C: 5 mg P/l. $D_s = 2.7.10^{-19}$ m^2 d^{-1} and the total sorption capacity is $3.3 * X_m$).

Long-term sorption experiments

Figure 5 illustrates the results of the long-term sorption experiments (averages of duplicates). They show, after a fast initial adsorption, a gradually decreasing sorption rate that can be described adequately by the solid-phase diffusion model (solid lines). The decreasing concentrations in the water have been taken into account. D_s was obtained by curve-fitting, and in all three cases a value of $2.7 \cdot 10^{-19}$ m^2 d^{-1} gives reasonably good results. The 90% confidence intervals for D_s, calculated with (13) are

series A: $2.7.10^{-19} - 3.8.10^{-19}$ m^2 d^{-1};
series B: $2.5.10^{-19} - 2.8.10^{-19}$ m^2 d^{-1};
series C: $2.0.10^{-19} - 3.0.10^{-19}$ m^2 d^{-1}.

The value of D_s, however, depends on the ratio between the thickness of the layer involved in the

surficial adsorption and the total thickness of the metal-oxide coating. The estimation of this ratio will be explained in the next section.

In the simulations the surficially adsorbed phosphorus was used as an initial condition. This was calculated from the Langmuir adsorption isotherm and the measured concentrations after 7 days, assuming that at this time the fast surficial adsorption step had reached equilibrium.

This is, however, an arbitrary choice, since no clear distinction in time can be made between the fast initial step and the slow step.

Total sorption capacity

The amount of iron extracted from the sand in 0.1 M HCl was 71 μg g^{-1} dry matter ($n = 4$, $\sigma = 6$

Table 2. Sensitivity analysis for simulation of aerated column experiment. C_{sim} is simulated concentration at $t = 69$ hr. Measured concentration at $t = 69$ hr was 0.74 mg P/l.

Parameter	$D_{dis, z=0}$ 10^{-4} m^2 d^{-1}	K m^3 g^{-1}	X_m μg g^{-1}	k_s m^3 g^{-1} hr^{-1} $* 10^6$	C_{sim} mg P l^{-1}
A (original)	0	0.57	101	2.59	1.53
B	0	0.57	126	2.59	1.48
C	0	1.03	101	2.59	1.43
D	0	1.03	126	5.18	1.36
E (optimal)	4.5	0.57	101	2.59	0.88

μg g^{-1} dry matter) and comparable to the oxalate-extractable amount (75 ± 24 μg g^{-1} dry matter, $n = 9$). The 0.1 M HCl extractable Al was 323 μg g^{-1} dry matter ($n = 15$, $\sigma = 18$ μg g^{-1} dry matter).

Lijklema (1980) found in homogeneous precipitation after addition of an aluminum solution to a phosphate solution a maximal molar ratio P/Al of 1.0 (pH 7). When assuming that the homogeneous sorption capacity of Al is twice that of Fe (Lijklema, 1991), it can be concluded that Al is the main P-sorbing element in the sediment considered (for about 95% of the total sorption capacity by Fe and Al). A maximal P/Al ratio of 1.0 and a maximal P/Fe ratio of 0.5 yields a total binding capacity of 391 μg P g^{-1} dry matter (371 μg g^{-1} dry matter Al-P and 20 μg g^{-1} dry matter Fe-P), which is 3.87 times the estimated maximal surficial sorption capacity X_m.

In Fig. 6 the amounts of P sorbed by the acid extracted and thereafter homogeneously precipitated Al-OH-phosphates are plotted versus the equilibrium concentrations and compared with the surficial Langmuir-adsorption isotherm. The plain sand without the metal-oxide coating (series 2) did not adsorb any phosphate and therefore is not displayed. The separated supernatant shows a larger P-sorption than the suspension that still contains sand grains. This may be caused by occupation of a part of the available sorption sites by the silicate surface. In the pure supernatant without phosphate addition the precipitation of the Al-hydroxides seems to be hampered. After three days only 60% of the dissolved Al was precipitated again. In samples with phosphate addition and/or in the presence of sand grains all Al was precipitated again.

The total sorption capacity of the supernatant + solid is about 3.3 times the surficial adsorption, and is slightly smaller than the potential sorption capacity calculated from the extractable Al + Fe content. The number of prints is insufficient to construct a 'total sorption isotherm', which displays the amount of sorbed P versus the equilibrium concentration of the slow step.

Simulation of vertical concentration profiles in the porewater

Figure 7 displays the concentrations of dissolved phosphorus in the water of the sand ditch with the highest P-loading. The monthly fluctuations result from the dosages. In the winter of 1990–1991 (days 550–650) no dosages could be given due to an ice cover. In Fig. 8 the simulated vertical profile of dissolved phosphorus in the interstitial water 660 days after the start of the dosages in May 1989, is fitted to the measured profile- $D_{\text{dis}, z=0}$ has been assigned values of 0, $1.5.10^{-3}$ and $1.8.10^{-3}$ m^2 d^{-1}. An average value of $D_{\text{dis}, z=0}$ of $1.8.10^{-3}$ m^2 d^{-1} gives a good agree-

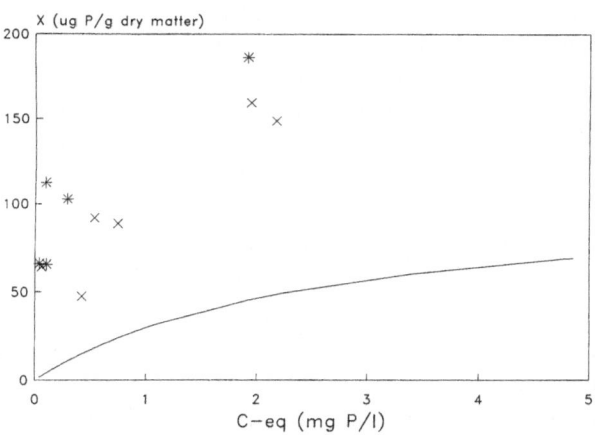

Fig. 6. Sorption of series 1 and 3 after three days in precipitation experiments compared to surficial adsorption isotherm. (———— = surficial adsorption, X = series 1, $*$ = series 3).

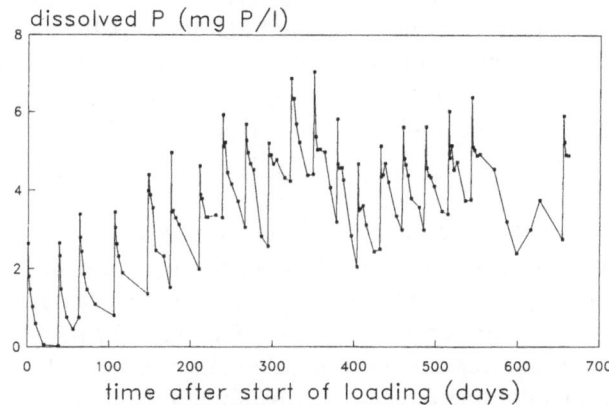

Fig. 7. Dissolved phosphorus concentrations in the ditch.

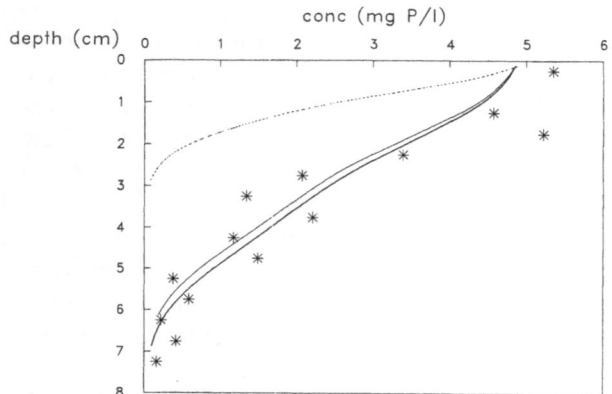

Fig. 8. Measured and simulated vertical concentration profile of phosphorus in the porewater after 660 days. (* = measured, – – – = simulated, $D_{dis, z = 0}$, thin solid line = simulated, $D_{dis, z = 0} = 1.5 . 10^{-3}$ m^2 d^{-1}, thick solid line = simulated, $D_{dis, z = 0} = 1.8 . 10^{-3}$ m^2 d^{-1}).

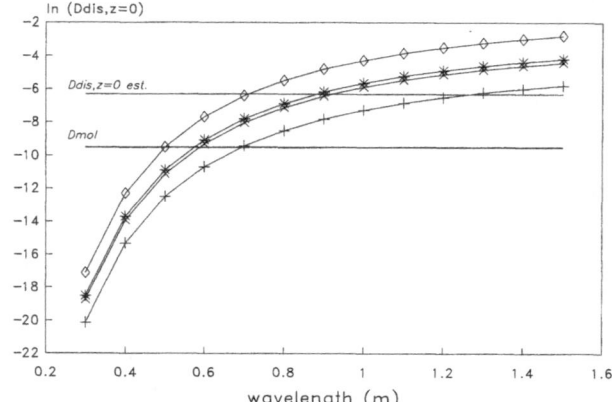

Fig. 9. In $(D_{dis, z = 0})$, as calculated according to Rutgers van der Loeff (1981), plotted versus wave length and compared to D_{mol} and $D_{dis, z = 0}$ estimated ($1.8 . 10^{-3}$ m^2 d^{-1}). (◆: $H = 10$ cm, $d = 1.5$ mm, *: $H = 5$ cm, $d = 1.5$ mm, X: $H = 10$ cm, $d = 1$ mm, +: $H = 5$ cm, $d = 1$ mm).

ment. This is about 3.9 times the value found for the aerated column experiment and indicates that on the average pressure gradients in the ditch are larger than in the column experiment. The simulated porewater profile at $D_{dis, z = 0} = 0$ shows bad agreement. The attenuation constant, k of equation (4), has been assigned a value of 50 m^{-1}. The sensitivity to k is quite low.

The estimated dispersion coefficient $D_{dis, z = 0}$ can be compared to theoretical values calculated as a function of water depth, wave length and amplitude and grain size. Two theories are used, the first (Rutgers van der Loeff, 1981), gives:

$$D_{dis, z = 0} = \pi k_p^2 \, T \, H^2 / 2 \, \lambda^2 \cosh^2 (h \, D) \quad (14)$$

with k_p the permeability coefficient (LT^{-1}), T the wave period, λ the wave length, H the wave height ($= 2 *$ amplitude), h the wave number ($= 2\pi/\lambda$) and D the water depth. k_p can be related to the grain size according to (Harrison, 1983):

$$k_p = c \, d^2 , \quad (15)$$

with d the grain size and c a constant. c has a value of $1.84 . 10^4$ (m s)$^{-1}$ for a porosity of 0.4 and at 20 °C. In Fig. 9 ln $D_{dis, z = 0}$ is plotted versus wave length for two values of H and d. The thick solid line represents molecular diffusion and the thin solid line the $D_{dis, z = 0}$ estimated from fitting to the measured phosphate porewater concentra-

tion profile (see Fig. 8). It is clear that $D_{dis, z = 0}$ strongly depends on wave length as well as grain size and wave amplitude.

In the second theory, Harrison (1983) used the term 'mechanical dispersion' for wave-induced dispersion in the sediment. This dispersion becomes significant when the Peclet number Pe ($= vd/D_{mol}$) $\gg 1$, with v the resulting velocity in the sediment. The mechanical dispersion $\alpha_{z = 0}$ ($= vd = $ Pe D_{mol}), can be calculated from

$$\alpha_{z = 0} = 0.5 \, c \, d^3 \, h \, H / \cosh (h \, D) . \quad (16)$$

The vertical component usually is about 10% of the total, and is assumed to represent $D_{dis, z = 0}$. Figure 10 presents ln $\alpha_{z = 0}$ plotted versus wave length. It again strongly depends on wave-length, grain size and wave amplitude, but exhibits higher values than calculated according to Rutgers van der Loeff (Fig. 9).

The wave length and wave amplitude required for a value of $D_{dis, z = 0}$ of $1.8 . 10^{-3}$ m^2 d^{-1} are in the range of 0.6–1.0 m and 0.025–0.05 m respectively. Visual observations showed that these values are on the average lower. $D_{dis, z = 0}$ estimated from fitting the simulated vertical phosphate profile in the porewater to measured data is higher than theoretical values derived from Rutgers van der Loeff (1981) and Harrison (1983).

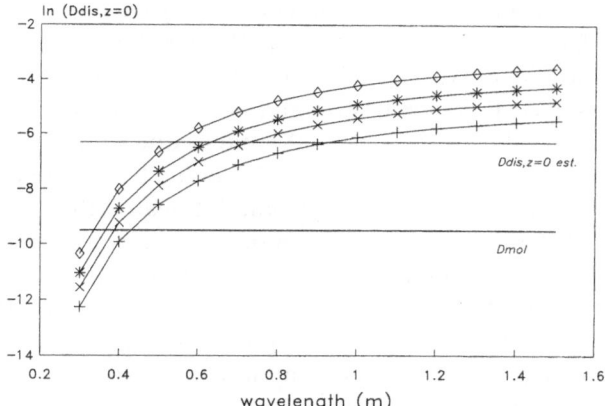

Fig. 10. In ($D_{dis, z=0}$), as calculated from Harrison (1983), plotted versus wave length and compared to D_{mol} and $D_{dis, z=0}$ estimated ($1.8.10^{-3}$ m^2 d^{-1}. (◆: $H = 10$ cm, $d = 1.5$ mm, *: $H = 5$ cm, $d = 1.5$ mm, X: $H = 10$ cm, $d = 1$ mm, + : $H = 5$ cm, $d = 1$ mm).

Due to strong non-linearity of the behaviour of the dispersion coefficient the variation in time will be considerable, but when considering longer periods the use of an average value seems, at least for the moment, justified.

Furthermore, the simulation indicated that after 660 days about 60% of the sorbed phosphorus in the sediment of the ditch had diffused into the interior of the aluminumhydroxide layer.

Conclusions and discussion

– Mixing of the overlying water has been demonstrated to enhance the flux of phosphate across the sediment-water interface. In column uptake experiments without mixing the overlying water, the simulation matched the data, when the dispersion coefficient at the sediment-water interface was about 65% of the molecular diffusion coefficient. Part of this 65% may result from underestimation of the tortuosity factor θ^{-2} in the top few millimeters of the sediment, which have the highest porosity (≈ 0.7) and play an important role in short-term column uptake experiments. From Ullman & Aller (1982) it can be calculated that θ^{-2} can be at maximum 0.89 in sandy sediments with a porosity of 0.7. In the simulations a constant value for θ^{-2} of 0.7 has been used.

Mixing by aeration strongly increased the initial fluxes. A sensitivity analysis for all parameters in the model showed that this could only be attributed to an enhanced effective diffusion coefficient. A dispersion coefficient that was one order of magnitude larger than the molecular diffusion coefficient gave the best fit.

The attenuation of dispersion in the sediment is described as an exponential decrease, with a relatively low sensibility of the model for the attenuation constant k. A k-value of 75 m^{-1} generates an intensity of pressure gradient induced mixing equal to molecular diffusion at a sediment depth of 6 cm. Hesslein (1980) found enhanced mixing rates of porewater up to a sediment depth of 10 cm at overlying water depths of 0.75 and 3.85 m without mixing of the solid phase of the sediments. In stagnant shallow waters the effect of wind action is an enhanced effective diffusion coefficient in the sediment. After enhanced external loading its effect is an increased penetration rate of phosphorus into the sediment and, thereby, an increased retention in the sediment.

– In the simulation of the phosphate fluxes across the sediment-water interface over a period of 660 days, an average dispersion coefficient, corrected for tortuosity, of $1.8.10^{-3}$ m^2 d^{-1} has been used and is satisfactory. This is about 35 times the molecular diffusion coefficient, corrected for tortuosity. Comparison with theoretical values calculated from wave characteristics indicated that the enhanced effective diffusion coefficient in the sediment could, for a substantial part, be attributed to wind induced pressure gradients. In shallow systems like the ditches, pressure gradients induced by congestion also might contribute to this enhancement.

When considering shorter time scales variations in wind velocity cannot be neglected. Wind velocity affects wave heights, which affect the pressure gradients in the sediment, and, as a result of that, affect D_{dis}. The attenuation constant of dispersion in the sediment, k, has been assumed to be constant over depth. It might increase slightly with depth because of decreasing porosity.

– A subsequent slow uptake process causes

immobilization of a considerable part of the externally supplied phosphorus. This slow process can well be described as a diffusion process in which adsorbed P penetrates into a coating of Al-hydroxides. In the situation considered, with aluminum as the main P-sorbing compound, the sensitivity for changing redox conditions is low, in contrast to sediments where iron is the dominating adsorbate. This allows a permanent storage of phosphate, gradually penetrating into deeper layers.

Considering the time scale of the solid phase diffusion process and the reversibility of the process suggested by the model, the release of phosphorus from the sediment after a reduction of the external loading with nutrients will be very long-lasting (see also Barrow, 1983). However, no direct evidence for the reversibility of the slow process has yet been obtained, and it may display other kinetics, although the model used assumes complete reversibility.

– Dissolution of the Al-hydroxide coating and subsequent precipitation in the presence of phosphate yielded a total sorption capacity that is about 3.3 times the maximal surficial adsorption. It may be that the polymeric structures of the precipitated Al-hydroxo-phosphates are still partly a function of the initial conditions in the experiment, but comparison with data from Lijklema (1980) showed a reasonable correspondence. The observation that precipitation of the dissolved Al-hydroxide coating in the presence of the sand particles sorbed less P than in the absence of the sand grains, suggests that the precipitation in the first case was not completely homogeneous. Bolan *et al.* (1985) found similar results after precipitation of Al- and Fe-hydroxides in the presence and absence of kaolinite.

Mineralization in the sediment has been neglected in the simulation of the vertical dissolved phosphorus profile in the interstitial water. This is based on the visual observation that a layer of benthic algae and dead organic matter on top of the sand in the ditch with the highest load does not mix into the sand to any appreciable extent. It may have caused variations in the dissolved phosphorus concentrations within this organic layer on top of the sediment, thereby affecting the flux. The living fraction in the benthic layer is continuously exposed to a relatively high dissolved P concentration and hence luxury uptake and other variations in P-content due to loading are small. The total P-content in the benthic material per m^2 is at its maximum towards the end of the simulated period, more or less equal to a monthly loading ($1 g P m^{-2}$). Besides this, the effects of luxury uptake have a short time-constant as compared to the dosing interval. Conversely, mineralization in the benthic layer continuously may produce dissolved P and thus maintain a gradient at the interface, so that PO_4-concentrations at $z = 0$ are underestimated, which results in overestimation of $D_{dis, z = 0}$. However, again the relative change induced by this process is small compared to the ambient concentration and the mixing rate will attenuate such concentration gradients. Finally, uptake and mineralization have opposite effects. The total error caused by assuming the overlying water concentrations as representative for the concentration at $z = 0$ will be small.

Acknowledgements

Martin Seidl of The Winand Staring Centre for Integrated Land, Soil and Water Research, Wageningen, the Netherlands, kindly supplied data on the vertical concentration profile of dissolved phosphorus in the interstitial water. Rein van Eck, Department of Soil Science and Plant Nutrition, Agricultural University, Wageningen, measured the Aluminum concentrations. The critical comments and suggestions of Wim van Raaphorst, Netherlands Institute for Sea Research, Bernie Boudreau and an anonymous reviewer are gratefully acknowledged.

References

Ahlgren, I., 1977. Role of sediments in the process of recovery of a eutrophicated lake. In: Golterman, H. L. (ed.), Interactions between sediments and freshwater. Dr W. Junk Publishers, The Hague: 372–377.

Barrow, N. J. & T. C. Shaw, 1975. The slow reactions between soils and anions: 2. Effect of time and temperature on the decrease in phosphate concentration in the soil solution. Soil Science 119: 167–177.

Barrow, N. J., 1983. A mechanistic model for describing the sorption and desorption of phosphate by soil. J. Soil Sci. 34: 733–750.

Berner, R. A., 1980. Early diagenesis, a theoretical approach. Princeton Univ. Press.

Bolan, N. S., N. J. Barrow & A. M. Posner, 1985. Describing the effect of time on sorption of phosphate by iron and aluminum hydroxides. J. Soil Sci. 36: 187–197.

Booij, K., 1989. Exchange of solutes between sediments and water. PhD Thesis, University of Groningen.

Brinkman, A. G. & W. van Raaphorst, 1986. De fosfaathuishouding van het Veluwemeer. PhD Thesis, Technical University Twente.

Brinkman, A. G., W. van Raaphorst, L. Lijklema & G. van Straten, 1987. Enkele experimentele technieken bij de bestudering van fosfaatuitwisselingsprocessen tussen meersediment en oppervlaktewater. H$_2$O 20: 664–668.

Draper N. R. & H. Smith, 1981. Applied regression analysis, 2nd edition. John Wiley & Sons, New York, pp. 458–474.

Golterman, H. L., 1977. Sediments as a source of phosphate for algal growth. In: Golterman, H. L. (ed.), Interactions between sediments and freshwater, Dr W. Junk Publishers, The Hague: 286–293.

Harrison, W. D., D. Musgrave and W. S. Reeburgh, 1983. A waveinduced transport process in marine sediments. J. geophys. Res. 88: 7617–7622.

Hesslein, R. H., 1980. *In situ* measurements of pore water diffusion coefficients using tritiated water. Can. J. Fish. aquat. Sci. 37: 545–551.

Kamp-Nielsen, L., H. Mejer & S. E. Jørgensen, 1982. Modelling the influence of bioturbation on the vertical distribution of sedimentary phosphorus in Lake Esrom. In P. G.

Sly (ed.), Sediment/Freshwater Interaction. Developments in Hydrobiology 9. Dr W. Junk Publishers, The Hague: 197–206. Reprinted from Hydrobiologia 91/92.

Lijklema, L., 1980. Interaction of orthophosphate with iron-(III) and aluminum hydroxides. Envir. Sci. Technol. 14: 537–540.

Lijklema, L., 1986. Phosphorus loading in sediments and internal loading. Hydrobiol. Bull. 20: 213–224.

Lijklema, L., 1991. Functional models describing fluxes of phosphorus across the sediment-water interface. In publ. (this workshop).

McDuff, R. E. & R. A. Ellis, 1979. Determining diffusion coefficients in marine sediments: a laboratory study of the validity of resistivity techniques, Am. J. Sci. 279: 666–675.

Murphy, J. & J. P. Riley, 1962. A modified single solution method for the determination of phosphate in natural waters. Anal. Chim. Acta 27: 31–36.

Novozamsky, I., R. van Eck, V. J. G. Houba & J. J. van der Lee, 1986. Use of inductively coupled plasma atomic emission spectrometry for determination of iron, aluminium and phosphorus in Tamm's soil extracts. Neth. J. Agric. Sci. 34: 185–191.

Rutgers van der Loeff, M. M., 1981. Wave effects on sediment-water exchange in a submerged sand bed. Neth. J. Sea Res. 15: 100–112.

Ullman, W. J. & R. C. Aller, 1982. Diffusion coefficients in nearshore marine sediments. Limnol. Oceanogr 27: 552–556.

Van der Zee, S. E. A. T. M., 1988. Transport of reactive contaminants in heterogeneous soil systems. PhD Thesis, Agricultural University Wageningen: pp. 35–64.

Van Riemsdijk, W. H., L. J. M. Boumans & F. A. M. de Haan, 1984. Phosphate sorption by soils I; A diffusion-precipitation model for the reaction of phosphate with metal-oxides in soils. Soil Sci. Soc. Am. J. 48: 537–540.

Hydrobiologia **253**: 263–274, 1993.
P.C.M. Boers, T.E. Cappenberg & W. van Raaphorst (eds),
Proceedings of the Third International Workshop on Phosphorus in Sediments.
© 1993 *Kluwer Academic Publishers.*

In situ phosphorus release experiments in the Warnow River (Mecklenburg, northern Germany)

Andreas Kleeberg[1] & Günter Schlungbaum
Rostock University, Department of Biology, Freiligrathstraße 7/8, Rostock DO-2500, Germany; [1] *Present address: Humboldt-University of Berlin, Department of Biology, Institute of Ecology, Luisenstr. 53, Berlin DO-1040, Germany*

Key words: River sediment, phosphorus release, phytoplankton collapse, pH, nitrate, oxygen

Abstract

Results are presented of *in situ* benthic phosphorus release experiments in an undercut bank of an impounded river. Due to high sedimentation of phytoplankton biomass high oxygen consumption rates between 259.4 and 947.0 mg O_2 m^{-2} d^{-1} developed, leading to almost anaerobic conditions and phosphorus releases between 175.2 and 236.3 mgP m^{-2} d^{-1} over a period of 18 days.

In a second series of experiments the water column overlying the sediment was aerated, resulting in much lower P release rates (1.1 to 32.9 mgP m^{-2} d^{-1}) over a period of 30 days. The influence of pH and nitrate was studied by adjusting pH and adding NO_3^- to the overlying water. Increasing pH positively affected P release rates and enhanced NO_3^- levels led to an increase of benthic P release, too.

Introduction

Although it is clear that dynamics of phosphorus in small rivers are influenced by land use, hydrology and soil geochemistry of the catchment area, climatic factors, interactions with biota, as well as with river sediments (Prairie & Kalff, 1988), the bihaviour mechanisms of phosphorus binding and mobilization processes at the sediment-water interface is not well understood. Maue (1989) reviewed the nutrient releases from sediments of running waters. However, the role of aerobic phosphorus release in impounded running waters is still largely unknown (Steinberg, 1989). For instance, in the Warnow soluble reactive phosphorus concentrations (SRP) increase with annual regularity from early spring to late summer (Kleeberg, 1992). What are the reasons for this increase? The underlying mechanisms are, how-

ever not quantified and may be related to benthic P releases.

There is evidence from other studies (e.g. Ripl & Lindmark, 1979), that phytoplankton dynamics may have a decisive influence in stimulating sedimentary P release. During the blooming of phytoplankton high pH values can develop, while sedimentation of algal material at the end of blooming can cause oxygen depletion in the sediment and the overlying water. P release can be enhanced at low concentrations of dissolved oxygen (Fillos & Molof, 1972; Fillos & Swanson, 1975; Bates & Neafus, 1980; Holdren & Armstrong, 1980; Nixon *et al.*, 1980 and Mawson *et al.*, 1980; Cerco, 1989) and high temperature (Hale, 1975; Fillos & Swanson, 1975; Nixon *et al.*, 1980; Boynten *et al.*, 1980; Sondergaard, 1989). High pH (pH > 8) in overlying water can also enhance SRP-release rates from sediments under aerobic conditions (Andersen, 1975). Nitrate depletion due to algal growth may also affect P-releases. Van Liere *et al.* (1983) found that in a continuous phosphorus release reactor phytoplankton growth can lead to nitrate consumption, thus lowering the sedimentary redox potential to a level were Fe(III) is reduced to Fe(II) and Fe(II) bound phosphorus may be mobilized. A continuous supply of nitrate is supposed to keep the redox potential high and lock phosphorus in the sediment (Pettersson & Boström, 1982). On the other hand, especially in aerobic sediments of shallow, eutrophic waters in which seasonal nitrate depletion occurs, nitrate is believed to enhance the phosphate release by stimulating the microbial mineralization (Boström *et al.*, 1982; Boström *et al.*, 1988).

In the present study *in situ* measurements are performed to determine the range of phosphorus release from river Warnow sediment and to assess the effects of settling phytoplankton, dissolved oxygen, temperature, pH and nitrate on the mobilization of sedimentary phosphorus.

Study area

The Warnow River rises at an altitude of 65 m, and flows for 161 km (mean depth 2.5 m), reaching the Baltic Sea east of Rostock in Northern Germany (Fig. 1). The river has a catchment area of 3230 km^2 with about 200 tributaries which is intensively used by agriculture. The river is divided by a lock nearby Rostock into the brackish, strongly enlarged lower part, the Unterwarnow, and the 148 km long Oberwarnow.

Of special interest for the present investigations is the 40 km long impounded part (Bützow – lock, Fig. 1) with a difference in height of only 0.19 m (= 0.005 ppt). The current velocity is 0.05 to 0.10 m s^{-1} and the mean monthly runoff amounts to 17.96 m^3 s^{-1} (1975–1990).

Due to high nutrient loadings to the Oberwarnow, seston concentrations and phytoplankton biomass, measured as Chlorophyll-*a*, are high. The minimal and maximal values of 1990 were

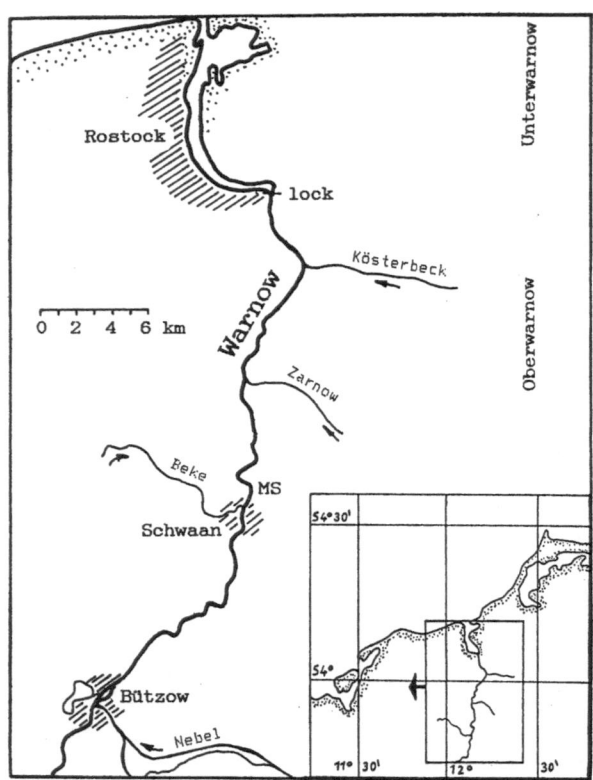

Fig. 1. Geographical location and the impounded part of the River Warnow.

19.2 and 50.0 mg d.w. l^{-1} (dry weight) and 9.6 and 205.3 µg Chl*a* l^{-1}, respectively.

Material and methods

Experiments on P release

A measuring station (MS) for water quality parameters is situated at river km 127 in a curve of the river. The measuring instruments were situated in the littoral zone of the undercut bank (Fig. 2).

Four glass tubes (height 1 m, diameter 0.2 m, open on top) were stuck 0.1 m in to the sediment, the top only a few centimeters above the water.

Two different series of experiments were carried out. In the first, running from 30/04/90 to 17/05/90 (18 days, phytoplankton collapse), the 4

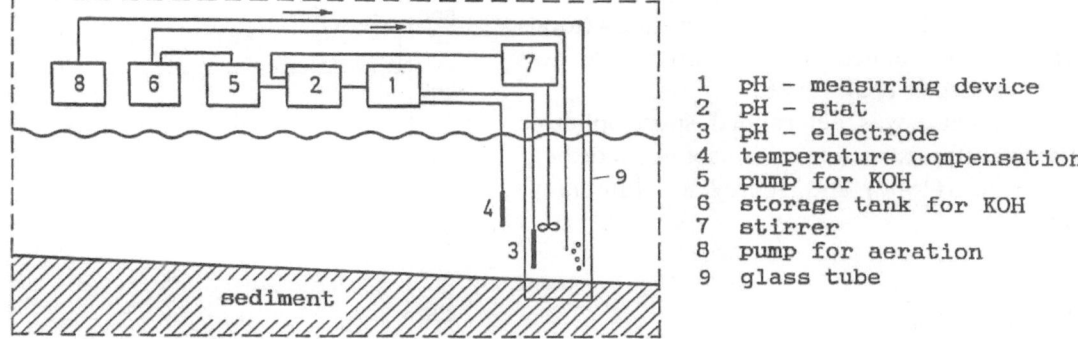

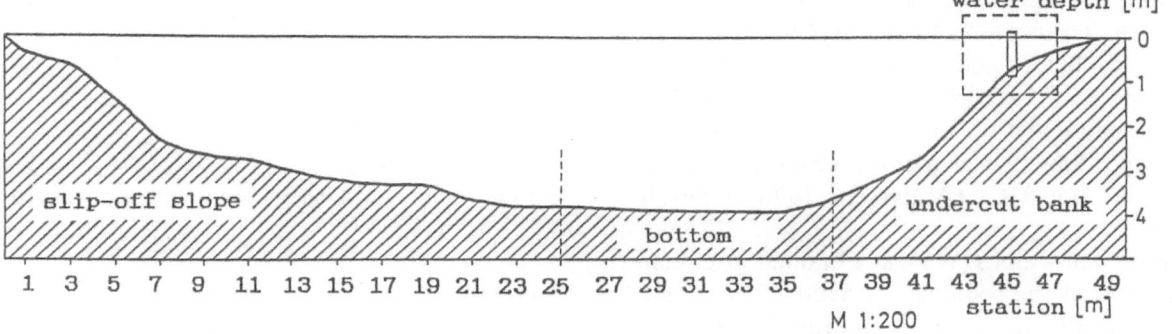

1	pH – measuring device
2	pH – stat
3	pH – electrode
4	temperature compensation
5	pump for KOH
6	storage tank for KOH
7	stirrer
8	pump for aeration
9	glass tube

Fig. 2. River cross section and locality of release experiments including equipment used (situation 12/03/1990).

tubes were not aerated or treated by addition of chemicals. In the second, running from 12/06/90 to 11/07/90 (30 days), nitrate was added to tube 1 and 2, and the pH of tube 3 was manipulated. Tube 4 was used as control. All 4 tubes were aerated and stirred gently. The average September/October nitrate concentrations at the MS are *ca.* 0.8 mg NO_3-N l^{-1} (period 1975–1989). We maintained an equivalent concentration in tubes 1 and 2 during the second experiment stirring in a few milliliters of a molar KNO_3 solution. The pH was manipulated according to Sordyl (1984) using a pH-measuring device with temperature compensation interconnected with pH-stat. The pH value was maintained by adding a 0.1 N KOH solution whenever the measured value decreased below the desired levels.

The gross phosphorus release (*i.e.* including the phosphorus portion liberated from algogenic material) and O_2 consumption rates (*i.e.* differences in the O_2 concentration in the overlying water after tube installment) were calculated using

the method described by Helbig *et al.* (1981). Every day 2 liters of water were removed from the tubes and were replaced by the same amount of nutrient poorer Warnow water. This intermittent removal reduces a build-up of high phosphorus concentrations in the overlying water, which otherwise would reduce the release rates.

Water chemistry

Samples from the river water were taken with a 21 Ruttner-reversing water bottle just below the water surface and directly above the sediment. Samples out of the tubes were taken with P.V.C. bottles by hand. Prior to chemical analysis, an aliquot was filtered (0.45 μm).

Concentrations of nitrate, dissolved oxygen (DO) and total phosphorus (TP) were determined according to Rohde & Nehring (1979). For the determination of SRP the molybdenum blue method (Murphy & Riley, 1962) in the modification of Vogler (1975) for flow-stream automats was used. The H_2S- and TFe-concentration (total iron) were determined with the procedure described by Legler *et al.* (1986).

Chlorophyll-*a* was determined spectrophoto-metrically after extraction with aceton according to Lorenzen (1967) and Jeffrey & Humphrey (1975).

Sediment chemistry

Sediment samples were taken with a sediment corer (similar to a KAJAK-corer) (Nausch, 1981) some days before and immediately after the tube experiments. Total phosphorus content of the mixed sediment (0–3 cm layer) was determined after digestion of the ignition residue (550 °C, 3 h) with hot HCl (Andersen, 1976; Nausch, 1981).

Interstitial water (IW) was obtained after centrifugation of the sediment at 5500 g for 15 minutes at *in situ* temperature of the sediment in closed centrifugation tubes immediately after sampling. All determinations were done in duplicates.

Results

Phytoplankton collapse and oxygen depletion

The installation of tubes without aeration or other treatments resulted in an expected stagnative phase. The Chl-*a* content decreased rapidly (30/04/90 = day 1, 200 μg l^{-1} = 100%) in the river itself by 56.4, and in tubes 1 to 4 by 58.1, 63.6, 80.1 and 31.2% at day 4. After the fourth day the phytoplankton growth started again in tubes 2 and 3, but not in the Warnow and tube 1 and 4 (Fig. 3).

The supply of easily degradable organic material via sedimentation of the algae (mainly diatoms) directly after the collapse (day 1 to 4) resulted in low dissolved oxygen concentrations (DO). Oxygen consumption was probably also favoured by an increase in the water temperature from 13.1 to 19.6 °C at the same time (Fig. 4).

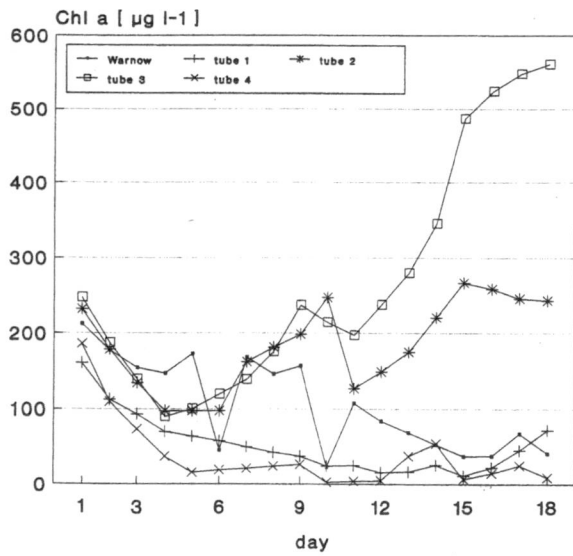

Fig. 3. Course of Chla concentration (μg l^{-1}) in the River Warnow in tubes 1 to 4, respectively, during the observation period 30/04/–17/05/1990.

Gross oxygen consumption rates were estimated at 947.0, 764.2, 936.4 and 259.4 mg O_2 m^{-2} d^{-1} for tube 1 to 4, respectively.

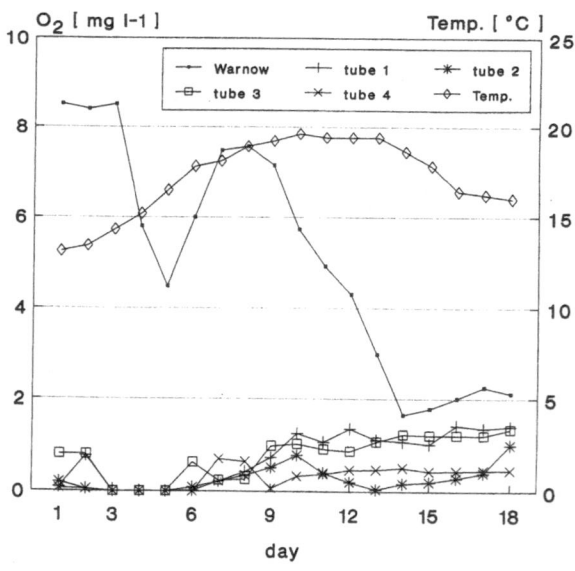

Fig. 4. Course of the concentration of dissolved oxygen (mg l^{-1}) and the water temperature (°C) in the River Warnow in tubes 1 to 4, respectively, during the observation period 30/04/–17/05/1990.

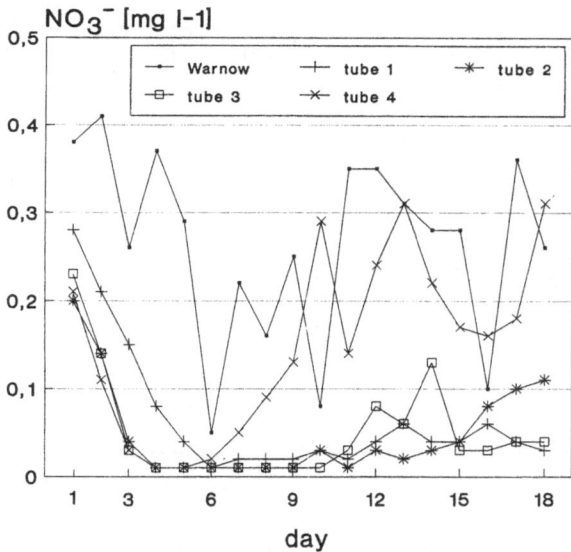

Fig. 5. Course of the concentration of nitrate (mg l⁻¹) in the River Warnow and tube 1 to 4, respectively, during the observation period 30/04/–17/05/1990.

The nitrate concentrations were decreasing rapidly within the first 4 days inside the tubes and kept constant around zero (except in tube 4 where the concentration rose again) in contrast to a fluctuating Warnow nitrate concentration (Fig. 5).

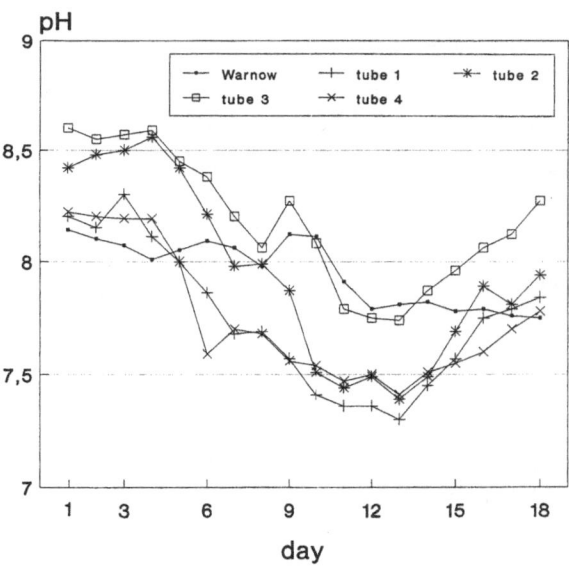

Fig. 6. Course of pH in the River Warnow and in tubes 1 to 4, respectively, during the observation period 30/04/–17/05/1990.

At the same time the pH value of the enclosed and river water decreased from the initial value around 8.5 towards around 7.5 at day 12 and then increased again (Fig. 6).

All conditions together induced a considerable uniform P-release in all tubes. Phosphorus concentration increased drastically when reaching 0.1 to 1.1 mg O_2 l⁻¹ just above the sediment at day 1–3. Hydrogen sulfide remained below the detection limit (*i.e.* 0.1 mg l⁻¹). The increase of TP concentration was initially slow but from day 7 (nitrate depletion) the concentration increased faster. The release reached its peak on day 16, then slowed down. The TP and SRP concentration outside the tubes increased continuously, but only slightly (Fig. 7). The pattern for release of SRP in the tubes resembled that of TP during the anaerobic and anoxic phase, respectively. The maximum SRP concentration was reached on day 15/16. The average phosphorus release rates in tube 1 to 4 were 99.1, 175.2, 236.3 and 201.3 mg P m⁻² d – 1, respectively. Regression analysis of the TP and SRP data indicated that DO has with 61% the greater influence compared to the water temperature (T_w) with 39%. The relations are expressed in the following equations:

$$SRP = -36.781 + 2.664 * T_w - 35.255 * DO,$$
$$(r = 0.938, n = 34, \alpha = 0.05),$$

$$TP = -42.332 + 3.332 * TW - 48.479 * DO,$$
$$(r = 0.936, n = 34, \alpha = 0.05).$$

During the time of the experiments, DO in the near bottom water of the river itself decreased from 8.52 mg O_2 l⁻¹ (115.9% saturation) at day 1 to 2.11 mg O_2 l⁻¹ (30.6% saturation) at day 18. The minimum was measured at 1.66 mg O_2 l⁻¹ (25.4% saturation) on day 14 (Fig. 4).

The phosphorus contents of the uppermost 3 cm sediment layer of tubes 1 to 4 before and after tube installment are compared with the values found for the Warnow sediment in Table 1.

The amounts of phosphorus released in the tube sediments vary between 2.2 and 18.3% of the initial contents. Outside the tubes TP content increased by 17.6% during the same period.

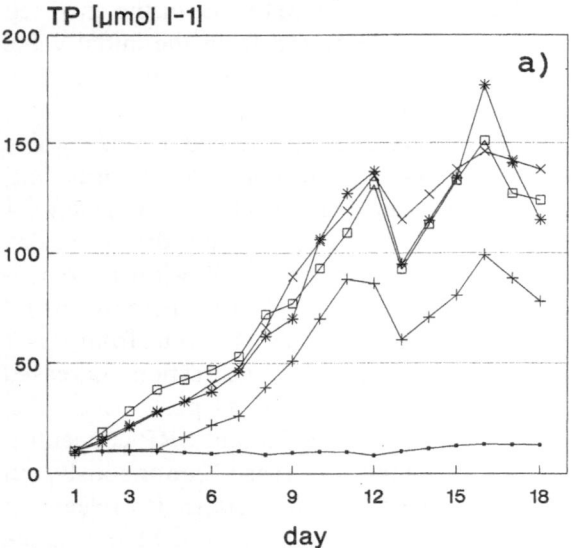

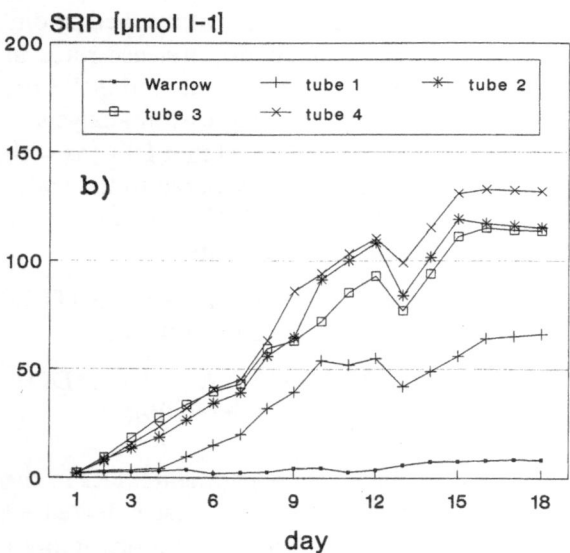

Fig. 7. Course of the concentration (μmol l^{-1}) of total phosphorus (a) and soluble reactive phosphorus (b) during the observation time 30/04–17/05/1990.

Table 1. Changes in concentration of interstitial water (IW) and total phosphorus (TP) during the observation period 30/04–17/05/90 in River Warnow sediment in tubes 1 to 4, respectively.

	TP [μg gdw^{-1}]		IW [μmol l^{-1}]	
	27/04/	17/05/	27/04/	17/05/90
Warnow	500.0	588.0	8.4	10.0
Tube 1	620.0	606.0	14.3	11.8
Tube 2	518.0	423.0	10.1	7.5
Tube 3	700.0	667.0	13.9	12.7
Tube 4	962.0	901.0	16.4	16.2

age TFe release was observed ranging from 480.7 to 927.6 mg Fe m^{-2} d^{-1}. Following the onset of anoxia, the TFe concentration decreased continuously and reached its maximum concentration at the end of the experiment (Fig. 8). The ambient TFe concentration remained constant. It should be noted, that TP, SRP and TFe were released rapidly in spite of low oxygen and nitrate concentrations at the beginning of the experiment indicating local anoxia at the sediment-water interface. Low oxygen concentrations could promote a redox dependent release of iron and phosphorus (Boström *et al.*, 1982).

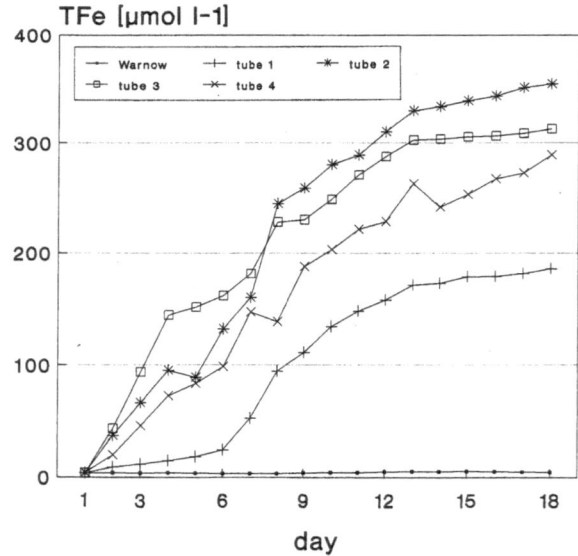

Fig. 8. Course of the concentration (μmol l^{-1}) of total iron during the observation period 30/04/–17/05/90.

A close relation was found between TP and IW-SRP:

$$IW\text{-}SRP = 1.66 + 0.017*TP$$
$$(r = 0.925, \; n = 24, \; \alpha = 0.05),$$

indicating rapid transitions between different phosphorus forms in the sediment.

Parallel with the SRP and TP release an aver-

Table 2. Molar TFe:TP relation in River Warnow sediment in tubes 1 to 4, respectively, during the observation period 30/04–17/05/90.

Days	Warnow	Tube 1	Tube 2	Tube 3	Tube 4
1.– 6.	2.2	5.9	15.5	16.8	14.7
6.–12.	2.5	12.0	18.3	16.9	12.7
12.–18.	2.4	13.7	16.4	15.2	11.9

With the exception of tube 1 (lowest phosphorus and iron release rates) the molar TFe:TP ratio of the sediment remained relatively constant over the incubation period (Table 2). On the average, molar TFe:TP ranges from 12 to 18:1, i.e. very close to ratios observed for P adsorption on iron hydroxides in laboratory experiments (Lijklema, 1977).

pH and nitrate

Because of the falling water level we moved the tubes towards the middle of the river for the second series of release experiments.

At a mean Chl*a* concentration in tubes 1 to 4 of 236.1, 99.2, 410.3 and 15.0 μg Chl*a* l^{-1}· different sedimentation rates are determined. They amounted in the same sequence to 136.1, 60.3, 37.8 and 9.2 μg Chl*a* l^{-1} d^{-1}. Contrary to the first series of experiments the Warnow Chl*a* increased by a rate of 43.8 μg Chl*a* l^{-1} d^{-1}. By a proportionate aeration of the tubes oxygen concentrations remained stable at 6.0 ± 2.7 mg O_2 l^{-1} above the sediment and 0.8 ± 1.7 mg O_2 L^{-1} at the water surface. During the experiments the water temperature increased from 16.2 °C at the beginning of the experiments to a maximum of 20.9 °C on the 20th day, and then decreased down to 18.2 °C at day 30. The pH in tube 3 was enhanced within the first 10 days of experiments and was then maintained at a level of 9.5 ± 0.03 over a period of 20 days (Fig. 9).

A daily addition of nitrate to tube 1 resulted in an average concentration of 0.81 ± 0.29 mg N O_3^--N l^{-1} ($= 0.69$ g m^{-2}) over the 30 days. The nitrate concentration in the river itself was 0.54 ± 0.25 mg N l^{-1} and decreased by

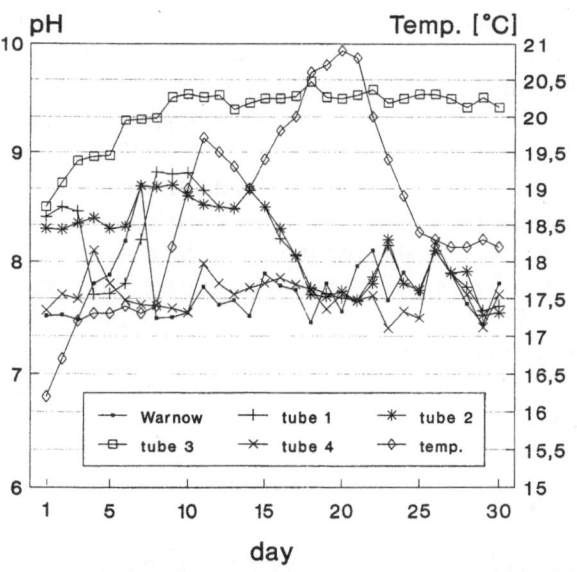

Fig. 9. Course of pH and the water temperature (°C) in the River Warnow in tubes 1 to 4, respectively, during the observation period 12/06/–11/07/1990. Tube 1: adjusted nitrate [0.69 mg NO_3^--N l^{-1}], daily addition. Tube 2: adjusted nitrate [0.98 mg NO_3^--N l^{-1}], two additions. Tube 3: pH controlled. Tube 4: control.

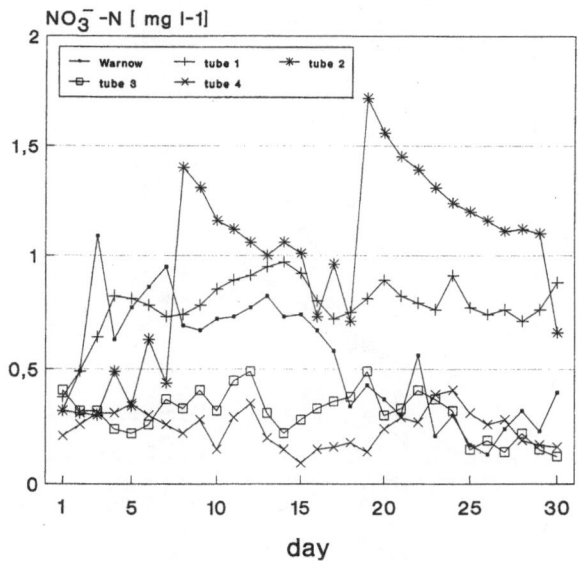

Fig. 10. Course of nitrate concentration (mg l^{-1}) in the River Warnow in tubes 1 to 4, respectively, during the observation period 12/06/–11/07/1990. Tube 1: adjusted nitrate [0.69 mg NO_3^--N l^{-1}], daily addition. Tube 2: adjusted nitrate [0.98 mg NO_3^--N l^{-1}], two additions. Tube 3: pH controlled. Tube 4: control.

0.32 ± 0.21 mg NO_3^--N l^{-1}. With the help of two additions of nitrate (1.0 mg l^{-1}) (firstly on the 6th and secondly on the 18th day of experiment) a mean concentration of 0.98 ± 0.40 mg NO_3^--N

Table 3. Phosphorus release rates (PRR) calculated for tubes 1 to 4, 12/06/–11/07/90. Tube 1: adjusted nitrate [0.69 mg NO_3^--N l^{-1}], daily addition. Tube 2: adjusted nitrate [0.98 mg NO_3^--N l^{-1}], two additions. Tube 3: pH controlled [pH 9.5]. Tube 4: control.

	PRR [mg P m^{-2} d^{-1}]			
	Tube 1	Tube 2	Tube 3	Tube 4
TP	4.0	32.7	106.1	− 0.1
SRP	1.1	32.9	49.4	0.8

l^{-1} (0.83 g m^{-2}) was achieved in tube 2 (Fig. 10).

Following the onset of increasing pH the TP and SRP concentration increased more rapidly in contrast to the tubes 1, 2 and 4, respectively (Fig. 11). The maximum occurred later for TP than for SRP. During the time of regulated pH the phosphorus concentration decreased and reached about half of the maximum concentration of SRP and about one-third of TP at the end of the experiment, respectively. In the tube which was supplied daily with nitrate, the phosphorus concentration increased slowly from the beginning of the experiment. Whereas in tube 2 (two additions of nitrate) the rate of phosphorus concentration increase (TP and SRP as well) was initially slow, but from day 6 the concentration increased faster (Fig. 11). The phosphorus release rates of both tubes are presented in Table 3, too.

Tube 4 (control) was only slightly influenced by settling phytoplankton and increasing water temperature and showed therefore the lowest release rates. The gross phosphorus release rates calculated in the experiment of phytoplankton collapse were much higher than those of the experiments with adjusted pH and nitrate conditions, respectively.

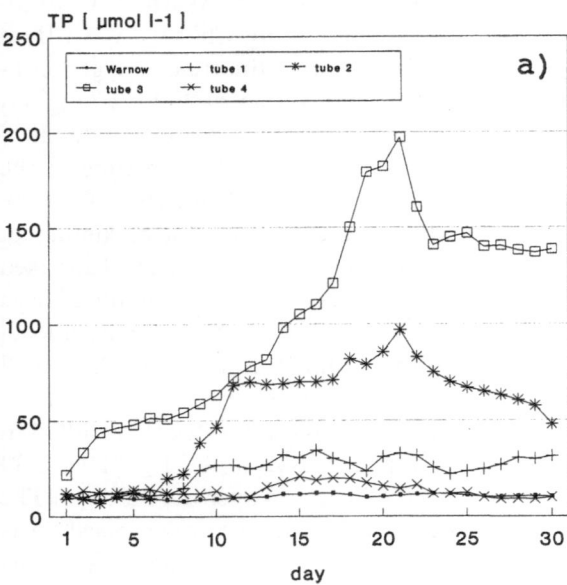

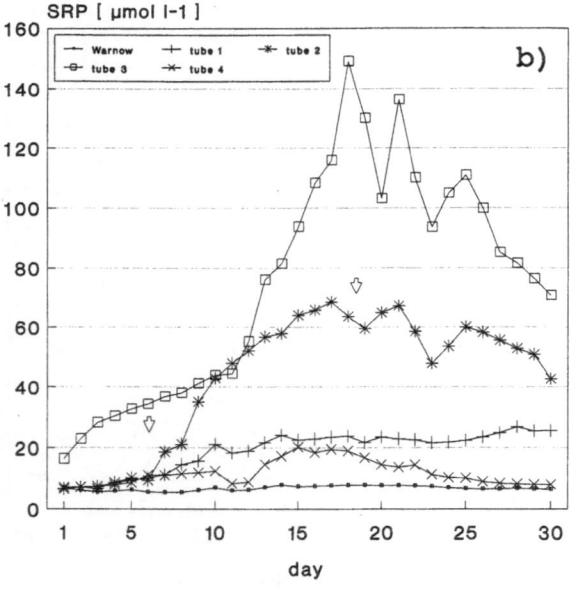

Fig. 11. Course of the concentration (μmol l^{-1}) of total phosphorus (a) and soluble reactive phosphorus (b) during the observation time 12/06/–11/07/1990. Tube 1: adjusted nitrate [0.69 mg NO_3^--N l^{-1}], daily addition. Tube 2: adjusted nitrate [0.98 mg NO_3^--N l^{-1}], two additions (arrows). Tube 3: pH controlled. Tube 4: control.

Discussion

Phytoplankton collapse and oxygen depletion

When a tube is pushed into the sediment, the sedimentary system is interfered with in different ways. Firstly, the tube favours the sedimentation of phytoplankton which enhances the supply of organic matter to the sediment, thus possibly stimulating benthic mineralization of the uppermost sediment layers and affecting the transport of solutes across the sediment-water interface. Sorensen et al. (1990) found a stimulation of decomposition processes, e.g. of the oxygen respiration and of denitrification in the sediment in a lowland river, which may have also led to an additional increase of the pH. But as shown in Fig. 6, the pH was decreasing from the beginning of the experiments, probably due to CO_2 formation during enhanced benthic respiration processes. Decreases in NO_3^- N are attributed mainly to denitrification. In addition, rising temperature as during the observation period, might have led to increased decomposition and release of phosphorus (Psenner, 1984) from the organic bound fraction of the sediment. The temperature dependence has been reported with undisturbed sediment cores by Sondergaard (1989). In these experiments the average release rate determined at 20 °C was about five times higher than the rate obtained at 10 °C. Another portion of the increase in SRP concentration in the water column likely resulted from mineralization of algae material.

Secondly, at the same time the phosphorus consumption by the growing phytoplankton probably caused a higher gradient of SRP between sediment and water.

Thirdly, due to the tube installment the actual in situ near bottom turbulence of running waters is interrupted, thus the supply of oxygen to the sediment was limited. Therefore, altogether the measured gross release rates are considered as the maximum possible.

We are not aware of existing in situ measurements on benthic phosphorus release in running waters and hence the release rates are compared to few literature data. With a mean phosphorus release rate of 178.0 mg P $m^{-2} d^{-1}$ (4 tubes) this phosphorus release is comparable with those of Lake Sobygard (Sondergaard, 1989; Sondergaard et al., 1990). Here the phytoplankton collapse resulted in an average of phosphorus release rates between 100 and 200 mg P $m^{-2} d^{-1}$, although the water column was well oxygenated by about 5 mg $O_2 l^{-1}$. By Helbig et al. (1981) a high phosphorus release rate of 155 mg P $m^{-2} d^{-1}$ was found during the transition from aerobic to anaerobic conditions, whereas at permanent anaerobic conditions the phosphorus release rates remained low between 3.1 and 9.3 mg P $m^{-2} d^{-1}$.

The pore water phosphorus concentrations are much higher than those in the river water, indicating a diffusional flux of phosphorus from the sediment to the overlying water. Thus, the observed phosphorus release is consistent with the pore water gradient of the uppermost 3 cm layer of the four tubes. Regarding the P release associated with sedimentation rate, Carignan & Lean (1991) stated, that regeneration of SRP is not well coupled with organic matter degradation and undefined anoxic P immobilization reactions seem to be taking place in the sediment. However, the relatively constant TFe:TP ratio found for the tubes with the highest release rates is an expression for a consistent, reductive dissolution of 'iron-phosphate-complexes' (Helbig et al., 1981). The microbial processes cause the development of reducing conditions due to oxygen consumption only when a sufficient supply of organic compounds is available. These allow iron oxide hydroxides to be reduced to soluble divalent ions under release of OH^- (Boström et al., 1982; Hoyer et al., 1982). The sorption of phosphorus to iron was not investigated, but the rapidly decreasing DO at the sediment surface may indicate that iron is reduced. Hence, dissolution of the phosphates and humic acids bound in the oxide hydroxide matrix occurs.

pH

The influence of pH on the phosphorus exchange between sediment and water has been discussed by several authors. E.g., Rippey (1977) observed a phosphorus release rate of 10.4 mg P $m^{-2} d^{-1}$ at pH of 10.4, Sanni (1982) found a phosphorus release rate of 20 mg P $m^{-2} d^{-1}$ when pH exceeded 9.8. Phosphorus release rates in a range of 2 (pH 5.9–6.3) to 75 mg P $m^{-2} d^{-1}$ (pH 10.5) were determined by Drake & Heany (1987).

In the Warnow River pH-values around 9 occur mainly in April and August. Therefore the effect of pH on the release rates was tested, but due to rising temperature at the same time, the obtained results are difficult to interpret. Boers & Van Hese (1988) showed that the temperature is the most important factor which controls the P-release in the shallow peaty Loosdrecht lakes.

In our experiments it was found that a higher pH positively affected the P release rates. In tube 3, having the same temperature as the control tube, release rates were substantially higher than in tube 4. Here the phosphorus release was only slightly influenced by temperature, however at the lowest level of phytoplankton sedimentation. These findings indicate that the combination of both factors may be very important factors for the shallow Warnow River.

In tube 3 the total amount of phosphorus released under alkaline conditions indicates a great reservoir of soluble phosphorus in the sediment. Under these conditions phosphorus is released through desorption from Fe and Al-complexes by substitution with hydroxyl ions (Andersen, 1975; Boström et al., 1982; Hoyer et al., 1982). This process is of importance in lakes where Fe, Al or organic-bound phosphorus predominates in the sediments (Andersson & Gahnström, 1985), and probably also in the iron rich (Kleeberg, 1992) Warnow River sediments.

However, the method of KOH addition used may overestimate the effect of increasing pH on benthic P release. Boers (1991) found in laboratory experiments with intact sediment cores that the addition of 0.1 M NaOH resulted in a 10-fold increase in release rate at pH 9.5 compared to the rate at pH 8.3, whereas the CO_2-stripping technique (CO_2-free air) yielded only a doubling of the release rate at pH 9.3. Increasing alkalinity in the water column may affect the pH within the near the sediment-water interface and hence the behaviour of phosphate in the sediment.

NO₃

The effect of nitrate on phosphorus release was tested by lowdose addition of nitrate (KNO_3). As shown in Fig. 6 the phosphorus release in tube 1 and 2 was slightly enhanced in comparison with the control. A relatively large difference in the release rates was recorded between the tube with daily addition and the one with two distinct additions, although the mean NO_3^--concentration was nearly similar (0.69 and 0.83 g m^{-2}) during the observation period. This concentration, however does not give sufficient information concerning the NO_3^- effect on phosphorus release. In most cases, no phosphorus release seems to occur if nitrate concentration remains high, i.e. > 0.5 mg N l^{-1} (Boström et al., 1988). The fact that the addition of nitrate in our experiments did not prevent or reduce substantially the release of phosphorus, although its concentration was always well above 0.5 mg NO_3^--N l^{-1}, favours the concept that the benthic P release depends on the sediment type (Boström & Petterson, 1982) including its biological activity.

So, the amounts of nitrate added were consumed probably by denitrification processes (Fig. 5). An enhanced denitrification is possible at a level $< 50\%$ oxygen saturation (Rheinheimer, 1985). Therefore, it is assumed that increased nitrate concentrations would increase the total mineralization rate by increased denitrification (provided that denitrification replaces other slower mineralization processes), thereby possibly promoting a more rapid transfer of phosphorus from degradable organic matter to the mobilized phosphorus pool.

Reffering to the ambient conditions in the impounded part of the Warnow River, rapid changes of phytoplankton biomass occur every spring. SRP concentration increases immediately after this collapse. Thus, the P release observed under nearly relevant environmental conditions may lead to a better understanding of internal loading and nutrient dynamics in the impounded river.

Conclusions

Both anaerobic (local anoxia) and aerobic release of phosphorus may occur in some sediments of the impounded part of Warnow River. The aerobic P release rates found should be tested in a P mass balance. Environmental conditions, such as a high decomposition rate (microbial and benthic activity) due to a considerable sedimentation (supply of easily degradable organic material) favoured by rising temperature and a high portion of iron-bound sediment phosphorus sensitive to changes in redox potential and pH, are regarded as major factors causing the benthic P release. Therefore, it is important for further *in situ* studies that the oxygen supply to the sediment during the observation time is maintained as close as possible to natural conditions. As far as the pH control used is concerned, a CO_2 regulation technique as proposed by Boers (1991) should be used in further experiments. Low additions of nitrate cannot inhibit the P release of shallow river sediment when the overlying water is oxygenated. Experimental maintenance of nitrate concentrations or additional nitrate loads during the time of seasonal nitrate depletion affects the P release positively.

Finally, it should be stressed that (1) shallow sediments may be important in regulating the trophic status of the impounded part of the river Warnow and (2) these sediments should not be ignored as a potential P source to rivers.

Acknowledgements

The authors wish to express their gratitude to Dr G. Nausch for many stimulating discussions on the subject of this work. Both reviewers are acknowledged for their many stimulating suggestions to improve the original drafts.

References

Andersen, J. M., 1975. Influence of pH on release of phosphorus from lake sediments. Arch. Hydrobiol. 76: 411–419.

Andersen, J. M., 1976. An ignition method for determination of total phosphorus in lake sediments. Wat. Res. 10. 329–331.

Andersson, G. & G. Gahnström, 1985. Effect of pH on release and sorption of dissolved substances in sediment-water microcosms. Ecol. Bull. 37: 301–318.

Boers, P. C. M. & O. Van Hese, 1988. Phosphorus release from the peaty sediments of the Loosdrecht Lakes (The Netherlands). Wat. Res. 22: 355–363.

Boers, P. C. M., 1991. The influence of pH on phosphate release from lake sediments. Wat. Res. 3: 309–311.

Boström, B. & K. Pettersson, 1982. Different pattern of phosphorus release from lake sediment. In P. G. Sly (ed.), Sediment/Freshwater Interaction. Developments in Hydrobiology 9. Dr W. Junk Publishers, The Hague: 415. Reprinted from Hydrobiologia 91/92.

Boström, B., M. Jansson & C. Forsberg, 1982. Phosphorus release from lake sediments. Arch. Hydrobiol. Beih. Ergebn. Limnol. 18: 5–59.

Boström, B., J. M. Andersen, S. Fleischer & M. Jansson, 1988. Exchange of phosphorus across the sediment–water interface. In G. Persson & M. Jansson (eds), Phosphorus in Freshwater Ecosystems. Developments in Hydrobiology 48. Kluwer Academic Publishers, Dordrecht: 229–244. Reprinted from Hydrobiologia 170.

Boynten, W. R., W. M. Kemp & C. G. Osborne, 1980. Sediment water nutrient fluxes in the sediment trap portion of the Patuxent estuary. In V. Kennedy (ed.), Estuarine Perspectives. Academic Press, New York, 93–109.

Carignan, R. & D. R. S. Lean, 1991. Regeneration of dissolved substances in a seasonally anoxic lake: The relative importance of processes occurring in the water column and in the sediments. Limnol. Oceanogr. 36: 643–707.

Cerco, C. F., 1989. Measured and modelled effects of temperature, dissolved oxygen and nutrients on sediment-water nutrient exchange. Hydrobiologia 174: 185–194.

Drake, J. C. & I. Heaney, 1987. Occurence of phosphorus and its potential remobilization in the littoral sediments of a productive English lake. Freshwat. Biol. 17: 513–523.

Fillos, J. & W. Swanson, 1975. The release rate of nutrients from river and lake sediments. J. Wat. Pollut. Cont. Fed. 44: 1032–1042.

Helbig, J., H. Reissig & R. Hölzig, 1981. Bestimmung von Nährstoff- und Metallfreisetzungsraten aus Sedimenten umweltbelasteter Fließgewässerstauhaltungen mit Hilfe von Inkubationsversuchen. Teil 1: Bestimmung von Phosphor- und Eisenfreisetzungsraten. Acta hydrochim. hydrobiol. 9: 203–212.

Hieltjes, A. H. M. & L. Lijklema, 1980. Fractionation of inorganic phosphates in calcareous sediments. J. envir. Qual. 9: 405–407.

Holdren, G. C. & D. E. Armstrong, 1980. Factors affecting phosphorus release from intact sediment cores. Envir. Sci. Technol. 14: 79–87.

Hoyer, O., H. Bernhardt, J. Clasen & A. Wilhelms, 1982. *In situ* studies on the exchange between sediment and water using caissons in the Wahnbach reservoir. Arch. Hydrobiol. Ergebn. Limnol. 18: 79–100.

Jeffrey, S. W. & G. Humphrey, 1975. New spectrophotometric equations for determining chlorophyll *a*, *b*, *c* in higher plants, algae and natural phytoplankton. Biochem. physiol. Pflanzen 167: 191–194.

Kleeberg, A., 1992. Untersuchungen zur Phosphorfreisetzung aus und Phosphorverteilung in Sedimenten der Oberwarnow, PhD Thesis, Rostock University, Rostock, 161 pp.

Legler, Ch., G. Breitig, G. Stepphuhn & V. Vobach, 1986. Ausgewählte Methoden der Wasseruntersuchung, 1., VEB Gustav Fischer Verlag, Jena, 517 pp.

Lijklema, L., 1977. The role of iron in the exchange of phosphate between water and sediments. In H. L. Golterman (ed.), Interaction between sediment and freshwater. Dr W. Junk Publishers, The Hague, 313–317.

Lorenzen, C. J., 1967. Determination of chlorophyll and phaeopigment – Spectrophotometric equation. Limnol. Oceanogr. 12: 343–347.

Maue, G. 1989. Literaturstudie zur Freisetzung von Nährstoffen aus Sedimenten in Fließgewässern. In Stoffbelastung der Fließgewässerbiotope. P. Parey, Hamburg, Berlin: 273–344.

Murphey, J. & J. F. Riley, 1962. A modified single solution method for determination of phosphate in natural waters. Anal. Chim. Acta 27: 31–36.

Nausch, G., 1981. Die Sedimente der Darß-Zingster-Boddengewässer – Zustandsanalyse und Stellung im Phosphorkreislauf. PhD Thesis, Rostock University, Rostock, 143 pp.

Nixon, S., J. R. Kelly, B. N. Furnas, C. A. Oviatt & S. S. Hale, 1980. Phosphorus regeneration and the metabolism of coastal marine bottom communities. In K. R. Tenore & B. C. Coull (eds), Marine Benthic Dynamics. Univ. South Carolina Press, Columbia: 219–242.

Pettersson, K. & B. Boström, 1982. A critical analysis of suggested methods for nitrate treatment of sediments. Vatten 38: 74–82.

Prairie, Y. T. & J. Kalff, 1988. Dissolved phosphorus dynamics in headwater streams. Can. J. Fish. aquat. Sci. 45: 200–209.

Psenner, R., 1984. Phosphorus release patterns from sediments of a meromictic mesotrophic lake (Piburger See, Austria). Verh. int. Ver. Limnol. 22: 219–228.

Rheinheimer, G., 1985. Mikrobiologie der Gewässer, 4. VEB Gustav Fischer Verlag, Jena, 262 pp.

Ripl, W. & G. Lindmark, 1979. The impact of algae and nutrient composition on sediment exchange dynamics. Arch. Hydrobiol. 86: 45–65.

Rippey, B., 1977. The behavior of phosphorus and silicon in undisturbed cores of Lough Neagh sediments. In H. Golterman (ed.), Interactions between Sediments and Freshwater. Dr W. Junk Publishers, The Hague; 349–353.

Rohde, K. H. & D. Nehring, 1979. Ausgewählte Methoden zur Bestimmung von Inhaltsstoffen im Meer- und Brackwasser. Geod. Geoph. Veröff. 27: 1–68.

Sanni, S., 1982. The effect of high pH on phosphorus release. In vitro quantification and in situ verification in lake Arungen. In I. Bergström, J. Kettunen & M. Stenmark (eds), Physical, chemical and biological dynamics in sediment, Proceedings of the 10th Nordic Symposium on Sediment 26: 21–30.

Sondergaard, M., 1989. Phosphorus release from an hypertrophic lake sediment: Experiments with intact sediment cores in a continuous flow system. Arch. Hydrobiol. 116: 45–59.

Sondergaard, M., E. Jeppesen, P. Kristensen & O. Sortkjaer, 1990. Interactions between sediment and water in a shallow and hypertrophic lake sediment: a study on phytoplankton collapses in Lake Sobygaard, Denmark. In P. Biró & J. F. Talling (eds), Trophic Relationships in Inland Waters. Developments in Hydrobiology 53. Kluwer Academic Publishers, Dordrecht: 139–148. Reprinted from Hydrobiologia 191.

Sordyl, H., 1984. Konzentrations- und zeitabhängige Untersuchungen zur Wirkung der Umweltfaktoren pH, NH_3, und Pb^{2+} auf das Blut von Regenbogenforellen (Salmo gairdneri RICH.). PhD Thesis, University Rostock, Rostock, 151 pp.

Sorensen, J., L. P. Nielsen, P. B. Christensen & N. P. Revsbech, 1990. Denitrification and oxygen metabolism in stream sediments. The danish N, P and organic matter research program 1985–1990 report C: on water courses, lakes and marine waters 2: 25–33.

Steinberg, Ch., 1989. Bioverfügbarkeit und Rolle des Phosphors im Gewässer. Münchener Beiträge 43: 1–29.

Teague, K. G., Ch. J. Madden & J. W. Day, 1988. Sediment-water oxygen and nutrient fluxes in a river dominated estuary. Estuaries 11: 1–9.

Vogler, P., 1975. Analysenautomation in Wasserlaboratorien mit flow-stream-Automaten. Acta hydrochim. hydrobiol. 3: 145–158.

Van Liere, L., J. Peters, A. Mutijin & L. R. Mur, 1983. Release of sediment-phosphorus, and the influence of algae growth. Hydrobiol. Bull. 16: 191–200.

Hydrobiologia **253**: 275–280, 1993.
P.C.M. Boers, T.E. Cappenberg & W. van Raaphorst (eds),
Proceedings of the Third International Workshop on Phosphorus in Sediments.
© 1993 *Kluwer Academic Publishers.*

Sulfate control of phosphorus availability in lakes

A test and re-evaluation of Hasler and Einsele's Model

N. F. Caraco, J. J. Cole & G. E. Likens
Institute of Ecosystem studies, The New York Botanical Garden, Millbrook, New York 12545, USA

Key words: phosphorus, iron, sulfate, lakes

Abstract

During summer stratification large amounts of phosphorus (P) accumulate in anoxic bottom waters of many lakes due to release of P from underlying sediments. The availability to phytoplankton of this P is inversely related to the Fe:P ratio in bottom waters. Using data from 51 lakes, we tested the hypothesis that sulfate concentration in lake water may be critical in controlling the Fe:P ratio in anoxic bottom waters. Results showed that Fe:P ratios in bottom waters of lakes were significantly ($p < 0.001$) related to surface water sulfate concentrations. The higher Fe:P ratios in low sulfate systems is due not only to higher iron concentrations in anoxic bottom waters but also to lower P concentrations in anoxic waters. Thus, our results suggest that anthropogenically induced increases in sulfate concentrations of waters (e.g. from fossil fuel burning) may have a double effect on P cycling in lakes. Higher sulfate concentrations can both increase the magnitude of P release from sediments as well as increase the availability of P released from sediments into anoxic bottom waters.

Introduction

When bottom waters of aquatic systems become anoxic there is a tendency for enhanced P release from sediments to these overlying bottom waters (Mortimer, 1941, 1942; Fonselius, 1970). Due in part to this enhanced release, the seasonally accumulated P in anoxic bottom waters can be extremely large, equalling or exceeding the annual loading from the watershed (Lawacz, 1985; Nurnberg, 1987). The availability to phytoplankton production of the P in anoxic bottom waters is, therefore, of critical importance to the trophic status of a water body.

There are two hurdles that phosphorus in anoxic bottom waters must pass before reaching trophogenic surface waters where it can be used by phytoplankton. First, the P must be mixed upward. During summer stratification this obstacle can limit severely the availability of P accumulated in bottom waters (Schindler *et al.*, 1980). In many lakes, however, complete mixing of the water column occurs in the fall. Thus, P accumulated in anoxic bottom waters during the summer stratified period potentially can be made available for phytoplankton growth. The second hurdle which can limit the availability of P accumulated in bottom waters comes into play during this water column turnover. Water column turnover oxidizes iron in bottom waters and the iron oxides which are formed during the vertical mixing process can precipitate P (Lean *et al.*, 1986). Hasler & Einsele (1948) discussed this as an unfortunate process in considering attempts to enhance the fertility of lakes. Presently, this co-precipitation process is considered a fortunate phenomenon as enhanced productivity of lakes is, generally, not desired. From either point of view, however, the factors that control the extent to which P in anoxic bottom waters can escape co-precipitation with iron-oxides are of interest both theoretically and for management considerations.

A proximate control on the availability of P in anoxic bottom waters is the iron to phosphorus ratio in these waters (Baccini, 1985; Lean et al., 1986). In 1948, Hasler and Einsele hypothesized that sulfate additions to lakes could lower the Fe:P ratio in bottom waters and, thus, increase availability of P. In the intervening decades, the burning of fossil fuels and fertilizer use have introduced large amounts of S into the environment (Berner & Berner, 1987) and have made the proposed link between S inputs to lakes and P availability an extremely interesting hypothesis. We know, however, of only one paper which has tested this proposed link directly (Curtis, 1989). In this paper we use data from 51 diverse lakes to test the hypothesis that sulfate is a critical variable which controls availability of P in bottom waters through its control of Fe:P ratios (Hasler & Einsele, 1948). This work on P availability expands on our earlier work on the quantity of P released from anoxic sediments (Caraco et al., 1989, 1991).

Methods

Our data set consists of samples from 51 different freshwater lakes located throughout North America. These lakes were seasonally anoxic, with some portion of the hypolimnion becoming anoxic during the stratification period. For each lake water column samples were taken with depth on 2 to 5 occasions. Samples were pumped from depth using a peristaltic pump and filtered in line through a Whatman GF/F filter without exposure to the atmosphere. The samples for the analysis of DIC were pumped directly into BOD bottles. DIC was measured according to Stainton (1973) using a Schimadzu GC-AIT equipped with a

thermal conductivity detector. Samples from 5 ml and 35 ml samples (depending on sulfide levels in system) for the analysis of sulfide and sulfate were pumped directly into 5 ml of a basic zinc mixture to prevent oxidation of the sulfide to sulfate (Gilboa-Garber, 1971). This oxidation can lead to serious overestimates of sulfate depletion in bottom waters and, obviously, severe underestimates of sulfide. The samples for the analysis of sulfide and sulfate were centrifuged and the precipitate was used for the colorimetric analysis of sulfide (Gilboa-Garber, 1971). Sulfate was analyzed on the supernatant on a DIONEX ion chromatograph. All standards and blanks were treated as the samples. Filtered samples for the analysis of Fe and P were acidified immediately to pH 2.1–2.3 using sulfuric acid and saved for later analysis. Dissolved iron was measured using the bipyridyl method (Lewis, 1954) dissolved P was measured using the molybdate method (Murphy & Riley, 1962). All blanks and standards for analysis were run on acidified water (as for samples). For each system, Fe:P, Fe:C and P:C accumulation ratios in bottom waters were calculated by linear regression (see Fig. 1 for Fe:P example). Phosphorus accumulation, iron accumulation and sulfate depletion are expressed for each lake system as the maximum value occurring above sediments at the end of stratification.

Results and discussion

Hasler & Einsele (1948) suggested that sulfate could affect the availability of P in bottom waters because sulfide (formed from microbial sulfate reduction) can remove, by the formation of iron sulfide, iron from solution in anoxic waters (Fig. 2, top panel). There are several points of agreement between our cross-system observations and the model of Hasler & Einsele (1948). First, we found that sulfate concentrations of surface waters (a measure of the potential for sulfate reduction) indeed were related negatively to Fe:P accumulation in bottom water (Fig. 2, bottom panel, Table 1). Second, we found that this relationship was at least in part due to lower iron accumula-

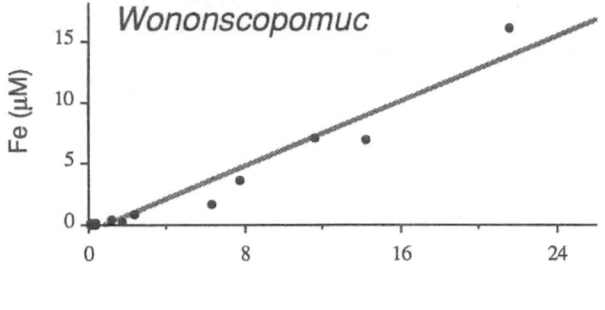

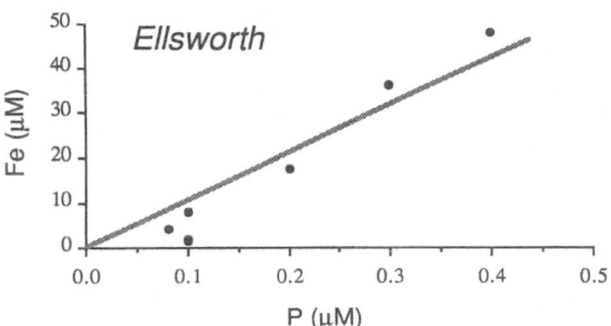

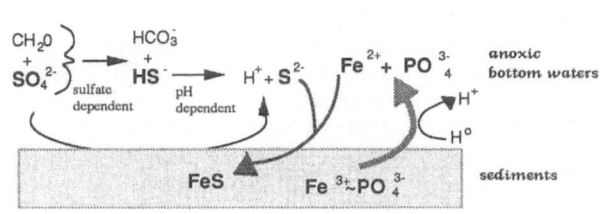

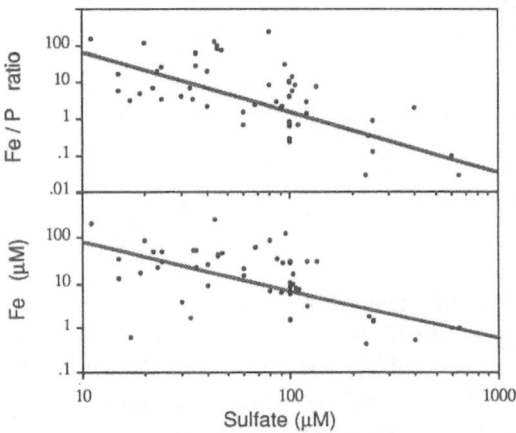

Fig. 1. Relationship between iron and phosphorus in the anoxic bottom waters of two lakes. The slopes on these graphs represent the Fe:P accumulation in bottom waters. Note that in Ellsworth Pond, N. H., U.S.A. (a low sulfate system) the Fe:P ratio is far greater than in Lake Wononscopomuc, CT, U.S.A. (a system with moderate sulfate concentration).

tion (presumably due to FeS formation) in high sulfate systems. Finally, we observed that the pH of bottom waters was related negatively to the Fe:P ratio in these waters (Table 1). Given equivalent amounts of Fe and S in solution FeS will more likely form at higher pH values (Fig. 2, top panel). A relationship between pH and Fe:P ratios is, therefore, in agreement with the model relating control of Fe:P ratio to precipitation of FeS.

Although several aspects of our data support the model of control of Fe:P ratios presented by Hasler & Einsele (1948), two results conflict with it. First, Hasler & Einsele proposed that the relationship between sulfate and Fe would be driven by precipitation of Fe as FeS, with sulfide in turn being created by sulfate reduction (Fig. 2, top panel). If FeS precipitation were driving the relationship between sulfate and Fe:P ratios, one would expect that there should be a good predic-

Fig. 2. Top Panel – Representation of the Hasler & Einsele model of the control of Fe:P ratios in anoxic bottom waters by sulfate concentration in those waters. In their model both Fe and P are released from sediments under anoxic conditions. After release, sulfate reduction to sulfide can cause precipitation of iron sulfides and can, therefore, decrease the Fe:P ratio in bottom waters. Note that although this sulfate reduction is shown to occur in the water column, reduction in the sediment would have the same consequence of sulfide production which could lead to decreased Fe in botton waters.

Bottom Panel – Our data showing a negative relationship between both Fe:P ratios in bottom waters and Fe concentrations in bottom waters support this model.

tive relationship between the degree of sulfate reduction (sulfide generation) in bottom waters of a system and iron accumulation in those waters. Further, this relationship should be as good or better than the relationship between sulfate and iron in bottom waters. Our data suggest, however, that sulfate reduction is related weakly to iron accumulation in bottom waters (Fig. 3, bottom panel, Table 1) and that the relationship between sulfate and iron is, in fact, a better one

Table 1. Statistical tests of the controls of P release from sediments into overlying anoxic waters and availability of released P (see text). All relationships are from linear regression analysis of data from 51 lakes. For the predictive variables tested: log (SO_4) is the concentration (μM) of sulfate in surface waters; log (SO_4 reduced) is the maximum observed sulfate depletion (μM) in anoxic bottom waters at the end of summer stratification; and pH is the measured pH in the deepest water of the lake at the end of summer stratification. For dependent variables: log (Fe/P) and log (P/C) are the ratios of Fe to P to C in atoms, respectively, in anoxic bottom waters; log (Fe) and log (P) are the maximum observed Fe and P concentrations, respectively, in anoxic bottom waters.

Predictive	Dependent	Slope	Intercept	r^2	P
Log (SO_4)	log (Fe/P)	-1.46	3.31	0.41	<0.001
Log (SO_4 reduced)	log (Fe/P)	-0.90	2.05	0.30	<0.001
pH	log (Fe/P)	-0.92	6.78	0.21	0.002
Log (SO_4)	log (Fe)	-0.97	2.87	0.32	<0.001
Log (SO_4 reduced)	log (Fe)	-0.30	1.49	0.05	0.17NS
pH	log (Fe)	-0.42	3.92	0.08	0.07NS
Log (SO_4)	log (P)	0.85	-1.24	0.19	0.001
Log (SO_4 reduced)	log (P)	1.09	-1.29	0.40	<0.001
pH	log (P)	0.82	-5.20	0.24	0.001
Log (SO_4)	log (P/C)	0.58	-0.64	0.12	0.015
Log ($SO4$ reduced)	log (P/C)	0.59	-0.43	0.18	0.004
pH	log (P/C)	0.51	-3.00	0.13	0.015

(Table 1). The reason for a lack of a good relationship could be the relatively complex interaction between sulfate reduction and iron concentration. That is, sulfate reduction potentially can have both positive and negative effects on Fe accumulation in bottom waters (Fig. 3, top panel). The reason that there is a relatively good relationship between surface water sulfate and iron is unclear. One possibility is that sulfate reduction over the past several years diminishes the pool of sediment iron oxides which would be available for release in any particular season. This possibility suggests that, over several years to decades, iron oxide pools can be diminished by enhanced S loading to surface waters (Giblin *et al.*, 1990; Carignan & Tessier, 1988).

The second way in which our results contradict the model of Hasler & Einsele (1948) is that, while their model presents control of Fe:P ratios only in terms of a relationship between S and Fe, our data show that the relationship between Fe:P and sulfate reduction (Table 1) is driven in part by a positive relationship between sulfate reduction and P accumulation in bottom waters (Table 1, Fig. 3, bottom panel). Sulfate could enhance P release from sediments by enhancing decomposition of organic matter in sediments (thus releasing P bound in organic matter). Additionally,

sulfate could alter the fate of P released from organic matter by altering post-decompositional binding of P. The P:C ratio in bottom waters is an estimate of the post-decompositional binding in sediments (Caraco *et al.*, 1991). Our data set shows that this ratio is related positively to sulfate and indicates that postdecomposition binding of P is greater in low sulfate systems. Some possible mechanisms for this lower binding capacity include:

1. Control mediated through pH (Curtis, 1989).
   – That is, systems with higher sulfate concentrations could have higher pH in bottom waters due to the fact that sulfate reduction is an alkalinity generating process while methane production (which predominates in many low sulfate systems) is not.

2. Control mediated through iron reduction (Krom & Berner, 1980; Caraco *et al.*, 1991) – Sulfide oxidation to sulfate may be coupled to the reduction of Fe(III) to Fe(II). If this reduction were faster or more complete than microbial reduction of Fe(III) using organics as the reductant, then S would act to decrease the pool of Fe(III) in the sediments and, thus, decrease the potential for P binding in sediments.

3. Control mediated through microbial P release

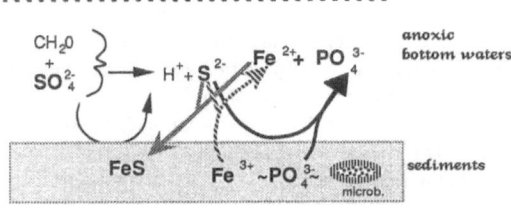

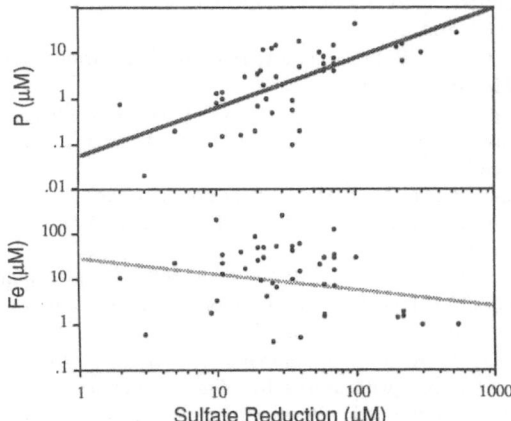

Fig. 3. Top Panel – A modified model of sulfate control of Fe:P ratios in bottom waters. In contrast to the original model of Hasler & Einsele (1948), this model emphasizes the direct control of P release by sulfur (black arrow). It also shows a more complicated relationship between iron concentration in bottom waters and sulfide generation (sulfate reduction) in bottom waters or sediments (see legend Fig. 2). Sulfide can interact with particulate Fe(III) and increase Fe(II) in solution (stippled arrow). Alternately, the interaction of sulfide with dissolved Fe(II) can lead to reduced concentrations of dissolved iron in bottom waters (stripped arrow).

Bottom Panel – Our data show a good relationship between sulfate reduction in bottom waters and P accumulation in those waters. Further, our data show that sulfate reduction is related poorly to Fe in bottom waters. These two observations suggested that modifications of Hasler & Einsele's model (Fig. 2.) were needed.

(Comeau *et al.*, 1986) – That is, data suggest that H_2S can increase the rate of P release from microbial polyphosphate pools.

Studies are needed to distinguish between these interesting potential mechanisms for regulating P release from sediments.

Implications

Sulfate concentrations of freshwaters have increased dramatically due to atmospheric inputs from fossil fuel burning and fertilizer use (Berner & Berner, 1987). It has been estimated that on a global scale this human impact has caused nearly a doubling of surface water sulfate concentrations from a mean pre-industrial value of 70 μM to a present mean value of 120 μM (Berner & Berner, 1987). This sulfur pollution of surface waters undoubtedly has multiple consequences for the functioning of aquatic systems. For example, within-system alkalinity generation may have increased (Schindler, 1985; Kelly *et al.*, 1982) and methanogenesis could have decreased (Kelly *et al.*, 1982). Another important consequence of increased sulfate input to surface waters may be a significant alteration of P cycling in aquatic systems.

An alteration of the P cycle in lakes due to changes in S loading could occur in two ways: 1. by affecting the availability of P released from sediments into anoxic bottom waters and 2. the actual P release from sediments may be affected by sulfate concentrations of overlying waters. According to our empirically derived relationships (Table 1) an increase in sulfate from 70 to 120 μM would increase P release from sediments by approximately 58%. Further, Fe:P accumulation ratios in bottom waters would have changed from 4.1 pre-industrially to 1.9 at present. Such a change is highly significant. Tessenow's (1974) results suggest that when Fe:P ratios in bottom waters are less than 2.2 bottom waters are far more likely to 'leak' P to surface waters. Thus, our results suggest that on a global scale S pollution may have been sufficient to change many lakes from systems which retain fully the P reaching sediments to systems where this P is 'leaked' back to surface waters where it can be reused by phytoplankton. This increased leakiness of systems coupled with the actual increase in P release from sediments suggest that humans may be enhancing the fertility of lakes indirectly through alteration of the S cycle.

Acknowledgements

We thank M. Filmer for laboratory assistance and P. Likens for field assistance. We also thank M. Pace and G. Steinhart for data and samples and the North Temperate Lakes – LTER as well. This work was supported by the National Science Foundation (BSR89-08855) and the Andrew W. Mellon Foundation. This paper was written, in part, during a visit by NFC and JJC to the Limnology Institute of the University of Uppsala. The support of the Limnology Institute is gratefully acknowledged.

References

Baccini, P., 1985. Phosphate interaction at sediment-water interfaces. In W. Stumm (ed.), Chemical Processes in Lakes. Wiley, New York: 189–205.

Berner, E. K. & R. A. Berner, 1987. The global water cycle: Geochemistry and environment, Prentice-Hall, Englewood Cliffs, 1–397.

Caraco, N. F., J. J. Cole & G. E. Likens, 1989. Evidence for sulphate-controlled phosphorus release from sediments of aquatic systems. Nature 341: 316–318.

Caraco, N. F., J. J. Cole & G. E. Likens, 1991. A cross-system study of phosphorus release from lake sediments. In J. J. Cole, G. Lovett & S. Findlay (eds), Comparative Analysis of Ecosystems. Springer-Verlag, New York: 241–258.

Carignan, R. & A. Tessier, 1988. The co-diagenesis of sulfur and iron in acid lake sediments of southwestern Quebec. Geochim. Cosmochim. Acta 52: 1179–1188.

Comeau, Y., K. Y. Hall, R. E. W. Hancock & W. K. Oldham, 1986. Biochemical model for enhanced biological phosphorus removal. Wat. Res. 20: 1511–1521.

Curtis, P. J., 1989. Effects of hydrogen ion and sulphate on the phosphorus cycle of a Precambrian Shield Lake. Nature 337: 156–158.

Fonselius, S. H., 1970. On the stagnation and recent turnover of the water in the Baltic. Tellus 22: 533–544.

Giblin, A. E., G. E. Likens, D. White & R. W. Howarth, 1990. Sulfur storage and alkalinity generation in New England lake sediments. Limnol. Oceanogr. 35: 852–869.

Gilboa-Garber, N., 1971. Direct Spectrophotometric determination of inorganic sulfide in biologic material and in other complex mixtures. Anal. Biochem. 43: 129–133.

Hasler, A. C. & W. G. Einsele, 1948. Fertilization for increasing productivity of natural inland waters. Trans. North. Amer. Wildr. Conf. 13: 527–555.

Kelly, C. A., J. N.. M. Rudd, R. B. Cook & D. W. Schindler, 1982. The potential importance of bacterial processes in regulating rate of lake acidification. Limnol. Oceanogr. 27: 868–882.

Lawacz, W., 1985. Factors affecting nutrient budget in lakes of the Jorka River watershed Masurian Lakeland Poland XI. Nutrient budget with special consideration to phosphorus retention. Ekol. Pol. 33: 357–382.

Lean, D. R. S., D. J. McQueen & V. A. Story, 1986. Phosphate transport during hypolimnetic aeration. Arch. Hydrobiol. 108: 269–280.

Lewis, G.. J. & E. D. Goldberg, 1954. Iron in marine waters. J. Mar Res. 13: 183–197.

Mortimer, C. H., 1941. The exchange of dissolved substances between mud and water in lakes (Parts I and II). J. Ecol. 29: 280–329.

Mortimer, C. H., 1942. The exchange of dissolved substances between mud and water in lakes (Parts III and IV). J. Ecol. 30: 147–201.

Murphy, J. & J. P. Riley, 1962. A modified single solution method for the determination of phosphate in natural waters. Analyt. chim. Acta 27: 31–36.

Nurnberg, G. K., 1987. A comparison of internal phosphorus loads in lakes with anoxic hypolimnia.: Laboratory incubation hypolimnetic phosphorus accumulation. Limnol. Oceanogr. 32: 1160–1164.

Riley, E. T. & E. E. Prepas, 1984. Role of internal phosphorus loading in two shallow, productive lakes in Alberta, Canada. Can. J. Fish. – aquat. Sci. 41: 845–855.

Schindler, D. W., 1985. Coupling of elemental cycles by organisms: evidence from whole-lake chemical perturbations. In W. Stumm (ed.), Chemical Processes in Lakes. Wiley & Sons, New York: 225–250.

Schindler, D. W., T. Ruszcynski & E. J. Fee, 1980. Hypolimnion injection of nutrient effluents as a method for reducing eutrophication. Can. J. Fish. aquat. Sci. 37: 320–327.

Stainton, M. P., 1973. A syringe gas-stripping procedure for gas-chromatographic determination of dissolved inorganic and organic carbon in freshwater and carbonates in sediments. J. Fish. Res. Bd Can. 50: 1441-1445.

Tessenow, V U., 1974. Solution, diffusion and sorption in the upper layer of lake sediments. IV. Reaction mechanisms and equilibria in the system iron-manganese-phosphate with regard to the accumulation of vivianite in Lake Ursee. Arch. Hydrobiol. Suppl. 47: 1–79.

Hydrobiologia **253**: 281–300, 1993.
P.C.M. Boers, T.E. Cappenberg & W. van Raaphorst (eds),
Proceedings of the Third International Workshop on Phosphorus in Sediments.
© 1993 *Kluwer Academic Publishers.*

Application of SWITCH, a model for sediment–water exchange of nutrients, to Lake Veluwe in The Netherlands

J. G. C. Smits[1] & D. T. van der Molen[2]
[1] *Delft Hydraulics, P.O. Box 177, 2600 MH Delft, The Netherlands;* [2] *Institute for Inland Water Management and Waste Water Treatment, P.O. Box, 8200 AA Lelystad, The Netherlands*

Key words: mathematical modelling, nutrients, sediment-water interaction, Lake Veluwe

Abstract

The formulations of SWITCH, a model for prediction of nutrient fluxes across the sediment-water interface, are presented. Results of the application to data on the sediment of Lake Veluwe are presented and discussed.

SWITCH calculates the thicknesses of the aerobic and denitrifying layers on the basis of a step-wise steady state approach. The concentrations of detritus, ammonium, nitrate and phosphate in the sediments and the pore water are simulated dynamically using mass balance equations.

Analysis of the data for Lake Veluwe show large spatial heterogeneity. This presents a major drawback for the calibration of SWITCH, which focused on the silty part of the lake. The results show that the model simulates realistically and consistently layer thicknesses, concentrations and mass fluxes connected with the transport and conversion processes. The model appears to have potential for describing both seasonal patterns and developments on the long term.

SWITCH calculates strongly increased phosphate return fluxes, following total reduction of the top sediments. An important hypothesis in the model is that phosphate precipitated in reduced sediment layers is transferred to the oxidized layer and dissolves instantaneously. This results in a decrease of the phosphorus content of the sediment, but also maintains high release rates of phosphorus after the reduction of the external phosphorus loading of Lake Veluwe. Model results and mass balance studies for the overlying water indicate that the removal of phosphorus to deeper sediment layers is underestimated or that dilution of the sediments occurs as the result of sedimentation.

Introduction

In water quality management studies the use of mathematical models is often beneficial by means of quantitative description of the water system and prediction of the effects of changes such as reduced nutrient loading. In the field of lake eutrophication a broad range of models have been developed for this purpose (Los, 1980; De Rooij, 1980; Los *et al.*, 1984; Jørgensen *et al.*, 1982; Jør-gensen, 1983; Beck, 1985). In most of these models the interaction of nutrients between water and sediment is a weak point. It is often omitted or sedimentation and release are described independently.

Relatively simple dynamic phosphorus models had already been developed elsewhere (Kamp-Nielsen, 1975; Jørgensen *et al.*, 1982; Van Raap-horst *et al.*, 1988; Van der Molen, 1991). An improved model for the exchange of nutrients between sediments and overlying water is needed for the predictive long-term modelling of aquatic systems with changing (improving) water quality. An attempt to construct such an improved model led to SEDMOD (Van Eck & Smits, 1986). This rather simple model followed the classic approach (Berner, 1974; Vanderborght *et al.*, 1977a) with steady state solutions of diffusion equations for nutrients and electron acceptors and dynamic mass balances for detritus (organic matter). This steady state approach delivered realistic results for equilibrated systems, but proved to be inflexible and unpractical at the same time.

SWITCH was developed to describe the exchange of nutrients between sediments and the overlying water, both annually and on a longer time scale. We aimed to model only those processes in mass transport and biological and chemical transformations which are relevant from a quantitative point of view. SWITCH simulates the exchanges between water and sediment and the concentrations in the sediments of organic matter, nutrients and dissolved oxygen, but in this paper we will focus on the fate of phosphorus.

The release of phosphorus from sediments is an important source of nutrients for algal growth during the summer period (Ryding & Forsberg, 1977; Hosper, 1984; Søndergaard *et al.*, 1989). An indication of the importance of the sediment release is an increase in the phosphate concentration in summer which is far greater than can be explained by external loading (see Sas (1989) for several examples). Since the pioneering work of Einsele (1936) and Mortimer (1941) the release of phosphate is generally attributed to a decrease in redox potential, followed by reduction of Fe (III) and consequently release of Fe(III) bound phosphorus. For a peaty sediment Boers (1986) found mineralization to be the driving force for release, and others demonstrated the importance of factors such as temperature (*e.g.* Holdren & Armstrong, 1980) and pH (Lijklema, 1980).

The choice of sediment schematization and process formulation is based on the 'classical' theory on release of phosphorus. They will be described briefly in the next section. Next, the results of the application of the model to Lake Veluwe, a shallow eutrophic lake in the Netherlands, is presented. Finally concluding remarks about the properties of the model will be made. The objective was to examine whether the most important processes were included in a satisfying way by comparing the results with measurements and mass balance studies.

Description of SWITCH

Schematization of the bottom

SWITCH computes concentrations in an 'active' layer of the bottom with a constant thickness, which may be deduced from the vertical pattern of bioturbation. In principle, the lower boundary of the active layer can be identified by a steep decline of the activity of the benthic fauna. Mass is removed from the active layer and transferred to the inactive part of the bottom as a consequence of net sedimentation. It is assumed that a particle buried in the inactive sediment never returns to the active layer, and that there is no net dispersive transport across the interface. The model complies with the demand of mass conservation, as all fluxes going into the sediment match the sum of accumulation in the sediment and the fluxes going out.

The active layer contains four sublayers, as is depicted in Fig. 1. The bottom is divided into a rather thin and relatively well mixed upper layer $(d_1 + d_2 + d_3)$ and a lower layer (d_4) in relation to vertical differences in the physical characteristics of the sediments. Porosity and mixing processes decrease with increasing depth. The thin top layer $(d_1 + d_2)$ is oxidized, the remaining part of the upper layer (d_3) is reduced just like the lower layer. The oxidized layer consists of an aerobic layer (d_1) at the sediment-water interface and a denitrifying layer (d_2). A further partition with respect to the reduction of iron, manganese, sulphate and carbon dioxide (methanogenesis) was rejected. It is thought that the incorporation of these sublayers would only complicate the model without improving its performance with respect to the nutrient budgets.

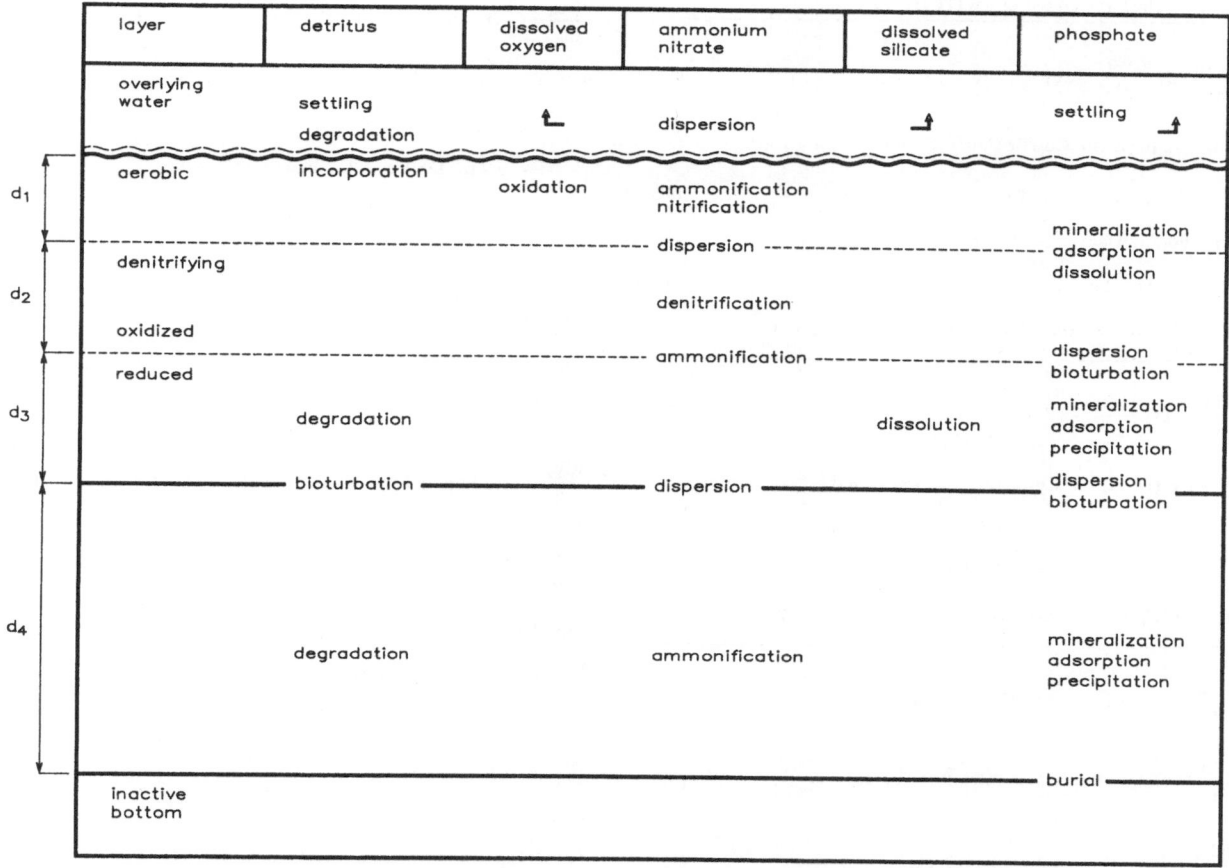

layer	detritus	dissolved oxygen	ammonium nitrate	dissolved silicate	phosphate
overlying water	settling degradation	⬇	dispersion	⬆	settling ⬆
d_1 aerobic	incorporation	oxidation	ammonification nitrification		
					mineralization adsorption dissolution
d_2 denitrifying		dispersion			
			denitrification		
oxidized		ammonification			dispersion
d_3 reduced					bioturbation
	degradation			dissolution	mineralization adsorption precipitation
	bioturbation	dispersion			dispersion bioturbation
d_4	degradation		ammonification		mineralization adsorption precipitation
					burial
inactive bottom					

Fig. 1. Schematization of the bottom and processes in SWITCH.

The oxidized layer may disappear when the degradation of detritus is substantial and when little nitrate is available in the overlying water. In order to avoid numerical problems, neither the aerobic layer nor the oxidized layer may become infinitely thin. A minimal thickness is therefore imposed on these layers. The denitrifying layer may disappear entirely. This happens in SWITCH when the nitrate concentration in the aerobic layer falls below a critical value.

Additionally, a meta-stable boundary layer is supposed to exist between the sediments and the overlying water. This layer contains detritus recently settled from the water column. The detritus in the boundary layer is gradually incorporated into the bottom through the succession of resuspension and sedimentation and through bio-turbation. In the model this is established via an overall incorporation rate. The boundary layer is very thin and does not actually affect the transfer of dissolved substances across the sediment-water interface. Local production of detritus by benthic phytoplankton is not (yet) considered.

Mathematical formulations and mechanisms

The processes included in SWITCH are collected in Fig. 1. The formulations are gathered in Table 1, but interested readers are referred to the research report, available from the authors.

Layers d_1 and d_2 are derived from the steady state solution of differential equations for the mass balances of oxygen and nitrate (according to

284

Table 1. The formulations of SWITCH.

$$\frac{dCo}{dt} = D \cdot \frac{d^2Co}{dz^2} - Ro/p_1 \tag{1}$$

$$d_1 = \sqrt{(2\, p_1 \cdot D \cdot fo \cdot Co_0/Ro)} \quad \text{if:} \quad Co = fo \cdot Co_0 \text{ for } z = 0 \tag{2}$$

$$Fo = \sqrt{(2\, p_1 \cdot D \cdot Ro \cdot fo \cdot Co_0)} \tag{3}$$

$$Ro = Fo_b/d_1 + ac \cdot kc_1 \cdot Cd_1 + p_1 \cdot an \cdot kn \cdot Ca_1 + Fo_c/d_1 \tag{4}$$

$$Fo_b = ac \cdot kc_b \cdot Cd_b \tag{5}$$

$$Fo_c = (1 - fro) \cdot ac \cdot (kc_1 \cdot Cd_1 \cdot (d_2 + d_3) + kc_4 \cdot Cd_4 \cdot d_4) \tag{6}$$

$$\frac{dCn_2}{dt} = D \cdot \frac{d^2Cn_2}{dz^2} - kd \cdot Cn_2 \tag{7}$$

$$d_2 = 2\sqrt{(D/kd)} \quad \text{if: } Cn_2 = Cn_1 \quad \text{for } z = d_1$$
$$Cn_2 = 0.0 \quad \text{for } z = do$$
$$Cn_2 = 0.5\, Cn_1 \quad \text{for } d_1 < z < do \tag{8}$$

$$d_2 = 2\,(Cn_1 - Cn_c)/Cn_1 \cdot \sqrt{(D/kd)} \,\Big|\, d_2 = 0.0 \text{ if } Cn_1 \le Cn_c \tag{9}$$

$$\frac{dCd_b}{dt} = Fd_s - Fd_b - kc_b \cdot Cd_b \tag{10}$$

$$\frac{dCd_1}{dt} = (Fd_b - Fb_3 \cdot Cd_1 + Fd_3)/dh - kc_1 \cdot Cd_1 \tag{11}$$

$$\frac{dCd_4}{dt} = (Fb_3 \cdot Cd_1 - Fd_3)/dl - kc_4 \cdot Cd_4 \tag{12}$$

with: $Fd_s = sc \cdot Cd_0$

$Fd_b = rc \cdot Cd_b$

$Fd_3 = 2\, Db \cdot (Cd_4/(1 - p_4) - Cd_1/(1 - p_1))/d$

$$\frac{dCn_1}{dt} = (-Fn_0 + Fn_1)/(p_1 \cdot d_1) + kn \cdot Ca_1 \tag{13}$$

with: $Cn_2 = 0.5\, Cn_1$

$Fn_0 = 2\, p_1 \cdot D \cdot (Cn_1 - Cn_0)/(1 + d_1)$

$Fn_1 = -p_1 \cdot kd \cdot Cn_2 \cdot d_2$

$$\frac{dCa_1}{dt} = (Fa_b - Fa_0 + Fa_1 + Fas_0 - Fas_1)/(p_1 \cdot d_1)$$
$$+ aa \cdot kc_1 \cdot Cd_1/p_1 - kn \cdot Ca_1 \tag{14}$$

$$\frac{dCa_2}{dt} = (-Fa_1 + Fa_3 + Fas_1 - Fas_3)/(p_1 \cdot (d_2 + d_3))$$
$$+ aa \cdot kc_1 \cdot Cd_1/p_1 \tag{15}$$

$$\frac{dCa_4}{dt} = (-Fa_3 + Fas_3 - Fas_4)/(p_4 \cdot d_4) + aa \cdot kc_4 \cdot Cd_4/p_4 \tag{16}$$

with: $Fa_b = aa \cdot kc_b \cdot Cd_b$

$Fa_0 = 2\, p_1 \cdot D \cdot (Ca_1 - Ca_0)/(1 + d_1)$

$Fa_1 = 2\, p_1 \cdot D \cdot (Ca_2 - Ca_1)/do$

$Fa_3 = (p_1 + p_4) \cdot D \cdot (Ca_4 - Ca_2)/(d - d_1)$

$$Cpp = fpp \cdot Cp \tag{17}$$
$$Cdp = fdp \cdot Cp/p$$
$$Cap = fap \cdot Cp$$
$$fap + fdp + fpp = 1$$

$$\frac{dCpp_3}{dt} = (-Fpp_1 + Fpp_3)/d_3 + p_1 \cdot kp \cdot (fdp_3 \cdot Cp_3/p_1 - Cdp_5) \tag{18}$$

$$\frac{dCpp_4}{dt} = Fpp_3/d_3 + p_4 \cdot kp \cdot (fdp_4 \cdot Cp_4/p_4 - Cdp_5) \tag{19}$$

with: $Fpp_1 = 2\, Db \cdot fpp_3 \cdot Cp_3/((1 - p_1) \cdot (do + d_3) \cdot d_3)$

$Fpp_3 = 2\, Db \cdot (fpp_4 \cdot Cp_4/(1 - p_4)$
$\quad - fpp_3 \cdot Cp_3/(1 - p_1))/((d_3 + d_4) \cdot d_3)$

$$Cap = Cam \cdot Cdp/(Ks + Cdp) \tag{20}$$
$$Cam = Cac \cdot (1 - p) \cdot Ws$$
$$fdp = [Cip - p \cdot Ks - Cam + \sqrt{(Cip - p \cdot Ks - Cam)^2 +}$$
$$4\,(Cip \cdot p \cdot Ks)]/(2\, Cp) \tag{21}$$

with: $Cip = (1 - fpp) \cdot Cp$

$$\frac{dCp_1}{dt} = (Fp_b + Fp_s - Fp_0 + Fp_2 + Fps_0 - Fps_2 - Fpd_2 -$$
$$Fb_2 \cdot Cp_1)/do + ap \cdot kc_1 \cdot Cd_1 \tag{22}$$

$$\frac{dCp_3}{dt} = (-Fp_2 + Fp_3 + Fps_2 - Fps_3 - Fpd_2 + Fpd_3 +$$
$$Fb_2 \cdot Cp_1 - Fb_3 \cdot Cp_3)/d_3 + ap \cdot kc_1 \cdot Cd_1 \tag{23}$$

$$\frac{dCp_4}{dt} = (-Fp_3 + Fps_3 - Fps_4 - Fpd_3 + Fb_3 \cdot Cp_3 -$$
$$Fb_4 \cdot Cp_4)/d_4 + ap \cdot kc_4 \cdot Cd_4 \tag{24}$$

with: $Fp_b = ap \cdot kc_b \cdot Cd_b$

$Fp_0 = 2\, p_1 \cdot D \cdot (fdp_1 \cdot Cp_1/p_1 - fdp_0)/(1 + do)$

$Fp_2 = 2\, D \cdot (fdp_3 \cdot Cp_3 - fdp_1 \cdot Cp_1)/(do + d_3)$

$Fp_3 = (p_1 + p_4) \cdot D \cdot (fdp_4 \cdot Cp_4/p_4 - fdp_3 \cdot Cp_3/p_1)/(d_3 + d_4)$

$Fpd_2 = 2\, Db \cdot ((fpp_3 + fap_3) \cdot Cp_3 - fap_1 \cdot Cp_1)/(1 - p_1)/(do + d_3)$

$$Fpd_3 = 2\,Db \cdot ((fpp_4 + fap_4) \cdot Cp_4/(1 - p_4) -$$

$$(fpp_3 + fap_3) \cdot Cp_3/(1 - p_1))/(d_3 + d_4)$$

$$D = Dm + (bt - 1) \cdot Dm \quad \text{for} \quad bt \geq 1 \tag{25}$$

$$k = k^{20} \cdot kt^{(T-20)} \tag{26}$$

aa	= stoichiometric constant for nitrogen in detritus (g N/g C)
ac	= stoichiometric constant for oxidation of detritus (g O_2/g C)
an	= stoichiometric constant for nitrification (g O_2/g N)
ap	= stoichiometric constant for phosphate in detritus (g P/g C)
bt	= amplification factor for bio-irrigation (–)
Ca	= ammonium concentration (g N m^{-3} water)
Cam	= maximal concentration of adsorbed phosphate (g P m^{-3} bottom)
Cac	= adsorption capacity (g P kg^{-1} dry matter)
Cap	= adsorbed phosphate concentration (g P m^{-3} bottom)
Cd	= detritus concentration (g C m^{-3} water or bottom)
Cd_b	= amount of detritus in the boundary layer (g C m^{-2})
Cdp	= dissolved phosphate concentration (g P m^{-3} pore water)
Cdp_s	= saturation concentration of ortho-phosphate (g P m^{-3} pore water)
Cn	= nitrate concentration (g N m^{-1} water)
Cn_c	= critical nitrate concentration (g N m^{-3} water)
Co	= dissolved oxygen concentration (g m^{-3} water)
Cp	= total inorganic phosphate concentration (g P m^{-3} bottom)
Cpp	= precipitated phosphate concentration (g P m^{-3} bottom)
D	= dispersion coefficient (m^2 d^{-1})
Db	= bioturbation dispersion coefficient (m^2 d^{-1})
Dm	= molecular diffusion coefficient (m^2 d^{-1})
Fa	= dispersive ammonium flux between layers (g N m^{-2} d^{-1})
Fa_0	= dispersive ammonium return flux to overlying water (g N m^{-2} d^{-1})
Fa_b	= ammonification flux from the boundary layer (g N m^{-2} d^{-1})
fap	= adsorbed phosphate fraction (–)
Fas	= seepage flux of ammonium between layers (g N m^{-2} d^{-1})
Fb	= burial flux based on bottom volume (m^3 m^{-2} d^{-1})
Fd	= bioturbation flux of detritus (g C m^{-2} d^{-1})
Fd_b	= flux of detritus incorporated in the upper layer (g C m^{-2} d^{-1})
Fd_s	= flux of detritus settled from overlying water (g C m^{-2} d^{-1})
fdp	= dissolved phosphate fraction (–)
fo	= correction factor for the oxygen concentration (–)
Fo	= sediment oxygen demand (g O_2 m^{-2} d^{-1})
Fo_b	= oxygen consumption flux in the boundary layer (g O_2 m^{-2} d^{-1})
Fo_c	= chemical oxygen demand (g O_2 m^{-2} d^{-1})
Fp	= dispersive phosphate flux between layers (g P m^{-2} d^{-1})
Fp_0	= dispersive phosphate return flux to overlying water (g P m^{-2} d^{-1})
Fp_b	= phosphate release flux from the boundary layer (g P m^{-2} d^{-1})
Fp_s	= sedimentation flux of adsorbed phosphate (g P m^{-2} d^{-1})
Fpd	= bioturbation flux of phoshate between layers (g P m^{-2} d^{-1})
fpp	= precipitated phosphate fraction (–)
Fpp	= bioturbation flux between layers (g P m^{-2} d^{-1})
Fps	= seepage flux of phosphate between layers (g P m^{-2} d^{-1})
Fn_0	= dispersive nitrate return flux to overlying water (g N m^{-2} d^{-1})
Fn_1	= nitrate flux to the denitrifying layer (g N m^{-2} d^{-1})
fro	= fraction reduced substances permanently removed or fixed (–)
kc	= degradation rate of detritus in the bottom (d^{-1})
kc_b	= degradation rate of detritus in the boundary layer (d^{-1})
kd	= denitrification rate (d^{-1})
kn	= nitrification rate (d^{-1})
kp	= phosphate precipitation rate (d^{-1})
Ks	= half saturation phosphate concentration (g P m^{-3} pore water)
l	= thickness of the water boundary layer (m)
p	= porosity (–)
rc	= rate of detritus incorporation in the sediments (d^{-1})
Ro	= oxygen consumption rate (g m^3 bottom d^{-1})
sc	= sedimentation rate for detritus (m d^{-1})
t	= time (d)
Ws	= specific weight of the sediments (kg m^{-3})
z	= sediment depth (m)

A subscript figure indicates either a layer number (0 = overlying water or an interface number (4 = lower boundary of active bottom).

Bouldin (1968). The assumption of steady state is allowed for these substances, because the change of their concentrations in the overlying water is slow compared to the mass fluxes of the diffusion and oxidation processes near the sediment-water interface.

The correction factor (fo) in equation 2 relates to the existence of a relatively stagnant boundary layer in the overlying water, which may contain a part of the oxygen gradient at the sediment-water interface (Jorgensen & Revsbech, 1985; Sweerts et al. 1989). Consequently, the oxygen concentration at the interface is a certain fraction of the average oxygen concentration in the water column.

The sediment oxygen demand can be derived from the first derivative of the steady state solution of equation 1. Multiplication with the dispersion coefficient (D) and filling in z = 0 yields equation 3. Oxygen is consumed in the degradation of detritus in the boundary layer and the aerobic layer, in nitrification and in chemical oxidation. The overall oxygen consumption rate (Ro) is formulated as the sum of these processes (Eq. 4). The oxygen consumption in the boundary layer connected with the degradation of detritus on top of the sediments is given by equation 5. Imposing this oxygen consumption flux on the aerobic layer (d_1) instead of spreading it evenly over the upper sediment layer ($d_1 + d_2 + d_3$) is necessary to reproduce a complete collapse of the oxidized layer at a high influx of detritus.

The chemical oxygen demand (Eq. 6) concerns the oxidation of reduced substances, such as iron(II), manganese(II), sulphide and methane originating from the degradation of detritus in the anaerobic part of the 'active' bottom layer. However, the reduced substances will not be oxidized completely. A part of the sulphide resulting from sulphate reduction precipitates with iron and may accumulate in the reduced sediments. Methane may escape from the sediments in gas bubbles. The extent to which all of this occurs differs from one system to the other and is poorly known in a quantitative sense. The removal and permanent fixation of a part of the reduced substances is simply taken into account by means of a constant fraction (fro), an empirical input parameter.

All organic matter is considered as detritus in SWITCH, regardless of its origin. Detritus is subject to settling, resuspension, incorporation into the sediments, degradation and burial (Berner, 1974). A certain small fraction is turned into refractory organic matter, which, besides carbon, also contains certain amounts of nutrients. This process of humification is ignored in SWITCH. The concentrations of detritus in the boundary layer and the bottom layers are described with the differential equations 10–12. It is assumed that all detritus has been degraded before it arrives at the lower boundary of the 'active' layer, so that burial does not affect the detritus concentration in the lower layer (Boers & Boon, 1988).

Nitrate is formed from ammonium through nitrification in the aerobic top layer. It is subject to vertical transport and denitrification in the zone just below this layer. The nitrate concentrations in the aerobic and denitrifying layers follow from equation 13. The ammonium concentrations in the aerobic top layer, the remaining part of the upper layer $(d_2 + d_3)$ and the lower reduced layer (d_4) are described with equations 14–16. Ammonium is released at the degradation of detritus and is nitrified by bacteria under aerobic conditions (Berner, 1974; Vanderborght et al., 1977b).

Bacterial activity liberates phosphate from organic matter just like ammonium. But as contrasted with ammonium, phosphate adsorbs strongly to several components of the sediments, the hydroxides of iron(III) and aluminum in particular. Iron(III) hydroxide is present in a rela-

tively high concentration in the oxidized layer, where it is stable. The concentration declines sharply at the interface of the oxidized and reduced layers and decreases further in the reduced layer under the influence of reduction processes. Consequently, the adsorption is much stronger in the oxidized layer than in the reduced layer (Berner, 1974; Lijklema, 1980; Van Raaphorst et al., 1988). Phosphate may also precipitate in minerals, the identity of which has not been determined unequivocally. Vivianite (iron(II) phosphate) has been mentioned as the main mineral (Emerson & Widmer, 1978). Co-precipitation with several carbonates and sulphides is also possible. The formation of apatite (calcium phosphate) in fresh water sediments in a stable form is not very likely.

Three phosphate fractions are distinguished in SWITCH; a dissolved fraction, an adsorbed fraction and a precipitated fraction (Eq. 17). The model assumes equilibrium for the adsorption process, whereas precipitation is formulated as a slow process. The assumption of equilibrium has the advantage, that only total inorganic phosphate and precipitated phosphate need to be calculated explicitly on the basis of mass balances. The dissolved and adsorbed phosphate concentrations follow from the equilibrium condition for adsorption. The precipitated phosphate concentrations are determined from mass balance equations 18 and 19. The driving force for precipitation is the difference between the actual concentration and the saturation concentration of ortho-phosphate dissolved in the pore water. In principle, the latter may be determined from the solubility product of the phosphate mineral. The precipitation rate is a linear function of the driving force, when diffusion to the surface of the mineral is the rate limiting process. In case the surface reaction is rate limiting, the function may be nonlinear. However, assumption of simple first order reaction kinetics seems reasonable.

Moreover, a hypothesis is required with respect to the stability of the phosphate mineral. It is supposed that the mineral can only permanently exist in the reduced sediments. It dissolves instantly when transported into the oxidized layer by means of bioturbation. The hypothesis can be justified assuming that the phosphate mineral consists mainly of vivianite, which is thermodynamically unstable under oxidized conditions.

The dissolved fraction is derived from a Langmuir adsorption isotherm (Eq. 20). A hyperbolic equation 21 is obtained when equations 17 are substituted. The adsorption capacity depends on the iron(III) and aluminum contents of the sediments. The oxidized layer has the largest capacity. The capacity of the lower reduced layer (d_4) is a certain fraction of the adsorption capacity of the upper reduced layer (d3). Dependencies on pH and salinity are ignored.

The mass balances for total inorganic phosphate in the oxidized layer, the upper reduced layer and the lower reduced layer are represented by equations 22–24. Some remarks must be made with respect to dispersive mass transport and temperature dependency. Dispersion in the pore water is the result of molecular diffusion and bio-irrigation. The dispersion coefficient follows from equation 25. The amplification factor for bio-irrigation is a sinus function with a period of one year and a maximum in the summer. The dispersion coefficient for bioturbation is a similar function. All the rates of conversion processes are temperature dependent according to equation 26.

Numerical aspects

The mass balances are given here as differential equations. They were reformulated according to the principle of finite differences for the computer code of SWITCH. The resulting algebraic equations describe layer average concentrations as functions of process rates and fluxes at the interfaces with adjacent layers. They are solved in sufficiently small time steps.

A complication arises from the variability of d_1, d_2 and d_3. A layer thickness in a certain time step will generally be different from the thickness in the previous time step. Violation of the mass conservation law will occur, if no corrective steps are taken with respect to the concentrations. In order to maintain mass conservation the following correction is carried out at the beginning of a time step: $C = C'.d'/d$. Corrections will be small when an appropriate time step is chosen. However, the method cannot be used for phosphate, because of the large quantities of phosphate in the sediments. Merely the shrinking of the oxidized layer would result in an unrealistically increased flux, as appeared at the first test of SWITCH. It was therefore decided to accept in the first version of SWITCH the inaccuracy of 'artificial' mixing, which pertains to redistribution of phosphate among the layers. The redistribution involves adding and subtracting of the quantities of phosphate present in the increments of the layers.

The nitrogen and phosphorus contents of detritus are assumed to be constant in the sediments. However, these contents vary in time in the boundary layer due to the changing of the phytoplankton species composition. The nutrient contents in detritus in the boundary layer must always be smaller than the contents in the sediment. SWITCH takes into account the discontinuity at the sediment-water interface by adjustment of nutrient release fluxes from the boundary layer in such a way that mass conservation is established.

Case study Lake Veluwe

Field situation

Lake Veluwe (The Netherlands) is an artificial lake located between the 'old' land and the Flevoland polder (Hosper, 1984). The lake has a total surface area of 32.8 10^6 m^2; approximately half of the lake is very shallow (depth less than 1 m) with a sandy sediment; the other half is deeper (average depth 2 m) with a more silty sediment (Fig. 2).

In the second half of the sixties the lake deteriorated, because of an increased nutrient loading. This resulted in high chlorophyll *a* concentrations and an almost permanent bloom of blue-green algae (predominantly *Oscillatoria agardhii* from 1970 onwards. In 1979 restoration measures were taken: reduction of the external phosphorus

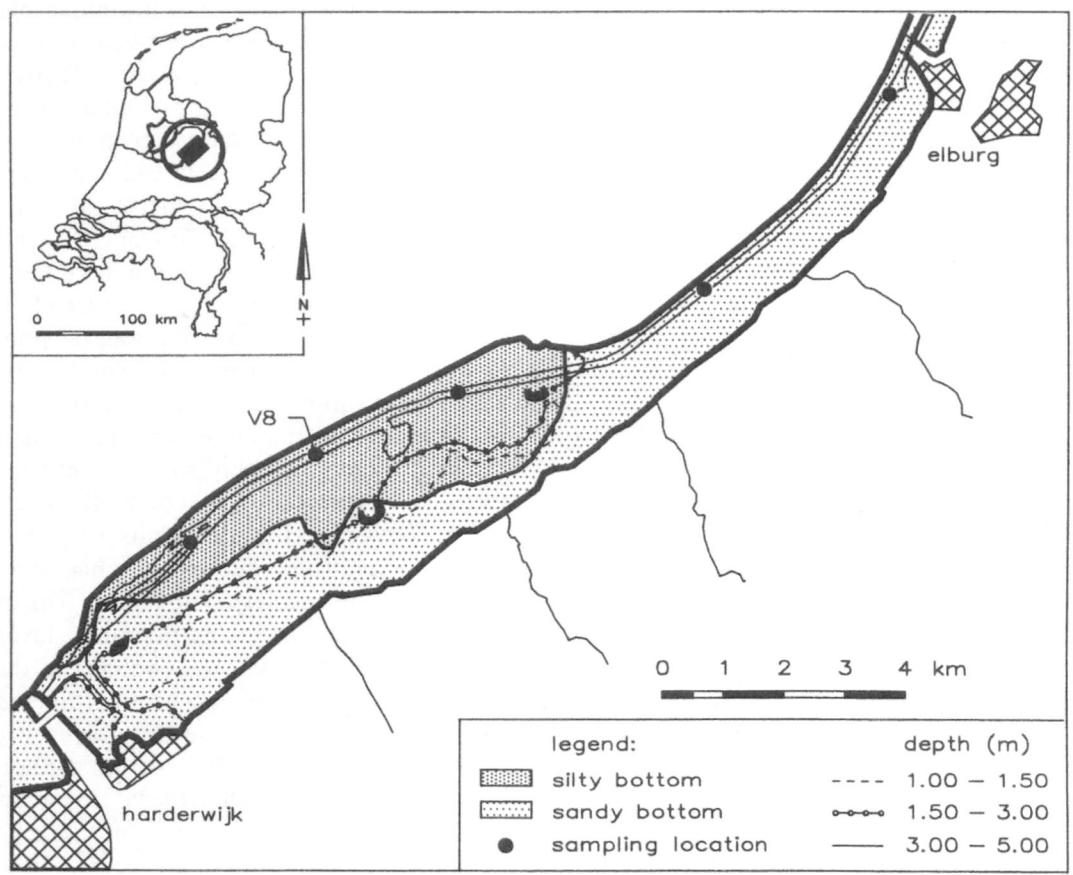

Fig. 2. Lake Veluwe and its surroundings.

loading from 3 to 1 gP m^{-2} y^{-1} by introduction of phosphorus elimination at the sewage treatment plant discharging its effluent to the lake, and flushing the lake with water from the Flevoland polder (poor in algae and phosphorus, but rich in calcium and nitrate) during the winter, decreasing the retention time of dissolved compounds from 0.35 to 0.15 y. In 1985 the lake was flushed in summer as well, decreasing the retention time of dissolved compounds from 0.50 to 0.25 y.

Consequently total phosphorus, total nitrogen and chlorophyll concentrations decreased in 1980. In 1982 and from 1985 onwards diatoms and green algae are dominant in summer (Jagtman *et al.*, 1992) and blue-green algae appear only in the late summer and in autumn.

Data analysis

Table 2 contains results for some sediment characteristics grouped for three different periods and two types of sediment. The periods are determined by data availability and the sediment types are defined by the water content of the samples (sandy and silty sediment defined by a water content less and more than 27% by weight respectively). Several conclusions can be drawn:
- Large differences exist in the composition of sandy and silty sediments; the contents of phosphate, iron, calcium and organic matter are about five times higher in silty sediments.
- The porosity of the sediment layers is based on the water content and the specific weight of the

Table 2. Sediment characteristics Lake Veluwe (all units in $g\,kg^{-1}$ DM, except for the water content which is expressed in % of wet weight). Data: Brinkman & Van Raaphorst (1986), Luttmer *et al.* (1992).

	Sand			Silt		
	1979–1981	1982	1990	1979–1981	1982	1990
Phosphorus						
mean	0.094	0.078	0.088	0.540	0.590	0.339
s.d.	0.048	0.033	0.036	0.181	0.095	0.106
n	14	10	10	21	12	29
Iron						
mean	4.6	5.2	3.4	21.6	19.8	10.2
s.d.	1.4	3.0	1.4	6.4	5.6	4.0
n	27	10	10	21	21	29
Calcium						
mean	6.3		8.1	33.3		32.3
s.d.	4.2		11.8	11.3		10.0
n	12		11	10		29
Water content						
mean	21		23	50		46
s.d.	2		3	15		9
n	18		11	50		29

sediment. Variations of porosity with depth in the sediments were only small.

– The total P and iron content in silty sediments probably decreased in 1990 compared to 1979–1982. However, this conclusion is tentative, given the relatively large standard deviations and the sometimes small number of data. Besides, analytical methods and sample locations vary for the different periods.

Pore water concentration profiles of soluble reactive phosphorus (SRP), available for 1983 only, show a rather irregular pattern (Brinkman & Van Raaphorst, 1986). The mean value of the observed concentrations is $0.198\,g\,P\,m^{-3}$, the standard deviation $0.192\,g\,P\,m^{-3}$ ($n = 94$). Nevertheless, a distinct maximum at a depth of 0.01–0.02 m and minimum values at a depth of 0.03–0.07 m can be observed. The presence of minimum values indicates the occurrence of a precipitation process.

The return fluxes of phosphate from the sediment determined with continuous flow column experiments (Boers & Van Hese, 1988) are available for 1983 and from 1987 onwards. The SOD was measured in 1983 *in situ* using benthic chambers (modification of the method described by Rutgers van der Loeff *et al.*, 1981). Results are gathered in Table 3.

The lake is characterized by a well mixed water column and two sediment types. Averaging the

Table 3. Sediment oxygen demand ($g\,m^{-2}\,d^{-1}$) and phosphorus release ($10^{-3}\,g\,m^{-2}\,d^{-1}$) in Lake Veluwe. Data: Brinkman & Van Raaphorst (1986), Boers & Van Hese (1988), Luttmer *et al.* (1992).

	Sand			Silt		
	1983	1987	1990	1983	1987	1990
Oxygen						
mean	2.2	−0.5		2.7	0.8	
s.d.	1.1	2.5		1.4	1.5	
n	5	12		12	29	
Phosphorus						
mean	<0.4		0.8	4	1.2	0.9
s.d.			0.1	5	0.5	0.7
n	1		6	2	5	13

sediment types is not correct because of several non-linear processes. The silty part of the lake is considered to be most important for the sediment – water interactions and therefore the characteristics of the silty sediment are used for model input.

Calibration method

Observed water quality parameters of Lake Veluwe were imposed on SWITCH as monthly average boundary conditions, derived from (bi)weekly data provided by the Institute for Inland Water Management and Waste Water Treatment. This concerns the concentrations of O_2, NH_4, NO_3, inorganic P, detritus (C, N, P) and water temperature. Detritus (C, N, P) is derived from chlorophyll *a* measurements and conversion factors depending on algal species composition and their growth limitation (Los *et al.*, 1988). Total inorganic P is derived from measured SRP and a conversion factor ($1/0.55$ g total inorganic P g^{-1} SRP). Both parameters are used to quantify the sedimentation fluxes, whereas the dissolved substances are used in the determina-

Table 4. Values of preset parameters.

Parameter	Value		References
Sedimentation			
Sed. rate detritus	0.1	$m\,d^{-1}$	Los *et al.*, 1988
Sed. rate adsorbed P	0.1	$m\,d^{-1}$	Los *et al.*, 1988
Adsorbed fraction of inorg. P in water	0.55	–	Los *et al.*, 1988
Sediment			
Thickness upper layer	0.02	m	
Thickness lower layer	0.08	m	
Thickness water-boundary layer	0.001	m	
Porosity upper layer	71	vol %	
Porosity lower layer	68	vol %	
Spec. weight of sediment	2500	$kg\,m^{-3}$	
Saturation conc. for phosphate precipitation	0.05	$g\,m^{-3}\,W$	Brinkman & Van Raaphorst, 1966
Mass transport			
Nett seepage velocity	-0.0035	$m\,d^{-1}$	PER, 1986
Molecular diff. coeff. oxygen	$5.5\,10^{-5}$	$m^2\,d^{-1}$	Sweerts *et al.*, 1989
Molecular diff. coeff. nitrate	$9.3\,10^{-5}$	$m^2\,d^{-1}$	Li & Gregory, 1974
Molecular diff. coeff. ammonium	$9.0\,10^{-5}$	$m^2\,d^{-1}$	Krom & Berner, 1980
Molecular diff. coeff. phosphate	$4.2\,10^{-5}$	$m^2\,d^{-1}$	Krom & Berner, 1980
(Average) bio-irrigation multiplier	3.0	–	Brinkman & Van Raaphorst, 1986
(Average) bioturbation dispersion coeff.	$1.0\,10^{-6}$	$m^2\,d^{-1}$	Brinkman & Van Raaphorst, 1986
Constants rate			
Degradation in boundary layer at 20 °C	0.075	d^{-1}	Los *et al.*, 1988
Temp. coeff. degradation	1.045	–	Bowie *et al.*, 1985
Temp. coeff. incorporation in upper layer	1.045	–	Bowie *et al.*, 1985
Temp. coeff. nitrification	1.07	–	Bowie *et al.*, 1985
Temp. coeff. denitrification	1.07	–	Bowie *et al.*, 1985
Temp. coeff. P precipitation	1.0	–	Bowie *et al.*, 1985
Stoch. constant detr. N	0.09	$gN\,g^{-1}\,C$	Los *et al.*, 1988
Stoch. constant detr. P	0.0065	$gP\,g^{-1}\,C$	Los *et al.*, 1988
Stoch. constant O_2 consumption detr. degr.	3.1	$g\,O_2\,g^{-1}\,C$	Los *et al.*, 1988
Stoch. constant O_2 consumption nitrification	4.57	$g\,O_2\,g^{-1}\,N$	Los *et al.*, 1988

tion of the return fluxes. The temperature in the sediments is supposed to be equal to the water temperature.

Several model parameters were excluded from calibration, because they could be estimated from field data and literature. These parameters are listed in Table 4. The net seepage velocity is calculated as the difference of infiltration in the western zone and seepage in the eastern zone as established for Lake Veluwe (PER, 1986). The PO_4 saturation concentration was determined as the lower boundary (precipitation process) of the observed SRP concentrations in pore water (Brinkman & Van Raaphorst, 1986). The temperature coefficients for the conversion processes were set at values well within the range given in literature (see Bowie *et al.* (1985) for an overview). The simulated bottom layer is 0.1 m thick. This layer agrees roughly with the layer affected by bioturbation.

Table 5. Values of parameters obtained by calibration.

Parameter	Value	
Detritus		
Degr. rate in upper layer at 20 °C	0.06	d^{-1}
Degr. rate in lower layer at 20 °C	0.0022	d^{-1}
Ammonium and nitrate		
Nitrification rate at 20 °C	50	d^{-1}
Denitrification rate at 20 °C	50	d^{-1}
Critical nitrate concentration	0.1	$g\,N\,m^{-3}$
Phosphate		
Adsorption cap. oxidized sediment	0.6	$g\,P\,kg^{-1}\,DM$
Adsorption cap. upper red. sediment	0.3	$g\,P\,kg^{-1}\,DM$
Ratio adsorption capacity upper and lower reduced sediment	0.25	–
Precipitation rate	1.2	d^{-1}
Half saturation conc. P adsorption	0.1	$g\,m^{-3}\,W$
Critical thickness oxidized layer	0.009	m

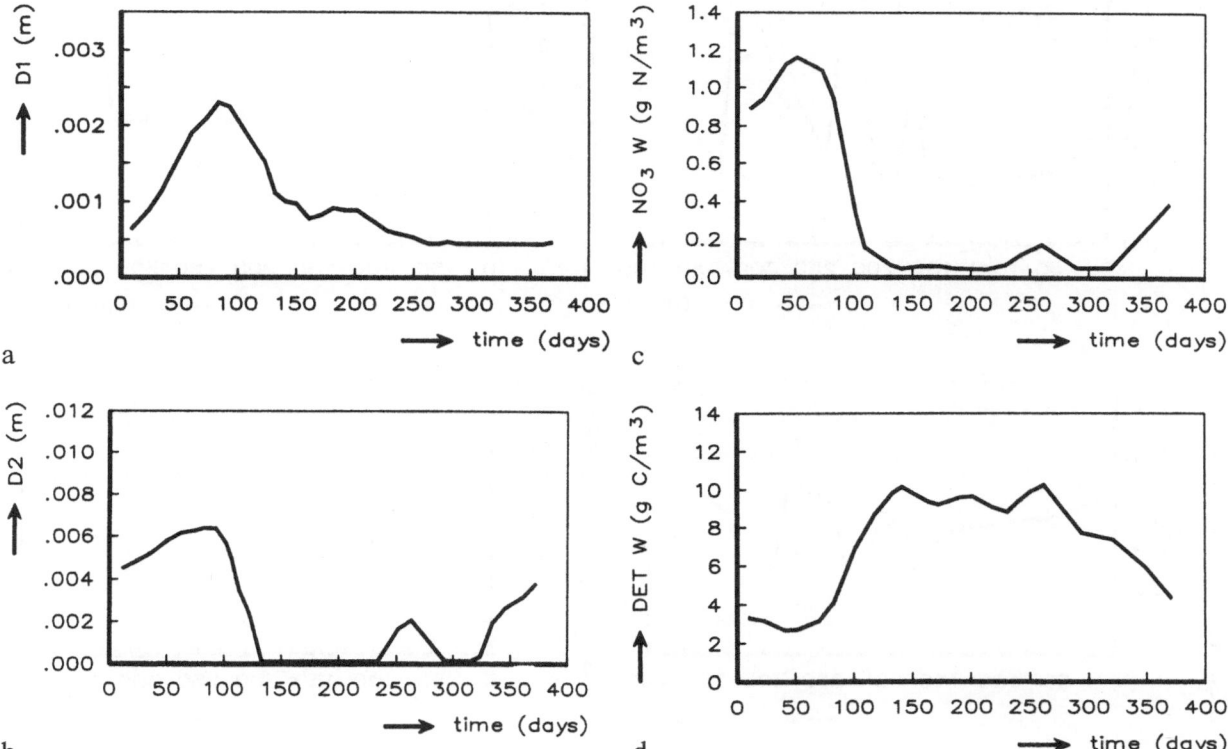

Fig. 3. The thicknesses of the aerobic and denitrifying layers (D1, D2), nitrate and detritus in the overlying water (NO3 W, DET W) for 1978.

The parameterization of SWITCH focused on 1978, when Lake Veluwe was very eutrophic. The bottom was still assumed to be in equilibrium with the external loadings. In the absence of measurements the main objective was to obtain the same concentrations and contents in the beginning and the end of the year. Next, the simulation results for the whole period 1978–1990 were checked on the available measurements on sediment content, pore water concentrations and fluxes between water and sediment.

Results and model performance

Results for 1978

Parameter values obtained by calibration are summarized in Table 5. Figure 3a shows that the thickness of the aerobic layer varies between 0.5 10^{-3}, the imposed minimum, and 2.5 10^{-3} m, which agrees with data from literature (Revsbech *et al.*, 1980; Sweerts *et al.*, 1989). The denitrifying layer (Fig. 3b) is absent during a large part of the summer as a consequence of the depletion of nitrate in the overlying water (Fig. 3c). In combination with the detritus influx of the spring algae bloom (Fig. 3d), this results in a collapse of the oxidized layer $(d_1 + d_2)$ between days 100 and 130. The aerobic layer remains very thin during the first winter months, due to the accumulation of detritus produced by autumn blooms.

The sediment oxygen demand varies within realistic bounds $(1.1–3.7 \, gO_2 \, m^{-2} \, d^{-1})$. The chemical oxidation constitutes about 60% of the oxygen consumption. Three times, the thickness of the oxidized layer falls below the critical value with respect to phosphate. The pattern of the

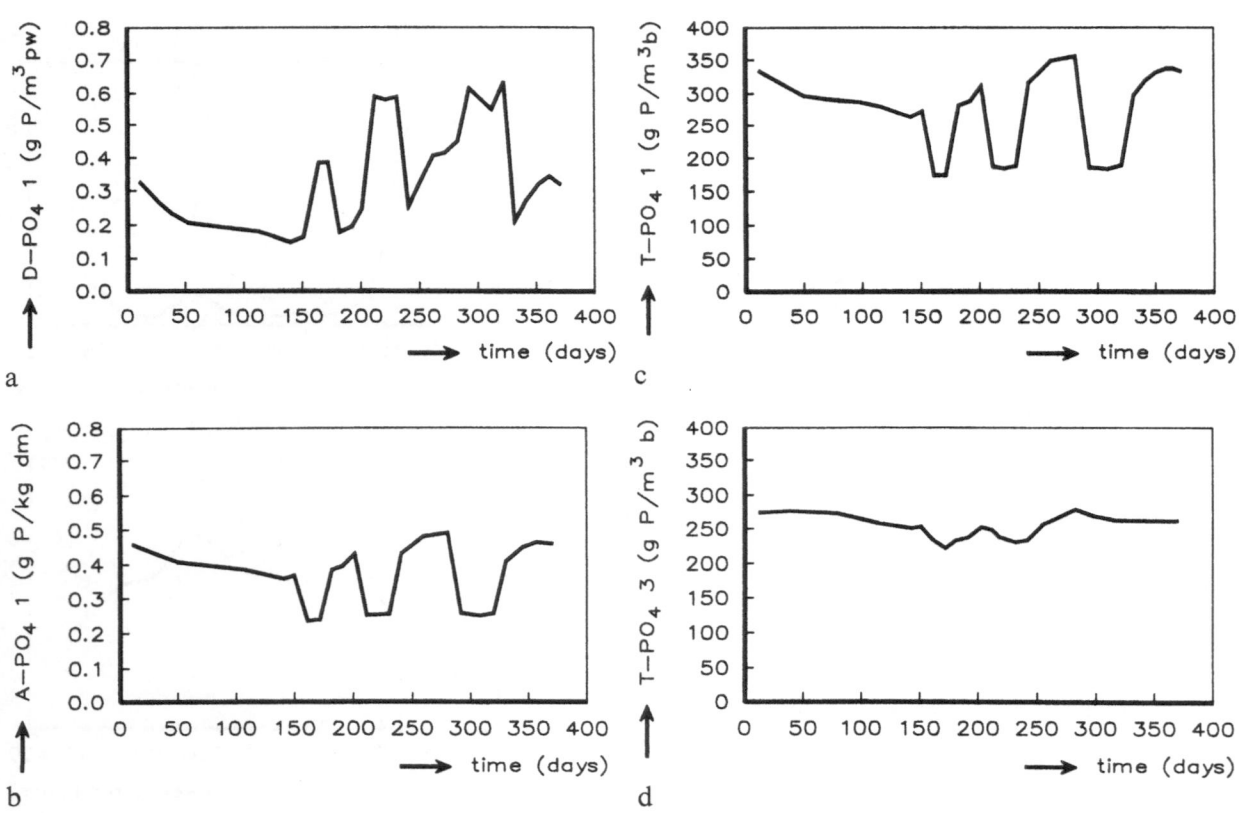

Fig. 4. Dissolved phosphate (D-PO4 1) and adsorbed phosphate (A-PO4 1) in the oxidized layer and total inorganic phosphate in the oxidized and upper reduced sediment layers (T-PO4 1, T-PO4 3) for 1978.

phosphate concentrations in pore water and sediment is entirely dominated by the sorption phenomena connected with the build-up or the collapse of the oxidized layer. The dissolved phosphate concentration in the top layer $(d_1 + d_2)$ shows maxima at a collapse of the oxidized layer (Fig. 4a). Consequently, 'explosive' phosphate return fluxes to the overlying water are simulated. The course of the adsorbed phosphate concentration is opposite to the dissolved phosphate concentration (Fig. 4b). Large quantities of phosphate are involved in the simulated fluxes as appears from Fig. 4c. The inorganic phosphate concentration in the overlying water and the total phosphate concentration in the sediment top layer are subject to large variations. Nevertheless, the total phosphate concentrations in the lower layers are hardly affected (Fig. 4d). Figure 5 shows the calculated release of phosphorus and SRP in the overlying water. The increase in SRP and total P in the overlying water during the summer of 1978 is almost $0.4 \, \text{g m}^{-3}$. This cannot be explained from the external phosphorus loading

(PER, 1986). To explain the increase in concentration by sediment release only, an average release of about $0.005 \, \text{gP m}^{-2} \, \text{d}^{-1}$ is necessary. Since the parameters of the model are chosen for the silty part of the lake, the magnitude of the calculated release may be overestimated. The variation in time of the calculated release, as a consequence of the repeatedly collapsing of the oxidized layer, is remarkable. Considerable fluctuations can also be observed in the SRP concentration in the overlying water.

The phosphate precipitation rate has been assigned a seemingly high value. It must be kept in mind that relatively high concentrations of coprecipitants such as iron are implicitly present in the first order rate. Precipitated phosphate is absent in the top layer, whereas a steep gradient occurs in the reduced layers. About 90% of the phosphate in the lower reduced layer is present in a precipitated mineral, the quantity of which changes only slowly. The gradient causes a constant upward transport of precipitated phosphate to the oxidized top layer, where it dissolves.

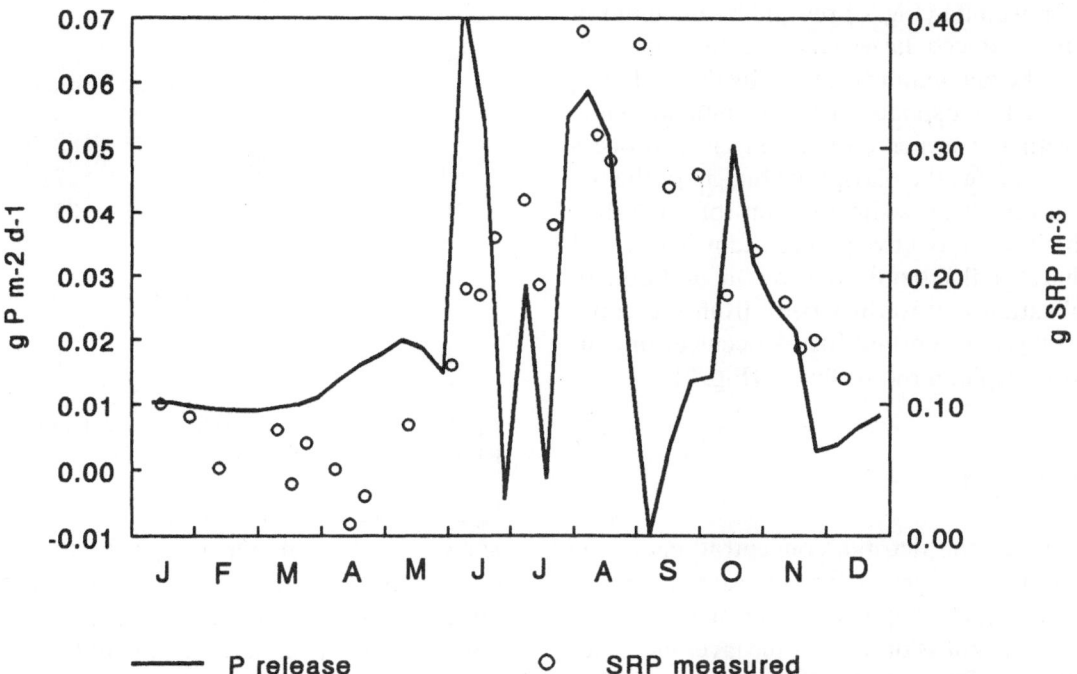

Fig. 5. Simulated phosphate release from the sediment and soluble reactive phosphorus in the overlying water in 1978.

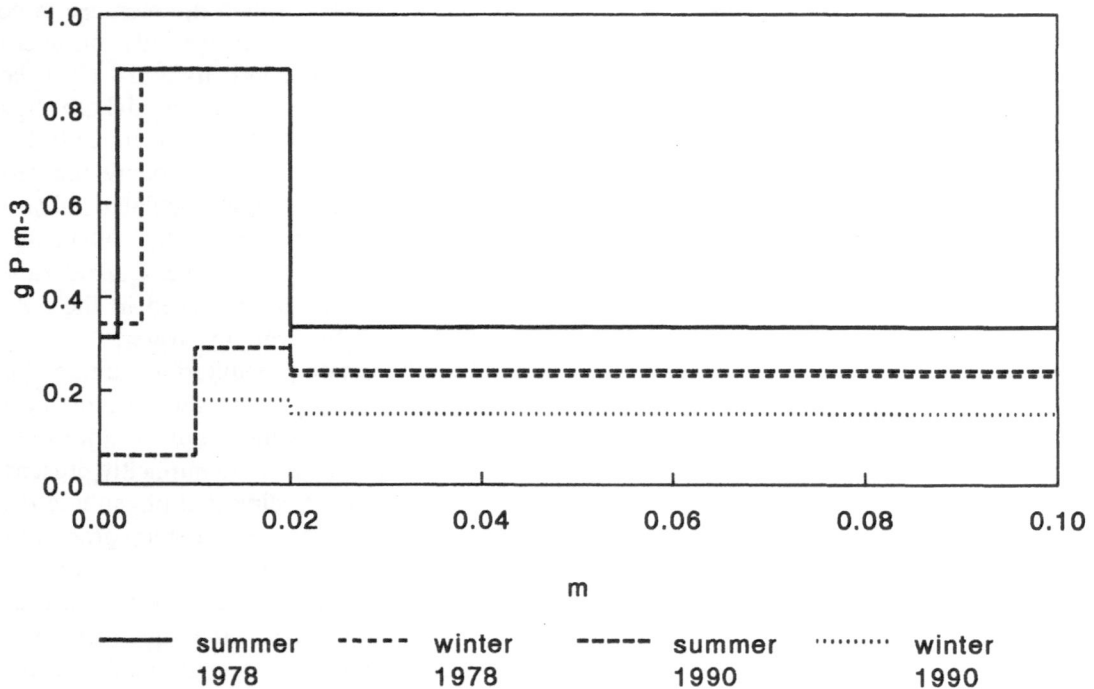

Fig. 6. Profiles of dissolved phosphorus in the sediment for summer and winter of 1978 and 1990.

The simulated concentration profile of dissolved phosphate (Fig. 6) reveals a maximum in the upper reduced layer just like the observed profiles. The maximum is caused by the exchange of adsorbed phosphate and precipitated phosphate with the upper reduced layer, a process which dominates the phosphate budget of the oxidized layer. The sedimentation of adsorbed phosphate is a relatively important source of phosphate in the oxidized layer in addition to mineralization and to dispersion from the upper reduced layer. Important losses occur during increased phosphate return fluxes (Fig. 7).

Results for 1978

In the model, the detritus concentrations in the sediment decrease gradually as a consequence of the decreasing chlorophyll *a* concentration as of 1980. The thickness of the aerobic layer increases from 1978 to 1985 (Fig. 8). From 1985 onwards, but with an exception in late 1988, the aerobic

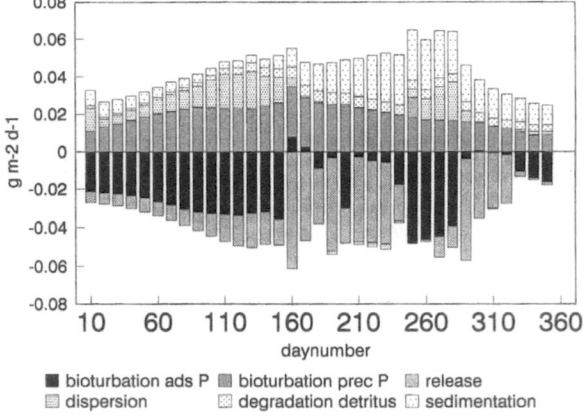

Fig. 7. Main processes of the mass balance for total inorganic phosphate in the oxidized layer for 1978.

layer is constantly thicker than 0.002 m. At the same time the summer dips in the thickness become narrower and less deep. Dips in the nitrate concentration in summer in the overlying water and in the top layer of the sediment hardly change over the period 1978–1990. Therefore, compared to the development of the aerobic layer, little

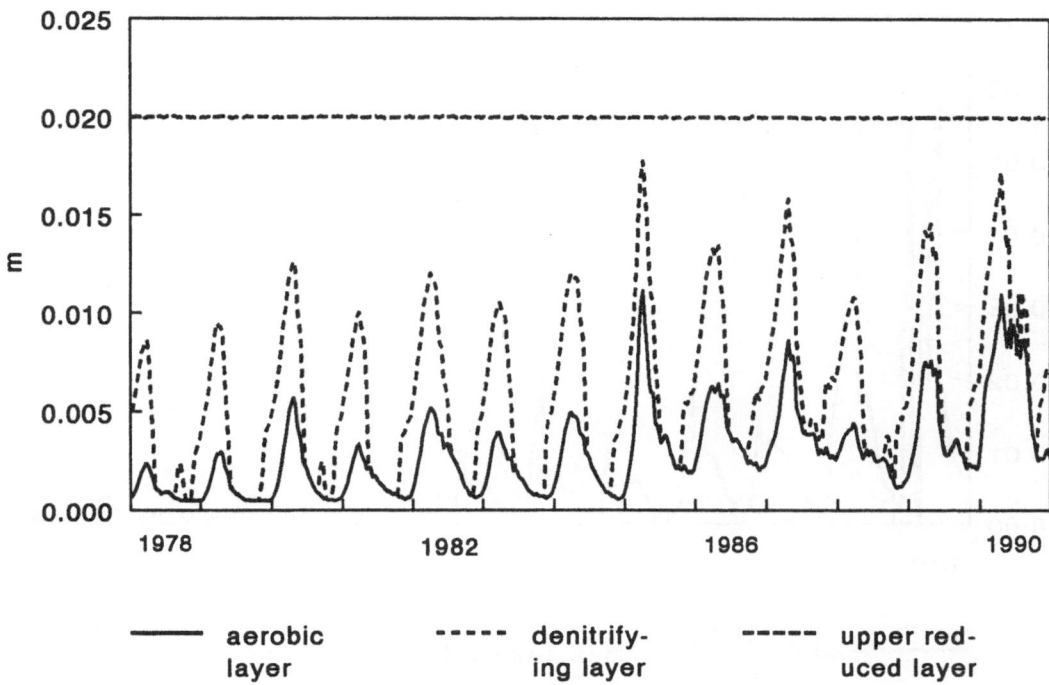

Fig. 8. Thickness of the aerobic, oxidized and upper – reduced layer (cumulative) of the sediment for 1978–1990.

changes in the thickness of the denitrifying layer. Due to the increased aerobic layer the denitrifying layer is displaced to greater depths. The total oxidized layer increases, which implies that the phosphate adsorption capacity per square meter increases.

The calculated sediment oxygen demand decreases gradually from 2.3 g m^{-2} d^{-1} in 1978 to 1.3 in 1983, 0.5 in 1985 and 0.4 g m^{-2} d^{-1} in 1990, which compares well with the measurements presented in Table 3. Mass balance studies revealed that from 1978 to 1983 there was almost continuously a (yearly averaged) nitrogen retention of 0.02–0.06 gN m^{-2} d^{-1} in the lake (PER, 1986), meaning that the sedimentation of organic nitrogen exceeded the release of ammonium and nitrate from the sediment. The retention calculated by the model is almost the same as the denitrification and varies between 0.01 and 0.05 gN m^{-2} d^{-1}.

The 'explosive' phosphate return fluxes, as simulated in 1978, becomes weaker each year and finally ceases to occur in 1985. Average release rates decrease from 6.8 10^{-3} g m^{-2} d^{-1} in 1983, to 2.7 10^{-3} in 1987 and 1.7 10^{-3} g m^{-2} d^{-1} in 1990, about twice as high as measured values (Table 3 and Fig. 9). It is reminded that the model uses parameters for the silty part of the lake, which is probably more active in the exchange of phosphorus between water and sediment.

The dissolved phosphate concentrations in the pore water remain relatively constant, which is caused by the buffering capacity of the adsorption and precipitation processes. The calculated dissolved phosphate concentrations in the upper and lower layer agree very well with measurements from 1983. In Fig. 6 calculated PO_4 profiles of 1978 and 1990 are compared. In 1978 calculated concentrations were higher and maxima were found closer to the sediment–water interface.

The simulated total inorganic phosphate content decreases by 40% (Fig. 10). A similar percentage of the phosphate bound in minerals has dissolved by 1990. The decreasing tendency of the phosphorus content is also visible in the observations (Table 2), although it cannot be statis-

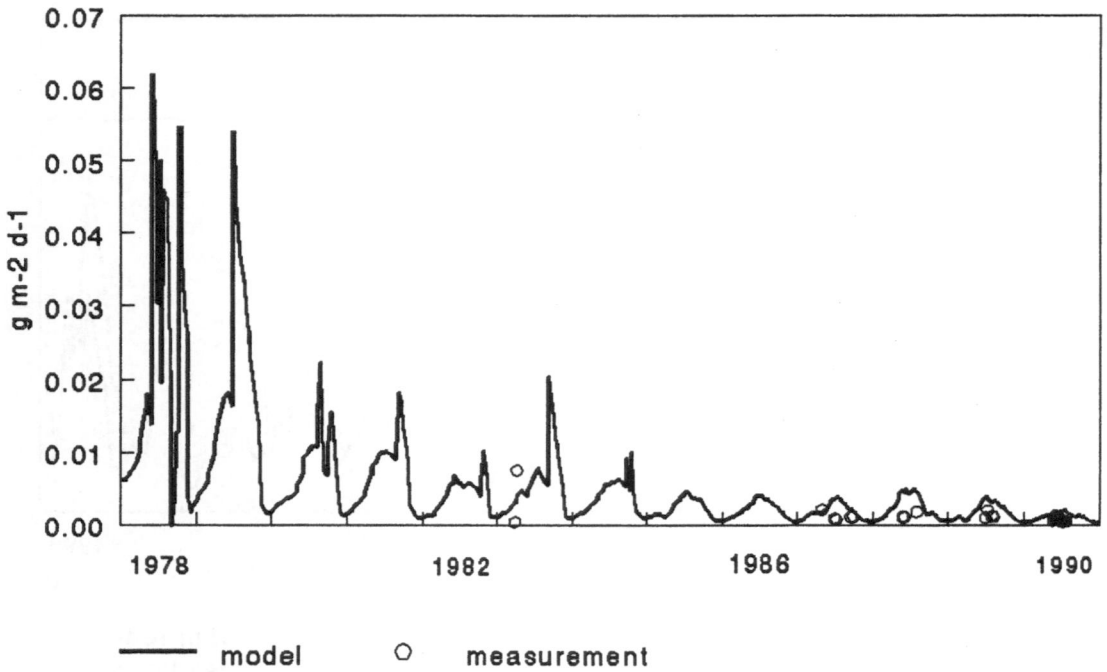

Fig. 9. Release of phosphate from the sediment for 1978–1990.

tically proved. However, mass balance calculations pointed out that on a yearly average there is net retention of phosphate in the sediment, which may be as high as 50% (PER, 1986). Sedimentation therefore exceeds the phosphate release, whereas the model produces the opposite.

Discussion

When applying a model describing sediment-water interactions one has to tackle the problem of spatial heterogeneity. Sediment characteristics, if available at all, are difficult to interpret in terms of average values. For this application it was decided to quantify those characteristics on the basis of data concerning the silty sediments of the lake, while both sedimentation and return fluxes were derived from lake average concentrations. The main consequence of this discrepancy is that the return fluxes predicted with SWITCH and the process parameters determined by calibration for

silty sediments are biased by the features of and the processes in the sandy sediments.

Little is known in literature about how to describe the net sedimentation of nutrients, organic matter and phosphate adsorbents. In this study we used measured chlorophyll and phosphorus in combination with conversion factors derived from an application of a water quality model calibrated for Lake Veluwe (Los *et al.*, 1988). It was assumed that the phosphate adsorption capacity did not change over the period 1978–1990. There are indications that the composition is changed as a result of the flushing (Brinkman & Van Raaphorst, 1986; Jagtman *et al.*, 1992), but the sediment phosphorus content did not show a significant tendency.

As appears from the simulations for Lake Veluwe the oxidized layer can become very thin, but it can never disappear entirely because of the high monthly averaged dissolved oxygen concentrations in the water column and the minimal layer thickness imposed on SWITCH. However, the dissolved oxygen concentration is subject to sub-

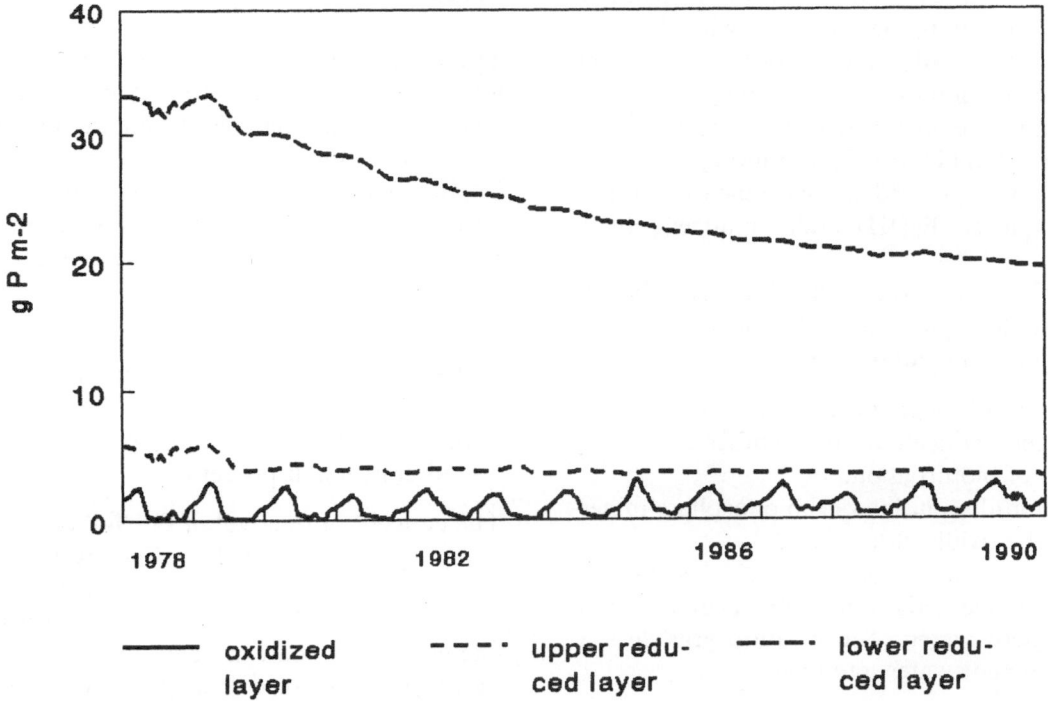

Fig. 10. Total inorganic phosphate (cumulative) in the sediment for 1978–1990.

stantial daily variations on the one hand and to spatial variations (patchiness) on the other. In order to reproduce the complete reduction of the sediment top layer and the increased phosphorus return flux according to the 'classical' theory, a critical thickness of the oxidized layer was introduced (0.9 mm for Lake Veluwe), below which the top layer is assumed to lose a large part of its phosphate adsorption capacity. As a result, a large part of the previously adsorbed phosphate is released into the pore water. The concentration gradient at the sediment-water interface becomes much steeper and, consequently, the dispersive flux increases suddenly.

This repeated reduction of the top layer of the sediments results in a remarkable peaked pattern of the simulated phosphorus release in 1978. Measured phosphorus concentrations in the overlying water neither confirm nor conflict with this pattern. Boers (1986) also reports a strongly fluctuating release over a summer period, but no information of frequently measured release over a

summer, nor information on repeatedly changing oxidation-reduction state of the top sediment could be found in literature. However, the transition between the adsorption capacities for oxidized and reduced sediments occurs instantly in SWITCH, whereas the reduction of the sediments will take some time in reality, so the fluctuations may be overestimated.

The vertical transport of phosphate in SWITCH is based mainly on the transfer of adsorbed phosphate to the reduced layers and on the transfer of precipitated phosphate to the oxidized layer through bioturbation. The adsorbed phosphate desorbs instantaneously in the (upper) reduced layer, whereas the precipitated phosphate dissolves immediately in the oxidized layer. These processes are 'smoothed out' by diffusive transport of dissolved phosphate, but nevertheless they result in maximum dissolved concentrations in the upper reduced layer. Although this shape of the dissolved phosphate concentration profile is confirmed by measurements, it is unlikely that desorption and dissolution proceed instantaneously. There must be a relation with the rates of the oxidation and reduction processes and the local accumulation of reduced and oxidized substances in the sediment, which is ignored in this version of SWITCH. Furthermore, Redshaw *et al.* (1990) suggested the existence of organically complexed Fe(III) under reduced conditions.

SWITCH uses the following hypothetical mechanism to explain the shift from eutrophic to more oligotrophic conditions:

1. transport of phosphate containing minerals from the reduced to the oxidized layer by means of bioturbation,
2. oxidation/dissolution of the phosphate minerals in the oxidized layer and
3. net release of phosphate from the sediments into the overlying water under influence of steep pore water concentration gradients at the sediment-water interface.

This mechanism leads to a gradual removal of the large stock of (precipitated) phosphate from the sediment, which might be expected, but cannot be unequivocally confirmed from the sediment phosphorus data of Lake Veluwe. On the other hand, mass balance calculations revealed that after the measures taken there is still net retention of phosphate in this lake of about 50% of the external loading (PER, 1986). As an increase in the sediment phosphorus content can be neglected, the phosphate content of the sediment in the silty area of Lake Veluwe may be determined to a large extent by dilution due to settling of material with a lower phosphate content than the sediment, or by an underestimation of the removal of phosphate to greater depths by infiltration and mixing processes. Also refractory phosphorus in the sediments should have more attention in this context. After the attention given to mass balances for the overlying water (Sas, 1989), mass balances over the active part of the sediment should be a subject of further study. A combined modelling of sediment and overlying water can be helpful in this respect.

Conclusions

The model SWITCH was applied to Lake Veluwe as a first attempt to test if the functional processes for nutrients in the interaction between water and sediment are described properly and to estimate values for the parameters in the model. Evaluation of the simulations learned that SWITCH can simulate in broad outline realistically and consistently:

- the thickness of the oxidized layer,
- the concentrations of nutrients in pore water and sediment and
- the mass fluxes connected with the transport and conversion processes.

The concept of removal of the 'lid' on sediments is reproduced and results in the increased phosphate release. The formulations used to describe long-term behaviour of the sediment-water exchange of phosphate are based upon interactions between a dissolved fraction and an adsorbed and a precipitated fraction. Incompleteness of the available data prohibited a full confirmation of the chosen formulations.

An underestimation of removal processes to deeper parts of the sediment resulted in too high release rates and, consequently, violation of the phosphorus mass balance of the overlying water.

Acknowledgements

We wish to thank P. C. M. Boers, G. Th. M. van Eck, W. van Raaphorst and N. M. de Rooij for their useful suggestions concerning the formulations, A. Hendriks for encoding SWITCH correctly and elegantly in FORTRAN 77 and A. M. Wagenaar-Hart for the grammatical corrections.

References

Beck, M. B., 1985. Water quality management: A review of the development and application of mathematical models. Springer Verlag, Berlin.

Berner, R. A., 1974. Kinetic models for the early diagenesis of nitrogen, sulfur, phosphorus and silicon in anoxic marine sediments. In E. D. Goldberg (ed.), The Sea: Marine Chemistry, Vol. 5. John Wiley & Sons, New York: 427–450.

Boers, P. C. M., 1986. Studying the phosphorus release from the Loosdrecht Lakes sediments, using a continuous flow system. Hydrobiol. Bull. 20: 51–60.

Boers, P. C.M. & J. J. Boon, 1988. Unmasking the particulate organic matter in a lake ecosystem: origin and fate of POM in the shallow eutrophic Loosdrecht Lakes (The Netherlands). Arch. Hydrobiol. 31: 27–34.

Boers, P. C. M. & O. Van Hese, 1988. Phosphorus release from the peaty sediments of the Loosdrecht Lakes (The Netherlands). Wat. Res. 22: 355–363.

Bouldin, D. R., 1968. Models for describing the diffusion of oxygen and other mobile constituents across the mud-water interface. J. Ecol. 56: 77–87.

Bowie, G. L., W. B. Mills, D. B. Porcella, C. L. Campbell, J. R. Pagenkopf, G. L. Rupp, K. M. Johnson, P. W. H. Chan and S. A. Gherini, 1985. Rates, constants and kinetics formulations in surface water quality modeling; second edition. Environmental research laboratory, Office of research and development, U.S. Environmental Protection Agency, Athens, Georgia, U.S.

Brinkman, A. G. & W. Van Raaphorst, 1986. De fosfaathuishouding van het Veluwemeer, Ph. D. Thesis. Technical University Twente, The Netherlands. 481 pp. (in Dutch with an English summary).

De Rooij, N. M., 1980. A chemical model to describe nutrient dynamics in lakes. In J. Barica & L. R. Mur (eds), Hypertrophic Ecosystems. Developments in Hydrobiology 2. Dr W. Junk Publishers, The Hague: 139–150.

Einsele, W., 1936. Uber die Beziehungen des Eisenkreislaufs zum Phosphatkreislauf im eutrophen See. Archiv für Hydrobiologie 29: 664–686.

Emerson, S. & G. Widmer, 1978. Early diagenesis in anaerobic lake sediments – II. Thermodynamic and kinetic factors controlling the formation of iron phosphate. Geochim. Cosmochim. Acta 42: 1307–1316.

Holdren, G. C. & D. E. Armstrong, 1980. Factors affecting phosphorus release from intact lake sediment cores. Envir. Sci. Technol. 14: 79–87.

Hosper, S. H., 1984. Restoration of Lake Veluwe, The Netherlands, by reduction of phosphorus loading and flushing. Wat. Sci. Technol. 17: 757–768.

Jagtman, E., D. T. Van der Molen & S. Vermij, 1992. The influence of flushing on nutrient dynamics, composition and densities of algae and transparency in Veluwemeer, The Netherlands. Hydrobiologia (in press).

Jørgensen, S. E., 1983. Eutrophication models of lakes. In S. E. Jørgensen (ed.), Application of ecological modelling in environmental management. Elsevier, Amsterdam.

Jørgensen, S. E., L. Kamp-Nielsen & H. F. Mejer, 1982. Comparison of a simple and a complex sediment phosphorus model. Ecol. Modelling 16: 99–124.

Jorgensen, B. B. & N. P. Revsbech, 1985. Diffusive boundary layers and the oxygen uptake of sediments and detritus. Limnol. Oceanogr. 30: 111–122.

Kamp-Nielsen, L., 1975. A kinetic approach to the aerobic sediment-water exchange of phosphorus in Lake Esrom. Ecol. Modelling 1: 153–160.

Li, Y-H. & S. Gregory, 1974. Diffusion of ions in sea water and in deep-sea sediments. Geochim. Cosmochim. Acta 38: 703–714.

Los, F. J., 1980. An algal bloom model as a tool to simulate management measures. In J. Barica & L. R. Mur (eds), Hypertrophic Ecosystems. Developments in Hydrobiology 2. Dr W. Junk Publishers, The Hague: 171–178.

Los, F. J., N. M. De Rooij & J. G. C. Smits, 1984. Modelling eutrophication in shallow Dutch lakes. Verh. int. Ver. Limnol. 22: 917–923.

Los, F. J., J. C. Stans & N. M. De Rooij, 1988. Eutrofiëringsmodellering van de randmeren. Delft Hydraulics, The Netherlands. 204 pp. (in Dutch).

Luttmer, W. J., P. C. M. Boers, D. T. Van der Molen & L. Van Ballegooijen, 1992. De nutriëntennalevering door de Veluwemeerbodem. RIZA report 92.010. 37 pp. (in Dutch).

Lijklema, L., 1980. Interaction of orthophosphate with iron(III) and aluminum hydroxides. Envir. Sci. Technol. 14: 537–541.

Mortimer, C. H., 1941. The exchange of substances between mud and water in lakes. J. Ecol. 29: 280–329.

PER, 1986. Bestrijding van de eutrofiëring van het Veluwemeer Drontermeer. Projectgroep Eutrofiëringsonderzoek Randmeren, Lelystad, The Netherlands. 296 pp. (in Dutch).

Redshaw, C. J., C. F. Mason, C. R. Hayes & R. D. Roberts, 1990. Factors influencing phosphate exchange across the sediment-water interface of eutrophic resevoirs. Hydrobiologia 192: 233–245.

Revsbech, N. P., J. Sørensen, Th. H. Blackburn & J. P. Lomholt, 1980. Distribution of oxygen in marine sediments measured with microelectrodes. Limnol. Oceanogr. 25: 403–411.

Rutgers van der Loeff, M. M., F. B. Van Es, W. Helder & R. T. P. De Vries, 1981. Sediment water exchanges of nutrients and oxygen on tidal flats in the Ems-Dollard Estuary. Neth. J. Sea Res. 15: 113–129.

Ryding, S. O. & C. Forsberg, 1977. Sediments as a nutrient source in shallow polluted lakes. In: H. L. Golterman (ed.), Interactions between sediments and freshwater. Dr W. Junk Publishers, The Hague: 227–234.

Sas, H. (ed.), 1989. Lake restoration by reduction of nutrient loading: expectations, experiences, extrapolations. AcademiaVerl. Richarz.

Søndergaard, M., E. Jeppesen, P. Kristensen & O. Sortkjaer, 1990. Interactions between sediments and water in a shallow and hypertrophic lake: a study on phytoplankton collapses in Lake Søbygard, Denmark. In P. Biró & J. F. Talling (eds), Trophic Relationships in Inland Waters. Developments in Hydrobiology 53. Kluwer Academic Publishers, Dordrecht: 247–258. Reprinted from Hydrobiologia 191.

Sweerts, J-P. R. A., V. St Louis & T. E. Cappenberg, 1989. Oxygen concentration profiles and exchange in sediment cores with circulated overlying waters. Freshwat. Biol. 21: 401–409.

Vanderborght, J. P., R. Wollast & G. Billen, 1977a. Kinetic models of diagenesis in disturbed sediments, Part I; Mass transfer properties and silica diagenesis. Limnol. Oceanogr. 22: 787–793.

Vanderborght, J. P., R. Wollast & G. Billen, 1977b. Kinetic models of diagenesis in disturbed sediments, Part II; Nitrogen diagenesis. Limnol. Oceanogr. 22: 794–803.

Van der Molen, D. T., 1991. A simple dynamic model for the simulation of the release of phosphorus from sediments in shallow eutrophic systems. Wat. Res. 25: 737–744.

Van Eck, G. Th. M. & J. G. C. Smits, 1986. Calculation of nutrient fluxes across the sediment-water interface in shallow lakes. In P. Sly (ed.), Sediments and Water Interactions. Springer-Verlag, New York: 289–301.

Van Raaphorst, W., P. Ruardij & A. G. Brinkman, 1988. The assessment of benthic phosphorus regeneration in an estuarine ecosystem model. Neth. J. Sea Res. 22: 23–36.

Hydrobiologia **253**: 301–309, 1993.
P.C.M. Boers, T.E. Cappenberg & W. van Raaphorst (eds),
Proceedings of the Third International Workshop on Phosphorus in Sediments.
© 1993 *Kluwer Academic Publishers.*

Measurement and verification of rates of sediment phosphorus release for a hypereutrophic urban lake*

Martin T. Auer[1], Ned A. Johnson[1], Michael R. Penn[1] & Steven W. Effler[2]
[1]*Department of Civil and Environmental Engineering, Michigan Technological University, Houghton, MI 49931, USA;* [2]*Upstate Freshwater Institute, P.O. Box 506, Syracuse, NY 13214, USA*

Key words: Phosphorus, sediment, lakes, eutrophication

Abstract

The contribution of sediment release to the phosphorus budget of hypereutrophic Onondaga Lake was determined through laboratory measurements made on intact cores. Rates ranged from 9–21 mg P m^{-2} d^{-1} with a mean of 13 mg P m^{-2} d^{-1}, values similar to those observed in other lakes of comparable trophic state. There was no statistically significant trend in rates in time (July versus September) or in space (location along the major N/S axis of the lake). Rates of sediment phosphorus release measured in the laboratory compared favorably with the observed rate of soluble reactive phosphorus accumulation in the lake's hypolimnion. The sediments are the second largest source of phosphorus for Onondaga Lake, contributing 24% of the overall phosphorus load to the system.

Introduction

Phosphorus availability controls phytoplankton growth in most freshwater systems (Hutchinson, 1973). The trophic state of a lake and the standing crop of phytoplankton which it supports are well correlated with water column phosphorus levels (Chapra & Dobson, 1981; Dillon & Rigler, 1974). Water quality reclamation efforts, aimed at reducing productivity through phosphorus management, generally focus on external sources, e.g. wastewater treatment plant effluents and agricultural runoff. Sediment inputs of phosphorus are less amenable to direct management for logistic and economic reasons, although a variety of approaches (e.g. dredging, chemical inactivation, and hypolimnetic aeration) have been successfully applied (Cooke *et al.*, 1986).

The role of sediments in mediating water column conditions must be considered in the development of lake restoration programs – a fact reflected in the attention which this topic receives within the scientific community (Golterman, 1977; Sly, 1986; Psenner & Gunatilaka, 1988). The contribution of sediment release to the total phosphorus load of a lake (3–35%, mean 19% for 22 lakes with anoxic hypolimnia; Nurnberg, 1984) may exceed that of some external sources. Several case studies (Larsen *et al.*, 1981; Marsden, 1989; Welch *et al.*, 1986) have demonstrated that sediment feedback can retard a lake's response to remediation efforts. Parallel programs of external loading reduction and in-lake manipulation of physical-chemical conditions (e.g. hypolimnetic aeration) can serve to reduce the trophic state of a lake and accelerate the rate of recovery (Gachter, 1987).

Contribution No. 116 of the Upstate Freshwater Institute, Inc. Contribution No. 23 of the New York State Fresh Water Research Institute.

Local, state, and federal authorities are currently considering remediation activities aimed at improving the water quality of Onondaga Lake, a hypereutrophic urban system. The interim goal of the program is to restore 'fishable' and 'swimmable' conditions in the lake. A variety of phosphorus control measures, including advanced waste treatment, diversion of wastewater effluent, and hypolimnetic aeration are being considered. The research reported here is intended to support the development of a phosphorus management plan for Onondaga Lake by defining the contribution of sediment inputs to the phosphorus budget of the lake. A companion paper (Driscoll et al., 1993) elucidates the mechanisms regulating phosphorus release in Onondaga Lake and comments on their significance to the suitability of hypolimnetic aeration as a remediation methodology.

Nurnberg (1987) recommends three approaches for estimating internal phosphorus loads:

- direct, laboratory measurement of release rates using intact sediment cores,
- calculation of the rate of phosphorus accumulation in the hypolimnion through examination of vertical profiles, and
- calculation by difference using the annual phosphorus budget, with all other sources and sinks determined independently.

Here, we utilize direct measurement, verifying the rates by comparison with observations of hypolimnetic phosphorus accumulation. Doerr (1991) has utilized the third approach, successfully applying measured rates in mass balance modeling of total phosphorus in Onondaga Lake. Together, these efforts constitute a rigorous and comprehensive testing of the direct measurement technique.

Methods

Study system

Onondaga Lake (area 11.7 km^2; mean depth 12 m), is a hardwater, dimictic lake located in metropolitan Syracuse, New York. The lake has great potential as a recreational resource, but severe water quality degradation limits present use (it has received the municipal and industrial wastes of the region for over 100 years; Effler, 1987). Onondaga Lake is hypereutrophic (Auer et al., 1990), receiving an abundant supply of phosphorus from seven tributaries and the Syracuse Metropolitan Treatment Plant. The lake experiences severe dissolved oxygen depletion as a result of its high productivity. Anoxic conditions are first detected in mid-May and by late July ± 50% of the lake volume is anoxic and 70% of the sediment surface is overlain by anoxic waters. Onondaga Lake is naturally enriched in sulfate (± 150 mg SO_4^{2-} l^{-1}); as a result of this and the lake's hypereutrophic state, high concentrations of H_2S (up to 16 mg S l^{-1}) develop in the hypolimnion. Scavenging by sulfide keeps iron concentrations low ($Fe^{2+} < 2$ mg l^{-1}).

Many of the lake's water quality problems (high algal standing crop, low transparency, oxygen depletion, and accumulation of toxic levels of ammonia and hydrogen sulfide in the hypolimnion) are linked to phosphorus inputs. Phosphorus loads to Onondaga Lake (1987: 350 kg P d^{-1}; 10.6 g P m^{-2} y^{-1}; Heidtke & Auer, 1992) exceed the bound for mesotrophy (92 kg P d^{-1}; OECD, 1982; Heidtke & Auer, 1992) by almost a factor of 4. The lake has been severely polluted with calcium (Ca ± 180 mg l^{-1}) by an adjoining soda-ash manufacturer (Effler, 1987; Driscoll et al., 1993) and the entire water column is oversaturated with calcium carbonate (Effler, 1987). Rates of phosphorus deposition (48.4 mg P m^{-2} d^{-1}; Wodka et al., 1985), influenced by the system's unusual chemistry, are extremely high. Approximately 30% of the phosphorus sedimented from the upper waters is in an inorganic form, probably co-precipitated with calcite (Effler et al., 1985), and the surficial sediment $CaCO_3$ content ranges from 30–92% DW (mean 66% DW; Johnson, 1989). Incorporation of this material into the sediment (Effler et al., 1979; Devan & Effler, 1984) has been thought to create an atypical chemical condition which may not favor mineralization and subsequent release (Murphy, 1978;

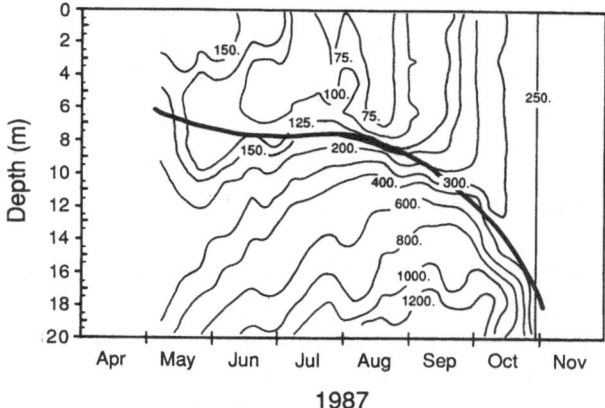

Fig. 1. Isopleths of soluble reactive phosphorus in Onondaga Lake for 1987 (mg P m^{-3}). Heavy line indicates approximate position of the thermocline.

Honstein, 1981; Effler *et al.*, 1985). Recent field measurements (Fig. 1) offer information to the contrary, however, showing that accumulation of soluble reactive phosphorus in the hypolimnion is

a prominent feature of the system during summer stratification.

Core collection

The distribution of phosphorus in the surficial sediments of Onondaga Lake is influenced by the nature of the depositional environment and by proximity to point source and tributary inputs (Johnson, 1989). The highest levels of sediment-P are generally found in a depositional basin which occupies much of the central portion of the lake (Fig. 2). This region is overlain by anoxic water from mid-May to late October and thus the sediments are exposed to physical and chemical conditions favorable for phosphorus release. Intact, undisturbed cores were collected on 31 July and 3 September 1987 at three locations representative of the deep-water depositional basin (Fig. 2) using a Wildco KB Core Sampler (Model

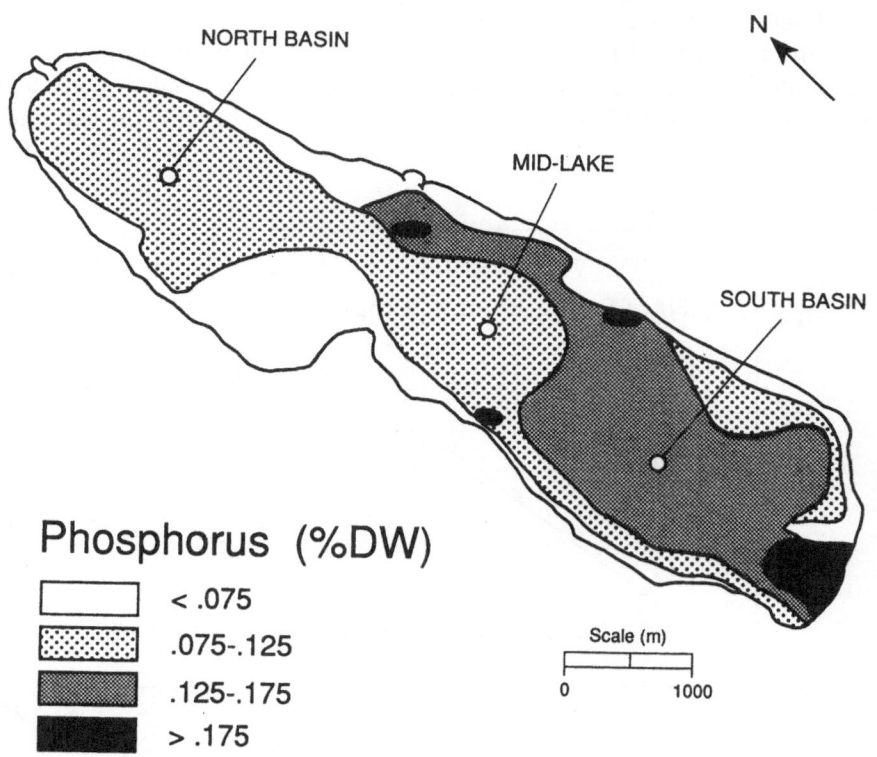

Fig. 2. Surficial sediment phosphorus distribution and the location of coring sites in Onondaga Lake.

No. 2404) fitted with (4.7 cm × 76 cm) CAB (cellulose-acetate-buterate) core liners. Cores were returned to the laboratory on ice, sealed at the top and bottom and stored at 8 °C. In some cases, cores were stored up to 21 days, however, most experiments were begun within 1–3 days of core collection.

Experimental design

Phosphorus release rates were measured on six cores from each of the three coring locations. Core liners containing approximately equal quantities (25 cm; 400 cm³) of sediment and overlying water were placed in a constant temperature incubator in the dark at 8 °C. Nitrogen gas was gently bubbled into the overlying water to maintain anaerobic conditions and to enhance mixing. Bubbling did not disturb the sediment–water interface. The initial pH of the overlying water was ± 7.0, essentially that of the lake's hypolimnion during summer (pH ± 7.3). Bubbling with nitrogen increased the pH of the overlying water to ± 9 within 1–3 days of the start of the experiment. Distilled water was added to correct evaporation losses and replace water removed for chemical analysis. Addition of distilled water, instead of phosphorus-rich lake water, helped to maintain phosphorus concentrations in the overlying water near the original level.

Sampling, analysis and rate calculation

A 50 ml aliquot of water was removed daily from each core over the three-week duration of the experiments. An equal amount of distilled water was returned to the core to maintain a constant volume. Water removed from the core was filtered through a 0.45 μm membrane filter (polycarbonate or nitrocellulose), placed in acid-washed glass bottles and frozen. Samples were analyzed for soluble reactive phosphorus (SRP; ascorbic acid technique) and total dissolved phosphorus (TDP; persulfate digestion) as prescribed by *Standard Methods* (APHA, 1985). SRP and TDP release rates were virtually identical as SRP accounted for more than 94% of the TDP fraction; only SRP release rates are presented here.

The rate of phosphorus accumulation (mg P d^{-1}) was determined through least squares, linear regression of daily measurements of the mass

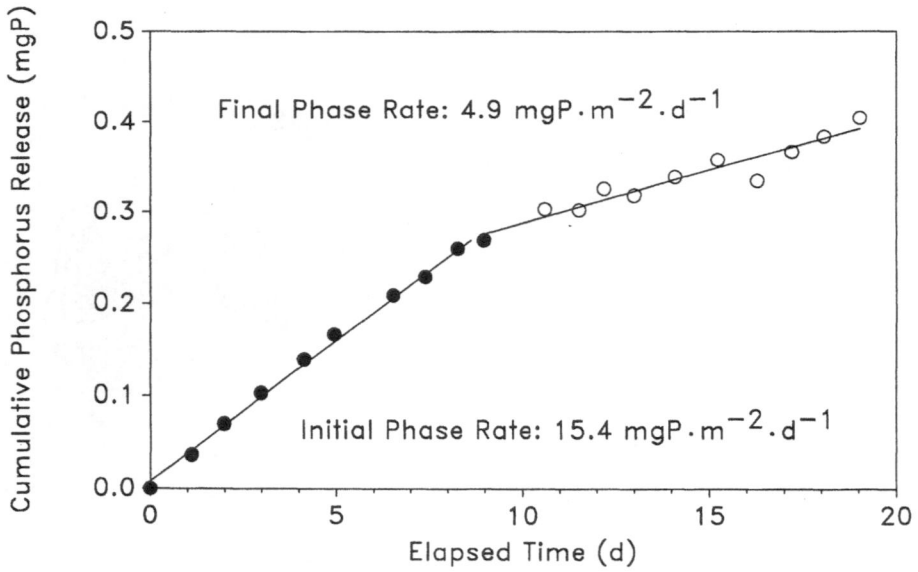

Fig. 3. Phosphorus release measurements: illustration of two-phase behavior.

of phosphorus in the overlying water, corrected for phosphorus removed through sampling. Areal release rates (mg P m^{-2} d^{-1}) were calculated as the rate of phosphorus accumulation divided by the surface area of the sediment in the core (m^2).

Results and discussion

Rate measurements

A two-phase pattern of phosphorus release was observed in all cores (Fig. 3). Rates observed during the initial 7–10 days were considerably higher than those observed over the remainder of the experiment. Phosphorus released to the overlying water includes that liberated from particulate matter freshly-deposited at the sediment-water interface as well as that originating from diagenetic processes within the sediment. The high, initial rates may reflect the result of processes occurring at the sediment surface (mineralization, dissolution, desorption), as well as diffusion of phosphorus from the sediment pore water. The lower rates, observed later in the incubation period when the pool of freshly-deposited phosphorus had been depleted, may represent exchange due to diffusion alone. The initial rates are considered representative of in-lake conditions at the sediment-water interface (where a supply of freshly sedimented phosphorus is readily available) and thus become the focus for further discussion.

Substantial phosphorus release was observed in all experiments. Rates ranged from 9–21 mg P m^{-2} d^{-1}, with a lakewide average of 13 mg P m^{-2} d^{-1} (Table 1). The reproducibility of the experiments, as reflected in the low standard error (Table 1), is good and compares favorably with results published for other systems (Nurnberg, 1987). While the release rates measured here for Onondaga Lake are of the general magnitude expected for eutrophic to hypereutrophic lakes (Fig. 4; Nurnberg, 1988), the rates are somewhat lower than may be expected for such a phosphorus-enriched, hypereutrophic system.

Nurnberg (1988) has shown that the release

Table 1. Summary of sediment phosphorus release rates.

Coring location	Release rate (mg P m^{-2} d^{-1})		
	Mean	Standard error	n
South Basin	12.9	0.7	6
Mid-Lake	15.5	2.2	5
North Basin	11.8	1.0	6
Overall	13.3	0.8	17

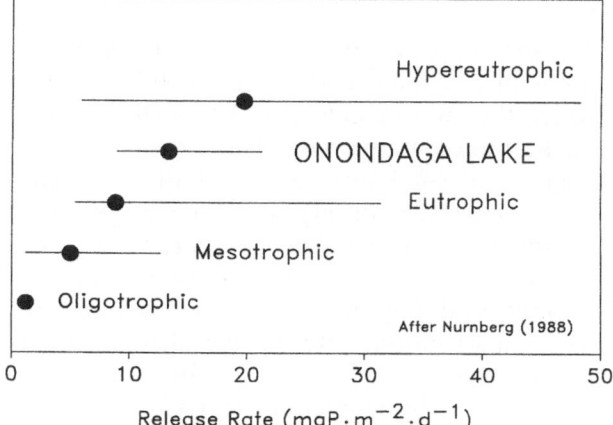

Fig. 4. Onondaga Lake sediment phosphorus release rates compared with those for other systems (after Nurnberg, 1988).

rate tends to increase with the total phosphorus content of the surficial sediment, although the relationship is not particularly strong over the range of phosphorus concentrations observed in the depositional basin of Onondaga Lake. The distribution of phosphorus in the surficial sediments of Onondaga Lake (Fig. 2) suggests that higher release rates might be associated with elevated phosphorus levels in the southern basin of the lake. However, there was no statistically significant (One Way ANOVA; Zar, 1984) trend in release rates with position along the major N/S axis of the lake. Variability in the downward flux of particulate phosphorus may be expected to influence phosphorus recycling if freshly deposited material plays a significant role in the overall release process. Phosphorus deposition varies dramatically over the summer in Onondaga Lake (Doerr, 1991) and thus some seasonality in release rates could be expected. However, statistical analysis (Two-Sample t Test; Zar, 1984) revealed no significant difference in rates for cores collected in July compared with those collected in September. This suggests that the rate of sediment-P release in Onondaga Lake is not limited by the supply of freshly sedimented phosphorus.

Rate verification

The reliability of rate estimates is influenced by the nature of the experimental conditions under which they were conducted. The ability to conduct a representative sampling of the lake's sediments, to avoid disturbance of the sediment-water interface during core collection and handling, and to maintain appropriate chemical conditions (e.g. redox, pH, and phosphorus concentration) during the course of the experiments can impart uncertainty to the determination. Some sense of the reliability of these estimates can be gained by comparing them with monitored changes in the phosphorus content of the lake's hypolimnion (Nurnberg, 1987).

The change in the mass of SRP in the hypolimnion of Onondaga Lake over the period 16 June to 8 September 1987 is illustrated in Fig. 5. Increases in SRP occur as a result of sediment phosphorus release and water-column mineralization of particulate phosphorus. Johnson (1989) has shown that water-column mineralization is negligible over the residence time of particles in the hypolimnion of Onondaga Lake. Loss of SRP occurs through vertical exchange (turbulent diffusion) with the comparatively phosphorus-poor waters of the epilimnion. Phosphorus flux across the thermocline due to vertical mass transport may be estimated by (Chapra and Reckhow, 1983):

$$J = v_t \cdot (P_h - P_e),$$

where:

J = phosphorus flux across the thermocline, mg P m^{-2} d^{-1};

v_t = vertical exchange coefficient, m d^{-1};

P_h = hypolimnion soluble reactive phosphorus, mg P m^{-3};

P_e = epilimnion soluble reactive phosphorus, mg P m^{-3}.

The summer average diffusive phosphorus flux (J) is calculated to be 2.6 mg P m^{-2} d^{-1} for $v_t = 0.01$ cm^2 s^{-1} (Wodka et al., 1983) and $(P_h - P_e) = 326$ mg P m^{-3} (Doerr, 1991). Correct-

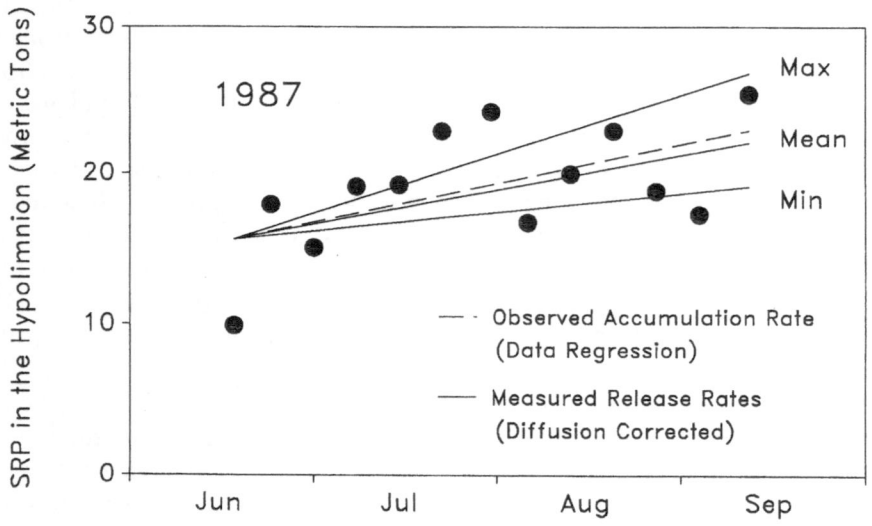

Fig. 5. Comparison of release rates determined using intact cores with the observed rate of hypolimnetic phosphorus accumulation.

ing the phosphorus release rates measured in the laboratory for losses due to vertical mixing:

$$R_{corr} = R_{meas} - J,$$

where:

R_{corr} = corrected rate of soluble reactive phosphorus release, mg P m^{-2} d^{-1};

R_{meas} = rate of soluble reactive phosphorus release measured in the laboratory, mg P m^{-2} d^{-1};

yields a mean predicted accumulation rate of 11 mg P m^{-2} d^{-1} (range 6–19 mg P m^{-2} d^{-1}; solid lines in Fig. 5). The mean observed rate of accumulation, calculated by linear regression of the measured mass of soluble reactive phosphorus in the hypolimnion versus time, is 12.4 mg P m^{-2} d^{-1} (dashed line in Fig. 5). The observed rate of increase agrees well with the diffusion-corrected laboratory estimates. Short-term dynamics in the observed mass of phosphorus in the hypolimnion (solid circles in Fig. 5) may reflect seasonal differences in release rates (not noted in laboratory measurements) or seasonality in mass transport (concentration gradient, degree of vertical mixing) not accommodated by the summer average approach utilized here.

Contribution to the lake's phosphorus budget

The importance of sediment release to the lake's overall phosphorus budget varies dramatically among lakes due to differences in phosphorus loading, productivity, morphometry, and the physical-chemical characteristics which mediate phosphorus cycling. The contribution of sediment phosphorus release to the overall phosphorus loading to Onondaga Lake for the 150-day summer period (mid-May to mid-October, 1987) is illustrated in Fig. 6. The discharge from the Syracuse Metropolitan Wastewater Treatment Plant (METRO WWTP) provides the largest loading at 64% (44.6 MT-P; 297 kgP d^{-1}; Heidtke & Auer, 1992). The internal load, calculated as the product of the lakewide mean release rate (13 mg P m^{-2} d^{-1}) and the surface area of the anoxic sediments (those below a depth of 7 m; 8.23×10^6 m^2), is the second largest source at 24% (16.4 MT-P; 109 kgP d^{-1}). Tributaries are the third largest source at 12% (8.4 MT-P; 56 kgP d^{-1}; Heidtke & Auer, 1992).

The true importance of sediment-derived phosphorus in mediating phosphorus-phytoplankton interactions is less clear. Although sediments contribute $\pm$ 16 MT-P to the water column over

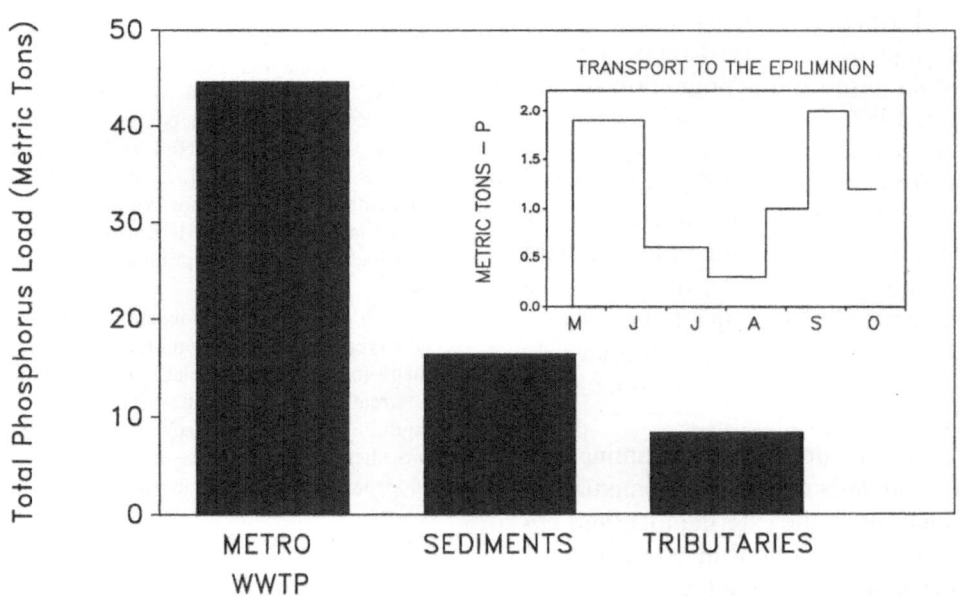

Fig. 6. Contribution of sediment phosphorus release to the overall phosphorus budget for Onondaga Lake.

308

the summer, transport to the epilimnion occurs primarily during the early (May) and late (September–October) stages of thermal stratification when vertical mixing is greatest (inset in Fig. 6; Doerr, 1991). Total transport to the epilimnion over the summer amounts to ± 7 MT-P (43 % of that released) and vertical transport is particularly low in June, July, and August when phytoplankton demand may be high. On the other hand, essentially all of the phosphorus released from the sediment is in the soluble reactive form, immediately available for uptake by the phytoplankton. Tributary phosphorus is generally less available (Young and DePinto, 1982). While resolution of this question requires a more sophisticated analysis (e.g. a coupled phosphorus – phytoplankton model), it is clear that vertical transport of sediment derived phosphorus may play a role in regulating primary production during summer stratification. At turnover, the accumulated phosphorous is mixed over the water column where its fate is governed by algal uptake, adsorption/precipitation phenomena and discharge from the lake.

Application, limitations, and future work

The sediment phosphorus release rates measured here are consistent with observed rates of phosphorus accumulation in the hypolimnion of Onondaga Lake in 1987. These rates have been applied in the development of a two-layer mass balance model for phosphorus in Onondaga Lake with good results (Doerr, 1991). The experiments reported here have served to quantify sediment phosphorus exchange and its contribution to the phosphorus budget of Onondaga Lake. Within the limitations of the experimental framework, these studies have satisfactorily characterized current conditions in the lake.

The next step is to determine the timing and degree of the sediment's response to remedial actions, e.g. reductions in the external load and hypolimnetic aeration. Reductions in external phosphorus loading can be expected to yield lower rates of sediment-P release, following a period of

equilibration (Lorenzen *et al.*, 1976). Particulate phosphorus flux from the epilimnion would decrease and newly deposited, low-phosphorus sediment would cover the high-phosphorus layers, reducing the release rate. If the phosphorus flux in Onondaga Lake is mediated by redox conditions, e.g. iron chemistry (Mortimer, 1971), then hypolimnetic aeration would be expected to inhibit phosphorus cycling and reduce internal loads. However, recent field observations and the results of preliminary modeling analysis and laboratory experimentation presented elsewhere in this volume (Driscoll *et al.*, 1993) suggest that Ca-P interactions may play an important role in regulating sediment phosphorus release in Onondaga Lake. This may make the system less sensitive to redox conditions (hypolimnetic aeration). Further study, supported by modeling analysis, is required to predict the impact of changes in external loading on rates of phosphorus recycle and to identify the degree to which internal loads will retard the improvement of water quality.

References

American Public Health Association, American Water Works Association and Water Pollution Control Federation. 1985. Standard Methods for the Examination of Water and Wastewater, 16th Edition. American Public Health Association, Washington, D.C., 1268 pp.

Auer, M. T., M. L. Storey, S. W. Effler, N. A. Auer & P. Sze, 1990. Zooplankton impacts on chlorophyll and transparency in Onondaga Lake, New York, USA. In R. D. Gulati, E. H. R. R. Lammens, M.-L. Meijer & E. van Donk (eds), Biomanipulation – Tool for Water Management. Developments in Hydrobiology 61. Kluwer Academic Publishers, Dordrecht: 603–617. Reprinted from Hydrobiologia 200/201.

Chapra, S. C. & H. F. H. Dobson, 1981. Quantification of the lake typologies of Naumann (surface quality) and Thienemann (oxygen) with special reference to the Great Lakes. J. Great Lakes Res. 7: 182–193.

Chapra, S. C. & K. H. Reckhow. 1983. Engineering Approaches for Lake Management. Volume 2. Mechanistic Modeling. Butterworth Publishers, Boston, Massachusetts, 492 pp.

Cooke, G. D., E. B. Welch, S. A. Peterson & P. R. Newroth. 1986. Lake and Reservoir Restoration. Butterworths, Boston, 392 pp.

Devan, S. P. & S. W. Effler, 1984. History of phosphorus

loading to Onondaga Lake. J. env. Eng. Div., ASCE 110: 93–107.

Dillon, P. J. & F. H. Rigler. 1974. The phosphorus-chlorophyll relationship in lakes. Limnol. Oceanogr. 19: 767–773.

Doerr, S. M., 1991. Development and application of a mass balance model for total phosphorus in a eutrophic lake. M.S. Thesis, Department of Civil Engineering, Syracuse University, Syracuse, New York, 159 pp.

Driscoll, C. T., S. W. Effler, M. T. Auer, S. M. Doerr & M. R. Penn, 1993. Supply of phosphorus to the water column of a productive hardwater lake: controlling mechanisms and management considerations. In P. C. M. Boers, T. E. Cappenberg & W. van Raaphorst (eds), Proceedings of the Third International Workshop on Phosphorus in Sediments. Developments in Hydrobiology 84. Kluwer Academic Publishers, Dordrecht: 61–72. Reprinted from Hydrobiologia 253.

Effler, S. W., 1987. The impact of a chlor-alkalai plant on Onondaga Lake and adjoining systems. Wat Air Soil Pollut. 33: 85–115.

Effler, S. W., M. C. Rand & T. A. Tamayo, 1979. Effects of heavy metals and other pollutants on the sediments of Onondaga Lake. Wat. Air Soil Pollut. 12: 117–134.

Effler, S. W., C. T. Driscoll, M. C. Wodka, R. Honstein, S. P-. Devan, P. Jaran & T. Edwards, 1985. Phosphorus cycling in ionically polluted Onondaga Lake, New York. Wat. Air Soil Pollut. 24: 121–130.

Gachter, R., 1987. Lake restoration. Why oxygenation and artificial mixing cannot substitute for a decrease in the external phosphorus loading. Schweiz. Z. Hydrol. 49: 170–185.

Golterman, H. L. (ed.) 1977. Interactions Between Sediments and Freshwater. Dr W. Junk Publishers, The Hague.

Heidtke, T. M. & M. T. Auer, 1992. Partitioning phosphorus loads: implications for lake restoration. J. Wat. Resour. Plan. Mgt. Vol. 118 No. 5: 562–579.

Honstein, R. L., 1981. An assessment of mechanisms by which phosphorus may be regulated within the sediments of Onondaga Lake, New York. M.S. Thesis, Department of Civil Engineering, Syracuse University, Syracuse, New York, 199 pp.

Hutchinson, G. E., 1973. Eutrophication: the scientific background of a contemporary practical problem. Am. Sci. 61: 269–279.

Johnson, N. A., 1989. Surficial sediment characteristics and sediment phosphorus release rates in Onondaga Lake, New York. M.S. Thesis, Department of Civil and Environmental Engineering, Michigan Technological University, Houghton, Michigan, 127 pp.

Larsen, D. P., D. W. Schults & K. W. Malereg, 1981. Summer internal phosphorus supplies in Shagawa Lake, Minnesota. Limnol. Oceanogr. 26: 740–753.

Lorenzen, M. W., D. J. Smith & L. V. Kimmel, 1976. A long-term phosphorus model for lakes: application to Lake Washington. In: R. P. Canale (ed.), Modeling Biochemical Processes in Aquatic Ecosystems. Ann Arbor, Science Publishers, Ann Arbor, Michigan: 75–91.

Marsden, M. W., 1989. Lake restoration by reducing external phosphorus loading: the influence of sediment phosphorus release. Freshwater Biology 21: 139–162.

Mortimer, C. H., 1971. Chemical exchanges between sediments and water in the Great Lakes – Speculations on probably regulatory mechanisms. Limnol. Oceanogr. 16: 387–404.

Murphy, C. B. Jr. 1978. Onondaga Lake. In: J. A. Bloomfield (ed.), Lakes of New York State. Volume II. Ecology of the Lakes of Western New York. Academic Press, New York: 224–336.

Nurnberg, G. K., 1984. The prediction of internal phosphorus load in lakes with anoxic hypolimnia. Limnol. Oceanogr. 29: 111–124.

Nurnberg, G. K., 1987. A comparison of internal phosphorus loads in lakes with anoxic hypolimnia: laboratory incubation versus in situ hypolimnetic phosphorus accumulation. Limnol. Oceanogr. 32: 1160–1164.

Nurnberg, G. K., 1988. Prediction of phosphorus release rates from total and reductant soluble phosphorus in anoxic lake sediments. Can. J. Fish. Aquat. Sci., 45: 453–462.

OECD, 1982. Eutrophication of waters: monitoring, assessment and control. Organization of Economic Cooperation and Development, Paris, France, 154 pp.

Psenner, R. & A. Gunatilaka (eds), 1988. Proceedings of the first international workshop on sediment phosphorus. Arch. Hydrobiol. 30: 1–115.

Sly, P. G. (ed.), 1986. Sediments and Water Interactions. Springer-Verlag, New York.

Welch, E. B., D. E. Spyridakis, J. I. Shuster & R. R. Horner, 1986. Declining lake sediment phosphorus release and oxygen deficit following wastewater diversion. J. Wat. Pollut. Cont. Fed. 58: 92–96.

Wodka, M. C., S. W. Effler, C. T. Driscoll, S. D. Field & S. P. Devan, 1983. Diffusivity based flux of phosphorus in Onondaga Lake. J. env. Eng. Div., ASCE 109: 1403–1415.

Wodka, M. C., S. W. Effler & C. T. Driscoll, 1985. Phosphorus deposition from the epilimnion of Onondaga Lake. Limnol. Oceanogr. 30: 833–843.

Young, T. C. & J. V. DePinto, 1982. Algal availability of particulate phosphorus from diffuse and point sources in the lower Great Lakes basin. In P. G. Sly (ed.), Sediment/Freshwater Interaction. Developments in Hydrobiology 9. Dr W. Junk Publishers, The Hague: 111–119. Reprinted from Hydrobiologia 91/92.

Zar, J. H., 1984. Biostatistical Analysis. 2nd Edition. PrenticeHall, Inc., Englewood Cliffs, New Jersey, 718 pp.

Hydrobiologia **253**: 311–317, 1993.
P.C.M. Boers, T.E. Cappenberg & W. van Raaphorst (eds),
Proceedings of the Third International Workshop on Phosphorus in Sediments.
© 1993 *Kluwer Academic Publishers.*

Iron content of sediment and phosphate adsorption properties

Heleen Danen-Louwerse, Lambertus Lijklema & Monique Coenraats
Agricultural University, Nature Conservation Department, P.O. Box 8080, 6700 DD Wageningen, The Netherlands

Key words: phosphorus, sediment, iron, aluminum, adsorption

Abstract

Phosphorus can occur in sediments in different forms and accordingly its availability varies. The distinction between the phosphorus fractions is made with two chemical extraction methods; an ammonium oxalate-oxalic acid extraction and an extraction according to Hieltjes & Lijklema (1980).

The iron and aluminum liberated with the ammonium oxalate-oxalic acid extraction method is linearly correlated ($r^2 = 0.73$) with the phosphorus liberated in the first two steps of the Hieltjes and Lijklema extraction by:

$P = 0.035 \, (Fe + Al) + 0.001$ (P, Fe and Al in mmol g^{-1}).

The iron and aluminum (hydr)oxides are very important fractions in the sediment adsorption capacity for phosphorus. The phosphorus sorption capacity (PSC) is 0.080 mol P (mol (Fe + Al))$^{-1}$ and the adsorption constant (k) is 11.9 μmol P l^{-1}. Here it is assumed that iron and aluminum (hydr)oxide have the same affinity for phosphorus.

Introduction

Reduction of the eutrophication of surface waters is an important issue in water quality management and research. Its prime purpose is to decrease algal biomass. The usual strategy is reduction of the external phosphorus loading. However, reducing the external loading alone would not guarantee a rapid recovery of the lake. In addition, the phosphorus content of the lake water is strongly influenced by the release of phosphorus from the sediments (Hosper & Meyer, 1986). Thus the recovery of the lake also requires measures to suppress internal loading.

Hence, it is necessary to determine the relationship between sediment composition and phosphate binding to assess the potential of the sediments to release phosphorus into the water phase.

Phosphorus can occur, as particulate and dissolved organic or inorganic P, in the water phase and in aerobic or anaerobic parts of the sediment and interstitial water. Phosphorus can move between compartments as a result of growth, degradation, sedimentation and resuspension, physico-chemical reactions, mixing, diffusion and burial.

The binding aspects studied here are the iron and aluminum phosphate ratio in the sediments in relation with the soluble inorganic P concentration for aerobic conditions.

Material and methods

Description of the lake area

Lake Veluwe is located between the coastline of the former Zuiderzee and the dikes of the new

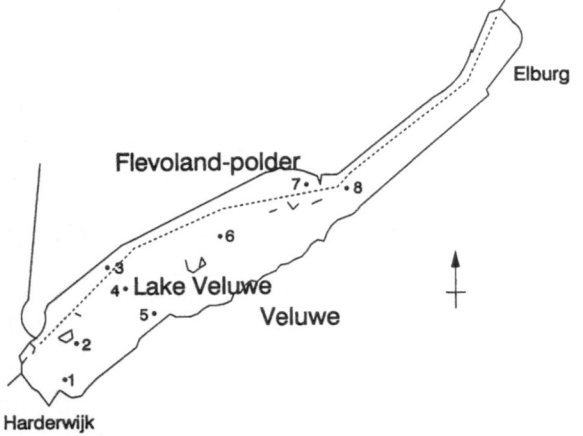

Fig. 1. The locations in Lake Veluwe.

Flevoland polder. The lake's hydrology is determined by its special location between the elevated 'old' land and the new polders, lying about 5 m below mean sea level. Hence, on one side of the lake there is an inflow of groundwater and a number of small streams, whereas along the other side water is lost by infiltration towards the polders. There is a possibility to pump the seepage water from the polders back into Lake Veluwe.

The lake has a surface area of 32 km^2, is very shallow (mean depth 1.3 m) except for a shipping channel and is of great importance for recreation and for nature.

The occurrence of both seepage and infiltration as well as intensive internal transport of sedi-

ments due to wind driven resuspension and sedimentation has caused a marked horizontal and vertical stratification in sediment composition and phosphate binding properties. This is a key feature in the potential for internal loading.

The locations 1 and 5 (Fig. 1) have a sandy sediment with a low water (20%) and organic matter (0.5%) content. In contrast with this the muddy sediment at location 2 contains 75% water and 10% organic matter (Table 1).

Phosphorus fractionation

The distinction between different chemical fractions of phosphorus in sediments is relevant in so far as they can be distinguished according to their ecological significance. This functional classification does not necessarily correspond to well defined chemical species, because extractions do not give a clear distinction between the various species. In fact the species are defined by the analytical procedure.

Several methods to characterize the different forms of bound phosphorus have been reported. The classical fractionation scheme of Chang & Jackson (1957) has some imperfections, particularly for calcareous sediments. The simple scheme of Hieltjes & Lijklema (1980) is used regularly, but does not identify organic phosphates.

In organic phosphate two parts can be differentiated: an inert and a degradable part. Organic

Table 1. The measured water and organic matter content and P, Fe and Al content extracted with ammoniumoxalate-oxalic acid.

Location sediment type	Water (%)	Organic matter (%)	Amoniumoxalate-oxalic acid extraction		
			Fe	Al (μmol g^{-1})	P
1 Sand	21	0.6	14.9	5.6	0.3
2 Mud	77	11.0	109.9	56.0	12.3
3 Clay	50	7.5	55.5	19.3	5.5
4 Sandyclay	48	7.4	63.4	19.3	5.5
5 Sand	20	0.5	13.3	3.7	0.6
6 Clay	34	3.2	19.3	8.9	1.6
7 Mud	28	2.6	14.5	6.7	1.3
8 Mud	20	0.7	1.3	35.2	0.3

phosphate consists of plant debris, living and dead algae, zooplankton, bacteria and detritus.

The C–O–P bond occurs frequently in the organic species but is (bio-)chemically instable. As a consequence, a major part of the phosphate in organic compounds will be degraded and regenerated into inorganic species via bacteria as an intermediate.

Clay minerals, Fe and Al (hydr)oxides and calcium components are the main inorganic P constituents. The adsorption of phosphate on clay minerals and Fe and Al (hydr)oxides is based on an exchange with OH-groups (Bolt, 1982), so the pH is important for the formation of the bonds. Further, reaction time and phosphate concentration control the adsorption.

Calcium can be important for the removal of phosphorus from the water. A calcium carbonate and phosphate (co)precipitate can be formed during photosynthesis.

There is a range of binding forms between phosphorus in precipitates and adsorbents with variable stoichiometry and frequently an amorphous character. Hence it is impossible to make an exact analytical separation of these different forms.

Therefore indirect characteristics are used, such as pH and redox sensibility. The separation of the fractions can be induced by water, acids, bases, salts, reducing, oxidizing and complexing substances.

To investigate the phosphorus binding characteristics in this study, two chemical extraction methods have been used: an ammonium oxalate-oxalic acid extraction (Schwertmann, 1964) and an extraction according to Hieltjes & Lijklema (1980), which are based on solubilization of iron by reduction and desorption and solubilization of phosphate by pH changes.

The ammonium oxalate-oxalic acid extract is considered to yield 'active iron and aluminum'. The sediments are shaken at a 1:20 weight to volume ratio with a solution of oxalic acid and ammonium oxalate at a pH of 3. The extraction solution reduces the poorly soluble iron (III) ions to the much more soluble iron (II) ions. As light influences the reducing action of oxalic acid

(Schwertmann, 1964), the extraction has to be performed in the dark.

The extraction according to Hieltjes & Lijklema (1980) is divided in three successive steps. First the 'labile' phosphate and the phosphate bound to calcium carbonate are liberated by a 1 M NH_4Cl solution at pH 7, in which $CaCO_3$ dissolves. In the second step iron and aluminum bound phosphate dissolves due to 0.1 M NaOH. The removal of $CaCO_3$ in the previous step prevents the precipitation of new calcium phosphates due to the high pH. In the last step calcium phosphates and occluded iron phosphates are extracted with 0.5 M HCl. In all extraction solutions P and Fe are determined. Organic P is not solubilized in this scheme, except a labile fraction that may decompose under alkaline or acid conditions.

The phosphate liberated in the first two steps of this extraction scheme is called 'reactive phosphate', because this fraction is mobile and may form other complexes (Moore *et al.*, 1991).

Phosphorus adsorption

The adsorption of phosphorus on sediment particles containing clay minerals and Fe and Al (hydr)oxides, is measured under aerobic conditions. The data are standardized on oxalate extractable Fe and Al, because Fe and Al (hydr)oxides are the main P adsorbing constituents (Van der Zee *et al.*, 1987). The data are fitted to a Langmuir equation in a form as used by Jensen *et al.* (1991):

$$SP + NAP = \frac{PSC * Ce}{k + Ce}$$

SP = Sorbed Phospahte (mol P (mol $(Fe + Al))^{-1}$)
NAP = Native Adsorbed Phosphate (mol P (mol $(Fe + Al))^{-1}$)
PSC = Phosphate Adsoption Capacity (mol P (mol $(Fe + Al))^{-1}$)
Ce = equilibrium concentration (μmol P l^{-1})
k = constant (μmol P l^{-1})
The parameters are explained in Fig. 2.

314

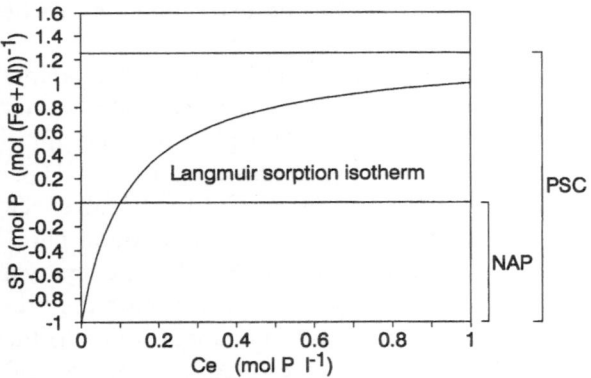

Fig. 2. The Langmuir sorption isotherm with the parameters used.

The shape of the sorption isotherm depends on pH and temperature.

Thus NAP is the phosphate adsorbed under the ambient ortho-phosphate concentration and can in principle be estimated by desorption in (pure) water (without phosphate). NAP is estimated via a Langmuir isotherm fit for each individual sediment sample, because the experimental determination of NAP is not yet completed. Reactive P (NH_4Cl-P + NaOH-P) cannot be used as NAP, because reactive P includes also other bound P components and overestimates NAP. *PSC* and *k* are estimated with the least square method, and the 90% reliability intervals for *PSC* and *k* are calculated.

Experimental methods

Sediment samples were taken in May 1990 and July 1991 using a Beeker sampler (Eijkelkamp, the Netherlands). A detachable perspex inner core containing the samples can be brought into the laboratory. The sediment core is cut into maximally 8 segments at 0–2, 2–4, 4–6, 6–8, 8–10, 10–14, 14–18 and 18–22 cm from the top (interface). Before further analysis the sediment slices are dried at 40 °C and stored at 4 °C in the dark.

For the determination of the ammonium oxalate-oxalic acid extraction, 2.500 g air dry sediment is shaken for 2 hours in the dark with 50 ml extraction solution. After centrifugation (10 minutes at 3000 rpm) Fe and Al are determined in the extracted solutions.

The total dissolved iron in the water is measured after reduction with hydroxyl ammonium chloride to Fe^{2+} with tripyridiltriazine as colour reagent on a Skalar SA-40 Autoanalyser. Aluminum is determined on a Spectro Analytical Instruments ICP type Spectroflame with AES.

To determine the Hieltjes & Lijklema extraction 50 mg air dry sediment is shaken for two hours with 40 ml 1 M NH_4Cl at pH = 7. After centrifugation the supernatant is saved and the extraction is repeated. The sediment is washed with 10 ml water and Fe and P are estimated in the combined supernatant and wash water.

The phosphate is determined on a Skalar SA-40 Autoanalyser or on a Vitatron Photometer, with the molybdenum blue method as modified by Murphy & Riley (1962).

The sediment remaining after step 1 is shaken for 17 hours with 50 ml 0.1 M NaOH. The supernatant after centrifugation is saved and the sediment washed with 20 ml water. The mixture of supernatant and washing water is brought to a pH of 6–8 with 12 N HCl and P and Fe are analysed.

In the last step the remaining sediment is shaken for 24 hours with 50 ml 0.5 M HCl. After centrifugation the supernatant is saved and the sediment is washed with 20 ml water. The combined solution is adjusted to a pH of 6–8 with 6 N NaOH and P and Fe are measured.

To determine the adsorption isotherms 50.0 ml of eight different phosphate solutions (between o and 125 μmol P l^{-1} in artificial Lake Veluwe water) are added to 0.200 g air dry sediment and shaken for 48 hours to achieve equilibrium. The pH of the solutions is 8, as a result of HCO_3^- presence in artificial Lake Veluwe water. The experiments are performed at room temperature. After centrifugation (15 minutes at 3000 rpm) the phosphate concentration (Ce) in the supernatant is measured and the adsorbed phosphate (SP) can be calculated.

Results and discussion

Sediment analysis

The segregation of fine materials due to internal transport and sieving of sediments is reflected in a proportionality between the Fe oxalate content at the different locations and the organic matter content. The linear relationship is:

$$\text{Org. matter } (\%) = 94.2 \text{ Fe (mmol g}^{-1})$$

$$(r^2 = 0.65).$$

Thus in principle there is no extraction procedure that can discern between iron bound or organic matter related phosphate by statistical means only.

In the first step of the Hieltjes & Lijklema (1980) extraction (NH_4Cl) only little phosphate is liberated: less than 1.5 μmol P g^{-1} (except location 2). The phosphate yield in the NaOH step is about 30% of the total extractable phosphate. In the last extraction step (HCl) 50% of the total extractable phosphate and most of the iron is liberated (Table 2).

The reactive aluminum content in the different sediments, is approximately proportional to the reactive iron content. The linear relationship is Fe = 1.9 Al (mmol g^{-1}), $r^2 = 0.80$. Aluminum can also contribute to phosphate adsorption (van der Zee, 1988). 70% (r^2) of the reactive phosphate (NH_4Cl + NaOH-step) can be explained by the reactive (oxalate extractable) iron and aluminum content.

The linear relationship found is (Fig. 3):

$$P = 0.035 \text{ (Fe + AL)} + 0.001$$
$$(n = 113, r^2 = 0.73)$$

with a standard deviation for P of 0.001 mmol P g^{-1}

Again the individual contributions of Fe and Al are difficult to distinguish.

Adsorption isotherms

The overall phosphate adsorption isotherm for all samples tested according to Langmuir is given in Fig. 4. The whole data set is fitted to the Langmuir equation with the least square method. The *PSC* and *k* values with their 90% confidence intervals are:

$$PSC = 0.080 \text{ mol P (mol(Al + Fe))}^{-1}$$
$$(0.074-0.087)$$
$$k = 11.9 \, \mu \text{ mol P l}^{-1} \quad (8.8-16.1)$$

It is expected that the variation will be reduced somewhat when NAP is measured experimentally, because the NAP value affects both the *PSC* and *k* value.

In this adsorption model of phosphate, equal affinities of P to iron and aluminum (hydr)oxide

Table 2. The location average of the iron and phosphate content extracted accroding to the Hieltjes & Lijkema (1980) extraction scheme.

Location	NH$_4$Cl		NaOH		HCl	
sediment type	P	Fe	P	Fe	P	Fe
	(μmol g^{-1})		(μmol g^{-1})		(μmol g^{-1})	
1 Sand	0.3	0.2	0.6	2.3	0.6	5.4
2 Mud	2.9	0.2	3.6	9.0	4.8	126.4
3 Clay	1.3	0.2	2.9	4.7	6.5	112.6
4 Sandyclay	1.0	0.5	2.9	3.8	5.8	108.3
5 Sand	0.6	0.4	0.3	0.2	0.3	15.8
6 Clay	0.6	0.5	1.3	2.0	1.3	27.2
7 Mud	1.3	0.4	0.6	5.1	2.9	24.7
8 Mud	0.6	0.9	0.3	3.4	0.3	10.7

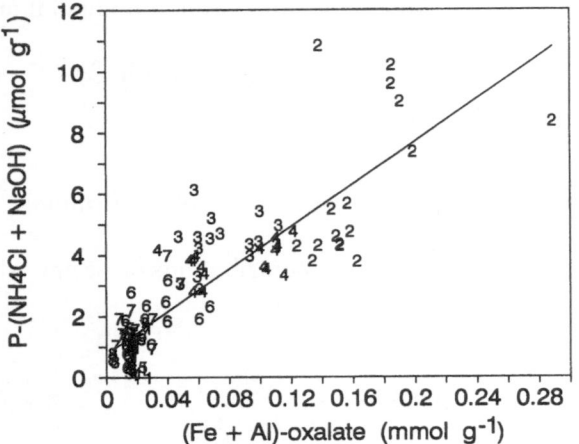

Fig. 3. Reactive iron and aluminum versus reactive phosphorus. Numbers refer to the sampling sites.

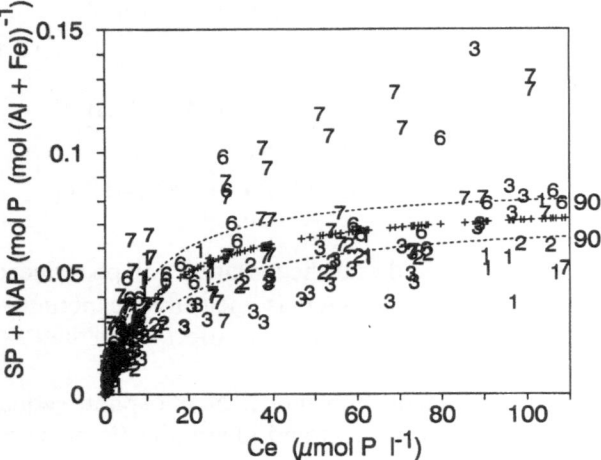

Fig. 4. The overall phosphate adsorption isotherm. Numbers refer to the sampling sites.

are assumed. However, the affinities of Fe and Al (hydr)oxide for P can differ because the zero points of charge of the two (hydr)oxides at a specific pH are not necessarily the same. Further, also the available specific surface area of the two species has an important influence on P adsorption and these may differ as well, depending on their diagenesis.

Another (1:2) affinity ratio of iron and aluminum (hydr)oxide for phosphorus is tested as well.

The basis for the assumed higher affinity of aluminum is the observation by Lijklema (1980) that the molar ratios P/Fe and P/Al respectively of adsorption to freshly precipitated hydroxide were 0.15 and 0.32 mol mol^{-1} respectively at pH = 7. This 1:2 ratio hardly improves the adsorption isotherm fit which is to be expected as a consequence of the correlation between iron and aluminum in sediments.

However, under anaerobic conditions this difference in affinity for P is important as iron (III) is readily reduced to iron (II). Whereas the adsorption capacity of iron is decreasing, aluminum maintains its adsorption capacity.

Conclusions

The phosphate extracted in the first two steps (NH$_4$Cl and NaOH) of the Hieltjes & Lijklema extraction is correlated with the iron and aluminum fraction in the sediment as determined with the ammonium oxalate-oxalic acid extraction.

The linear relationship is P = 0.035 (Fe + Al) + 0.001 r^2 = 0.73; σ = 0.001 mmol P g^{-1}

Thus reactive iron and aluminum explain over 70% of the P binding and so play a dominant role in the phosphate sorption in Lake Veluwe. This concurs with experiences elsewhere. Ku *et al.* (1978) reported that surfaces of iron and aluminum in clays, oxides and hydroxides are of great importance for phosphate adsorption.

The phosphate adsorption isotherm can be described reasonably well with the Langmuir adsorption equation in the form as used by Jenssen *et al.* (1991). The least square fit on the whole data set, with NAP values evaluated from curves for individual samples, resulted in:

$$PSC = 0.080 \text{ mol P (mol (Fe + Al))}^{-1}$$
$$(0.074-0.087)$$

$$k = 11.9 \, \mu\text{mol P l}^{-1} \quad (8.8-16.1)$$

() = 90% confidence interval.

Iron and aluminum are the primary constituents controlling the adsorption isotherm, but substantial variation remains.

Acknowledgements

This research is carried out in conjunction with the Institute for Inland Water Management and Waste Water Treatment in Lelystad, the Netherlands.

The support and comments by Dr. P. C. M. Boers of this Institute are gratefully acknowledged. R. v. Eck, Department of Soil Science and Plant Nutrition, Agricultural University of Wageningen, measured the aluminum concentrations.

References

Bolt, G. H., 1982. Soil Chemistry B; Physico-chemical models. Elsevier, Amsterdam, 527 pp.

Chang, S. C. & M. J. Jackson, 1957. Fractionation of soil phosphorus. Soil Sci. 84: 133–144.

Hieltjes, A. H. M. & L. Lijklema, 1980. Fractionation of inorganic phosphates in calcareous sediments. J. envir. Qual. 9: 405–407.

Hosper, H. & M.-L. Meyer, 1986. Control of phosphorus loading and flushing as restoration methods for Lake Veluwe, The Netherlands. Hydrobiol. 20: 183–194.

Jensen, H. S., P. Kristensen, E. Jeppesen & A. Skytthe, 1991. Iron:Phosphorus ratio in surface sediment as an indicator of phosphate release from aerobic sediments in shallow lakes. Hydrobiol. In press.

Ku, W. C., F. A. DiGiano & T. H. Feng, 1978. Factors affecting phosphate adsorption equilibria in lake sediments. Wat. Res. 12: 1069–1074.

Lijklema, L., 1980. Interaction of orthophosphate with iron(III) and aluminum hydroxides. Envir. Sci. Technol. 14: 537–540.

Moore, P. A., K. R. Reddy & D. A. Graetz, 1991. Phosphorus geochemistry in the sediment-water column of a hypereutrophic lake. J. Envir. Qual. 20: 869–875.

Murphy, J. & J. P. Riley, 1962. A modified single-solution method for the determination of phosphate in natural waters. Analyt. chim. Acta 27: 31–36.

Schwertmann, U., 1964. Differenzierung der Eisenoxide des Bodens durch Extraktion mit Ammoniumoxalat-Lösung. Z. Pfl. Ern. Düngung Bdk. 105: 194–202.

Van der Zee, S. E. A. T. M., L. G. J. Fokkink & W. H. van Riemsdijk, 1987. A new technique for assessment of reversibly adsorbed phosphate. Soil Sci. Soc. Am. J. 51: 599–604.

Hydrobiologia **253**: 319–322, 1993.
P.C.M. Boers, T.E. Cappenberg & W. van Raaphorst (eds),
Proceedings of the Third International Workshop on Phosphorus in Sediments.
© 1993 *Kluwer Academic Publishers.*

Abstracts

Phosphorus sedimentation in seven basins of Lake Lucerne, Switzerland

Jürg Bloesch
Swiss Federal Institute of Technology (EAWAG), 8600 Dübendorf, Switzerland

In 1989 sedimentation in seven basins of oligotrophic Lake Lucerne was measured by means of cylindrical traps below the epilimnion (30–50 m) and in the hypolimnion 5 m above lake bottom. The aim of this study was to evaluate horizontal P sedimentation differences and the possible causes thereof. By means of measuring autochthonous primary production, mean sestonic PP concentration, major allochthonous P inputs and the P output, and increments in bottom sediments, the fate of P in the different lake basins might be elucidated, despite the complicated morphological structure of Lake Lucerne.

The annual PP flux in the seven basins ranged between 0.6 and 2.0 g $PP \cdot m^{-2} \cdot yr^{-1}$. PP fluxes varied not only horizontally, but also seasonally and vertically between 0.4 and 11.7 mg $PP \cdot m^{-2} \cdot d^{-1}$, with one extreme sedimentation rate of 32 mg $PP \cdot m^{-2} \cdot d^{-1}$ in a sediment trap 5 m above lake bottom, measured after a rainfall event in one basin and indicating a turbidity current. The ratio between maximum and minimum flux value of 6.6 for seasonal variances, 3.1 for vertical differences, and 21.3 for horizontal differences showed that differences from basin to basin were far the largest. This is basically caused by differences in allochthonous PP input rather than autochthonous PP production by phytoplankton, since the horizontal differences in primary production were negligible. Because horizontal seston differences were much smaller than those in settling flux, a fast settling velocity of inorganic particles with high density is assumed. (As usual, calculations dividing flux by mean concentration, yielding 0.18–1.35 $m \cdot d^{-1}$ in the different basins, underestimated true settling velocity considerably.) Typically, the PP flux was lower in winter than in summer, reflecting again the high summer discharge of glacier fed rivers. The vertical differences in PP fluxes indicated usually a greater hypo flux by about 68% in all basins indicating bottom sediment resuspension or P uptake during settling. Evidence of resuspension events is limited due to the great depth of the basins; however, density currents in basins with large river input cannot be excluded. There was no evidence, however, that such currents stratified in any depth of the upper hypolimnion, since additional traps exposed at various hypolimnetic depths in two basins did never show increased settling flux. On the other hand, P uptake experiments with settling particles revealed that bacteria attached at settling particles can incorporate significant amounts of phosphorus during sedimentation.

Phosphorus dynamics at the sediment–water interface in marine and freshwater environments during positive redox-turnover

Anneli Gunnars[1] & Sven Blomqvist[2,3]
[1] *Dept. of Physical, Inorganic and Structural Chemistry, Stockholm University, S-106 91 Stockholm, Sweden;* [2] *Dept. of Systems Ecology, Sect. Marine Ecology;* [3] *Dept. of Geology and Geochemistry, Stockholm University, S-106 91 Stockholm, Sweden*

Comparative studies on the exchange processes of phosphorus over the sediment-water interface in marine and freshwater environments are reported. The study is based on laboratory experiments with sediment cores collected at three soft bottom stations; two brackish-marine and one freshwater. One of the marine stations was reduced and azoic, whereas the freshwater and the other marine station had well oxygenated conditions in the bottom waters. Positive redox-turnovers, including anaerobic incubation followed by reaeration, were generated in the cores and the supernatant water.

In cores from the oxygenated freshwater and marine stations, dissolved phosphate and ferrous ions were released from the sediment during the anaerobic incubation. At the positive redox-turnover, the concentration of dissolved phosphate in the supernatant water decreased sharply due to scavenging by rapidly formed colloidal ferric hydroxide. Dissolved phosphate was also released during the incubation of the marine sediment from the reduced station. However, in these cores the concentration of iron in the supernatant water was low throughout the experiment and after the redox-turnover phosphate remained dissolved. In a parallel experiment in which iron was added to the supernatant water, dissolved phosphate was scavenged by ferric hydroxide at the positive redox-turnover in a similar way as observed for the two oxygenated stations. The low abundance of dissolved iron in the reduced marine system could be due to a rich supply of sulphide.

In freshwater systems the concentration of dissolved phosphate is effectively diminished after a positive redox-turnover due to interaction with iron. In marine systems, which have had prevailed reduced conditions in the bottom waters, iron is immobilised. Consequently, a potent retention mechanism for phosphorus is eliminated. Our results imply that the cycling of phosphorus, in this aspect, differs in fresh and saltwater systems. This difference might have large effects on the availability of phosphorus as a nutrient.

Manganese showed a consistent redox-dependent behaviour in all systems, but it did not interact with iron or phosphorus.

The phosphorus cycle in coastal marine sediments

Bjørn Sundby[1], Charles Gobeil,[1] Norman Silverberg[1] & Alfonso Mucci[2]

[1] *Department of Fisheries and Oceans, Maurice-Lamontagne Institute, P.O. Box 1000, Mont-Joli, Québec G5H 3Z4, Canada;* [2] *Department of Geological Sciences, McGill University, Montréal, Québec H3A 2A7, Canada*

The concentration of dissolved phosphate in sediment box-cores from the Laurentian Trough in the Gulf of St. Lawrence increased sharply across the sediment-water interface from 2.0 μmol/PO$_4$/l in the bottom water to 6 ± 3 μmol PO$_4$/l in the top cm, remained almost constant at this value down to 5–15 cm depth, and then increased rapidly with further depth. In the region of constant concentration, phosphate is buffered by sorption equilibria with the sediment. The production rate of phosphate, the sorption capacity of the sediment, and the thickness of the diffusive boundary layer at the sediment-water interface appear to control the shape of the pore water profile. Even though the buffering places an upper limit on the concentration gradient across the sediment–water interface, and hence on the flux, the phosphate flux to the overlying water is controlled by the production rate of phosphate within the sediment. A model is proposed to relate sediment chemistry to phosphorus fluxes.

Approximately half of the net sedimentation flux of phosphorus is not buried but is mobilized within the sediment and returned to the water column. The phosphorus mobilized in this way from the suspended sediment load carried by the St. Lawrence River adds 40% to the load of dissolved phosphate carried by this river to the ocean, demonstrating that phosphorus diagenesis in coastal sediments can increase significantly the flux of dissolved phosphate from the continents to the ocean.

Sediment–water exchange of phosphorus in Wadden Sea and North Sea

W. van Raaphorst, H. T. Kloosterhuis, A. J. M. Gieles
Netherlands Institute for Sea Research (NIOZ), PO 59 1790 AB Den Burg, The Netherlands

Sediment-water exchange rates of SRP were measured over full annual cycles at 3 tidal flats in the western dutch Wadden Sea and at 2 stations (30 and 40 m depth) in the southern North Sea. The fluxes were estimated from experiments with intact sediment cores, using the general mass transfer equation:

$$\text{FLUX} = K_m * (Cb - Co),$$

where: K_m = mass transfer coefficient (m/s), Cb = parameter representing a characteristic SRP concentration in the pore water (mmol/m^3), Co = concentration in the water column (mmol/m^3).

The annual variation and the differences between the 5 stations concerning fluxes and underlying parameters (K_m, Cb) are discussed and the ecological interpretation is indicated. Factors included are e.g. overall mineralization activity (oxygen respiration), oxygen penetration depth, Fe contents of the sediments, abundance of macrofauna, influence of benthic diatoms. To enable this analysis data on SRP pore water concentrations and results of sequential extractions according to the Hieltjes-Lijklema scheme are presented also. In addition, experiments were performed to measure sorption characteristics (isotherms, kinetics) of the upper 5 mm of the Wadden Sea sediments. Results of these will be related to the data of the extractions.

Integration of all results is done by simple models based on general early diagenesis modelling concepts, leading to a further parameterization of the above flux equation. In this parameterization K_m is a function of e.g. diffusion, sorption kinetics, oxygen penetration depth and macrofauna abundance (bio-irrigation), and Cb is determined by mineralization rate, aerobic sorption isotherms and kinetics, anaerobic reaction characteristics etc. Finally benthic phosphorus exchange rates as observed in both North Sea and Wadden Sea are evaluated according to the models.

The release of phosphorus during sediment resuspension

Ryszard N. Wisniewski
Copernicus University, Institute of Biology, Dept. of Zoology, ul. Gagarina 9,87-100 Toruń, Poland

The release of phosphorus from sediments to overlaying water was studied during laboratory and in situ experiments.

The in situ investigations were conducted in Włocławek dam reservoir and in two lakes differing in morphology and trophy.

Lake Druzno is a shallow (3 m max depth) and highly eutrophied while lake Partęczyny is deep (28.5 m) and mesotrophic.

It was established, that rapid and enhanced release of phosphorus occured during sediment resuspension.

In lake Partęczyny greatest increase of phosphorus release was noted for the deep profundal deposits.

In littoral zone (4 m depth) it changed from 25 mg in undisturbed sediments to 40 mg during resuspension – (per sq m, per hour), while in profundal (27 m depth) from 30 mg to 90 mg respectively.

The degree of enhancement was correlated with the SOD increase and the rate of resedimentation – i.e. share of fine-grained sediment particles.

It seems that for this reason shallow deposits from accumulation zones of lake Druzno, physically very like to those from profundal of lake Partęczyny, showed similar character of phosphorus release dynamics.

For the more precision laboratory measurements a micro-cosm was constructed. Modular construction consists of teflon rings, having different functions, inserted between glass tube segments.

It allows to observe the release of phosphorus at different physical and chemical properties of sediment and water, for example different kind of deposits, desired temperature profile and oxygen regime.

The detailed description of the micro-cosm and preliminary results are presented.

Application of new functional models for calculation of phosphorus fluxes from sediments

Natalia V. Ignatieva
Institute for Lake Research Ac. Sci. USSR, Sevastyanov Str. 9, St.-Petersburg, 196199, Russia

Diffusive fluxes of dissolved phosphate, the rates of oxygen consumption and sediment accumulation were calculated using new formulas of Dr I. B. Misandrontsev (1990). All the formulas are based on the equations of macrokinetics processes near the sediment–water interface under the steady-state conditions. There are two independent methods to calculate the diffusive fluxes of phosphate using the formulas. One is based on rates of sediment accumulation and organic phosphorus content in surficial sediment layers, another on the shape of concentration profile of inorganic phosphorus in interstitial waters.

To test the Misandrontsev's formulas other methods have been used for different sediments of several lakes. Data obtained during short period laboratory measurements with undisturbed sediment cores in chambers and calculated from the Misandrontsev's formulas (two independent methods) are similar, while the fluxes calculated from Fick's first law of diffusion are different from ones measured in chambers. This indicates that Misandrontsev's formulas give accurate results.

These formulas were applied for calculation of oxygen consumption and sediment accumulation rates and phosphorus fluxes from sediments of the largest European Lake Ladoga. Sediments of this lake are of various types from sands to fine-grained silts and clays. There are considerable bottom areas with one or several layers a few centimetres beneath the sediment-water interface, which are enriched with Fe or Mn. These layers prevent vertical diffusion of phosphate within sediment. The overlying water is usually well-aerated during the year, thus the exchange processes take place mainly under aerobic conditions.

Concentrations of phosphorus and organic carbon in the solid phase of sediment, inorganic and total phosphorus in interstitial waters, dissolved oxygen in overlaying water, porosity and some other parameters were determined in samples from 46 sites of bottom area. All the calculated parameters vary widely. Patterns of space distribution of these characteristics are presented. The least values are connected with sandy sediments of southern shallow areas of the lake, and the greatest with fine-grained sediments of deep region and some littoral areas. Mean value of phosphorus flux is 0.45 mg P/sq. m day. The calculated internal phosphorus load is about 2900 ton/year.

These calculative methods may be used for research of both shallow and deep lakes particularly where measurements in situ are difficult or impossible.

Hydrobiologia **253**: 323–335, 1993.
P.C.M. Boers, T.E. Cappenberg & W. van Raaphorst (eds),
Proceedings of the Third International Workshop on Phosphorus in Sediments.
© *1993 Kluwer Academic Publishers.*

Effectiveness of Al, Ca, and Fe salts for control of internal phosphorus loading in shallow and deep lakes

G. Dennis Cooke[1], Eugene B. Welch[2], Angela B. Martin[1], Donald G. Fulmer[1], James B. Hyde[1] &
Garin D. Schrieve[2]
[1]*Department of Biological Sciences and Water Resources Research Institute, Kent State University, Kent,*
OH 44242, USA; [2]*Department of Civil Engineering, University of Washington, Seattle, WA 98195,*
USA

Key words: alum, phosphorus inactivation, internal loading, eutrophic

Abstract

Internal P loading can maintain high P concentrations and delay eutrophic lake recovery following abatement of external loading. Sediment P inactivation with Al salts has been shown to provide long-term (5–14 years) control of sediment P release; long-term effectiveness of Fe and Ca salts has not been reported. Al toxicity problems are possible unless pH is maintained in the 6–8 range. Vertical transport of hypolimnetic P is unlikely in small, deep, dimictic lakes ($\overline{Z}\sqrt{A_0} > 8$), and effectiveness of P inactivation in lowering their mid-summer epilimnetic P has not been demonstrated. To date, P inactivation has been found to be most effective in improving trophic state in shallow, softwater, polymictic lakes where control of sediment P release affects the entire water column. Abatement of external loading, where necessary, is essential for a successful P inactivation treatment.

Introduction

Eutrophication remains a major world environmental problem. Lakes and reservoirs with algal blooms, siltation and shoaling, nuisance macrophytes, and periods of low or zero dissolved oxygen are likely to have impaired recreational activities, to be a source of potable water with taste, odor and perhaps unhealthful levels of chlorination by-products, and to have altered biological diversity. These problems are costly, may pose health risks, and in some geographic areas may contribute to developing shortages of clean freshwater (Cooke & Carlson, 1989).

The eutrophic state is the product of excessive loading of dissolved and particulate organic and inorganic materials. When loading is greatly reduced, some lakes recover to a previous and more acceptable trophic state, whereas others recover slowly and may not reach a significantly better condition. Lakes with high internal loading and a close proximity of productive zones to nutrient regenerative zones are among those which may be slow to improve (Cullen & Forsberg, 1988; Marsden, 1989).

The net release of phosphorus (P) from sediments probably occurs in every eutrophic lake, especially those with extended periods of hypolimnetic anoxia leading to anaerobic bacterial metabolism and iron reduction. These processes can provide a large subsidy of P to surface waters when the lake is polymictic or when wind fetch and lake morphometry combine to produce mixing between hypolimnetic and epilimnetic waters.

P released from anoxic sediments is primarily biologically available phosphate (Nürnberg, 1988). In some lakes, horizontal P transport from littoral and wetland areas may be important (James & Barko, 1991).

Mass balance studies of small lakes indicate that 65–100% of summer increase in lake P content is from internal loading (Cooke *et al.*, 1977; Perkins, 1983). Such lakes may be the rule rather than the exception, particularly in the common small, shallower (< 15 m) lakes and/or those with extensive wetland and littoral areas.

In dimictic, eutrophic lakes, most of this P increase occurs in the hypolimnion. Vertical P entrainment must occur in order for hypolimnetic sediment P release to be significant to algal production, a process not equally important in every lake. It is most likely to occur in larger lakes exposed to a series of summer cold fronts, rather than in sheltered lakes or in deep lakes with a small surface area.

Recovery of lakes with extensive internal loading and vertical P entrainment requires both the reduction of external loading and treatment or removal of P-rich sediments. There are several methods to accomplish the latter objective, including sediment removal or skimming, sediment oxidation, hypolimnetic withdrawal, and P inactivation by the addition of iron, calcium, or aluminum salts (Cooke *et al.*, 1993). This report examines this use of these salts, emphasizing some unpublished comparative data about alum treatments in shallow and deep lakes.

The phosphorus inactivation process

The purpose of iron, calcium, or aluminum applications is to produce initial lowering of lake P content (P precipitation) and to provide long-term (years) control of sediment P release (P inactivation). P inactivation/precipitation is not the same as an algicide treatment, and its classification as such is erroneous. P inactivation, in conjunction with reduction of external loading, improves trophic state through the limiting nutrient concept, and can be effective for years, whereas the brief effect of an algicide is through its toxicity to cells.

Reviews of aluminum chemistry include Burrows (1977) and Driscoll & Schecher (1990), from which the following discussion is partially drawn. When aluminum sulfate (alum) is added to water it dissociates, aluminum ions appear, and a hydrolysis series occurs leading to formation of aluminum hydroxide ($Al(OH)_3$) and a decrease of pH. $Al(OH)_3$ is a colloidal, amorphous floc with high coagulation and P adsorption properties. $Al(OH)_3$ settles through the water column, removing particles and growing in size and weight as it falls to the sediment surface. Within hours, transparency increases dramatically.

The pH of lake water determines which aluminum hydrolysis products will dominate, and their solubilities. At pH 6–8, insoluble polymeric $Al(OH)_3$ dominates, whereas at pH 4–6, various soluble intermediate forms occur, and at pH less than 4, hydrated and soluble Al^{3+} dominates. When alum ($Al_2(SO_4)_3 \cdot 14\,H_2O$) is added to soft waters, acid neutralizing capacity (ANC) is quickly exhausted and soluble aluminum species dominate. At higher pH levels (> 8.0), the amphoteric nature of aluminum hydroxide forms the aluminate ion. High pH will also lead to the liberation of P sorbed to the floc. In lakes with low or moderate ANC (0.6–1.0 meq l^{-1}), alum addition produces a decline in pH at a low or moderate alum dose, leading to the appearance and increasing concentration of toxic, soluble aluminum forms, including $Al(OH)_2^+$ and Al^{3+}. This limits the amount of aluminum sulfate which can be added to soft water lakes. This problem has been addressed by adding buffers, usually sodium aluminate, to the lake or to the alum slurry as it is applied. Though more costly than aluminum sulfate, sodium aluminate ($Na_2Al_2O_4 \cdot N\,H_2O$), a high alkalinity compound, has the additional benefit of having aluminum content. This approach will maintain a pH in the 6–8 range and provides excellent P sorption. Details of dose determination are described in Cooke *et al.*, 1986, 1993, and in Kennedy & Cooke (1982).

The important properties of $Al(OH)_3$ to P inactivation treatments are its low or zero toxicity to lake biota under normal treatment conditions (pH > 6.0), its ability to adsorb large amounts of inorganic P, and the essentially irreversible binding of P to the floc (except at high pH). Low or zero dissolved oxygen conditions do not solubilize the floc and allow P release. $Al(OH)_3$ is not effective in removing or sorbing dissolved organic P compounds. This could be of significance because P is released from phosphomonoesters and other organic molecules through autolytic and enzymatic activities.

Phosphorus also forms precipitates and complexes with iron and calcium. Inorganic iron exists in the oxidized ferric (Fe^{3+}) or the reduced ferrous (Fe^{2+}) forms, depending upon pH and oxidation-reduction potential. $Fe(OH)_3$ strongly sorbs P and forms part of an oxidized 'microzone' over the sediment surface, providing high sediment P retention. $FePO_4$ also forms, but the primary P removal and retention in sediments is by solid $FeOOH-PO_4$ complexes, and is maximal at pH 5–7 (Lijklema, 1977). In the reduced state (Fe^{2+}), iron becomes soluble and P is released, a change which occurs rapidly so that even brief periods of thermal stratification and dissolved oxygen consumption leads to P release. It appears to occur on a diurnal basis in littoral areas of some lakes. P is also released from iron complexes during periods of high pH, as may occur through photosynthesis (Boers, 1991a). This high sensitivity of iron to redox and pH conditions means that its use as a P inactivant might have to be accompanied by aeration or artificial circulation.

Calcium carbonate (calcite) and calcium hydroxide can be added to a lake or be produced during periods of photosynthesis. Calcite sorbs P, especially when pH exceeds 9.0 (Koschel et al., 1983), and can result in significant P removal (Gardner & Eadie, 1980). At high levels of pH, Ca^{3+}, and P, hydroxyapatite forms. Unlike $Fe(OH)_3$, it has its lowest solubility at pH > 9.5, and P sorbs to it (Andersen, 1975). The solubility of calcite and hydroxyapatite increases as CO_2 concentration increases and pH falls, as would occur in a hypolimnion or dark littoral zone with intense respiration, leading to P release. As with iron, effective P removal and inactivation is possible with calcium but conditions for continued P sorption can be lost without an additional management step.

Alum case histories

These experiences with P inactivation illustrate typical lake responses and important problems. The technique has been used in deep and shallow lakes, poorly buffered lakes, reservoirs, and ponds. Comprehensive analyses of alum, calcium, and iron case histories have been reported elsewhere (Cooke et al., 1986, 1993).

Mirror and Shadow Lakes, Wisconsin

Mirror (A = 5 ha, Z_{max} = 13.1 m, $\overline{Z}$ = 7.8 m) and Shadow (A = 17 ha, Z_{max} = 11.6 m, $\overline{Z}$ = 5.3 m) lakes are located in the rich glacial till and rolling terrain of east-central Wisconsin. Mirror Lake in particular, is surrounded by steep bluffs along the shoreline. The following case history data have been reported by Garrison & Knauer (1983) and Garrison & Ihm (1991a, b).

Urban storm drainage, prior to diversion in 1976, accounted for 65% of external P load to Mirror and 58% to Shadow and produced severe blue-green algal blooms after summer storms. Internal P loading was significant. During summer stratification in 1977, the P mass of Mirror increased by 15 kg, of which 2 kg were from external loading. In Shadow, P mass increased 40 kg during this interval and only 8 kg came from external loading.

Alum was applied to each hypolimnion in 1978, at a dose of 6.6 mg Al l^{-1} to Mirror and 5.7 mg Al l^{-1} to Shadow. Figures 1 and 2 illustrate the total (TP) and soluble reactive P (SRP) concentrations of their water columns from 1977 through 1990. Diversion and alum together produced a sharp drop in TP from a post-stormwater diversion concentration of about 90 μg l^{-1} to about 20 μg l^{-1} in Mirror. In Shadow Lake, post-diversion whole-lake TP ranged from 40–80 μg l^{-1}; post-alum concentration were about 20 μg

326

MIRROR LAKE

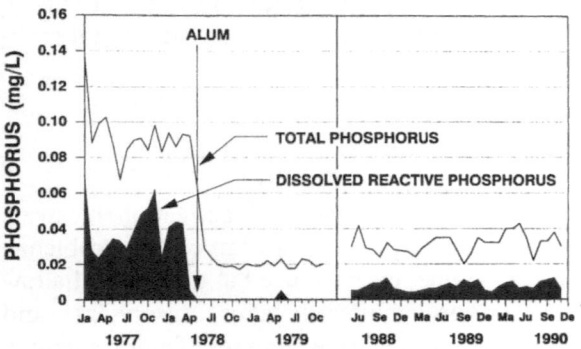

Fig. 1. Volume weighted mean of phosphorus concentration in Mirror Lake, WI, before, immediately following, and a decade after completion of the restoration work (from Garrison & Ihm, 1991a, b).

SHADOW LAKE

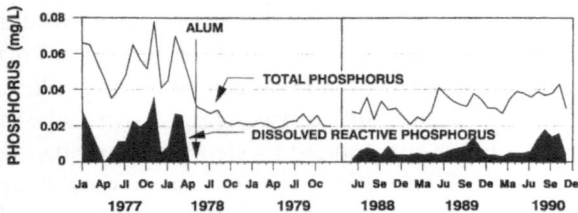

Fig. 2. Volume weighted mean of phosphorus concentrations in Shadow Lake, WI, before, immediately following, and a decade after completion of the restoration work (from Garrison & Ihm, 1991a, b).

l^{-1}. SRP was undetectable in either lake in 1978–79 and then increased to 5–10 μg l^{-1} in 1988–1990. Volume-weighted TP in both lakes increased about 50% between 1979 and 1988. Epilimnetic TP prior to a stormwater diversion from Mirror Lake average 56 μg l^{-1}, declined to 28 μg l^{-1} in the year following diversion (1977), and declined after alum application to 1978–1990 annual means ranging from 13–18 μg l^{-1}.

Sediment P release was determined *in situ*, from 1977–1981, and again in 1990, using chambers sampled by SCUBA divers. Two chambers (1 m^2 area exposed to sediments) were placed in each lake at depths of 13 m in Mirror and 9 m in Shadow. In Shadow Lake, one chamber was placed on the sediment prior to 1978 treatment and another just afterwards. P release in Shadow Lake was 9 times less after treatment than before.

P release rates increased between 1981 and 1990 to 0.20 for Mirror and 0.30 mg m^{-2} day^{-1} for Shadow (Table 1). Sediment cores taken in 1991 by us clearly show the Al(OH)$_3$ layer in both lakes about 5–8 cm below the sediment-water interface. The effectiveness of the Al(OH)$_3$ layer in retarding P release has lasted through 1991 (13 yrs), though some increase in P release from the sediment layer overlying the Al(OH)$_3$ has occurred (Figs 1 and 2).

Summer blue-green algal blooms ended after diversion and alum application, chlorophyll and transparency have remained at a greatly improved level, and epilimnetic P has remained at the post-alum treatment level through 1991. Garrison & Ihm (1991a, b) conclude that effectiveness of stormwater diversion was enhanced by the role of Al(OH)$_3$ in retarding P release and P transport to the epilimnion.

West Twin Lake, Ohio

The evaluation of the alum treatment at West Twin Lake (WTL), OH had the advantage of a nearby (200 m) control or reference lake (East Twin, ETL). The lakes are small(WTL = 35 ha, ETL = 27 ha) shallow (WTL Z_{max} = 11 m, $\overline{Z}$ = 4.3 m; ETL Z_{max} = 12 m, $\overline{Z}$ = 5 m), dimictic, and somewhat sheltered from prevailing winds by low bluffs. WTL drains into ETL, providing 40–70% of ETL's annual water income but usually

Table 1. Phosphorus Release Rates from *in situ* P Release Chambers (data from Garrison & Knauer, 1983 and Garrison & Ihm, 1991a,b)

Year	Release Rates mg TP m^{-2} d^{-1}	
	Mirror Lake	Shadow Lake
1977	1.30	–
1978 (control)	–	1.27
1978 (alum)	0.03	0.14
1979	0.13	0.18
1980	0.04	0.07
1981	0.07	0.08
1990	0.20	0.30

little of its summer water load. The lakes had received septic flows and were highly eutrophic, but waste water diversion was completed in 1972. WTL's hypolimnion was treated with 26 mg Al l^{-1} in July, 1975, using the maximum dose criterion of Kennedy & Cooke (1982). It was predicted that after alum treatment and diversion, internal loading would be less in WTL and its recovery would be faster than ETL's.

The alum application to WTL was effective in retarding sediment P release from anoxic hypolimnetic sediments. Based on mass balance data, mean summer pretreatment (1972–74) internal P release rates were 1.8 and 2.2 mg m^{-2} day^{-1} for WTL and ETL respectively. The mean release rates for summers 1976 and 1978, after alum application to WTL, were 0.4 and 1.2 mg m^{-2} day^{-1} for WTL and ETL respectively (Cooke *et al.*, 1982). Other studies, including the use of seepage meters on treated and untreated hypolimnetic sediments, also demonstrated alum effectiveness (Kennedy, 1978).

Despite the control of hypolimnetic sediment P release, the alum treatment of WTL may not have been the major factor in the lake's improved trophic state. Figures 3 and 4 illustrate mean summer (June through August) surface (0.3 m) and 10 m (just above sediments) TP concentrations in WTL and ETL prior to diversion (1972), before alum application to WTL (1972–74), and after alum (1975–1991). Sampling frequency was

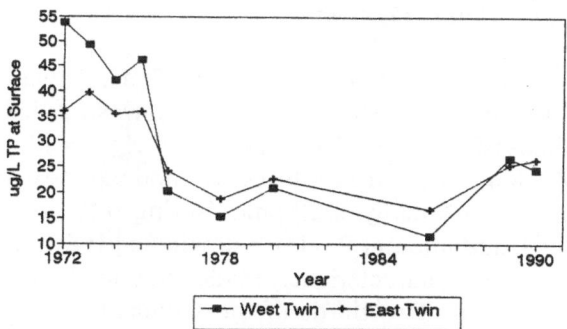

Fig. 4. Summer (June–August) mean (*n* = 5–15 samples per summer) surface total P concentrations in West and East Twin Lakes, OH. Restoration treatments as in Fig. 4.

high (N = 12–14) during 1972–78, and lower (N = 4–7) in later years. The persistent (at least through 1980) low 10 m TP in WTL demonstrates the effectiveness of alum, but between 1980 and 1986 (no monitoring during that interval), treatment effectiveness diminished and by 1988 the 10 m concentrations of WTL and ETL were similar. The $Al(OH)_3$ layer could be seen in 1989 and 1991 WTL sediment cores at a depth of about 9–10 cm, overlain with a layer of new sediments. Despite the alum treatment, the mean summer surface TP concentrations of the two lakes were nearly identical and declined in an identical fashion after the 1975 treatment, suggesting that the primary factor in this decline was diversion. The hypolimnetic alum treatment, while effective in retarding P release from anoxic sediments, may not have played a major role in lowering surface P concentrations.

The low surface TP in both lakes led to a sharp reduction in epilimnetic algal biomass. Chl *a* declined from a pre-diversion summer mean of about 20 $\mu g\,l^{-1}$ in both lakes to a 1991 mean of 4 $\mu g\,l^{-1}$. This change was correlated with the decrease in TP ($r^2 = 0.68$). Transparency increased accordingly. The trophic state of both lakes improved to borderline mesotrophic-oligotrophic after diversion and alum treatment.

The prediction of an accelerated recovery of WTL through a hypolimnetic alum application was apparently not realized.

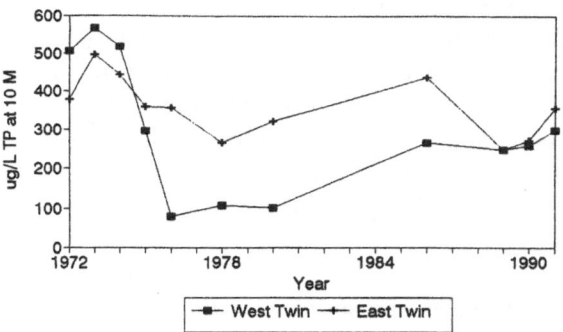

Fig. 3. Summer (June–August) mean (*n* = 5–16 samples per year) 10 meter total P concentration in West and East Twin Lakes, OH. Waste water diversion completed in 1972; alum addition to West Twin occurred in 1975.

Lake Morey, Vermont

The treatment of Lake Morey, VT, a soft water lake (alkalinity = 0.6–1.0 meq l^{-1}), was partially successful. The lake (220 ha, $Z_{max} = 13$ m, $\overline{Z} = 8$ m), located in a forested, mountainous region, was mildly eutrophic (spring TP ~ 40 μg l^{-1}), and internal P release dominated P dynamics. Two characteristics made internal loading significant. Two-thirds of the sediment area occurs in the anoxic hypolimnion, and a low hypolimnetic iron: TP ratio (about 0.5), perhaps related to high sulfide concentrations and the precipitation of FeS, made precipitation of P by Fe minimal during turnover. Abatement of cultural loading occurred many years earlier. Additional Lake Morey characteristics are in Smeltzer (1990).

Because the lake waters are poorly buffered, Na aluminate and alum were added together at an empirically determined ratio of 1.4 gm Al from aluminate per gm of Al from alum, in order to maintain pH above 6.5. A dose of 12 mg Al l^{-1} was added to the $5 \times 10^6 m^3$ hypolimnion in June 1986.

Treatment sharply lowered the P content of Lake Morey, an effect which has endured through August, 1991 (Smeltzer, 1990; this study). Before application epilimnetic concentrations were 20–30 μg l^{-1}, and values in the 30–40 μg l^{-1} range were found at turnover. Concentrations above 500 μg l^{-1} were found near the hypolimnetic sediments. Following treatment, and in 1991, epilimnetic and spring-fall concentrations were less than 10 μg l^{-1} and hypolimnetic P was less than 50 μg l^{-1} (Table 2).

Summer chlorophyll a levels declined from 5–20 μg l^{-1} to less than 5 μg l^{-1} and Secchi disc transparency increased from pre-treatment depths of 2–4 m to post-treatment values (through 1991) of 7–8 m. Dissolved oxygen in the hypolimnion greatly increased because of the improved water clarity (Smeltzer, 1990).

The treatment was successful in breaking up a summer pattern of an onset of anoxia, followed by P release and vertical transport to epilimnetic waters, followed by algal production and sedi-

Table 2. Lake Morey, VT, August Vertical Profile of Total P μg l^{-1}

Meters	1981*	1986* (post-treatment)	1991**
Surf	20	10	2
1	20	10	2
2	20	10	2
3	20	10	2
4	20	10	2
5	20	10	2
6	20	10	2
7	20	10	5
8	20	20	13
9	30	30	27
10	70	30	20
11	500	30	36
12	500	30	–

* data from Smeltzer (1990)
** this study

mentation of cells and additional anoxia and P release. Iron concentrations were unaffected by alum, and the Fe:P ratio increased to 3.3. This has apparently facilitated P precipitation during turnover. The effect of the application on the P concentration has lasted through August, 1991.

The Lake Morey alum application produced a temporary increase in dissolved aluminum concentration in waters above the application depth, an event associated with a significant decrease in condition factor of adult yellow perch (*Perca flavescens*) and a temporarily decreased density and species richness of benthic invertebrates. This problem, described in detail by Smeltzer (1990), and Cooke *et al.* (1993), was unexpected because pH was maintained at a level which maximized Al(OH)$_3$ formation and precipitation. Similar alum impacts have been noted in some other softwater lakes, suggesting that extra precautions are needed for their treatment.

Eau Galle Reservoir, Wisconsin

Reservoir alum treatments provide documentation of the response of waterbodies where external loading is not reduced or eliminated. Eau

Galle Reservoir is a small (0.6 km^2), weakly stratified impoundment which receives very heavy external and internal P loading (Barko *et al.*, 1990). Dose was calculated to be equivalent to five times the average summer internal load, assuming that P-aluminum complexes would attain a stoichiometric ratio of one (Kennedy *et al.*, 1987). The treatment was effective for most of one summer in controlling hypolimnetic P, but there was no effect on algal biomass or epilimnetic P because high external loading continued. Sediment P inactivation is ineffective when external loading remains high.

Shallow lakes

Internal P loading can be significant in shallow unstratified lakes. Mechanisms of P release can involve temporary anoxia and iron redox phenomena, increased pH (from photosynthesis) and ligand exchange, and microbial enzymatic hydrolyses. These reactions are enhanced by the higher temperatures which may be found across the entire basin of mixed lakes (Boström *et al.*, 1982; Ryding, 1985). Internal loading may also occur via migration of blue-green algae from the sediment surface to the upper waters (Barbiero & Welch, 1992).

P inactivation treatments of shallow lakes were initially considered to be ineffective because early applications were failures. In two of these cases (Lake Langsjön, SWE (Jernelov, 1971) and Lake Lyngby SØ, DEN (Norup *et al.*, 1975, cited in Welch *et al.* (1988)), high external loading and sediment disturbance prohibited control of lake P content. In the third case (Pickerel Lake, WI, USA; Garrison & Knauer, 1984)), the alum floc was redistributed to the lake's center by wind mixing. P inactivation, however, should be highly effective in improving the trophic state of polymictic lakes dominated by internal P loading because the entire water column will be affected by the control of sediment P release.

Welch *et al.* (1988) reviewed the effectiveness of the 12 published alum treatments in shallow lakes. Of the 9 for which there are data, 5 had a

reduced P content, decreased algal abundance, and increased transparency for at least one year, and were labelled as successes. One recent treatment (Lake Wapato) and the 3 just mentioned were failures.

The five successes plus Lake Wapato are located in Washington state. Each lake was greatly dominated by internal loading. Two, Pattison and Long (Thurston Co.), are separated by narrow isthmuses so that, for analysis purposes, they are divided into north (N) and south (S) halves. The S halves of Pattison and Long (Thurston Co.) are dominated by dense beds of submersed macrophytes which appeared to influence internal P release through increased pH and tissue decomposition. Effectiveness has been dramatic and long-lasting (up to 7 years) in all the lakes except Wapato (Fig. 5). Corresponding decreases in chlorophyll and an increase in transparency have also occurred. In Wapato Lake, areal macrophyte coverage (*Ceratophyllum* sp.) increased from 20% to 75% in the year after treatment. TP doubled over the previous year's concentration, apparently from an increase in pH to 10.1. Also *Ceratophyllum* may have released P throughout the growing season from living and dead tissues.

Long Lake (Kitsap Co., Washington) is the most thoroughly studied of these shallow lakes (Welch *et al.*, 1988; Welch & Kelly, 1990). In the fifth year (1985) following treatment, P concen-

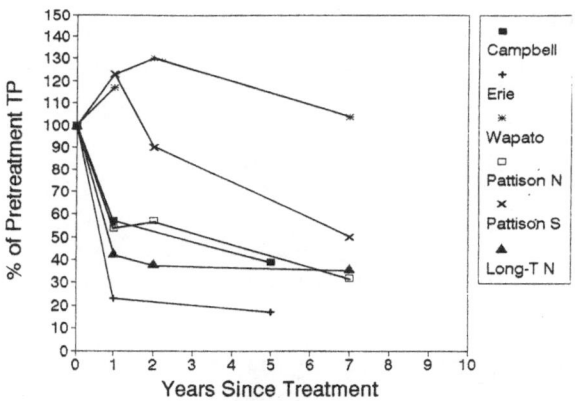

Fig. 5. Mean summer total P concentrations, as percent of pre-alum treatment TP, in five shallow lakes in Washington state, USA.

330

tration abruptly returned to pre-treatment levels and then declined sharply again in years 6 and 7. By 1990, 10 years after alum application, concentration had gradually returned to the pre-treatment point (Fig. 6). The 1985 increase appears to have been the result of a sharp decline in macrophyte (*Eloda densa*) biomass, allowing wind mixing to stir sediment P into the water column. Unlike *Ceratophyllum* in Wapato Lake, *Elodea* appears to be a plant which acts as a sink for P in summer and releases large amounts of nutrients during fall-winter die-back (Rørslett *et al.*, 1986). *Elodea* infestations in Long Lake thus produced P release only when pH increased (Jacoby *et al.*, 1982), a process apparently of little importance after alum was applied. The alum treatment of Long Lake (Kitsap Co.) was highly effective in controlling internal P release for 4 years and was partially effective for 9. Summer mean chlorophyll *a* showed a corresponding improvement. The pre-treatment mean chlorophyll *a* was 22, the mean for years 1–4 post-treatment was 9, and for years 5–9 was 18 μg l^{-1} (Welch & Kelly, 1990). A re-application of alum occurred in Autumn, 1991. These results, when compared to dimictic lakes (e.g. Mirror, Shadow, West Twin) show that alum treatments of shallow, polymictic lakes are effective in lowering the P content of the entire water column with corresponding and long-lasting improvements in trophic state. Wind mixing of exposed sediments, however, may produce increases in P concentration.

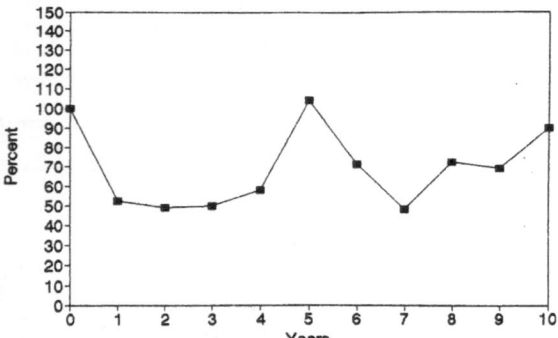

Fig. 6. Mean summer total P concentration, as percent of pre-alum treatment TP, in Long Lake (Kitsap Co), Washington, USA.

Iron

Iron salts alone have rarely been used successfully to inactivate sediment P because low redox values develop, releasing iron-bound P. For example, Hayes *et al.* (1984) added 3.5 mg l^{-1}, as ferric sulfate, to polymictic Foxcote Reservoir, England. A floc formed over the sediments, and mean SRP fell from 7 to 3 μg l^{-1}, and TP from a mean of 30 to 16 μg l^{-1}. The floc layer, however, was gone within 30 days, apparently from sediment anoxia. Algal biomass was reduced during one summer. The authors recognized that the iron dose was too small and that an aeration device should have been installed. Laboratory studies have suggested the use of iron doses as high as 100 g Fe^{+3} m^{-2} (Wijesooriya, 1989). Boers (1991b) also determined this rate to be optimal, even under anoxic conditions, and it was applied as FeCl$_3$ to the upper 15–20 cm of Lake Groot-Vogelenzang (Netherlands) as a 40% solution diluted 100–150 fold with lake water. The treatment was effective in controlling sediment P release for 3 months. The return to pre-treatment P conditions was attributed to high external loading and storm-related sediment disturbance.

The operation of a hypolimnetic aerator alone has not always produced declines in hypolimnetic P, possibly because of low iron content or because iron has been lost from the cycle as FeS or FeS$_2$. When iron is added, aeration devices can effectively retard sediment P release (Cooke *et al.*, 1986, 1993; Lean *et al.*, 1986). An example is Vadnais Lake, MN where iron chloride was continuously injected into hypolimnetic aeration units just above the diffuser rings at a rate (100 kg Fe day^{-1}) which would produce a removal ratio of 0.07–0.17 gm orthophosphorus per gm of iron (Walker *et al.*, 1989). This procedure was needed to supplement the aerators because the hypolimnetic Fe:P ratio was < 0.5 whereas ratios > 3.0 are optimal to promote FePO$_4$ precipitation at turnover (Stauffer, 1981). There was a large response to iron injections (Fig. 7) throughout the hypolimnion. However, metalimnetic P increased and algal blooms developed as the metalimnion eroded in late August. Because P began to accu-

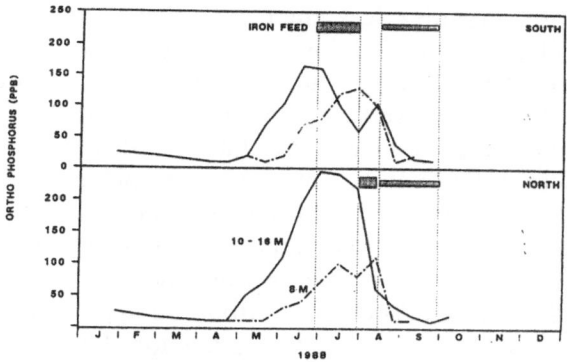

Fig. 7. Orthophosphorus responses to iron chloride injections in northern and southern regions of Vadnais Lake, MN (USA) (from Walker et al., 1989).

mulate in the metalimnion prior to initiation of iron injections on 30 June, Walker et al. (1989) recommended that aerator operation and iron additions begin in early May. They also suggested that iron additions could be terminated in several years when the iron content of the sediments has been replenished.

The effects of hypolimnetic aeration and iron injections on epilimnetic P and algal biomass, like the hypolimnetic application of alum, will depend upon the significance of vertical P transport in a particular lake.

Calcium

Reactions of calcium compounds with P have been used recently to precipitate and possibly inactivate P in eutrophic hard-water lakes, ponds, and stormwater detention basins. Murphy et al. (1988) described the addition of 23 metric tons (summer, 1983) and 16 metric tons (summer, 1984) of slaked lime ($Ca(OH)_2$) to the surface of dimictic, eutrophic Frisken Lake, B.C. (Canada) (34 ha, $Z = 11$ m, $\overline{Z} = 5.5$ m) to precipitate $CaCO_3$. P precipitated from the epilimnion, *Aphanizomenon flos-aquae* was flocculated, and transparency increased. The precipitate dissolved in the hypolimnion, limiting the treatment's long-term effectiveness. Prepas et al. (1990) added $CaCO_3$ and $Ca(OH)_2$ over two summers to the surface of hypereutrophic Figure Eight Lake (Al-

berta, Canada), a weakly stratified lake which develops hypolimnetic anoxia, TP concentrations of 1900 μg l^{-1} near the sediments, and pH values near 10 during *Aphanizomenon* blooms. TP decreased sharply. Prepas et al. believe this to be due to apatite formation and increased P-binding capacity of the sediments. Unlike other area lakes, diatoms and green algae replaced *Aphanizomenon*. Longevity of the treatment of Figure Eight Lake is unknown. The studies of Babin et al. (1989) suggest that slaked lime is more effective than calcite, perhaps because smaller, higher surface area crystals form.

Calcium compounds for P inactivation should receive additional study. Even if long-term effectiveness is less than with aluminum, costs are lower, application is simpler, and potential for toxicity is less as long as pH is kept below 10 (Murphy et al., 1988). These compounds could also be used effectively in shallow water sediments, perhaps in combination with aluminum application to hypolimnetic sediments.

Phosphorus inactivation problems

The use of aluminum salts has potential to create toxic conditions like those in acidified lakes. All forms of aluminum have greater or lesser potential for harmful or toxic effects. Soluble forms are toxic to cells whereas colloidal-sized $Al(OH)_3$ (s) may accumulate on cell surfaces following prolonged contact and produce gill hyperplasia. In nearly every case of an aluminum treatment, there have been major chemical (e.g. lowered P concentration) and physical (e.g. increased transparency) changes, and sometimes negative (e.g. reduced condition factor, decreased invertebrate species diversity) biotic changes. Algal biomass almost always declines. There has never been a report of a significant amount of mortality to fish, nor a demonstration of aluminum accumulation in tissues. Only in soft-water Lake Morey, in laboratory studies where prolonged exposure occurs, and in acidified lakes, have there been documented problems with aluminum toxicity. Surface and hypolimnetic treatments of well-buffered lakes have been without toxicity problems (Cooke et al., 1986, 1993; Narf, 1990).

There are several factors which reduce or eliminate potential for aluminum toxicity in well-buffered lakes. Driscoll et al. (1980) demonstrated that organic ligands (e.g. humics), common in lakes with high total organic carbon (e.g. eutrophic), complex with aluminum, even in soft water, and can eliminate toxicity to fish. Birchall et al. (1989) found that water high in silicon, even at low pH levels, eliminated aluminum toxicity to fish fry. Increased silicic acid may be as important in preventing toxicity as maintaining circumneutral pH. Finally, when the dose guidelines of Kennedy & Cooke (1982) are followed, pH will not fall to the level where soluble aluminum forms appear, and the treatment of only hypolimnetic waters or directly to sediments should eliminate the continuous exposure to $Al(OH)_3$ floc required for fish gill hyperplasia to occur.

Some lakes are unsuited for P inactivation with aluminum, including all lakes with high external P loads and lakes with very low (< 0.4 meq l^{-1}) alkalinity and/or acidic epilimnetic pH. Even if the treatment of a poorly buffered lake were successful, future acidification events could lead to elevated aluminum levels.

Discussion

Successful sediment P inactivation is determined by two criteria: (1) did the treatment reduce sediment P release? and (2) did the treatment lower photic zone P concentration? The first question is answered by comparing pre- and post-treatment hypolimnetic P concentrations and by determining P release rates. The answer to the second question is obvious for polymictic lakes. In dimictic lakes, vertical entrainment of P-rich hypolimnetic waters must be shown to have been a significant subsidy to surface water algae.

A useful estimate of the probability of vertical P transport via deep mixing events is the Osgood Index (Osgood, 1988), the ratio of mean depth ($\overline{m}$) to the square root of surface area (km^2) ($\overline{Z}/\sqrt{A_0}$). Osgood found that Minnesota lakes

with an index less than 6–7, which includes polymictic and weakly stratified dimictic lakes, were lakes in which surface water P may greatly increase from vertical P entrainment in excess of summer external loading. Lakes with an index near 7 may not exhibit a summer epilimnetic P increase from vertical transport, but will have a late summer-early autumnal increase in TP associated with thermocline erosion, and the possibility of fall algal blooms. P from the hypolimnion is not likely to reach the epilimnion during the height of summer stratification in dimictic lakes with index numbers above 8. External loading alone likely determines their summer surface P concentrations. This scale of index numbers may vary somewhat at latitudes and climates different from Minnesota. Also, shoreline physiography such as hills and trees may provide shelter from the effects of summer winds.

Dimictic lakes with demonstrably high P loading from hypolimnion to epilimnion are Lake Mendota, WI (Stauffer & Lee, 1973), and the large west basin of Shagawa Lake, MN (Stauffer & Armstrong, 1984), lakes with Osgood Index numbers of 2.0 and 2.3, respectively (Table 3). In contrast, Third Sister Lake, WI, a dimictic lake with an Osgood Index of 36.7, has been shown to

Table 3. Morphometry and Osgood Index (Osgood, 1988) Number for Case History Lakes

Lake	A_0 (km^2)	$\overline{Z}$ (m)	$\overline{Z}/\sqrt{A_0}$
Third Sister, MI	0.04	7.2	36.7
Mirror, WI	0.05	7.8	34.98
Shadow, WI	0.17	5.3	12.96
East Twin, OH	0.27	5.0	9.7
West Twin, OH	0.30	4.3	7.9
Morey, VT	2.20	8.4	5.7
Wapato, WA	0.11	1.5	4.5
Pattison, WA	1.10	4.0	3.8
Long, WA (Thurston Co.)	1.34	3.7	3.2
Erie, WA	0.45	1.8	2.7
Shagawa, MN (West Basin)	6.76	5.95	2.3
Campbell, WA	1.50	2.4	2.0
Mendota, WI	39.40	12.4	2.0
Long, WA (Kitsap Co.)	1.37	2.0	1.7

have very low or no vertical P transport (Lehman & Naumoski, 1986). Comparison of these Osgood Index values to those of WTL (7.9), ETL (9.7), Mirror (34.9), and Shadow (12.9) strongly suggests that vertical P entrainment in them during mid-summer months was unlikely as a significant P subsidy to epilimnetic algae. As well, all of these lakes, especially Mirror, are surrounded by bluffs and trees which shelter them from wind mixing. Even though hypolimnetic sediment P release was controlled with alum, improvement of these lakes was likely due primarily to diversion. Hydromechanical studies are needed at each lake to support or reject this conclusion.

The alum treatment of Lake Morey may be more successful in controlling algal blooms because vertical P transport is probably significant (Osgood Index 5.7; Table 3). Additional years of monitoring will be needed at this recently (1986) treated lake.

Alum treatment of polymictic lakes has been very successful in reducing photic zone P concentration (normally the entire water column), and there has been longevity to the treatments of 4–9 years (Figs 5 and 6). Stratification of them is absent or temporary and P released from sediments can be mixed throughout the water column (Table 3).

Smaller, deep lakes, which have little summer vertical P transport, may benefit from P inactivation if nuisance blooms occur during autumn destratification. P-rich hypolimnia of untreated lakes eventually mixes with the upper waters as the lake cools, and may lead to substantial late summer-autumn algal blooms. Such bloom events should be far less in an alum treated lake. Also, P concentration during spring mixing may be reduced by the $Al(OH)_3$ layer, leading to reduced spring algal blooms and to reduced early summer epilimnetic P.

We propose that the most successful P inactivation treatments for control of surface water P will be to polymictic lakes, or to those dimictic lakes where significant vertical P entrainment can be shown prior to alum application. We hypothesize that these lakes will have an Osgood Index less that 6. A recent (1991) U.S. alum application

was to Delavan Lake, WI, a large, deep eutrophic lake with a flat surrounding terrain. The Osgood Index of Delavan is 2.8 (Table 3), suggesting that both its wastewater diversion and P inactivation will play major roles in controlling summer algal blooms. What is needed are whole lake experiments in which vertical P transport is determined before application, followed by monitoring and post-treatment assessment. The use of control lakes such as ETL, a rare opportunity in whole-lake experiments, would be ideal, as well as long-term monitoring of loading and trophic state.

We conclude that P inactivation will improve trophic state in those lakes which have had sufficient diversion of external loading and are either polymictic or dimictic with an Osgood Index $(\bar{Z}\sqrt{A_0})$ of 6 or less. We encourage continued research with calcium and iron salts, especially in conjunction with other management procedures.

Acknowledgements

Portions of this research were supported by Phase III grants from the U.S. Environmental Protection Agency's Clean Lakes Program to Kent State University (KSU) and the University of Washington, and by a USEPA research grant (R801936) to KSU. We thank Paul Garrison, Douglas Knauer, William James, Eric Smeltzer, Jody Connor, Tom Gordon, and Richard Burton for providing access to their alum-treated lakes and generously sharing data. We also thank Ellie Prepas and Curt Forsberg for their helpful comments on the manuscript.

References

Andersen, J. M., 1975. Influence of pH on release of phosphorus from lake sediments. Arch. Hydrobiol. 76: 411–419.

Babin, J., E. E. Prepas, T. P. Murphy & H. R. Hamilton, 1989. A test of effects of lime on algal biomass and total phosphorus concentrations in Edmonton stormwater retention lakes. Lake Reserv. Manag. 5: 129–135.

Barbiero, R. P. & E. B. Welch, 1992. Contribution of benthic blue-green algal recruitment to lake populations and phosphorus. Freshwater Biol. 27: 249–260.

334

Barko, J. W., W. F. James, W. D. Taylor & D. G. McFarland, 1990. Effects of alum treatment on phosphorus and phytoplankton dynamics in Eau Galle Reservoir: A synopsis. Lake Reserv. Manag. 6: 1–8.

Birchall, J. D., C. Exley, J. S. Chappell & M. J. Phillips, 1989. Acute toxicity of aluminum to fish eliminated in silicon-rich acid waters. Nature 338: 146–148.

Boers, P. C. M., 1991a. The influence of pH on phosphate release from lake sediments. Wat. Res. 25: 309–311.

Boers, P. C. M., 1991b. The release of dissolved phosphorus from lake sediments. Ph. D. Dissertation. Limnological Institute and the Institute for Inland Water Management and Waste Treatment. Lelystad, The Netherlands.

Boström, B., M. Jansson & C. Forsberg, 1982. Phosphorus release from lake sediments. Arch. Hydrobiol. Beih. Ergebn. Limnol. 18: 5–59.

Burrows, H. D., 1977. Aquatic aluminum chemistry, toxicology, and environmental prevalence. CRC Crit. Rev. Environ. Control 7: 167–216.

Cooke, G. D., M. R. McComas, D. W. Waller & R. H. Kennedy, 1977. The occurrence of internal phosphorus loading in two small, eutrophic, glacial lakes in northeastern Ohio. Hydrobiol. 56: 129–135.

Cooke, G. D., R. T. Heath, R. H. Kennedy & M. R. McComas, 1982. Change in lake trophic state and internal phosphorus release after aluminum sulfate application. Wat. Res. Bull. 18: 699–705.

Cooke, G. D., E. B. Welch, S. A. Peterson & P. R. Newroth, 1986. Lake and Reservoir Restoration. Butterworth Pub., Stoneham, MA.

Cooke, G. D. & R. E. Carlson, 1989. Reservoir Management for Water Quality and THM Precursor Control. Amer. Water Works Assoc. Research Foundation. Denver, CO.

Cooke, G. D., E. B. Welch, S. A. Peterson & P. R. Newroth, 1993. Restoration and management of Lake and Reservoirs. Lewis Pub. (CRC Press, Inc.), Boca Raton, FL.

Cullen, P. & C. Forsberg, 1988. Experience with reducing point sources of phosphorus to lakes. In G. Persson & M. Jansson (eds), Phosphorus in Freshwater Ecosystems. Developments in Hydrobiology 48. Kluwer Academic Publishers, Dordrecht: 321–336. Reprinted from Hydrobiologia 170.

Driscoll, C. T. & W. D. Schecher, 1990. The chemistry of aluminum in the environment. Env. Geochem. Health 12: 28–49.

Driscoll, C., J. Baker, J. Bisogni & C. Schofield, 1980. Effects of aluminum speciation on fish in dilute, acidified waters. Nature. 184: 161–164.

Gardner, W. S. & B. J. Eadie, 1980. Chemical factors controlling phosphorus cycling in lakes. *In:* Nutrient Cycling in the Great Lakes: A Summarization of Factors Regulating the Cycling of Phosphorus. D. Scavia and R. Moll (eds). Special Report No. 83, Great Lakes Research Division, The University of Michigan, Ann Arbor: 13–34.

Garrison, P. J. & D. R. Knauer, 1984. Long term evaluation

of three alum treated lakes. *In:* Lake and Reservoir Management. EPA 440/5–84–001: 513–517.

Garrison, P. J. & D. M. Ihm, 1991a. First annual report of long-term evaluation of Wisconsin's Clean Lake projects. Part B. Lake Assessment. Wisconsin Department of Natural Resources, Madison, WI.

Garrison, P. J. & D. M. Ihm, 1991b. Long-term effectiveness of a lake restoration project involving nutrient diversion and a hypolimnetic alum treatment. Abstract, North American Lake Management Society, Annual Meeting, Denver, CO.

Hayes, C. R., R. G. Clark, R. F. Stent & C. J. Redshaw, 1984. The control of algae by chemical treatment in a eutrophic water supply reservoir. J. Instit. Wat. Eng. Sci. 38: 149–162.

Jacoby, J. M., D. D. Lynch, E. B. Welch & M. A. Perkins, 1982. Internal phosphorus loading in a shallow eutrophic lake. Wat. Res. 16: 911–919.

James, W. F. & J. W. Barko, 1991. Estimation of phosphorus exchange between littoral and pelagic zones during nighttime convective circulation. Limnol. Oceanogr. 36: 179–187.

Jernelöv, A., 1971. Phosphate reduction in lakes by precipitation with aluminum sulphate. 5th International Water Pollution Research Conference. Pergamon Press, New York.

Kennedy, R. H., 1978. Nutrient inactivation with aluminum sulfate as a lake reclamation technique. Ph. D. Dissertation, Kent State University, Kent, OH.

Kennedy, R. H. & Cooke, G. D., 1982. Control of lake phosphorus with aluminum sulfate. Dose determination and application techniques. Wat. Res. Bull. 18: 389–395.

Kennedy, R. H., J. W. Barko, W. F. James, W. D. Taylor & G. L. Godshalk, 1987. Aluminum sulfate treatment of a reservoir: Rationale, application methods, and preliminary results. Lake Reserv. Manag. 3: 85–90.

Köschel, R., J. Benndorf, G. Proft & F. Recknagel, 1983. Calcite precipitation as a natural control mechanism of eutrophication. Arch. Hydrobiol. 98: 380–408.

Lean, D. R. S., D. J. McQueen & V. A. Story, 1986. Phosphate transport during hypolimnetic aeration. Arch. Hydrobiol. 108: 269–280.

Lehman, J. T. & T. Naumoski, 1986. Net community production and hypolimnetic nutrient regeneration in a Michigan lake. Limnol. Oceanogr. 31: 788–797.

Lijklema, L., 1977. The role of iron in the exchange of phosphate between water and sediments. In: H. L. Golterman (ed.), Interactions between Sediment and Freshwater. Dr W. Junk Publishers, The Hague: 313–317.

Marsden, M. W., 1989. Lake restoration by reducing external phosphorus loading: The influence of sediment phosphorus release. Freshwat. Biol. 21: 139–162.

Murphy, T. P., K. J. Hall & I. Yesaki, 1983. Coprecipitation of phosphate with calcite in a naturally eutrophic lake. Limnol. Oceanogr. 28: 58–69.

Murphy, T. P., K. G. Hall & T. G. Northcote, 1988. Lime

treatment of a hardwater lake to reduce eutrophication. Lake and Reservoir Manage. 4: 51–62.

Murphy, T. P., E. E. Prepas, J. T. Lim, J. M. Crosby & D. T. Walty, 1990. Evaluation of calcium carbonate and calcium hydroxide treatments of prairie water dugouts. Lake Reserv. Manag. 6: 101–108.

Narf, R. P., 1990. Interaction of Chironomidae and Chaoboridae (Diptera) with aluminum sulfate treated lake sediments. Lake Reserv. Manag. 6: 33–42.

Norup, B., 1975. Lyngby SØ feltundersogelser efter fosfatfaeldeperioden sommeren 1974: Bundfaunaundersogelse fysisk-kemisk and planteplanktonproduction. Lyngby Taarboek Kommune. Copenhagen, Denmark.

Nürnberg, G. K., 1988. Prediction of phosphorus release rates from total and reductant-soluble phosphorus in anoxic lake sediments. Can. J. Fish. aquat. Sci. 45: 453–462.

Osgood, R. A., 1988. Lake mixes and internal phosphorus dynamics. Arch. Hydrobiol. 113: 629–638.

Perkins, M. A., 1983. Limnological characteristics of Green Lake, WA. Final Rept., Phase I, Restoration Analysis. Dept. of Civil Engineering, Univ. Washington, Seattle, WA.

Prepas, E. E., R. P. Murphy, J. M. Crosby, D. T. Walty, J. T. Lim, J. Babin & P. A. Chambers, 1990. Reduction of phosphorus and chlorophyll a concentrations following CaCO₃ and Ca(OH)₂ additions to hypereutrophic Figure Eight Lake, Alberta. Env. Sci. Technol. 24: 1252–1258.

Rørslett, B., D. Berge & S. W. Johnson, 1986. Lake enrichment by submersed macrophytes: A Norwegian whole-lake experience with Elodea canadensis. Aquatic Bot. 26: 325–340.

Ryding, S. O., 1985. Chemical and microbiological processes as regulators of the exchange of substances between sedi-

ments and water in shallow eutrophic lakes. Int. ges. Hydrobiol. 70: 657–702.

Smeltzer, E., 1987. Lake Morey Restoration Project. Interim Progress Report. January 1987, Vermont Dept. Water Resources and Environ. Eng., Watersburg, Vt.

Smeltzer, E., 1990. A successful alum/aluminate treatment of Lake Morey, Vermont. Lake Reserv. Manag. 6: 9–19.

Stauffer, R. E., 1981. Sampling strategies for estimating the magnitude and importance of internal phosphorus supplies in lakes. USEPA 600/3–81–015. Washington, D.C.

Stauffer, R. E. & G. F. Lee, 1973. The role of thermocline migration in regulation of algal blooms. In: Modeling the Eutrophication Process. E. J. Middlebrooks, D. H. Falkenborg & T. E. Maloney (eds), Utah State University, Water Resources Center, Logan: 73–82.

Stauffer, R. E. & D. E. Armstrong, 1984. Lake mixing and its relationship to epilimnetic phosphorus in Shagawa Lake, Minnesota. Can. J. Fish. aquat. Sci. 41: 57–69.

Walker, W. W. Jr., C. E. Westerberg, D. J. Schuler & J. A. Bode, 1989. Design and evaluation of eutrophication control measures for the St. Paul water supply. Lake Reserv. Manag. 5: 71–83.

Welch, E. B. & T. S. Kelly, 1990. Internal phosphorus loading and macrophytes: An alternative hypotheses. Lake Reserv. Manag. 6: 43–48.

Welch, E. B., C. L. DeGaspari, D. E. Spyridakis & T. J. Belnick, 1988. Internal phosphorus loading and alum effectiveness in shallow lakes. Lake Reserv. Manag. 4: 27–33.

Wijesooriya, W. A. D. D., 1989. Phosphorus inactivation in sediments of shallow lakes. M.S. Thesis. Internat. Instit. Environ. Engin. Delft, The Netherlands.

Hydrobiologia **253**: 337–344, 1993.
P.C.M. Boers, T.E. Cappenberg & W. van Raaphorst (eds),
Proceedings of the Third International Workshop on Phosphorus in Sediments.
© 1993 *Kluwer Academic Publishers.*

A new technique to reduce internal phosphorus loading by in-lake phosphate fixation in shallow lakes

Marinus Quaak [1], Joop van der Does [2], Paul Boers [3] & Jo van der Vlugt [4]
[1] *HAM International Dredging Contractors, P.O. Box 8574, 3009 AN Rotterdam, The Netherlands;*
[2] *Waterboard of Rijnland, P.O. Box 156, 2300 AD Leiden, The Netherlands;* [3] *Institute for Inland Water*
Management and Waste Water Treatment, P.O. Box 17, 8200 AA Lelystad, The Netherlands; [4] *National*
Institute of Public Health and Environmental Protection, P.O. Box 1, 3720 BA Bilthoven, The Netherlands

Key words: lake restoration, eutrophication, phosphates, iron-dosing, lake-bottom treatment, water-jet

Abstract

A new, flexible, fast, robust and economic technique was developed to treat sediment in shallow lakes with phosphate binding chemicals. The upper 0.15 m of the sediment is thoroughly mixed with ferric chloride using a water-jet manifold coupled to a dosing pump and a navigation control system. Its logistics were tried out in a small, shallow and hypertrophic peat lake, Lake Groot Vogelenzang.

Introduction

Abating the deleterious effects of eutrophication has been the goal of many attempts at lake restoration but success is still variable. Removal of excessive available phosphorus is one of the key factors for successful lake restoration and many ways to do this have been tried. It appears that we cannot as yet predict the effect of a single restoration measure in a single lake sufficiently accurate for a convincing cost-benefit analysis. This means that lake restoration is, after several decades of experience, still very much an empirical science, not an established technique. This suggested that there is still room for improvement of the toolbox of applicable actions. Lake restoration is treated in the books by Edmondson, 1991; Cooke *et al.*, 1986; Sas, 1989; the interrelations of dredging techniques with environmental problems in surface water by Donze *et al.*, 1990; the interrelations of sediment and water were reviewed by Golterman, 1984.

A first step always must be reduction of external phosphorus loading as far as possible. Sometimes this is enough to restore water quality. In many cases improvement appeared only after a lag time of several years. Often this first step is only partly effective or not at all due to internal loading of the lakewater. After reduction of external sources, phosphates, that over time have accumulated in the top-layer of the lake sediment are released again into the water during the growth season of algae. In Fig. 1 are shown two examples of depth profiles by Siebers (1985) from two peat lakes.

To overcome this problem in several lakes the phosphate-rich top layers were removed by dredging, leading to successful restoration in all cases (Bengtsson *et al.*, 1975). However, dredging is expensive and sometimes difficult to do because a large area is temporally necessary to store and dewater the dredged spoils before it can be used or dumped. If the sediments are contaminated special storage facilities must be built at

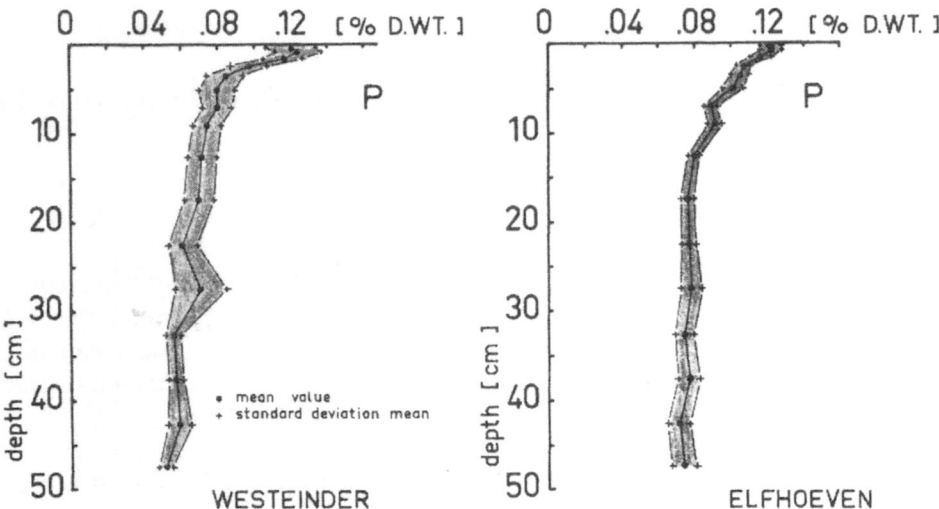

Fig. 1. Depth profiles of total phosphate with standard deviations in two shallow Dutch lakes (from Siebers, 1985). Severe eutrophication started in about 1960; between that time and 1982 phosphate accumulated in the top 30 cm; below this the 'old' concentration was found. Lake Elfhoeven is part of the same complex of lakes as the one studied here (Fig. 2).

considerable costs. Especially in shallow lakes with very loose sediments of a dry solids content of 5% or less dredging is difficult. Minor disturbances of the top-layer induce a turbidity cloud that spreads over the freshly dredged area, redistributing the phosphate-rich sediments.

If dredging is not necessary for nautical reasons or for quantitative water management an economic alternative is offered by techniques that, by chemical treatment, improve the binding capacity of the sediment for phosphates, thereby lowering the release rate. In deep, stratified lakes treatment of the water column with phosphate binding agents has been successful in, among others, Dollar Lake, U.S.A. (Cooke *et al.*, 1986; Kennedy & Cooke, 1982). It did not work well in shallow lakes due to resuspension and dispersion of bottom material by wind induced waves.

A further development was application of phosphate binding agents not only to the water column but also in the top-layer of the sediment. In Sweden the Riplox method (reviewed by Cooke *et al.*, 1986) has been applied. Oxidizing or phosphorus binding substances were introduced into the lake bed by a harrow type tool. Another variation is to cover the lake bottom with cheap (waste) materials in bulk that simultaneously bind

phosphates and cover the layer below of high release rate. Fly-ash, alum and lime have been successfully applied this way as discussed by Cooke *et al.* But the use of aluminium is suspect due to its potential toxicity, fly-ash because it can introduce pollutants to the lake.

The technique of phosphate fixation described here is a new robust method allowing fast treatment of a lake bottom in a controlled manner.

Materials and methods

The study area

The pilot project was carried out in Lake Groot Vogelenzang, part of the Reeuwijk lake system (Fig. 2) near Gouda in the Netherlands. Some characteristics of the sediment are given in Table 1. A long term water quality monitoring program is in progress. (Van der Vlugt & Klapwijk, 1990).

Iron dosing technique

In-lake phosphate fixation is developed to reduce the release of phosphates from shallow lake bottoms to overcome problems caused by resuspen-

Fig. 2. Map of the Reeuwijk lake area near Gouda in the western part of the Netherlands with Lake Groot Vogelenzang located in the centre. It is level with and directly connected by small thoroughfares with the surrounding lakes Ravensberg, 's Gravenkoop and 's Gravenbroek.

sion of loose sediments of the top-layer by wind-wave action, outboard engines and bioturbation by fish scavenging for food and by bottom dwelling animals.

In Dutch peat lakes the lake bed consists of very loose sediments with 5 to a maximum of 15% dry matter. A low estimate of the thickness of the layer sensitive to resuspension and reworking is 0.15 m. Often the highest concentrations of phosphorus are found in this layer (Fig. 1).

Table 1. Average ($n = 5$) composition of the 0–5 cm top-layer of Lake Groot Vogelenzang sediment.

Parameter	Quantity
Dry weight	6.2% of wet weight
Loss on ignition	49% of dry weight
$CaCO_3$	13% of dry weight
pH	7.4
Total P	1.6 mg g^{-1}
Total Fe before treatment	24 mg g^{-1}
Total Fe after treatment	40 mg g^{-1}

To achieve optimal results, the top 0.15 to 0.20 m of the lake sediment were mixed with an amount of iron (III) as determined from laboratory experiments. This was done by stirring up the top-layer by water-jets, at the same time injecting a diluted iron-chloride solution. The top-layer of the sediment is temporarily transformed in a turbulent suspension where intensive mixing of sediment and iron solution occurs.

Water-jets are commonly used in the dredging industry, especially to disintegrate material being dredged by means of a suction pump. A newer application is water injection dredging. Surface water is injected directly into the bottom generating a turbidity or density cloud. The suspended material moves outside the area being dredged by hydrodynamic and gravitational forces (Estourgie, 1988). The technique is schematically illustrated in Fig. 3. Our phosphate fixation method is derived from this technique.

It was calculated that low pressure jets at a limited stand-off distance only resuspend about 0.15 m of the loose peat. The turbidity cloud has a sufficiently low density not to cause significant sediment transport.

To guarantee full coverage of the treated area the Artemis high precision electronic distance-bearing positioning system was used. The iron solution was accurately dosed by coupling the rate of the dosing pump to the sailing speed of the craft. In Fig. 4 the relationships between the main components of the equipment are shown.

Technical grades of iron(III)-chloride contain low concentrations of heavy metals, which was considered unacceptable. Environmental regulations required the use the same quality as used as a flocculant in the preparation of drinking water.

Lake Groot Vogelenzang is, with some difficulty, only accessible over water. Narrow passages at bridges have to be negotiated. For the pilot project a special craft was constructed from locally available equipment. The injection vessel

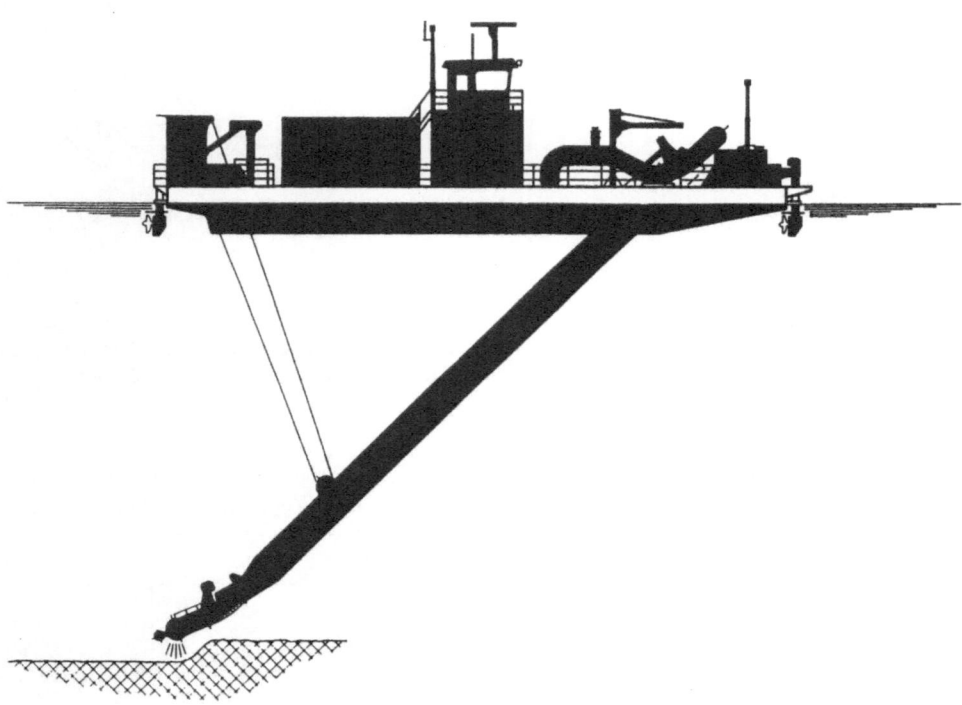

Fig. 3. Schematic view of the water injection method. Just above the sediment the manifold with water-jets is moved forward by the ship, generating a density cloud. The water injection vessel 'Jetsed' was built in 1987, manifold width 14 m, total jet-capacity 12 000 m^3 h^{-1}.

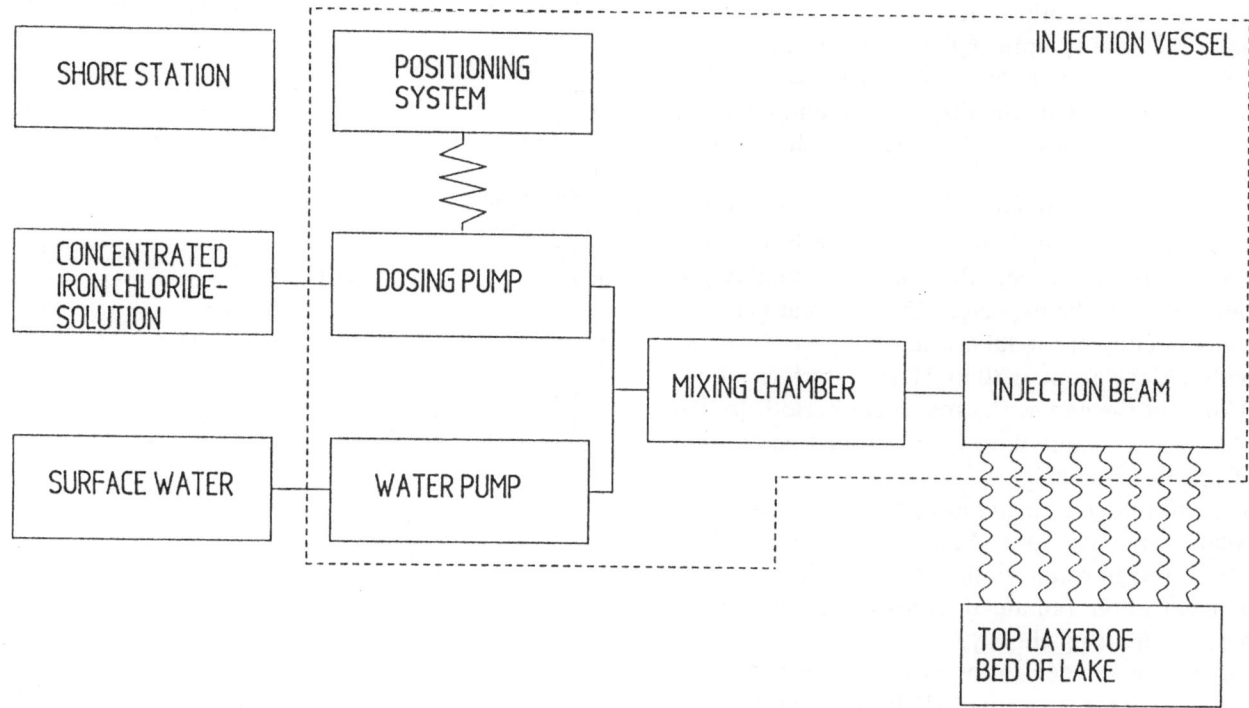

Fig. 4. The relationships between the main components of the equipment built for phosphate fixation.

consisted of two coupled pontoons and a push unit each 2.60 m wide. The injection beam or manifold with closely spaced jets and a width of 2.50 m was mounted on one of the pontoons. With this small improvised craft between 2 and 3 ha per day were treated. Larger equipment like the water injection vessel Jetsed (Fig. 3) a rate of 25 ha per day or more can be reached.

Laboratory experiments

Reduction of the phosphorus release rate of sediment was studied by continuous flow measurement in experimental columns according to Boers & Van Hese (1988). Intact cores were moved to the laboratory and treated. The overlying water was flushed and phosphate release from the sediment surface was measured in the outflowing water. This method was used before the in-lake experiment to estimate the dose of iron needed,

and during and after the experiment to follow the reactions of the lake sediment over time.

The release rate was also studied under oxic and anoxic conditions, using both iron and aluminium as a binding agent (Boers *et al.*, 1991a). At the same molar dose, iron(III) and aluminium showed similar effects under oxic as well as anoxic conditions, but anoxic conditions required about a double dose for both metals to reduce the phosphorus release rate by a fixed amount.

Bio-availability of phosphorus to algal cultures was measured according to Bruning & Klapwijk (1984). It was expressed as a percentage of the total amount of phosphorus in the sediment sample being tested.

Results

By column experiments the appropriate dosage of iron was determined at 100 g of iron(III) per

342

square metre. Both under oxic or anoxic conditions this would appear sufficient. Treatment of the lake bottom in November 1989 took three weeks. Technical details are summarized in Table 2. 120 Tons of 40% FeCl$_3$ solution was dosed.

Since not only the top-layer of the sediment but also the lake water is mixed intensively with the iron solution, a direct effect on water quality parameters is to be expected. During treatment the transparency of the lake water began to increase within days from 0.30 m at the start to about 0.70 m at the end of the treatment period. In Table 3 water quality data are given just before and after iron dosing.

Total phosphorus, chlorophyll and suspended solids concentrations (Fig. 5) dropped to values below the normal winter averages. The oxygen concentration remained constant during treatment, increasing shortly afterwards to total saturation. The concentration of total phosphorus increased again in spring 1990. In January and February two peaks in total phosphate were caused by heavy storms of up to 30 m s^{-1} which temporarily resuspended much sediment.

Together with the iron also chloride was added to the lake: dosing of iron-chloride offers the possibility to measure accurately the residence time using the chloride balance. Also the measurement of exchange coefficients between lakes if it is traced where the chloride goes.

Chloride concentration increased from 80 mg

Table 2. Summary of technical details of the phosphate fixation pilot project in Lake Groot Vogelenzang.

Lake area	18 ha
Average water depth	1.75 m (max. depth 2.50 m)
Applied dosage	100 g Fe m^{-2}
Treatment depth	0.15–0.20 m
Amount of FeCl$_3$ used	120 tons or 85 m^3, 40% solution, high purity quality
Dilution of FeCl$_3$	100–150 times with local surface water before injection
Manifold width	2.50 m, restricted due to local conditions
Production	Up to 2 ha d^{-1}
Treatment period	3 weeks (23 Oct.–9 Nov. 1989)
Cost of treatment	About Dfl 250,000 (US $ 140,000)

Table 3. Summary of water quality parameters determined just before and shortly after the phosphate fixation treatment of Lake Groot Vogelenzang.

Parameter		Before	After
pH	–	7.7	7.4
Secchi depth	cm	35	65
Suspended solids	mg l^{-1}	23	15
COD	mg l^{-1}	69	49
BOD	mg l^{-1}	10	6
O$_2$	mg l^{-1}	9	9
Cl$^-$	mg l^{-1}	82	150
P–PO$_4$	μg l^{-1}	7	6
Total P	μg l^{-1}	180	80
Chlorophyll-*a*	μg l^{-1}	160	60

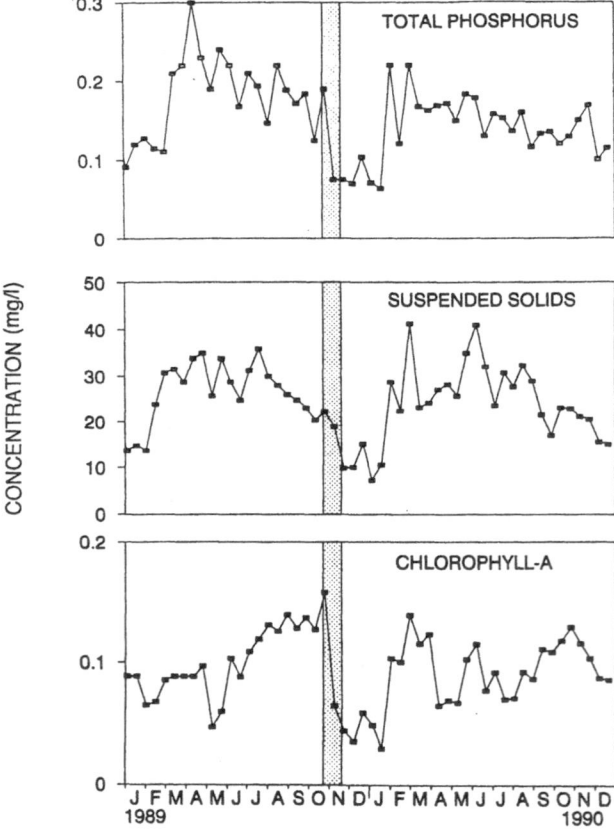

Fig. 5. Total phosphorus, suspended solids and chlorophyll-a concentrations of the lake water in Groot Vogelenzang during 1989–1990. The shaded field indicates the treatment period. Notwithstanding continuing external loading, total P is considerably reduced compared with the pre-treatment period.

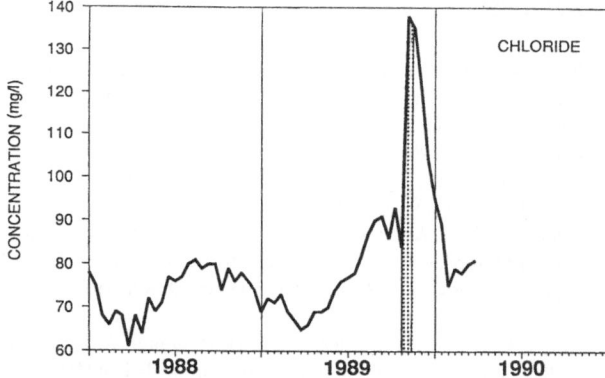

Fig. 6. Chloride concentration of the lake water in Groot Vogelenzang. The sharp increase in November 1989 is caused by the phosphate fixation treatment, the subsequent decrease by exchange with water from the surrounding lakes. From the decrease rate of the chloride concentration a residence time of the lake water of about 5 weeks is found.

l^{-1} to 150 mg l^{-1}. The dilution rate after this increase (Fig. 6) shows that the residence time during this period in Lake Groot Vogelenzang was about 5 weeks. Continuous exchange of water from the surrounding lakes takes place. These lakes have comparable phosphorus concentrations as Groot Vogelenzang before treatment. So this lake is subject to a considerable external loading which is also reflected by the high rate of increase of the total phosphorus level of the lake water after treatment. Total P during the summer of 1990 was still about 20% lower that during the summer of 1989, but it cannot be excluded to reflect normal annual fluctuations. The exchange of Lake Groot Vogelenzang water with the adjacent lakes made it impossible to study the effect of treatment using the response of the lake.

During treatment no adverse ecological side effects were observed. The sensitive vegetation of reeds bordering the lake was unharmed. A few times some larvae of Chironomidae and similar animals were disengaged and seen drifting at the surface. Next summer a local fisherman reported more and healthier eel (Anguilla anguilla) than before. The density of benthic macro-invertebrates and macrophytes in the profundal zone remained low. In the littoral zone both zoobenthos and macrophytes remained well developed, increasing somewhat in biomass after

treatment (Boers *et al.*, 1991b; Van Gemeren, 1990).

Discussion

In-lake phosphate fixation is a simple, robust method, large surfaces can be treated within a short period of time and is economic, especially when compared to dredging with the aim to remove the phosphorus-rich top-layer as well. In the case of Lake Groot Vogelenzang dredging would probably have cost 5 to 10 times as much as the phosphate fixation treatment. An important reason for this is the long distance between dredging location and the nearest potential land disposal site.

The main limitation of phosphate fixation seems to be the minimum water depth required to apply the method. A practical but solvable problem is how to move around with the necessary equipments, more difficult to solve is how to supply the equipment with sufficient surface water for the injection of the phosphorus binding agent. So the minimum water depth in which phosphate fixation is applicable will be about 0.30 to 0.50 m.

The maximum water depth in which phosphate fixation seems useful will be in the range of 5 to 8 m, increasing with the size of the lake. Shipping channels are an exception since considerable resuspension is caused by the action of ships propellers.

From the logistic point of view we have shown in-lake phosphate fixation feasible. It is unfortunate that no sufficiently small suitable lake (or sufficiently large budget) could be found to carry out the pilot project in such a way that a real contribution to water quality improvement could be made. It need not, however, be demonstrated that phosphates do bind to iron.

The injection technique is not limited to the use of iron compounds as phosphate-binding agent; also calcium and aluminium solutions can be applied.

The pilot project has shown that the phosphate fixation method offers advantages over other techniques used to abate eutrophication in shallow lakes:

– disposal of the dredged material is not necessary;
– phosphate fixation is performed in a fraction of the time required to dredge a lake and therefore more economic.

Compared to the Riplox method, phosphate fixation uses simpler and more robust equipment and the injection of the phosphate binding agent by water jets results in a more controlled and more thorough mixing with the lake sediment.

No undesirable side effects of this technique were observed in the pilot project, but in future applications an eye should be kept on this possibility. In particular, effects on zoobenthos and macrophytes during treatment and in the longer term should be monitored.

Long-term stability of the formed iron-phosphorus complexes, particularly in an anoxic environment could not yet be determined univocally. To prove this another pilot project should be set up shortly, preferably at a well isolated location.

An advantage of phosphate fixation (and of dredging if executed properly) over biomanipulation is that excess phosphorus is removed from the system, while biomanipulation leaves the 'stock' of phosphorus in the lake bottom intact. As yet very little is known about how long beneficial effect of such measures will keep.

If dredging is necessary to increase water-depth in silted up areas, this might be done cheaper when combined with phosphate fixation. The dredging operation can be laid out to reach the target depth and need not to bother to much about removal of the semi-liquid top-layer, where remaining phosphorus is neutralized by phosphate fixation. Applying this combination the need to remove all phosphorus-rich sediment no longer exists. Final dredging depth can be adjusted to the proposed function of the lake, and for instance be limited to about 1.50 to 2.00 m to give (submerged) water plants a better chance to recolonize at the lake bottom. Another positive effect is a substantial reduction of the amount of dredged material to be stored or deposited.

References

Bengtsson, L.,S. Fleischer, G. Lindmark & W. Ripl, 1975. Lake Trummen restoration project. I. Water and sediment chemistry. Verh. Int. Ver. Limnol. 19: 1107–1116.

Boers, P. C. M & O. Van Hese, 1988. Phosphorus release from the peaty sediments of the Loosdrecht Lakes (the Netherlands). Wat. Res. 22: 355–363.

Boers, P. C. M. & W. A. D. D. Wijesooriya, 1989. Phosphorus fixation using iron: a method to reduce internal phosphorus loading? H_2O 22: 501–504. In Dutch with an English summary.

Boers, P. C. M., J. Van der Does, M. P. Quaak & J. C. Van der Vlugt, 1991a. Phosphorus fixation with iron(III)chloride: a new method to combat internal phosphorus loading in shallow lakes? In: P. C. M. Boers, thesis, Univ. of Wageningen, The Netherlands.

Boers, P. C. M., J. Van der Does, M. P. Quaak, J. C. Van der Vlugt & P. Walker, 1991b. Fixation of phosphorus in lake sediments using iron(III)chloride, experiences, expectations. Submitted to Hydrobiologia.

Bruning, C. & S. P. Klapwijk, 1984. Application of derivative spectroscopy in bioassays estimating algal available phosphates in lake sediments. Verh. int. Ver. Limnol. 22: 172–178.

Cooke, G. D., E. B. Welch, S. A. Peterson & P. R. Newroth, 1986. Lake and Reservoir Restoration. Butterworth, Boston.

Donze, M., C. Nieuwendijk, A. M. J. V. Van Boxtel & M. P. Quaak, 1990. Shaping the environment: aquatic pollution and dredging in the EC. Delwel, The Hague, The Netherlands.

Edmondson, W. T., 1991. The uses of ecology. Lake Washington and beyond. Univ. Washington Press, Seattle, USA.

Estourgie, A. L. P., 1988. Theory and Practice of Water Injection Dredging. Terra et Aqua 38: 21–28.

Golterman, H. L., 1984. Sediments, modifying and equilibrating factors in the chemistry of freshwater. Verh. int. Ver Limnol. 22: 23–59.

Kennedy, R. H. & G. D. Cooke, 1982. Control of lake phosphorus with aluminium sulphate: dose determination and application techniques. Wat. Resour. Bull. 18: 389–396.

Sas, H. (coordinator), 1989. Lake restoration by reduction of nutrient loading. St. Augustin: Akademia Verl., Richardz.

Siebers, H. H., 1985. Patterns and variability of phosphate and heavy metals in sediments of two shallow lakes. Thesis, Delft University of Technology, Delft, the Netherlands.

Van der Vlugt, J. C. & S. P. Klapwijk, 1990. The Reeuwijk Lakes: a five years water quality study in an eutrophic ecosystem. Hydrobiol. Bull. 24, in press.

Van Gemeren, J., 1990. The development of aquatic and littoral vegetations in some Reeuwijk lakes in relation to phosphorus fixation and biomanipulation. Report Waterboard of Rijnland (in Dutch).

Hydrobiologia **253**: 345–356, 1993.
P.C.M. Boers, T.E. Cappenberg & W. van Raaphorst (eds),
Proceedings of the Third International Workshop on Phosphorus in Sediments.
© 1993 *Kluwer Academic Publishers.*

Eight years of internal phosphorus loading and changes in the sediment phosphorus profile of Lake Søbygaard, Denmark

Martin Søndergaard, Peter Kristensen & Erik Jeppesen
National Environmental Research Institute, Department of Freshwater Ecology, Vejlsøvej 25, P.O. Box 314, DK-8600 Silkeborg, Denmark

Key words: phosphorus, retention, sediment, release, fractionation, iron:phosphorus ratio

Abstract

During each of the first 8 years following an 80–90% reduction in external phosphorus loading of shallow, hypertrophic Lake Søbygaard, Denmark in 1982, phosphorus retention was found to be negative. Phosphorus release mainly occurred from April to October, net retention being close to zero during winter. Net internal phosphorus loading was 8 g P m^{-2} y^{-1} in 1983 and slowly decreased to 2 g P m^{-2} y^{-1} in 1990, mainly because of decreasing sediment phosphorus release during late summer and autumn. The high net release of phosphorus from Lake Søbygaard sediment is attributable to a very high phosphorus concentration and to a high transport rate in the sediment caused by bioturbation and gas ebullition. Sediment phosphorus concentration mainly decreased at a depth of 5 to 20 cm, involving sediment layers down to 23 cm. Maximum sediment phosphorus concentration, which was 11.3 mg P g^{-1} dw at a depth of 14–16 cm in 1985, decreased to 8.6 mg P g^{-1} dw at a depth of 16–18 cm in 1991. Phosphorus fractionation revealed that phosphorus release was accompanied by a decrease in NH$_4$Cl-P + NaOH-P and organic phosphorus fractions. HCl-P increased at all sediment depths. The Fe:P ratio in the superficial layer stabilized at approximately 10. Net phosphorus release can be expected to continue for another decade at the present release rate, before an Fe:P ratio of 10 will be reached in the sediment layers from which phosphorus is now being released.

Introduction

Recovery of eutrophic lakes following a reduction in external phosphorus loading can be severely delayed by the release of phosphorus from sediment (Marsden, 1989; Sas, 1989; Jeppesen *et al.,* 1991). The duration of net annual sediment release is usually considered to be relatively short, i.e. not more than 5 years (Sas, 1989). However, shallow, highly polluted lakes tend to have a longer recovery period (Boström *et al.,* 1985; Cullen & Forsberg, 1988; Jeppesen *et al.,* 1991), this being mainly attributable to high loading (relative to the lake area) prior to the reduction in external phosphorus loading. Recent results furthermore indicate that the iron phosphorus ratio may play an important role in the control of phosphorus release from sediments, and that iron can suppress phosphorus release when the ratio (by weight) is higher than 15 to 20 (Jensen *et al.,* 1992a).

The present article describes persistent internal loading, expressed as negative phosphorus retention, in shallow, hypertrophic Lake Søbygaard, Denmark. Changes in phosphorus retention during the first 8 years of recovery following an 80–

90% reduction in external phosphorus loading were determined from the monthly mass balance calculated by means of inlet-outlet measurements. In addition we describe how the internal loading of Lake Søbygaard is reflected by changes of sediment phosphorus profile and composition in the upper 30 cm of the sediment. Finally we demonstrate that phosphorus and iron profiles may give valuable information concerning the role of the sediment in future internal loading of the lake.

Methods

Lake description

Lake Søbygaard is a small (40 ha) and shallow (mean depth 1.0 m) hypertrophic lake situated in the central part of Jutland, Denmark. 80–90% of the total hydraulic loading is derived from one main inlet, the remainder coming from groundwater springs along the lake shore. The lake has two outlets. Mean hydraulic residence time varies from 15–20 days in winter to 25–30 days in summer.

Prior to 1982, when phosphorus reduction was implemented at the sewage plant, Lake Søbygaard received large amounts of poorly treated waste water which caused accumulation of a phosphorus-rich sediment (Søndergaard, 1989; Søndergaard, 1990). From 1976 (when biological treatment of sewage was started) to 1982 most of the phosphorus entering the lake was inorganic. Since 1982, the total phosphorus concentration at the inlet has usually been between 0.1 and 0.3 mg P l^{-1}. Outlet phosphorus concentration, which was alike lake water concentration, varied with the season, winter concentration being roughly the same as inlet concentration, and mean summer concentration being 0.5 to 1.0 mg l^{-1} (Jeppesen et al., 1990; Søndergaard et al., 1990). Mean summer chlorophyll a concentration, which was about 800 μg l^{-1} in 1984 and 1985, has declined, and has been 200–300 μg l^{-1} since 1987 (see also Jeppesen et al., 1990; Jeppesen et al., 1992).

Mass balance

Hydraulic flow in main inlet and outlet was determined from stage discharge rating curves. Stage was measured continuously and discharge by a propeller as a minimum with monthly intervals. Daily flow rates were calculated.

Phosphorus concentration in the inlet and outlet was determined weekly or biweekly in 1983 and 1984, and by automatic sampling every 6 hours since 1985. Point samples from all the main springs along the lake shore, which represent approx. 20% of the hydraulic loading, were collected 2–15 times during the investigation. The latter displayed low variability in phosphorus concentration (mean phosphorus loading was 0.28 g P m^{-2} y^{-1}), and were included in the mass balance calculations. The tot-P concentration of inlet and outlet water was analyzed as molybdate reactive phosphorus following persulphate digestion (Koroleff, 1970).

Total phosphorus input and output were calculated by linear interpolation of the measured concentrations between sampling dates and by multiplying the concentration by the daily discharge volume. This was done on an yearly or quarterly basis.

Sediment

Data concerning sediment in 1985 are based on samples collected on 6 occasions during the period July 1985 to February 1986. Subsequently sediment was only sampled once every two years: in March 1987, in September 1989 and in April 1991. The sediment cores were sampled with a Kajak bottom sampler (Kajak et al., 1965) or a modification of this (piston-corer). Inner core diameter was 52 mm and core length 50 cm. Sampling was each time conducted at the same position ($\pm$ 20 m) in the center of the lake.

Each analysis was made on a composite sample of three cores sectioned into 1 or 2 cm thick slices for determination of the sediment profiles. Dry weight (DW) was determined by drying at 105 °C for 24 h, and loss on ignition (LI) was subsequently determined by drying to constant weight at 550 °C. Total phosphorus (tot-P) was analyzed spectrophotometrically as molybdate reactive phosphorus after extraction of ash-free sediment with 1 M HCl (modified from Andersen, 1976). Total iron (tot-Fe) was measured after bipiridyl complex formation in the same solution. Phosphorus fractionation, carried out (with minor modifications) according to the method of Hieltjes & Lijklema (1980), gave the following fractions: NH_4Cl-P, NaOH-P and HCl-P. Residual phosphorus is expressed both as org-P (determined by analysis of total phosphorus in the sediment remaining after the HCl extraction) and res-P (calculated as tot-P (NH_4Cl-P + NaOH-P + HCl-P). Org-P is considered to represent refractory organic bound phosphorus, while res-P includes easily dissolvable organic phosphorus that is released during the preceding extractions. Sediment phosphorus fractionation was only carried out in 1985 and 1991.

As Lake Søbygaard is characterized by a distinct seasonal variation in NH_4Cl-P in the upper sediment (Søndergaard, 1988), and as the sediment cores were sampled at different times of the year, NH_4Cl-P has been included in the NaOH-fraction when presenting the data.

Volumetric phosphorus concentration was calculated by means of tot-P, DW and LI, assuming an inorganic matter density of 2.6 g cm^{-3} and an organic matter density of 1.05 g cm^{-3}.

Sediment profiles from 1985, 1987 and 1989 were adjusted, so as to take into account a net sediment increase of 0.6 cm y^{-1}. Sediment profiles shown have thus been adjusted supposing that the level of the sediment surface was zero in 1991; accordingly, the 1985 sediment surface is located at 3 cm on figures. The net sedimentation rate was calculated on the basis of a mean iron retention since 1984 of 36 g Fe m^{-2} y^{-1} (S.D. = 7.4); a mean iron concentration of 60 mg Fe g^{-1} dw and a mean DW of 10% in the upper 20 cm of the sediment. Due to compaction, however, especially the uppermost 5 to 10 cm of the sediment are believed to undergo changes in DW over time. This complicates the direct comparison of sediment profiles of different years in the surface sediment. Changes in sediment DW below 20 cm depth are believed to have been negligible during the period from 1985 to 1991.

Results

Towards the end of 1982 external phosphorus loading of Lake Søbygaard was reduced from about 30 g P m^{-2} y^{-1} to less than 5 g P m^{-2} y^{-1} (Fig. 1). Phosphorus retention had been high (and positive) for decades prior to this reduction; afterwards it immediately became negative due to sediment phosphorus release, and has remained so, although the amount released has gradually decreased (Fig. 1). During the last 4 years retention has largely remained unchanged between −2

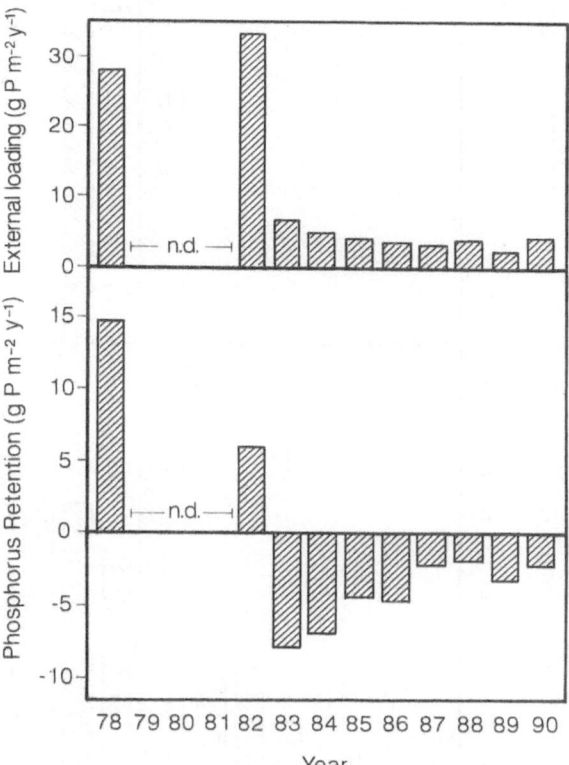

Fig. 1. External phosphorus loading (upper) and phosphorus retention (lower) in Lake Søbygaard from 1978 to 1990 (g P m^{-2} y^{-1}) Measures to severely reduce external phosphorus loading were implemented from the end of 1982. Data from 1978 are from Aarhus County (1979). n.d. = no data.

348

and $-3.3 \, \mathrm{g \, P \, m^{-2} \, y^{-1}}$, which corresponds to a net annual phosphorus release from the lake of 0.8 to 1.3 tons phosphorus.

Quarterly phosphorus retention was usually close to zero during the 4th and 1st quarters (i.e.

autumn/winter), whereas during the 2nd and 3rd quarters (i.e. spring/summer) retention was highly negative, varying between -0.7 and $-2.7 \, \mathrm{g \, P \, m^{-2}}$ quarter^{-1} (Fig. 2). This corresponds to a net internal phosphorus loading of 8–30 mg P m^{-2} d^{-1}.

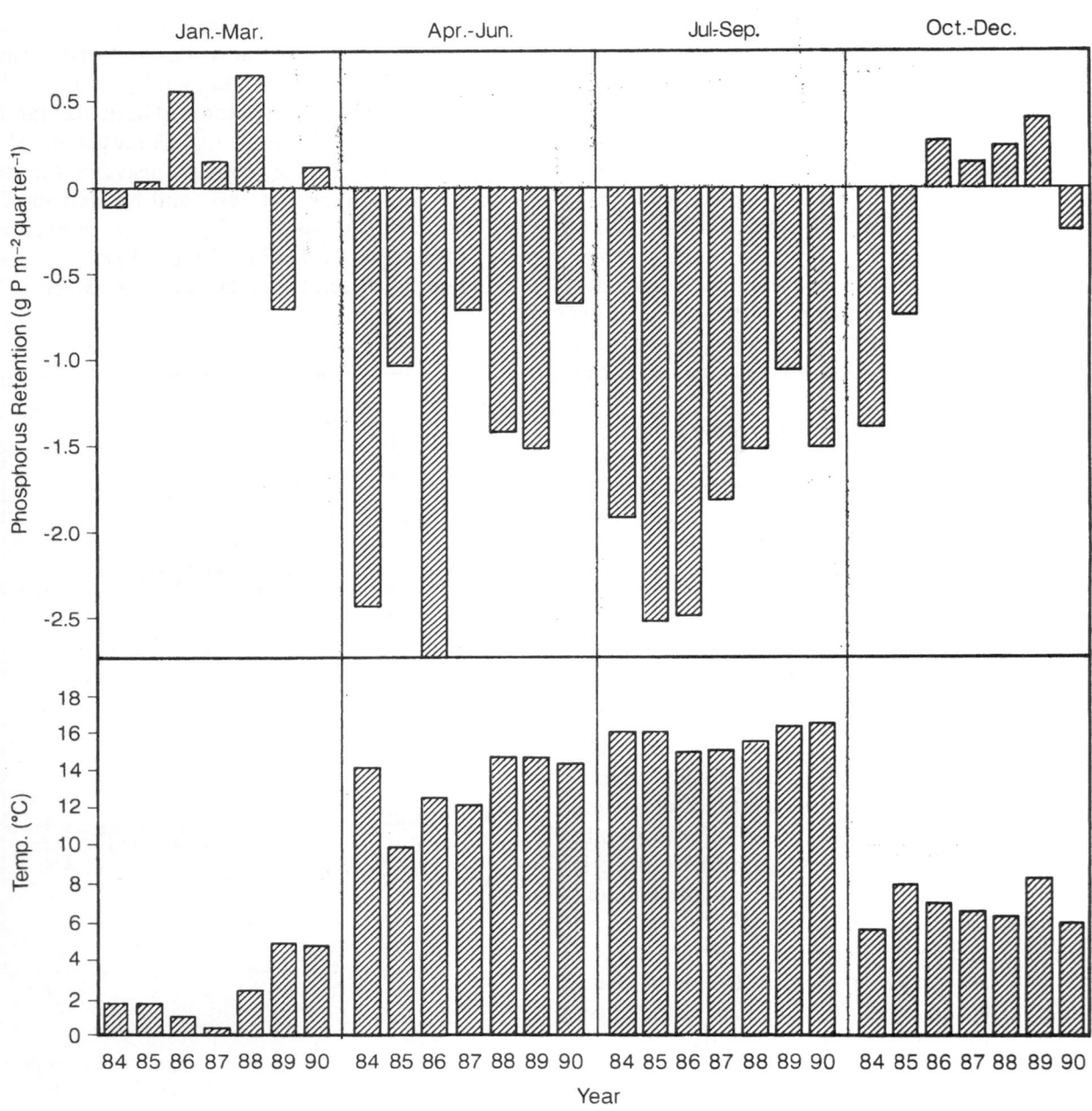

Fig. 2. Quarterly variation in phosphorus retention in Lake Søbygaard from 1984 to 1990 (g P m^{-2} quarter^{-1}). The calculations are based on mass balances corrected for changes in the phosphorus pool accumulated in the lake water.

In 1984 and 1985 retention was also markedly negative during the 4th quarter. Inter-annual variation in retention was most pronounced in the 2nd quarter from April to June. Retention in the 3rd quarter has tended to become less negative since 1986 ($p < 0.02$). Inter-annual variation in quarterly mean lake water temperature was up to 5 °C (Fig. 2). Apart from general seasonal changes from winter to summer, no clear relationship could be established between different temperature and different net phosphorus retention recognized within the quarters.

Sediment dry weight (DW) and loss on ignition (LI) changed only slightly during the period 1985 to 1991 when comparing corresponding sediment depths (relative to sediment surface). DW was

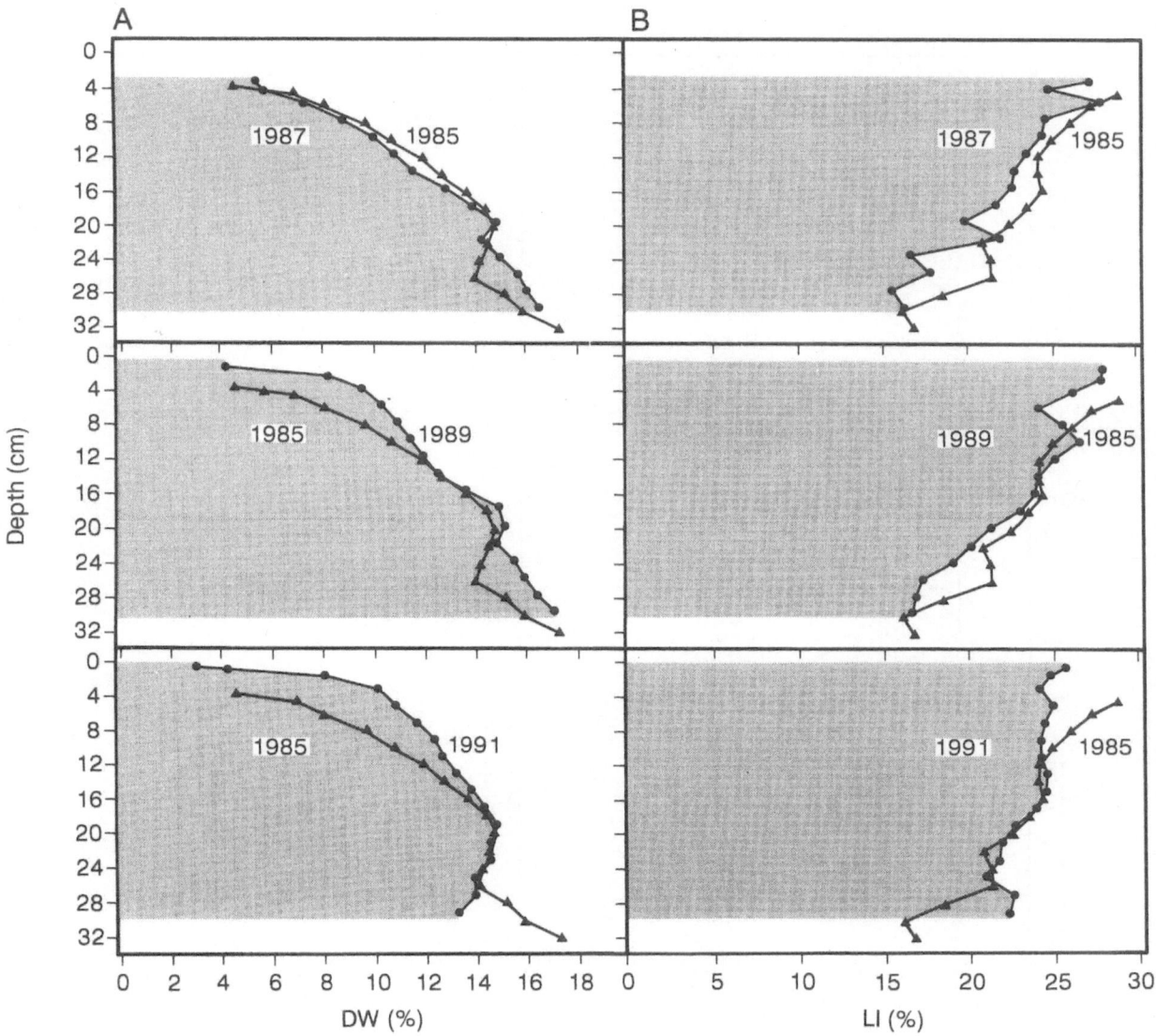

Fig. 3. Profiles of Lake Søbygaard sediment in 1985, 1987, 1989, and 1991 of A: dry weight (DW) and B: loss on ignition (LI). All profiles have been adjusted relative to the surface sediment being zero in 1991 using a net sediment increase of 0.6 cm y^{-1}. 1985 surface is 3 cm.

3–5% in the superficial sediment, increasing to 15% at a depth of 20 cm. LI was 24–27% in the superficial sediment, decreasing to 20% at a depth of 20 cm (Fig. 3). Tot-Fe was almost homogeneously distributed throughout the sediment, the concentration ranging from 58–65 mg Fe g^{-1} dw in the upper 30 cm of the sediment in 1991, and from 50–70 mg Fe g^{-1} dw in 1985.

Sediment tot-P per g dry sediment decreased markedly during the period 1985 to 1991 (Fig. 4),

the decrease occurring in surface sediment and at depths from 10 to 23 cm, but in particular at depths from 12 to 20 cm (sediment surface = 0 cm in 1991). Maximum tot-P, which was 11.4 mg P g^{-1} dw at a depth of 16–20 cm in 1985 decreased to 10.0 mg P g^{-1} dw in 1987, and to 8.5 mg P g^{-1} dw in 1989 and 1991 (Fig. 4A). This corresponds to a reduction in volumetric tot-P from 1.9 to 1.5 mg P cm^{-3} between 1985 and 1991 (Fig. 4B). At sediment depths from

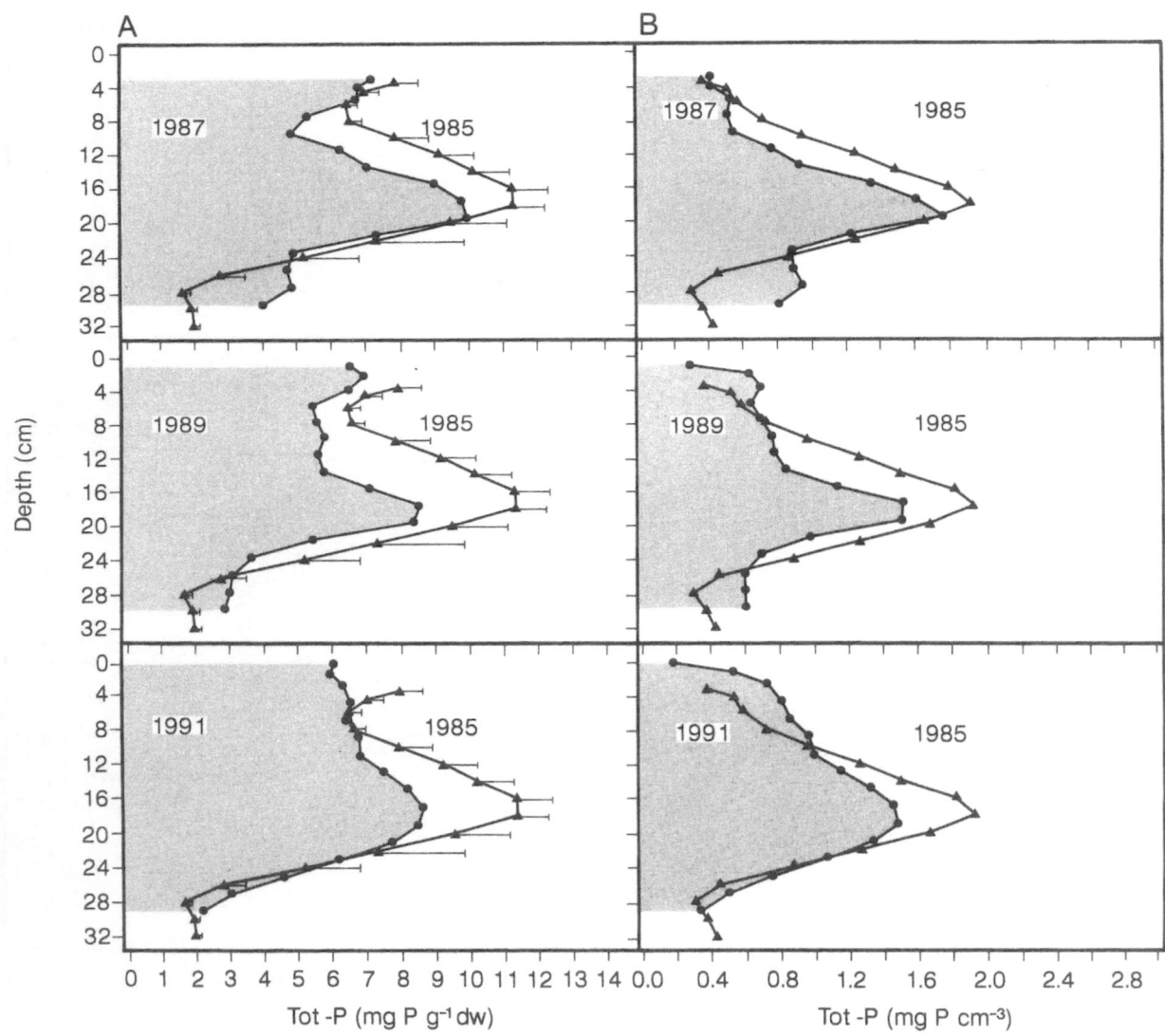

Fig. 4. Profiles of Lake Søbygaard sediment in 1985, 1987, 1989, and 1991 of total phosphorus concentration. A: on a dry weight basis (mg P g^{-1} dw) and B: on a volume basis (mg P cm^{-3}). 1985 error bars indicate 1 SD of mean value of 6 samples. Sediment profiles adjusted as in Fig. 3.

0–10 cm the volumetric tot-P content increased between 1985 and 1991, especially in the newly settled surface sediment (Fig. 4B). Maximum tot-P in the surface sediment decreased from 7.8 mg P g^{-1} dw in 1985 to 6.1 mg P g^{-1} dw in 1991.

According to an almost unchanged and distinct phosphorus gradient from approx. 23 to 26 cm depth, the upper 23 cm of the sediment seems to represent the sediment part subject to changes in phosphorus concentration. This part contained a total of 242 g P m^{-2} in 1985 that was reduced by 3 g P m^{-2} to 239 g P m^{-2} in 1991. In the same period inlet-outlet measurements revealed a net phosphorus release from the sediment of 14.5 g P m^{-2}. Inconsistency between measured net phosphorus release and recorded changes in the sediment concentrations should be attributed to local sediment variations and to difficulties in obtaining exact sediment profiles.

Changes in tot-P are primarily reflected in a decrease in NH$_4$Cl + NaOH-P, which constituted between 60 and 80% of tot-P in 1985, and in the organic fractions. From 1985 to 1991 NH$_4$Cl + NaOH-P decreased from a maximum of 9.2 mg P g^{-1} dw to 7.2 mg P g^{-1} dw (Fig. 5). Both org-P and res-P decreased at all sediment depths, but especially so in the superficial layer in which org-P decreased from 0.8 to 0.5 mg P g^{-1} dw and res-P decreased from 2.4 to 1.1 mg P g^{-1} dw between 1985 and 1991. HCl-P increased at all sediment depths by between 0.1 and 0.4 mg P g^{-1} dw in the same period.

Calcium is homogeneously distributed in the sediment with a concentration of approx. 150 mg Ca g^{-1} dw.

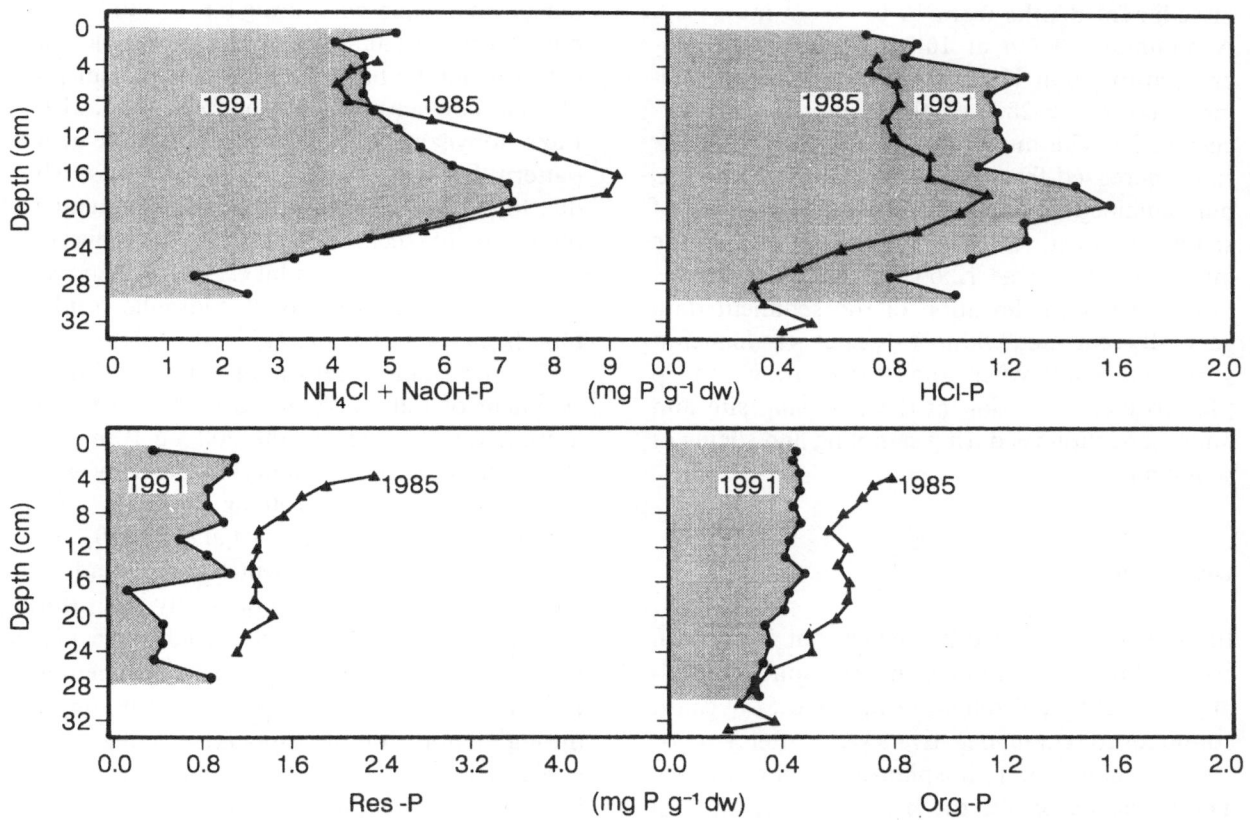

Fig. 5. Profiles of Lake Søbygaard sediment showing changes in NH$_4$Cl + NaOH-P, HCl-P, org-P and res-P (calculated as tot-P − (NH$_4$Cl-P + NaOH-P + HCl-P)) from 1985 to 1991. Sediment profiles adjusted as in Fig. 3.

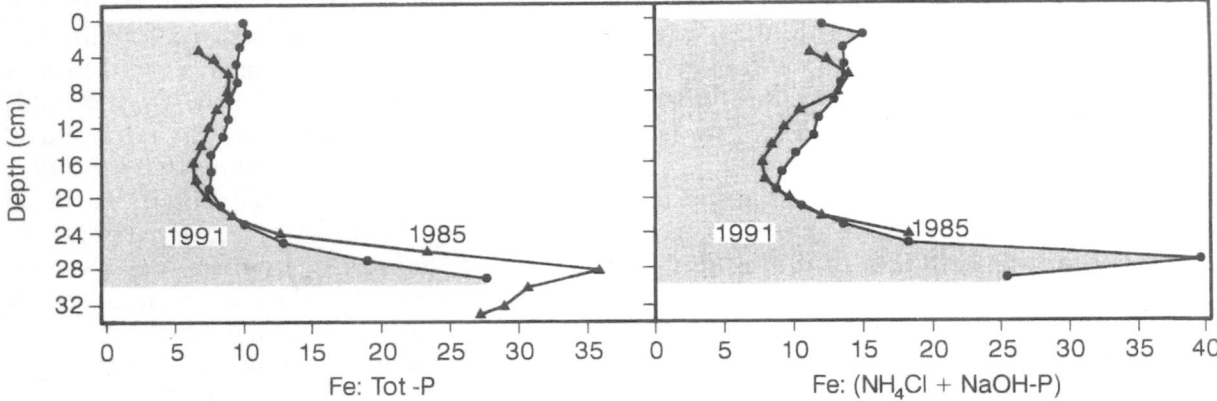

Fig. 6. Profiles of Lake Søbygaard sediment showing iron phosphorus ratio (Fe:P) and iron $NH_4Cl + NaOH$-P ratio (Fe:$NH_4Cl + NaOH$-P) by weight in the upper 30 cm in 1985 and 1991. Sediment profiles adjusted as in Fig. 3.

While the Fe:tot-P ratio remained relatively constant from 6 to 10 cm at a level of approx. 10 (by weight), the ratio in the surface sediment and at sediment depths from 10 to 22 cm increased. A minimum of 6.4 at 16 cm depth in 1985 increased to 7.6 in 1991 (Fig. 6). In the deep sediment layers (> 28 cm), the Fe:tot-P ratio exceeded 25. The minimum Fe:$NH_4Cl + NaOH$-P ratio increased from 8.0 in 1985 to 9.0 in 1991, but remained at about 13 in the upper 8 cm of sediment. Mean (1984 to 1990) Fe:P ratio in the inlet to the lake was 12.6 (S.D. = 2.0).

The standard deviation of the sediment data generally remained within 10%, but was higher at sediment depths where the P-gradient was steep (due to local variation at the sampling site and minor inaccuracies during sampling and slicing of sediment).

Discussion

In contrast to most other lakes that have been reported to suffer from internal phosphorus loading (Sas, 1989), P-retention in Lake Søbygaard continues to remain negative 8 years after a major decrease in external phosphorus loading (Fig. 1). The basic reason for the persistence of internal loading is the large phosphorus pool that has accumulated in the lake sediment as a consequence

of previously excessive and prolonged external phosphorus loading, and concomitant high iron loading. The high phosphorus affinity of iron explains why, when comparing lakes with the same phosphorus loading, those with iron rich-sediments generally have the highest sediment phosphorus concentration (Jeppesen *et al.*, 1991). Lake Søbygaard thus seems to reflect the general pattern that rapidly flushed Danish lakes, which due to a large phosphorus pool accumulated in the sediment, seem to have a longer recovery period than slowly flushed lakes, even when recovery is expressed in terms of hydraulic residence time (Jeppesen *et al.*, 1991).

The net summer phosphorus release from the sediment of Lake Søbygaard is the net result of a gross release rate of 100–200 mg P m^{-2} d^{-1} that is more or less counterbalanced by a sedimentation rate of 100–150 mg P m^{-2} d^{-1} (Søndergaard, 1989; Søndergaard *et al.*, 1990). Thus, net phosphorus release, which was found to be 8–30 mg P m^{-2} d^{-1} on a quarterly basis during summer in the present study, would be expected to be very sensitive to changes in sedimentation rate. This has been confirmed by the finding that during periods of phytoplankton collapse and subsequently low sedimentation rate, net release increases to 100–200 mg P m^{-2} d^{-1} due to a sustained high gross release rate (Søndergaard *et al.*, 1990).

Although the composition of the phytoplankton community changed towards dominance by larger species during the study period, thus increasing the species-specific sedimentation rate, there was a simultaneous 4-fold decrease in primary production and enhanced algal grazing by zooplankton (Jeppesen *et al.*, 1992a). Thus, the overall sedimentation rate probably did not increase and hence would not have influenced net phosphorus release. The effect of reduced sedimentation is most likely restricted to the relatively short-lasting periods of phytoplankton collapse. The decrease in yearly phosphorus release from Lake Søbygaard sediment since 1983 is therefore unlikely to be attributable to an increase in sedimentation, although the differences in duration and frequency of phytoplankton collapse may account for some of the inter-annual variability.

Instead, the decreasing net phosphorus release in Lake Søbygaard must be due to a decrease in the gross release from sediment that can be expected to occur as the pool of releasable phosphorus in the upper sediment becomes exhausted, and the distance to the lake water increases. The reduction in sediment phosphorus concentration (Fig. 4) indicates that phosphorus was mainly released from the upper 6–18 cm in the period 1985 to 1987, whereas deeper layers down to 23 cm were included from 1989 to 1991, in turn indicating that the increasing transport distance may be partly responsible for the decreasing net release of phosphorus from the sediment.

Factors influencing phosphorus release from sediment

Phosphorus release from lake sediments is facilitated by a variety of mechanisms (Boström *et al.*, 1988; Marsden, 1989), some of which are discussed below. The distinct seasonal variation in P-retention in Lake Søbygaard, which was highly negative in spring/summer and approximately zero in autumn/winter, implies that phosphorus release from the sediment is strongly associated with biological activity, either directly through mineralization, or indirectly through an increase in upwards transport. Mineralization processes are believed to be crucial for nutrient cycling especially in the upper sediment layers, to which large amounts of sedimenting algae are constantly being added (Jensen & Kristensen, 1987). Changes in the fractionation profile i.e. the decrease in organically bound phosphorus in the surface sediment emphasize the importance of mineralization. However, the net phosphorus release is probably to some extent only indirectly coupled to mineralization in the deeper parts of the sediment since up to 85% of the decrease recorded in total phosphorus could be referred to a decrease in the inorganic NaOH-P fraction. In contrast, the reduction of org-P only accounts for 10–20% of phosphorus lost from the sediment at a depth of 10–20 cm, i.e. from sediment layers in which the reduction in phosphorus is greatest. These indirect effects can probably be ascribed to bioturbation and gas ebullition, both of which are known to enhance phosphorus release (Skogheim & Erlandsen, 1979; Wisniewski & Planter, 1985; Löfgren & Boström, 1989). The existence of such an indirect, biologically-mediated transport mechanism is supported by the finding that simple chemical diffusion driven solely by a phosphate gradient is not rapid enough to explain the high gross release rates recorded from deeper parts of Lake Søbygaard sediment (Søndergaard, 1990).

Sulphate reduction and subsequent H_2S production lead to inactivation of iron through the formation of FeS and FeS_2, thereby disabling or reducing its phosphorus sorption capacity (Schindler, 1985; Boström *et al.*, 1988). Sulphate reduction activity was not measured in the present study. But a distinct summer sulphate gradient in the porewater from 70 mg SO_4 l^{-1} in the surface sediment to 5 mg SO_4 l^{-1} a depth of 20 cm, and an almost complete absence (<0.3 mg l^{-1}) of Fe^{+2} in the porewater in the upper 12 cm of the sediment due possibly to the precipitation of iron sulfide (Søndergaard, 1990), indicate that sulphate reduction might be important. The coupling between sulphate reduction and phosphorus release would be expected to be of particular importance in sediments like Søbygaard that have a low Fe:P ratio and a large proportion of iron-bound phosphorus.

The presence of nitrate can both enhance and reduce phosphorus release from sediments due to either increased decomposition through denitrification or improved redox conditions in the presence of nitrate (Boström et al., 1982). The summer period during which the nitrate concentration of Lake Søbygaard was less than 1 mg NO_3 l^{-1} increased from 2.5–3.0 months in 1984–1986 to 4.7–5.6 months in 1987–1990 (unpublished observation), probably as a result of a 36% decrease in the external nitrogen loading in August 1987 (Jensen et al., 1992b). Although it is difficult to assess the effect of this change in nitrate concentration, it seems that nitrate depletion does not affect phosphorus release.

Finally, resuspension and high pH are also able to enhance sediment release rates (Boström et al., 1982; Hamilton & Mitchell, 1988; Søndergaard, 1991). However, the influence of these two factors in Lake Søbygaard seems to be restricted to the upper 5 to 10 cm of the sediment. Thus, pH has been shown to increase to between 9 and 10 during summer in the upper 5 cm of the sediment, while wind-induced resuspension seems to provoke major changes in porewater profiles down to 6–8 cm depth (Søndergaard, 1990). Furthermore, pH and resuspension are mainly believed to be important at the beginning of the spring/summer season. At this time phosphorus derived from the lake water and deeper parts of the sediment, and which has been retained in a relatively thick oxidized sediment layer during the winter period, is released when the thickness of this layer is reduced by increasing decomposition. The phosphorus release pattern, expressed in terms of mass balance, can thus be interpreted as being determined by two events: early seasonal release from the superficial sediment which does not lead to any inter-annual differences in the superficial sediment phosphorus concentration, and late seasonal release which can be attributed to net phosphorus transport from deeper sediment layers. This would explain why phosphorus release in the 2nd quarter has not tended to decrease since 1983, whereas the release rate in the 3rd quarter has been diminishing.

Future perspectives

Taking into account that the sediment profiles shown in Figs 3–6 represent only one location in the lake, and that the amount of phosphorus accumulated in the sediment (240 g P m^{-2} in the upper 23 cm) is much higher than the net release of phosphorus (2–8 g P m^{-2} y^{-1}), then the sediment P-budget is in reasonable agreement with the mass balance calculated from inlet-outlet measurements. We thus consider it justifiable to make use of the changes in P-profile from 1985 to 1991 to clarify the overall transport and mobility of phosphorus in the sediment.

In comparison with most other investigations (Shapiro et al., 1971; Tessenow, 1972; Boström, 1982), the sediment layers subject to changes in phosphorus concentration in Lake Søbygaard are very deep. This is probably due to the very loose nature of the sediment, and the presence of easily-degradable organic material in deeper parts of the sediment, which enables decomposition to enhance phosphorus release and transport. Most of the phosphorus released seems to be derived from the $NH_4Cl + NaOH$-P and organic fractions.

That there was considerable organic decomposition is supported by the fact that organically bound phosphorus (both org-P and res-P) decreased at all sediment depths during the period 1985-1991. The decrease was most pronounced in the superficial sediment which would be expected to contain the largest amount of easily-degradable organic material. However, part of the decrease in organically bound phosphorus in the surface sediment may have resulted from seasonal variation in the organic P fraction (Boström et al., 1985). This may cause an artifact since most of the 1985 samples were collected during phytoplankton growth season, whereas the 1991 sample was collected in February. Org-P had also decreased, which indicates that this fraction cannot be totally considered as refractory and non-releasable.

HCl-P increased at all sediment depths between 1985 and 1991, hence confirming the assumption that HCl-P is generally immobile, and

355

suggesting the occurrence of diagenetic and transformation processes which involve HCl-P. A possible explanation is that the continuously high porewater phosphate concentration, combined with a high calcium content, leads to steady adsorption of phosphate by $CaCO_3$, and the formation of apatite or other Ca complexes. This would in turn lead to a general increase in the calcium-bound phosphorus fraction at increasing sediment depth as a consequence of early diagenetic processes (Boström et al., 1988). In addition, increased pH, such as the high levels found in the surface sediment of Lake Søbygaard (Søndergaard, 1990), may also stimulate apatite formation (Stumm & Leckie, 1971).

That the very deep sediment layers take part in the sediment–water interactions is important when considering the future role of the sediment as a net source of phosphorus. In this respect the sediment iron:phosphorus ratio may be important in the control of phosphorus release (Jensen et al., 1992a).

The Fe:tot-P ratio in the upper 6 to 10 cm of sediment changed only slightly between 1985 and 1991. If this stability is interpreted as reflecting a steady-state condition in which the Fe:tot-P ratio is such that no further phosphorus release can take place, then the Fe:tot-P ratio in deeper parts of the sediment would have to increase to 9–10 (by weight) before phosphorus release would cease. This is about the same as the ratio in the uppermost 3 cm of the sediment, which has been formed during the last 5 years. For the sediment at a depth of 7 to 22 cm to reach a Fe:tot-P ratio of 9.6, a total of 26 g P m^{-2} would have to be released (Fig. 4). Thus, as the net release rate in recent years has been 2–3 g P m^{-2} y^{-1}, it can be concluded that at the present release rate, the sediment contains enough phosphorus to support net release for another 10 years or so. The total duration of net internal phosphorus loading in Lake Søbygaard would then be approximately 20 years. The steady decrease in annual net release, which is probably due to the increased transportation distance, has not been taken into account.

Conclusion

Phosphorus mass balance and sediment profiles indicate that the net internal phosphorus loading in Lake Søbygaard, at present continuing for the 9th year, is likely to continue for another decade. The decrease in net phosphorus release from the sediment since cessation of external loading is mainly attributable to a decrease in the gross release rate caused by an increase in the transport distance within the sediment. Indirect effects of biological activity, such as bioturbation and gas ebullition, are believed to be the main factors controlling the rate of net phosphorus release. Net phosphorus release mainly occurs at a sediment depth of 10–23 cm. However, decomposition in the superficial sediment causes a high flux of phosphorus from sediment to water i.e. a high gross release, and occasionally, when sedimentation rate is low, a high net release. The majority of the phosphorus lost from the sediment was derived from the NaOH-P fraction, and to some extent, from the organically-bound fraction. The HCl-P fraction increased in all sediment depths during the years of internal phosphorus loading.

Acknowledgements

The assistance of the staff of The National Environmental Research Institute, Department of Freshwater Ecology, Silkeborg, in particular Birte Laustsen, Lissa Hansen and Lone Nørgaard (field and laboratory), Kathe Møgelvang, Inge Nielsen and Pia Nygaard (layout and manuscript preparation) is gratefully acknowledged. Financial support was provided by the Aarhus County Council.

References

Aarhus County, 1979. Søbygard Sø 1978 (in Danish). Aarhus County Council.
Andersen, J. M., 1976. An ignition method for determination of total phosphorus in lake sediments. Wat. Res. 10: 329–331.
Boström, B., M. Jansson & C. Forsberg, 1982. Phosphorus release from lake sediments. Arch. Hydrobiol. Beih. Ergebn. Limnol. 18: 5–59.

356

Boström, B., I. Ahlgren & R. Bell, 1985. Internal nutrient loading in a eutrophic lake, reflected in seasonal variations of some sediment parameters. Verh. int. Ver. Limnol. 22: 3335–3339.

Boström, B., J. M. Andersen, S. Fleicher & M. Jansson, 1988. Exchange of phosphorus across the sediment water interface. In G. Persson & M. Jansson (eds), Phosphorus in Freshwater Ecosystems. Developments in Hydrobiolgy 48. Kluwer Academic Publishers, Dordrecht: 229–244. Reprinted from Hydrobiologia 170.

Cullen, P. & C. Forsberg, 1988. Experience with reducing point sources of phosphorus to lakes. In G. Persson & M. Jansson (eds), Phosphorus in Freshwater Ecosystems. Developments in Hydrobiolgy 48. Kluwer Academic Publishers, Dordrecht: 321–336. Reprinted from Hydrobiologia 170.

Hamilton, D. P. & S. F. Mitchell, 1988. Effects of wind on nitrogen, phosphorus, and chlorophyll in a shallow New Zealand lake. Verh. int. Ver. Limnol. 23: 624–628.

Hieltjes, A. H. M. & L. Lijklema, 1980. Fractionation of Inorganic Phosphates in Calcareous Sediments. J. envir. Qual. 9: 405–407.

Jensen, P. & P. Kristensen, 1987. Sedimentation og resuspension i Søbygaard Sø. Mc. thesis from the Botanical Institute, University of Aarhus, 150 pages (in Danish).

Jensen, H. S., P. Kristensen, E. Jeppesen & A. Skytthe, 1992a. Iron:Phosphorus ratio in surface sediment as an indicator of phosphate release from aerobic sediments in shallow lakes. In B. T. Hart & P. G. Sly (eds), Sediment/Water Interactions. Developments in Hydrobiology 75. Kluwer Academic Publishers, Dordrecht: 731–743. Reprinted from Hydrobiologia 235/236.

Jensen, J. P., E. Jeppesen, P. Kristensen, P. B. C. Christensen & M. Søndergaard, 1992b. Nitrogen loss and denitrification as studied in relation to reductions in nitrogen loading in a shallow, hypertrophic lake (Lake Søbygaard, Denmark). Int. Revue ges. Hydrobiol 77: 29–42.

Jeppesen, E., M. Søndergaard, E. Mortensen, P. Kristensen, B. Riemann, H. J. Jensen, J. P. Müller, O. Sortkjær, J. P. Jensen, K. Christoffersen, S. Bosselmann & E. Dall, 1990. Fish manipulation as a lake restoration tool in shallow, eutrophic lakes 1: cross-analysis of three Danish case-studies. In R. D. Gulati, E. H. R. R. Lammens, M.-L. Meijer & E. van Donk (eds), Biomanipulation – Tool for Water Management. Developments in Hydrobiology 61. Kluwer Academic Publishers, Dordrecht: 205–218. Reprinted from Hydrobiologia 200/201.

Jeppesen, E., P. Kristensen, J. P. Jensen, M. Søndergaard, E. Mortensen & T. Lauridsen, 1991. Recovery resilience following a reduction in external phosphorus loading of shallow, eutrophic Danish lakes: duration, regulating factors and methods for overcoming resilience. Mem. Ist. ital. Idrobiol., 48: 127–148.

Jeppesen, E., O. Sortkjær, M. Søndergaard & M. Erlandsen, 1992. Impact of a trophic cascade on heterotrophic bacterioplankton production in two shallow fish-manipulated lakes. Arch. Hydrobiol., in press.

Kajak, Z., K. Kacprzak & R. Polkowski, 1965. Tube bottom sampler for taking samples og micro and macro benthos, and for sampling of undisturbed structures of mud samples for experimental purposes. Ekol. pol. 11: 159–165.

Koroleff, F., 1970. Determination of total phosphorus in natural waters by means of persulphate oxidation. An interlab. Report No. 3 Cons. int. Explor. Mer.

Löfgren, S. & B. Boström, 1989. Interstitial water concentrations of phosphorus, iron and manganese in a shallow, eutrophic Swedish lake – implication for phosphorus cycling. Wat. Res. 23: 1115–1125.

Marsden, W. M., 1989. Lake restoration by reducing external phosphorus loading: the influence of sediment phosphorus release. Freshwat. Biol. 21: 139–162.

Sas, H., 1989. Lake restoration by reduction of nutrient loading. Expectations, experiences, extrapolation. Academic Verlag St, Augustin, 497 pp.

Schindler, D. W., 1985. The coupling of elemental cycles by organisms: evidence from whole-lake chemical perturbations. In W. Stumm (ed.), Chemical Processes in Lakes. John Wiley and Sons, New York: 225–250.

Shapiro, J., W. T. Edmondson & D. E. Allison, 1971. Changes in the chemical composition of sediments of lake Washington, 1958–1970. Limnol. Oceanogr. 16: 437–452.

Skogheim, O. K. & A. H. Erlandsen, 1979. Transport mechanisms of solutes in lake sediments. Seventh Nordic sediment symposium: 124–134.

Stumm, W. & J. O. Leckie, 1971. Phosphate exchange with sediments; its role in the productivity of surface water. Eidgen. Techn. Hochschulen, Separatum Nr. 406. Dübendorf, Schweiz.

Søndergaard, M., 1988. Seasonal Variations in the Loosely Sorbed Phosphorus Fraction of the Sediment of a Shallow and Hypereutrophic Lake. Envir . Geol. Wat. Sci. 11: 115–121.

Søndergaard, M., 1989. Phosphorus release from a hypertrophic lake sediment: Experiments with intact sediment cores in a continuous flow system. Arch. Hydrobiol. 116: 45–59.

Søndergaard, M., 1990. Pore water dynamics in the sediment of a shallow and hypertrophic lake. Hydrobiologia 192: 247–258.

Søndergaard, M., E. Jeppesen, P. Kristensen & O. Sortkjær, 1990. Interactions between sediment and water in a shallow and hypertrophic lake: a study on phytoplankton collapses in Lake Søbygård, Denmark. In P. Biró & J. F. Talling (eds), Trophic Relationships in Inland Waters. Developments in Hydrobiology 53. Kluwer Academic Publishers, Dordrecht: 139–148. Reprinted from Hydrobiologia 191.

Tessenow, U., 1972. Lösungs-, Diffusion- und Sorptionsprozesse in der Oberschicht von Seesedimenten. 1. Ein Langzeitexperiment unter aeroben und anaeroben Bedingungen im Fliessgleichgewicht. Arch. Hydrobiol. Suppl. 38: 353–398.

Wisniewski, R. J. & M. Planter, 1985. Exchange of phosphorus across sediment water interface (with special attention to the influence of biotic factors) in several lakes of different trophic status. Verh. int. Ver. Limnol. 22: 3345–3349.

Hydrobiologia **253**: 357–366, 1993.
P.C.M. Boers, T.E. Cappenberg & W. van Raaphorst (eds),
Proceedings of the Third International Workshop on Phosphorus in Sediments.
© 1993 *Kluwer Academic Publishers.*

A comparison of sedimentary and diatom-inferred phosphorus profiles: implications for defining pre-disturbance nutrient conditions

N. J. Anderson[1], B. Rippey[2] & C. E. Gibson[3]
[1] *Geobotany Division, Geological Survey of Denmark, Thoravej 8, DK-2400 Copenhagen NV, Denmark;*
[2] *Freshwater Laboratory, University of Ulster, Traad Point, Ballyronan BT45 6LR, N. Ireland;* [3] *Dept. of Agriculture (NI), Aquatic Sciences Research Division, Newforge Lane, Belfast BT9 5PX, N. Ireland*

Key words: weighted averaging, calibration, phosphorus, diatoms, palaeolimnology, Northern Ireland

Abstract

Interpreting sedimentary phosphorus profiles in terms of changes in the historical P load is difficult due to variable retention and post-depositional diagenesis. An alternative approach is to use diatom assemblages in surface sediments and derive a transfer function for epilimnetic SRP and total P concentrations using weighted average regression and calibration. The obtained relationship can then be applied to down-core changes in sedimentary diatom assemblages and diatom-inferred P (DI-P) used to assess historical changes in epilimnetic P-concentrations. A diatom-phosphorus calibration data set for 43 eutrophic lakes in Northern Ireland has been constructed and applied to two small eutrophic lakes (Lough Mann, White Lough). DI-total P (i.e. predicted) is highly correlated with observed TP ($r^2 = 0.75$) for the surface-sediment training data-set. The resultant changes in DI-P derived from application of the transfer function to down-core changes in diatom assemblages are compared to sedimentary P concentrations. The latter are highly variable, presumably due to redox-derived effects, while DI-P profiles are more readily interpretable, and agree with other stratigraphic records of lake eutrophication. The method offers a good possibility of defining pre-disturbance (i.e. natural) phosphorus concentrations in lakes with associated implications for lake-restoration programmes.

Introduction

Pre-eutrophication phosphorus concentrations in lakes are unknown: many lakes in north-west Europe have been subject to perturbation over a variety of timescales, nearly all of which are longer than current monitoring programmes. However, for both ecological and lake-management strategies it is necessary to have information concerning the natural situation before disturbance, prior to the start of lake restoration (Schindler, 1987).

Realistic estimates of background conditions can often only be made by using the sediment record. As lake eutrophication primarily results from increased phosphorus loading, it is often assumed that sedimentary P provides a suitable record of enrichment (e.g. Birch *et al.* 1980). Unfortunately, factors controlling phosphorus retention in lake sediments are complex and sedimentary P profiles may have only a vague relationship to the timing of increased P inputs, and even less to the whole-lake P load (Engstrom & Wright, 1985; Carignan & Flett, 1981; Marsden, 1989; Mayer *et al.*, 1982; Bengtsson & Persson, 1978).

Algal communities change as nutrient concentrations increase (e.g. Hörnström 1981; Rosén,

1981). Until recently the diatom record in lake sediments has been interpreted subjectively using such criteria; e.g. phosphorus-rich (eutrophic) lakes are dominated by *Stephanodiscus* and P-poor (oligotrophic) lakes by *Cyclotella*. However, with the development of weighted averaging regression and calibration (ter Braak & van Dam, 1989; ter Braak & Prentice, 1988; Birks *et al.*, 1990; Line & Birks, 1990) it is possible, theoretically, to determine quantitatively the relationship of a species to a given nutrient (e.g. total phosphorus [TP]), and then use these relationships to infer past changes in the chosen parameter from down-core changes in diatom assemblages in lake sediments. This approach has been used with considerable success to infer past changes in lakewater pH and salinity (Birks *et al.*, 1990; Fritz *et al.*, 1991).

We present here results from a diatom calibration or training data-set for eutrophic lakes in Northern Ireland and compare diatom-inferred phosphorus (DI-P) to the geochemical phosphorus record, and assess its implications for determining the pre-enrichment nutrient status of lakes.

Study sites

Training set

The lakes used for the training data-set were selected to provide a phosphorus gradient (Table 1), and all are meso- to eutrophic by OECD guidelines (Forsberg & Ryding, 1980). The lakes are located in Counties Down (16 lakes), Armagh

Table 1. Summary characteristics for 43 training lakes.

	Range
Lake area:	1–30 ha
Maximum depth	0.6–20 m
Total P	25–800 μg l^{-1} #
SRP*	2–470 μg l^{-1} #

\# Mean of all available chemistry.
* Soluble reactive phosphorus.

(25 lakes), and Tyrone (2 lakes), along a broad S.W. – N.E. transect of increasing conductivity and P concentrations. The limnological characteristics of these lakes have been summarized by Gibson (1986, 1988, 1989). The lakes in Co. Down are primarily situated on Silurian greywacke sandstones, siltstones and shales, and a large granitic intrusion, the Newry granodiorite. The geological setting of the Co. Armagh lakes is more complex, but this area too consists of substantial areas of Silurian greywackes (see Gibson, 1989; Wilson 1972 for further details). The two lakes from Co. Tyrone are located on Carboniferous shales and sandstones.

The lakes are small (< 50 ha) with depths ranging from ~ 1 to 20 m (Table 1). Many of them are located within drumlin belts and are, therefore, quite sheltered. Land use at most sites, especially in Co. Armagh, is unimproved grassland, but there is a N.W. trend towards increasing soil fertility and agricultural productivity in Co. Down.

Stratigraphic study sites

To assess the applicability of the diatom-P calibration to down-core changes in diatom assemblages and its comparison with the geochemical record, two eutrophic lakes were chosen (Table 2). Lough Mann has a small, mainly pastoral, catchment with no apparent disturbance history. White Lough was chosen as it was the site of a chemical manipulation exercise (addition of ferric aluminium sulphate) in 1978–1982 (Foy, 1985). Although the effect on epilimnetic P-concentrations was limited, it was hoped that the addition of Fe (~ 3 times the normal load) would be clearly identified in the stratigraphic record, but unfortunately this was not possible.

Methods

For the training set samples surface sediment samples (~ 1 cm depth) were taken using a modified Brinkhurst-Kajak corer from the deepest part of the lakes during April 1990 and May 1991.

Table 2. Summary of water chemistry and sediment core chemistry for White Lough and Lough Mann.

Lake			White		Mann
Area (ha)			6		12
Max. depth (m)			11.2		8
Total P* (μg l^{-1})			19–43		51–73
SRP* (μg l^{-1})			9–16		9–12
Chl. a* (μg l^{-1})			17–26		18–41

Sediments	Mean*	Max	Min	Mean	Max	Min
		(mg g^{-1})			(mg g^{-1})	
Mg	4.15	5.62	2.94	4.51	4.87	4.02
Na	1.96	2.56	1.86	2.02	2.17	1.88
K	11.82	15.1	10.5	2.45	3.10	2.24
Ca	9.53	21.2	5.44	8.18	10.5	6.55
Fe	40.9	60.6	27.9	32.6	61.1	10.8
Mn	1.85	3.51	0.76	1.03	1.65	0.57
P	2.62	4.51	1.16	2.33	4.38	1.36

\# Range 1991.
* Mean whole-core.

For the diatom and geochemical stratigraphic analyses short cores (~ 30 cm sediment recovery) were taken using similar methods (Table 2), sealed in the field, and transported to the laboratory for extrusion. The core from White Lough was extruded at 0.5 cm intervals over the top 11 cm, and at 1 cm intervals below this depth. The Lough Mann core was extruded at 1 cm intervals down to 20 cm, and at 2 cm intervals below this depth.

Water chemistry

The lakes were sampled for water chemistry 2 to 4 times between April 1990 and May 1991, although some lakes were only sampled once in May 1991. Sampling has, however, continued in an effort to characterize better the chemistry of the lakes. Total phosphorus (TP) was determined by persulphate digestion of an unfiltered sample (Eisenreich *et al.*, 1975); and soluble reactive phosphorus (SRP) as phosphorus detectable in an undigested, filtered sample (Millipore

0.45 μm), with final determination by the method of Murphy & Riley (1962).

For the statistical analyses the log-transformation of the arithmetic mean of all samples was used (34 lakes) or the log of the single available sample (8 lakes). A comparative analysis was made using the December 1990 sample ($n = 31$), to assess the possible predictive capacity of winter chemistry prior to depletion by algal growth.

Diatom and geochemical methods

Dry weight and loss-on-ignition measurements were made on the cores using standard methodology. Diatom analyses followed standard methodology (Battarbee, 1986) with ~ 500 valves counted for each surface sediment sample. For the sediment cores a reduced count size was used (~ 300 per level for Lough Mann, and ~ 400 for White Lough). Only species with an occurrence of 1% in any one of either the training or core samples were retained for use in the statistical analyses (112 species from 189 identified). Bulk geochemical analyses for the major ions follow Anderson & Rippey (1988), with total phosphorus determined by colorimetry (phosphomolybdate method: Golterman *et al.*, 1978) on the sediment digest after a twenty- or fifty-fold dilution.

Statistical methods

Preliminary data analysis was undertaken using the canonical correspondence analysis (CCA) routine in the program CANOCO (ter Braak, 1987). Weighted averaging (WA) regression (to estimate taxa P optima) and WA-calibration (to estimate diatom-inferred lake P concentrations) were then done using the program WACALIB version 2.1 (Line & Birks, 1990). Inverse de-shrinking was used during WA calibration as this reduces the predictive error, as measured as Root Mean Standard Error (RMSE; Birks *et al*, 1990; Fritz *et al.*, 1991). WA calibration with tolerance down weighting [WA(tol)] was also undertaken (Birks *et al.*, 1990). The underlying statistical rationale and theory, together with further details of the methodology are given elsewhere (ter Braak & van Dam, 1989; ter Braak & Prentice, 1988; Birks *et al.*, 1990).

Apart from routine error screening of both the diatom and chemistry data no species or lakes were removed from the data set prior to calibration. The present data set is not sufficiently large to support screening or splitting, and the ecological basis is not entirely obvious. However, retaining lakes with unusual diatom assemblages at given P concentrations reduces, therefore, the predictive ability of the model (see discussion).

Results

Diatom inferred P using weighted averaging calibration

A CCA (ter Braak, 1986) of the full chemistry data set and diatom data indicated that TP is the first significant chemical variable ($P < 0.05$) in a forward selection of all variables and is highly correlated with CCA axis 1 ($r = 0.86$). SRP is similarly significant and correlated with axis 1 ($r = 0.80$).

That TP is the dominant variable with a larger constrained eigenvalue than any other variable in a series of CCAs with single environmental variables is a requirement 'for the calibration to be useful' (ter Braak, 1987). Even though TP only explains 8% of the variance in the weighted averages of the diatom abundance data, this percentage is similar to that for pH in the Surface Water Acidification Programme (Stevenson *et al.*, 1991) data set (8.1% $\lambda_1 = 0.495$, $\lambda_2 = 0.394$). For the data set represented here, TP is the dominant variable even though $\lambda_1 < \lambda_2$.

A complete analysis of diatom distribution in relation to the full chemistry data set will be published elsewhere (Anderson, Gibson & Rippey, ms in prep.). Both WA and WA(tol) diatom-inferred TP have a good relationship with the observed values ($r^2 = 0.75$ and 0.77 respectively; Table 3; Fig. 1a), with no substantial outliers for TP. RMSE and r^2 for TP are marginally better for

Table 3. Summary statistics for Weighted Averaging Phosphorus calibration.

	WA			WA (tol)		
	n	r^2	RMSE	*n*	r^2	RMSE
Mean TP	43	0.75	0.172	43	0.77	0.164
Mean TP*	41	0.67	0.157	41	0.71	0.156
December TP	31	0.69	0.166	31	0.75	0.150
MeanSRP	43	0.64	0.304	43	0.62	0.309
Mean SRP#	42	0.71	0.267	42	0.73	0.257
December SRP	31	0.70	0.287	31	0.73	0.269

* Excluding Islandderry & Monlough (cf. Fig. 1a).
# Excluding Carrigullian (cf. Fig. 1b).

WA(tol) than for WA (Table 3). Removing the two shallow hypertrophic lakes (Monlough and Islanderry) lowers the r^2 value slightly. The RMSE values for December TP are lower than of those for the mean of all measurements (Table 3), suggesting that winter chemistry may prove to provide a more reliable basis for prediction in a larger data set. Here, although the number of lakes is reduced, r^2 is not substantially lower.

Predicted mean SRP concentrations are also well correlated with observed values (Table 3; Fig. 1b), with only one outlier, but the fit is not as good as for TP (see discussion). Removing the outlier (Carrigullian, a very shallow [z mean: ~1.5 m] eutrophic lake; measured TP range 100–180 μg l^{-1}; September 1991 170 μg l^{-1} Chl. *a*) improves the relationship (Table 3).

These data suggest that there is a reasonable basis for the use of the relationship for inference of TP from changes in down-core diatom assemblages. Although the summary statistics suggest that WA(tol) is a better predictive tool, WA was used for the down-core reconstructions because the ecological implications of the WA(tol) approach have yet to be assessed.

Core applications

Lough Mann. Cation profiles are stable indicating minimal catchment disturbance, although zinc and copper increase above 3 cm (data not shown),

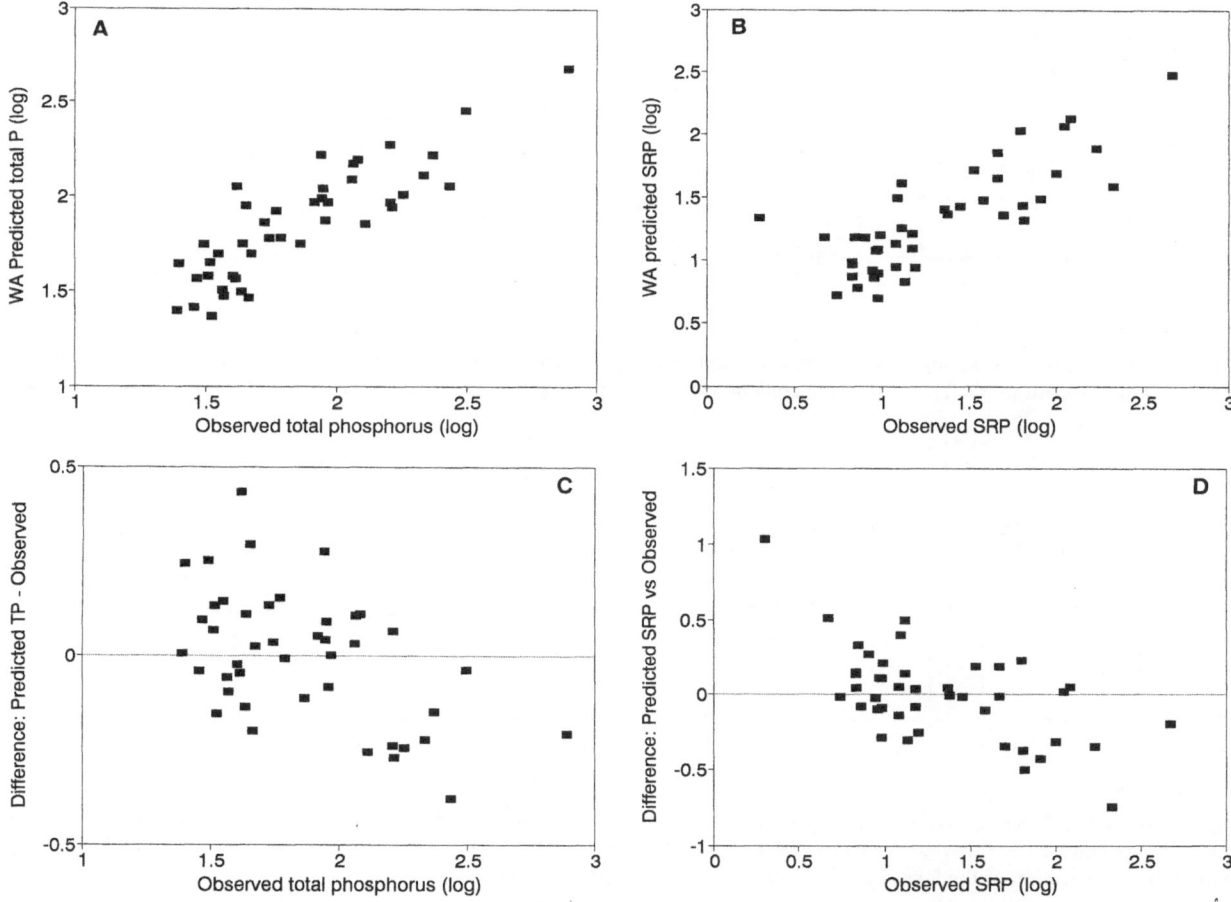

Fig. 1. Parts a & b: Scatter plots for diatom-inferred phosphorus vs observed phosphorus concentrations ($\log_{10}$ μg TP l^{-1}); a: TP; b: SRP; both plots use all available chemistry – see text. Parts c & d: Differences between predicted and observed values for diatom-inferred total phosphorus (c) and soluble reactive phosphorus (d), both using Weighted Averaging model (WA).

perhaps due to surface enrichment (Fig. 2; Table 2). The sedimentary P concentration increases a little up-core, but this steady increase is interrupted briefly by a high value (12 mg g^{-1}) at 8.5 cm (Fig. 2). This feature coincides with temporary highs in Fe and particularly Mn concentration (Fig. 2).

Diatom assemblages show few major changes throughout the core. The assemblage is dominated by planktonic species and although there are no species replacements, percentages fluctuate. DI-P values are ~ 100 μg l^{-1} at the base of the core and decline steadily up-core to around 67 μg l^{-1} at the surface. The measured mean TP

concentration ($n = 2$) for Lough Mann in 1991 was 62 μg l^{-1}.

White Lough. Again cation and loss-on-ignition profiles indicate stable catchment conditions (i.e. limited shifts in erosional intensity; Fig. 3, Table 2). Mn increases steadily up-core from < 1 mg g^{-1} to > 3 mg g^{-1} at ~ 2 cm, apart from a decrease between 10 and 12 cm depth (Fig. 3). Fe concentrations are ~ 40 mg g^{-1} to 10 cm, after which there is a variable decline. The Fe:Mn ratio declines from a basal maximum to a pronounced peak at 10–12 cm (Fig. 3), but is then stable to the surface. Minimum sedimentary P concentrations

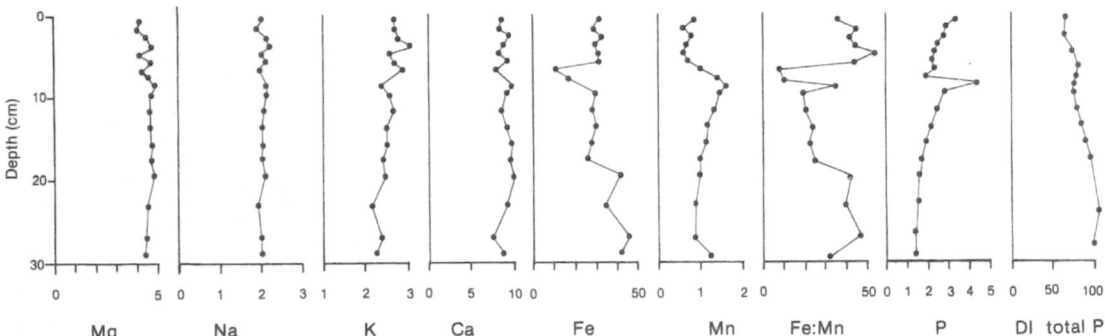

Fig. 2. Lough Mann core stratigraphy showing the differences between sedimentary P and diatom-inferred total phosphorus (DI-P). DI-P concentrations (i.e. for the lake epilimnion) are in $\mu g \, l^{-1}$; geochemical values are mg $g[DW]^{-1}$; Fe:Mn is a ratio.

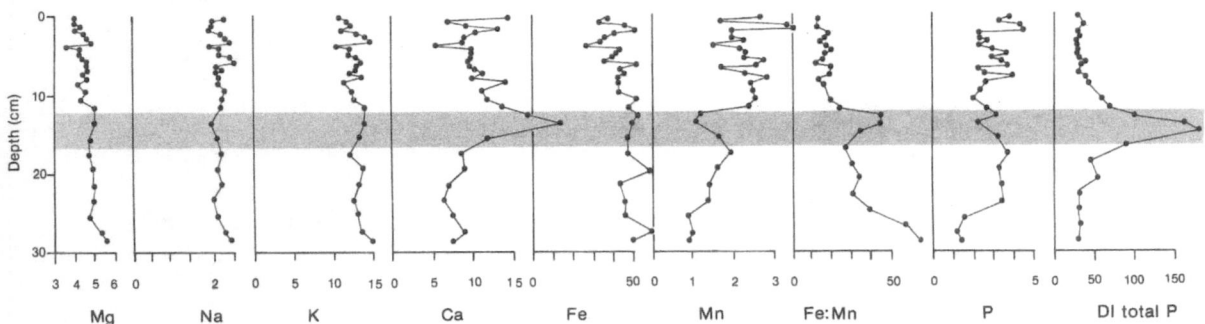

Fig. 3. As for Fig. 2, but for White Lough. The stipple indicates the period of good agreement between the diatom predicted increase in epilimnetic phosphorus concentrations and increases in sedimentary Ca and changes in Mn, Fe:Mn ratios associated with the eutrophication of the lake caused by the increased phosphorus availability.

(~ 1 mg g^{-1}) occur at the base of the core, after which there is a pronounced increase to ~ 3 mg g^{-1}. Above 12 cm values oscillate with no major trend. Sedimentary P concentrations are, however, correlated with Mn ($r = 0.60$) indicating the importance of redox cycling on the sedimentary P record at this site.

Despite changes in the diatom assemblages (Anderson, unpubl. data) DI-P is constant throughout the length of the core at $\sim 35 \, \mu g \, l^{-1}$, apart from a major increase to $160 \, \mu g \, l^{-1}$ between 12 and 15 cm (Fig. 3). The surface DI-P value ($35 \, \mu g \, l^{-1}$) agrees well with the measured mean value ($31 \, \mu g \, l^{-1}$). The peak in DI-P at 12–15 cm, apart from coinciding with reduced Mn and the peak in Fe:Mn (Fig. 3; both parameters are indicators of increased anoxia due to, presumably, increased P-loading and eutrophica-

tion), is also associated with a major peak in Ca, due to increased photosynthetic activity and precipitation of $CaCO_3$.

Discussion

Increased concern about the quality of inland waters has resulted in the widespread application of a variety of measures (tertiary wastewater treatment, diversion schemes, chemo- and biomanipulation) to reduce nutrient (especially phosphorus) loadings to lakes (Sas, 1989). The possible results of these reductions in nutrient load are normally assessed by either OECD-type regression relationships or site-specific models (Dillon & Rigler, 1974; Forsberg & Ryding, 1980; Kamp-Nielsen, 1986). Few lakes, however, have been extensively monitored, and prior to lake restoration attempts, monitoring may last only a few years with associated difficulties in determining intra-annual variability and mean nutrient and chlorophyll levels. The specific aim of lake restoration methods is normally not site-specific, but determined by either government legislation or 'subjective' criteria about clean lakes. The approach taken here is an attempt to develop a reliable method of determining pre-enrichment P levels for individual lakes.

Diatom-inferred P: Predictive ability

Diatom inferred-TP is well correlated with observed TP concentrations (Table 3), although there is some scatter associated with the relationship ($\sim 30-40$ μg TP l^{-1}). This range around the predicted values should, however, be compared both to the range in the data set ($25-800$ μg TP l^{-1}) and the intra-annual range of P concentrations that occurs in many eutrophic lakes (e.g. White Lough; Foy, 1985).

Errors associated with DI-P and its predictive ability, i.e. reasons for the scatter in the relationship between diatom-inferred and observed phosphorus, relate to a number of factors, in particular, diatom ecology, heterogeneity of diatom assemblages in lake sediments and chemical variability. The errors are, however, primarily due to the fact that the factors controlling algal abundance in eutrophic lakes are multivariate (Reynolds, 1984; Heyman & Lundgren, 1988).

There is a systematic bias – to underestimate at high P concentrations and vice versa – for both predictions of TP and SRP (Fig. 1c & d), and for both WA and WA(tol) models (data not shown). For the moment it is unclear what causes these trends, but at high P concentrations they are probably related to the inability of diatom communities to respond to increasing P concentrations by structural change. Diatom assemblages are quite uniform above ~ 300 μg l^{-1} TP due to the limited number of diatom species physiologically adapted to high P and low Si:P ratios (e.g. Stephanodiscus). In the mid-TP range, tolerance weighting probably provides a better fit be-

cause it downweights those species which have a wider tolerance and are found across a range of nutrient concentrations (e.g. Asterionella formosa and Aulacoseira ambigua).

The range of r^2 values and RSME (nb. RSME values are only comparable for each variable; not between TP and SRP) for the different models (Table 3) suggests that the method is quite robust, in terms of its predictive ability. Interestingly, the December SRP model (WA[tol]), despite a reduced data set is as good as the full data set. On ecological grounds one would expect winter SRP concentrations to be well correlated with diatom assemblages. The overall agreement of the different models (Table 3) suggests that as the data set is expanded to include low P lakes and the chemistry set refined, statistically and ecologically sound predictions of TP will be obtained.

A more stringent test of the predictive ability of DI-P is to test the training-set model against a set of lakes (test set) not included in the training set (cf. Birks et al., 1990). This approach permits a statistically rigorous estimate of the RMSE to be made. The model can also be tested by application to a lake with a known history of epilimnetic P concentrations, reconstructions in a dated sediment core can be compared to monitored data.

A measure of the usefulness of diatom P-calibration models can be gained by comparison with OECD-type regression relationships and the error associated with them (e.g. Dillon & Rigler, 1974). Criticism of the chlorophyll-phosphorus models has been extensive recently because of their poor predictive ability (e.g. Reynolds, 1987). Compared to the predictive ability of these regression relationships, the errors associated with DI-P are as good or better, and indicate that there is an acceptable future for the method in lake management.

Diatom-inferred P: interpretation and error sources

Given that diatoms are deposited in the sediments shortly after death or the collapse of a crop (Jewson et al., 1981; Reynolds et al., 1982) DI-P is a measure of the mean epilimnetic P concentration at the time of the growth of the diatom assemblage. The DI-P from sediment cores thus provides a record of changing epilimnetic P concentrations.

The problem of mixing of diatom assemblages in lake sediments by bioturbation and resuspension, although of possible importance, is reduced in eutrophic systems because of the higher sediment accumulation rates. Equally, however, some mixing of the surficial sediment assemblage smooths out interannual variability in the diatom assemblages, not all of which may be related to nutrient concentrations. Importantly, the smoothed record is acceptable for management purposes where only major trends are required.

Diatom errors. The assemblages in this training set are dominated by planktonic species, but there are clear relationships between P concentrations and some benthic/epiphytic species (e.g. *Achnanthes minutissima*; Anderson unpubl.). It is possible, however, that some benthic diatoms do not relate directly to epilimnetic chemistry, due to the interactions between motile species and interstitial water nutrient sources and epiphytes with macrophytes (e.g. Jørgensen, 1957). These factors will result in a poorer predictive ability.

Factors controlling diatom assemblages in lake sediments are complex (Reynolds, 1984; Heyman & Lundgren, 1988), and apart from variability associated with depositional processes, lake morphometry influences the diatom assemblage and may result in underrepresentation of certain communities. However, where planktonic diatoms are dominant, it is reasonable to assume that a single sample will be representative (Anderson, 1990). Relative frequency data have the advantage over diatom accumulation rate data in that they are less variable (Anderson, 1992; Whitmore, 1991).

Where diatoms are totally dissolved (e.g. some carbonate systems) the approach cannot, of course, be applied. More importantly, however, are the possible effects of differential dissolution (i.e. loss of lightly silicified species; e.g. *Rhizosolenia*), biasing the assemblage and creating poorer fits between predicted and observed phosphorus concentrations. Fortunately, severe differential dissolution is rare in freshwater lakes and therefore diatom assemblages in sediments should reflect water chemistry changes (Battarbee, 1986; Whitmore, 1991). This assumption is particularly true for surface sediments, but down-core diagenetic losses in some systems should not be discounted.

Lake water chemistry derived errors. A major problem with diatom training data-sets is the variability in lake-water chemistry. This stems from two sources: first, the limited number of samples that are commonly available when more than 40 lakes are being sampled. Resources do not generally permit a rigorous assessment of inter- and intra-annual variability at all sites. The second problem is the natural inter- and intra-annual variability of TP concentrations in lakes: what mean should be used: winter, spring or annual mean? P concentrations can vary considerably throughout the year, due to internal and external loading variations and biological depletion.

It is often assumed that winter chemistry provides the best predictive basis for mean summer chlorophyll. However, for the SRP model the poor predictive ability may be due to the fact that in many of the lakes in the data set, overwintering populations of *Oscillatoria* (Chl. $a \sim 20$–$40 \, \mu a$ l^{-1}) deplete SRP prior to the spring diatom crop (cf. Foy & Fitzsimons, 1987). Similarly, some of the diatoms in the data set commonly occur during summer (*Fragilaria crotonensis*, *Aulacoseira granulata*). It is, then, difficult to expect a better fit than that obtained whether the annual mean or winter mean chemistry is used.

For these reasons it may be difficult to reduce the errors associated with DI-P. A reasonable aim would be to get the total predictive error below the expected seasonal variation in measured epilimnetic TP.

Interpretation of sedimentary P

The problems of interpreting sedimentary P profiles in terms of historical P loads have been discussed extensively elsewhere (Carignan & Flett, 1981; Engstrom & Wright, 1985). Although unambiguous sedimentary P profiles do occur (Shapiro *et al.*, 1971), P mobility associated with redox changes (due to increased organic loadings) means that sedimentary P profiles need to be interpreted with caution (e.g. Bengtsson & Persson 1978). Sedimentary fluxes, as well as concentrations, need to be used to clarify the changes in lake P concentration and loading, and the lack of fluxes for the two sediment cores presented here limits the possible interpretation that can be made.

At both Lough Mann and White Lough it appears that redox changes have influenced the sedimentary P record. Neither sites have P profiles that are readily interpretable in relation to other stratigraphic records related to eutrophication. At White Lough the sedimentary P profile varies considerably, perhaps due to redox control (cf. P relationship with Mn). The DI-P profile, however, agrees well with other indicators of eutrophication (increased Fe:Mn ratio, Ca peak) and suggests a major eutrophication event in the recent past due to increases in epilimnetic TP concentrations to $\sim 160 \ \mu g \ l^{-1}$ (Fig. 3).

Sedimentary P profiles record gross changes in the lake water phosphorus concentrations, particularly when the change is from oligotrophic to eutrophic conditions, e.g. in Lake Washington (Shapiro *et al.*, 1971). However, extracting finer detail is much more difficult and it appears that this is particularly difficult when sediment P concentrations are high. Factors such as redox potential, year-to-year variations in lake conditions, e.g. hydraulic retention time and bioturbation, may have a more pronounced effect on sedimentary P concentrations when these factors are high, compared to when they are low.

Conclusions and management implications

The possibility of inferring epilimnetic phosphorus concentrations from diatom assemblages is very good, and its implications are clear: to provide important background information (i.e. pre-disturbance) on P concentrations for individual lakes prior to the initiation of restoration schemes.

An important benefit of this approach is that it integrates complex diatom stratigraphic data into a form which is more readily interpretable by water managers (i.e. TP concentrations). The approach is quantitative and gives reliable error estimates, together with enhanced precision in interpreting the sediment record. As part of a multidisciplinary approach (dating, diatoms, pigments and geochemistry) DI-P would increase the possibility of interpreting sediment records unambiguously.

Acknowledgements

NJA acknowledges support from the Swedish Natural Science Research Council (NFR) and the J.C. Kempe Memorial Fund; and Prof. H. John B. Birks for discussions and comments on the manuscript. Prof. Brian Wood willingly made available the facilities at Traad Point. Field assistance by Paul Raven, Steven Smith and Mat Quinn, and the technical support of the FBIU Chemistry Laboratory staff are gratefully appreciated.

References

Anderson, N. J., 1990. Variability of diatom concentrations and accumulation rates in sediments of a small lake basin. Limnol. Oceanogr. 35: 497–508.

Anderson, N. J., 1992. Inferring diatom paleoproduction and lake trophic status from fossil diatom assemblages. Proc. Calif. Acad. Sci., in press.

Anderson, N. J. & B. Rippey, 1988. Diagenesis of magnetic minerals in the recent sediments of a eutrophic lake. Limnol. Oceanogr. 33: 1476–1492.

Battarbee, R. W., 1986. Diatom analysis. In B. E. Berglund (ed.), Handbook of Holocene palaeoecology and palaeohydrology. Wiley, Chichester: 527–570.

Bengtsson, L. & T. Persson, 1978. Sediment changes in lake used for sewage reception. Pol. Arch. Hydrobiol. 25: 17–33.

Birch, P. B., R. S. Barnes & D. E. Spyridakis, 1980. Recent sedimentation and its relationship with primary productivity in four western Washington lakes. Limnol. Oceanogr. 25: 240–247.

Birks, H. J. B, J. M. Line, S. Juggins, A. C. Stevenson & C. J. F. ter Braak, 1990. Diatoms and pH reconstruction. Phil. Trans r. Soc., Lond. B 327: 263–278.

Carignan, R. & R. J. Flett, 1981. Postdepositional mobility of

366

phosphorus in lake sediments. Limnol. oceanogr. 26: 361–366.

Dillon, P. J. & F. H. Rigler, 1974. The phosphorus-chlorophyll relationship in lakes. Limnol. Oceanogr. 19: 767–773.

Eisenreich, J. J., R. T. Bannerman & D. E. Armstrong, 1975. A simplified phosphorus analysis technique. Environmental Letts. 9: 43–53.

Engstrom, D. R. & H. E. Wright Jr., 1985. Chemical stratigraphy of lake sediments as a record of environmental change. In J. W. G. Lund & E. Y. Haworth (eds), Lake sediments and environmental history. University of Leicester Press, Leicester: 11–67.

Forsberg, C. & S. -O. Ryding, 1980. Eutrophication parameters and trophic state indices in 30 Swedish lakes. Arch. Hydrobiol. 89: 189–207.

Foy, R. H., 1985. Phosphorus inactivation in a eutrophic lake by the direct addition of ferric aluminium sulphate: impact on iron and phosphorus. Freshwat. Biol. 15: 613–629.

Foy, R. H. & A. G. Fitzsimons, 1987. Phosphorus inactivation in a eutrophic lake by the direct addition of ferric aluminium sulphate: changes in phytoplankton populations. Freshwat. Biol. 17: 1–13.

Fritz, S. C., S. Juggins, R. W. Battarbee & D. R. Engstrom, 1991. Reconstruction of past changes in salinity and climate using a diatom-based transfer function. Nature 352: 706–708.

Gibson, C. E., 1986. A contribution to the regional limnology of Northern Ireland: the lakes of County Down. Record of Agricultural Research (Dept. Agric. NI) 34: 75–83.

Gibson, C. E., 1988. Contributions to the regional limnology of Northern Ireland (ii): the lakes of County Fermangh. Record of Agricultural Research (Dept. Agric. NI) 36: 121–131.

Gibson, C. E., 1989. Contributions to the regional limnology of Northern Ireland: (3) the lakes of Co. Armagh. Ir. Nat. J. 23: 41–47.

Golterman, H. L., R. S. Clymo & M. A. M. Ohnstad, 1978. Methods for the physical and chemical examination of freshwaters. Blackwell Sci. Pub., Oxford.

Heyman, U. & A. Lundgren, 1988. Phytoplankton biomass and production in relation to phosphorus: some conclusions from field studies. In G. Persson & M. Jansson (eds), Phosphorus in Freshwater Ecosystems. Developments in Hydrobiology 48. Kluwer Academic Publishers, Dordrecht: 211–227. Reprinted from Hydrobiologia 170.

Hörnström, E., 1981. Trophic characterization of lakes by means of qualitative phytoplankton analysis. Limnologica 13: 249–261.

Jewson, D. H., B. H. Rippey & W. K. Gilmore, 1981. Loss rates from sedimentation, parasitism, and grazing during the growth, nutrient limitation, and dormancy of a diatom crop. Limnol. Oceanogr. 26: 1045–1056.

Jørgensen, E. G., 1957. Diatom periodicity and silicon assimilation. Dansk Bot. Arkiv 18: 1–54.

Kamp-Neilsen, L., 1986. Modelling the recovery of hypertrophic L. Glumsø (Denmark). Hydrobiol. Bull. 20: 245–255.

Line, J. M. & H. J. B. Birks, 1990. WACALIB version 2.1 – a computer program to reconstruct environmental variables from fossil assemblages by weighted averaging. J. Paleolimnol. 3: 170–173.

Marsden, M. W., 1989. Lake restoration by reducing external phosphorus loading: the influence of sediment phosphorus release. Freshwat. Biol. 21: 139–162.

Mayer, L. M., F. P. Liotta & S. A. Norton, 1982. Hypolimnetic redox and phosphorus cycling in hypertrophic lake Sebasticook, Maine. Wat. Res. 16: 1189–1196.

Murphy, J. & J. P. Riley. 1962. A modified single solution method for the determination of phosphate in natural waters. Analyt. Chim. Acta. 27: 31–36.

Reynolds, C. S., 1984. The ecology of freshwater phytoplankton. Cambridge University Press.

Reynolds, C. S., 1987. The response of phytoplankton communities to changing lake environments. Schweiz. Z. Hydrol. 49: 220–236.

Reynolds, C. S., H. R. Morrison & C. Butterwick, 1982. The sedimentary flux of phytoplankton in the south basin of Windermere. Limnol. Oceanogr. 27: 1162–1175.

Rosén, G., 1981. Phytoplankton indicators and their relations to certain chemical and physical factors. Limnologica 13: 263–290.

Sas, H., (ed.) 1989. Lake restoration by reduction of nutrient loading: expectations, experiences, extrapolations. Academia Verlag., St. Augustin.

Schindler, D. W., 1987. Detecting ecosystem response to anthropogenic stress. Can. J. Fish. aquat. Sci. 44, Suppl. 1: 6–25.

Shapiro, J., W. T. Edmondson & D. E. Allison, 1971. Changes in the chemical composition of sediments of Lake Washington, 1958–1970. Limnol. Oceanogr. 16: 437–452.

Stevenson A. C. [and others] 1991. The Surface Waters Acidification Project Palaeolimnology Programme: modern diatom/lake-water chemistry data set. Ensis Ltd., London 86 pp.

ter Braak, C. J. F., 1987 CANOCO – A FORTRAN program for Canocial Community Ordination. Manual. Wageningen., TNO Insitute of Applied Computer Science.

ter Braak, C. J. F. & I. C. Prentice, 1988. A theory of gradient analysis. Adv. ecol. Res. 271–317.

ter Braak, C. J. F. & H. van Dam, 1989. Inferring pH from diatoms: a comparison of old and new calibration methods. Hydrobiologia 178: 209–223.

Whitmore, T. J., 1991. Sedimentary diatom concentrations and accumulation rates as predictors of lake trophic status. In J. P. Smith; P. G. Appleby, R. W. Battarbee, J. A. Dearing, R. Flower, E. Y. Haworth, F. Oldfield & P. E. O'Sullivan (eds), Environmental History and Palaeolimnology. Developments in Hydrobiology 67. Kluwer Academic Publishers, Dordrecht: 163–169. Reprinted from Hydrobiologia 214.

Wilson, H. E., 1972. Regional geology of Northern Ireland. HMSO, Belfast.

Hydrobiologia **253**: 367–372, 1993.
P.C.M. Boers, T.E. Cappenberg & W. van Raaphorst (eds),
Proceedings of the Third International Workshop on Phosphorus in Sediments.
© 1993 *Kluwer Academic Publishers.*

Release rates and biological availability of phosphorus released from sediments receiving aquaculture wastes

Liam A. Kelly
Institute of Aquaculture, University of Stirling, Stirling FK9 4LA Scotland

Key words: algae, aquaculture, phosphorus, sediments

Abstract

Aquaculture is an increasingly significant user of freshwater resources in Scotland. In 1989, the total fish biomass produced in Scottish freshwater amounted to 7000 t. 50% of this total was reared in floating cage systems situated in lochs (lakes). Both solid (mainly in the form of uneaten feeds and faecal matter) and dissolved byproducts of the production cycle enter the limnetic environment untreated. Much solid waste material accumulates directly on the sediments beneath the cage systems. This leads to a localised enrichment in nutrient elements of the sedimentary environment. The experiments served to quantify rates of total phosphorus (TP) and dissolved reactive phosphorus (DRP) release from undercage and control sites, and to relate such releases to the biological availability of the released P. Results indicate significantly higher levels of NH_4Cl-extractable P in sediments affected by waste deposition from fish cages. TP and DRP release, and greater growth of Chl$a$ are obtained from undercage cores compared with control sites. No link between extractable-P content of sediments, or release rate and Chl$a$ production was established.

Introduction

Over the past ten years, freshwater cage rearing of Atlantic salmon (*Salmo salar*) juveniles in Scotland has undergone a rapid increase in terms of biomass produced. Together with rainbow trout (*Onchorhynchus mykiss*) the total biomass produced in cages is approximately 3500t, or 50% of the total freshwater production in Scotland (NCC, 1990). In recent years, concern has arisen that waste products from such aquaculture activities may stimulate change in the nutrient status of the lakes being used for this purpose, the main concern being how to maintain the pristine nature of Scottish freshwater lakes. Two forms of nutrient loss are normally attributed to cage aquaculture; firstly, losses of particulate material in the

form of uneaten feeds and faeces, and secondly, losses of dissolved nutrients during the operation of cages, mainly through metabolic wastes of the fish (Trojanowski *et al.*, 1982). Losses of dissolved nutrients include products of leaching from feeds as they pass through the water column, together with enhanced release of nutrients from waste materials deposited in the lakes (Skogheim *et al.*, 1982; Weismann *et al.*, 1988; Colman & Jacobson, 1991).

Fish held in cages are fed artificial diets, which are produced in dry or moist pellet form. Diets used in freshwaters usually contain between 1.0–1.8% D.W. phosphorus (P). P added to diets may appear in many different forms, ranging from phytin phosphorus to defluorinated rock phosphate (Crampton, 1987). Various estimates of losses of feed from cages have been produced: Beveridge (1987), in a review of such losses, estimates that up to 50% of the total solid waste generated by a cage farm comes from uneaten or egested feed (NCC, 1990).

Past studies have cited various environmental effects attributable to the operation of fish cages in freshwater. Included in these are increased nutrient content of lake water near the cages (Trojanowski et al., 1985; Korzeniewski et al., 1985), shifts in benthic population structure (Adams et al., 1976; Dobrowolski, 1987) and increased nutrient status of the water body overall (Penczak et al., 1982; Phillips, 1985). Few studies in freshwaters have considered the impact of cage wastes on bottom sediments, or their potential as a source of nutrients to the overlying water column and biota. Very few studies of the bioavailability of aquaculture wastes exist (Carr & Goulder, 1990).

Algae are known to be able to utilise P released from both sediments in suspension (Long & Cooke, 1978; Logan et al., 1979; DePinto et al., 1981; Marengo & Premazzi, 1985; Persson, 1990) or those already deposited in lake basins (Golterman, 1977; Böstrom et al., 1988). In much of the reported literature, studies have focused on sites which have enhanced nutrient loading extending throughout the site; in the case of lakes used to support cage aquaculture, enhanced sedimentation and therefore nutrient loading to the sediments is usually restricted to areas close to the cages themselves (Weston, 1991). A consequence of this may therefore be the existence of areas of sediments in an essentially oligotrophic system with physical characteristics which may be more normally found in eutrophic systems, i.e. high in % organic and P content.

In this project, microcosm experiments were undertaken in order to attempt preliminary assessments of release rates of P from sediments affected by waste deposition. As a secondary objective, bioavailability of P releases from lake bottom sediment impacted by cage aquaculture wastes, compared to unaffected sediments in the same lake, was assessed by exposing sediment microcosms to a monoculture of the green test alga Selenastrum capricornutum Printz.

Experimental procedures

Sediment sampling and site details

Samples of bottom sediments from three lakes supporting cage aquaculture were obtained from beneath operational juvenile Atlantic salmon (Salmo salar) cages and at control sites, established at a distance of 200 m from the cages. Replicate samples were obtained at both points using an Ekman grab. Initial attempts to obtain samples using a Jenkins corer were discontinued as the corer penetrated too far into soft undercage sediments. Samples of the upper 0.05 m of sediment were taken and placed immediately into the microcosms used for the experiments. The cylindrical microcosms were made from clear, acid washed perspex, 0.25 m high, 0.04 m internal diameter, and were tightly sealed with bungs at each end for transport. A large volume (40 l) water sample was also obtained at the same time. Of the three sites, Lake 3 has had cages operational for the longest period (7 years), whilst Lakes 1 (2 years) & 2 (5 years) have been operational as cage farm sites only intermittently.

Laboratory analysis

In the laboratory, the large volume water sample was analysed for total phosphorus (TP), dissolved reactive phosphorus (DRP) and Chla, then filtered through a 0.45 μm Millipore filters, and autoclaved for 20 minutes at 120 °C. Initial TP levels were all < 10 μg P l^{-1}. After cooling, it was then carefully added to the sediment microcosms, replacing the initial water, and the microcosm resealed with a stopper. Once resealed, the microcosm contained 10 ml wet sediment and 215 ml overlying water. This stopper incorporated a glass U-shaped tube to allow continuous aeration of the water column via a scrubbed air supply. For each lake, a total of 24 microcosms were established, allowing for six replicates of each treatment, i.e. undercage and control site with test algae added, together with six replicates for each control i.e. undercage and control site without algae.

DRP analysis was carried out following the methodology of Murphy & Riley (1962), and corrected for colour interference by subtracting water colour at 690 nm. TP was obtained in the same manner following digestion of water samples with a persulphate-sulphuric acid mixture, autoclaved for 20 minutes at 120 °C. Chlorophyll *a* was determined by methanol extraction (APHA, 1985) at 663 nm and 750 nm. All determinations were made on a Uvikon 810/820 twin beam spectrophotometer.

The upper 0.05 m of the sediments from cage and control sites were also examined for chemical speciation of P. This was done using a NH_4Cl extraction of P according to the scheme of Hieltjes & Lijklema (1980), and subsequent NaOH and HCl extractions to determine non-reactive P. Sediment water content was obtained by drying samples from each site at 105 °C, until no further weight loss was observed. The percentage of inorganic content was determined by ashing sediment samples in a muffle furnace at 550 °C for 24 hrs. All sediment analyses were performed in triplicate.

Algal culture and experimental conditions

An axenic stock of *Selenastrum capricornutum* Printz, (CCAP No. 27814) was cultured in SANM medium (Miller *et al.*, 1978), in an enclosure lit constantly at 95 μE s^{-1} m^{-2} PAR at benchtop, (measured by an LI 170 quantum sensor, 400–700 nm range), produced by four 75–85W cool white fluorescent tubes. Temperature of the enclosure was maintained at 22 °C $\pm$ 2 °C by a thermistor controlled fan and heater. Algae from this culture were taken during day 6 of the growth period, and placed in P-free SANM for 24 hrs. After this period, algae were introduced to the water columns of six undercage and six control site sediment microcosms, at an initial concentration of 1000 cells ml^{-1}. Cell counts were made in triplicate using a haemocytometer and microscope. pH (Phillips PW9409 meter and probe) and oxygen content (YSI Type 52 meter and probe) of the microcosm water was measured at the beginning and end of each experiment. Neither parameter was found to vary significantly during the experimental period. The sediment microcosms were maintained in the same light enclosure as the algal medium for a period of 10 days. After this period, the water column was drained and analysed for DRP, TP, and Chl*a*. Comparisons of results from controls and algal additions for each lake were made using the Mann-Whitney U-test on STATGRAPHICS, a PC-based statistical package.

Results and discussion

Sediment analyses

Sediments from Lakes 1 & 2 demonstrated similar basic physical and chemical characteristics, and were found to be low in organic content, with only slight enrichment at the undercage location in Lake 2 (Table 1). Sediments from the undercage site in Lake 3 contained considerably more organic matter than sediments from the other two lakes, because of the greater period of cage operation and subsequent waste accumulation that has taken place in Lake 3. In all three lakes as expected, significantly higher levels of NH_4Cl-extractable P and non-reactive P were found in the undercage sediments. Sediments from Lake 3 again demonstrated significant differences with the other two lakes, with NH_4Cl-P levels substantially higher in undercage sediments. The range of NH_4Cl-P at sites 1 and 2 (0.11–0.51 mg P g^{-1}) are comparable with results reported for a range of lake types (van Liere *et al.*, 1982; Mayer & Manning, 1989; Böstrom *et al.*, 1982), whilst NH_4Cl-P content of undercage sediments from Lake 3 represents a considerable enrichment over background levels.

NaOH- and HCl-extractable P fractions, with the exception of the undercage sample from Lake 3, accounted for over 50% of extractable sediment P (Table 1). Variations in the relative proportion of either fraction, as a percentage of total extractable P, were less marked than for the NH_4Cl-extractable fraction. Non-reactive P,

Table 1. Chemical and physical characteristics of the upper 0.05 m of sediments at control and cage sites ($n = 3$, 95% c.l. error on average in parentheses).

Site	Site 1		Site 2		Site 3	
	Cage	Control	Cage	Control	Cage	Control
Organic content (%)	13 ± 1.3	13 ± 1.1	18 ± 1.5	15 ± 1.4	38 ± 2.3	15 ± 0.7
Water content (%)	72 ± 1.5	72 ± 1.2	82 ± 2.3	71 ± 1.3	90 ± 2.1	70 ± 1.5
NH_4Cl-P (mg P g^{-1} D.W.)	0.35 ± 0.2	0.11 ± 0.0	0.51 ± 0.1	0.13 ± 0.01	2.77 ± 0.6	0.17 ± 0.03
NaOH-P (mg P g^{-1} D.W.)	0.20 ± 0.1	0.10 ± 0.1	0.60 ± 0.3	0.10 ± 0.1	1.41 ± 0.2	0.20 ± 0.1
HCl-P (mg P g^{-1} D.W.)	0.23 ± 0.1	0.16 ± 0.1	0.42 ± 0.1	0.20 ± 0.1	1.14 ± 0.3	0.27 ± 0.1

however, represents a considerable portion of sediment P content of all the sites examined, including those where enhanced waste deposition occurred.

Algal addition results and release rates

With the exception of Lake 2 ($n_1 = 6$, $n_2 = 6$, $U = 7.1$, $p = 0.07$) the experiments demonstrated a significant enhancement in Chl*a* derived from the inoculated sediment microcosms containing undercage sediments, when compared with inoculated control sites (Table 2). For all lakes, and for microcosms with or without algae added, TP and DRP release was significantly greater for undercage sediments as well ($n_1 = 6$, $n_2 = 6$, minimum $U = 6.1$, $p = 0.05$). Unlike the findings of van Liere *et al.* (1982), no significant differences in rates of TP and DRP release were observed as a result of microcosms being inoculated with algae, *i.e.* release in these enclosures was found to be driven primarily by sediment processes. During the experiment involving Lake 3 sedi-

Table 2. TP and DRP release rates from sediments with and without algae added to the water, together with Chl*a* generated during the experimental period. ($n = 6$, 95% c.l. of average in parentheses).

Parameter	Site 1			Site 2			Site 3		
	TP (mg m^{-2} d^{-1})	DRP (μg l^{-1})	Chl*a* (μg l^{-1})	TP (mg m^{-2} d^{-1})	DRP (μg l^{-1})	Chl*a* (μg l^{-1})	TP (mg m^{-2} d^{-1})	DRP (μg l^{-1})	Chl*a* (μg l^{-1})
Undercage + algae	0.63 (0.2)	0.53 (0.3)	54.8 (13.5)	7.6 (1.3)	4.3 (0.5)	12.7* (12.6)	58.6 (2.3)	44.5 (9.6)	4.2 (1.9)
Control + algae	0.04 (0.06)	0.03 (0.05)	0.3 (0.2)	1.3 (0.1)	0.5 (0.3)	1.1 (1.1)	2.8 (0.7)	1.6 (0.4)	0.18 (0.16)
Undercage	1.5 (1.0)	0.9 (0.6)	–	7.8 (0.5)	5.3 (0.9)	–	57.6 (1.5)	57.2 (5.5)	–
Control	0.2 (0.06)	0.15 (0.04)	–	1.3 (0.2)	0.1 (0.02)	–	2.3 (0.4)	1.5 (0.4)	–

All differences in undercage and control experiments with algae added significant at $\geq 95\%$ c.l., excluding results marked*.

ments, the water column in the microcosms containing undercage sediments became extremely stained, indicating a release of humic compounds to the water column.

Chl*a* increases were not found to be positively related to TP or DRP release rates, nor to the NH_4Cl-P fraction within the sediments. This result runs counter to those commonly observed (Golterman, 1977, Klapwijk *et al.*, 1982; Böstrom *et al.*, 1988). Phosphorus in all forms was plentiful in the water column during each of the experiments and it must be concluded that other limitations, such as lack of nitrogen (N) (Barica, 1990) may have occurred in the microcosms. Another potential explanation of the very low production of Chl*a* from the P-rich water column of undercage sediments in Lake 3 is the chemical limitation of P availability for algae resulting from the presence of large quantities of dissolved humic matter (Jones, 1990) in the water column.

Release rates of TP and DRP from the sediments examined were highly variable (Table 2). Sediments from Lake 1 clearly had the lowest production rate of TP and DRP, and it would appear that rate of release from sediments affected by cage wastes is linked at least in part to the length and intensity of the cage operations taking place, with Lake 3 being the site having the longest record of cage aquaculture operations. Rates of release from undercage sediments are comparable with reported values of aerobic release from eutrophic lakes elsewhere (Böstrom *et al.*, 1982, Premazzi & Provini, 1985, Sinke & Cappenberg, 1988) whilst rates of release at control sites are comparable with ranges reported for mesotrophic sites (Marsden, 1989).

The area over which sediments are affected by the impact of wastes for cage culture is of considerable importance in determining the overall impact of such deposits upon the water column nutrient and biological status. Following these preliminary results, it will be an aim of future work to include an assessment of the likely extent and influence of such waste deposits on the overall sediment nutrient budget for lakes supporting cage aquaculture, together with investigations of N cycles in the sediments.

Conclusions

The results of preliminary microcosm experiments with sediments from three lakes indicate that rates of nutrient release from sediments affected by wastes derived from freshwater cage aquaculture are significantly enhanced (1.5-57.6 mg TP m^{-2} d^{-1}) over unaffected sediments within the same lake (0.2-1.3 mg TP m^{-2} d^{-1}). Levels of extractable P were found to be consistently higher in the upper 0.05 m of sediments obtained from undercage sites, when compared with controls. Experiments using *S. capricornutum* as the test alga indicated that enhanced levels of algal-available P are produced from undercage sediments. However, results obtained from the present series of experiments did not establish a strong link between growth of algae and P release from, or P content of sediments. Other possible factors, *e.g.* N limitation or the presence of humic matter in the water column may also influence P uptake by algae. In order to fully understand the effect of such waste loadings on lake biotic and trophic status, an additional survey of the areal extent of waste dispersion will be necessary.

Acknowledgements

The author wishes to thank the anonymous referees and editors for their constructive comments and suggestions on the original draft of this paper.

References

Adams, J. C., R. V. Kilambi, W. A. Wickizer & A. V. Brown, 1976. Macrobenthos population changes in Crystal Lake, Arkansas, subsequent to cage culture of fish. Ark. Acad. Sci. Proc. 3: 12–13.

APHA, 1985. Standard methods for the examination of waters and wastewaters (16th Edition). American Public Health Association, Washington DC.

Barica, J. 1990. Seasonal variability of N:P ratios in eutrophic lakes. In P. Biró and J. F. Talling (eds), Trophic Relationships in Inland Waters. Developments in Hydrobiology 53. Kluwer Academic Publishers, Dordrecht: 97–103. Reprinted from Hydrobiologia 191.

Beveridge, M. C. L., 1987. Cage Aquaculture. Fishing News Books Ltd, Farnham, U.K., 352 pp.

372

Böstrom, B., M. Jansson & C. Forsberg, 1982. Phosphorus release from lake sediments. Arch. Hydrobiol. Beih. 18: 5–59.

Böstrom, B., G. Persson & B. Broberg, 1988. Bioavailability of different phosphorus forms in freshwater systems. In G. Persson & M. Jansson (eds), Phosphorus in Freshwater Ecosystems. Developments in Hydrobiology 48. Kluwer Academic Publishers, Dordrecht: 133–155. Reprinted from Hydrobiologia 170.

Carr, O. J. & R. Goulder, 1990. Fish farm effluents in rivers-II. Effects on inorganic nutrients, algae, and the macrophyte Ranunculus pencillatus. Wat. Res. 24: 639–647.

Colman, J. A. & A. R. Jacobson, 1991. Review and development of aquaculture models for predicting solute flux at the sediment-water interface. In Brune, D. E. and J. R. Tomasso (eds), Aquaculture and Water Quality, Advances in World Aquaculture, Vol. 3. pubs. World Aquaculture Society, Baton Rouge, Louisiana, pp. 460–488.

Crampton, V, 1987. How to control phosphorus levels. Fish Farmer. July/August: 38–39.

DePinto, J. V., T. C. Young & S. C. Martin, 1981. Algal-available phosphorus in suspended sediments from lower Great Lakes tributaries. J. Great Lakes Res. 7: 311–325.

Dobrowolski, Z., 1987. The application of benthic indices for evaluation of caged trout culture's influence on the littoral of the Łetowskie Lake. Pol. Arch. Hydrobiol. 34: 579–591.

Golterman, H. L., 1977. Sediments as a source for algal growth. In H. L. Golterman (ed.), Interactions between sediments and freshwater. Dr W. Junk Publishers, The Hague: 286–293.

Hieltjes, A. H. M. & L. Lijklema, 1980. Fractionation of inorganic phosphates in calcareous sediments. J. envir. Qual. 9: 405–407.

Jones, R. I., 1990. Phosphorus transformations in the epilimnion of humic lakes: biological uptake of phosphate. Freshwat. Biol. 23: 323–337.

Klapwijk, S. P., J. M. W. Kroon & M.-L. Meijer, 1982. Available phosphorus in lake sediments in The Netherlands. In P. G. Sly (ed.), Sediment/Freshwater Interaction. Developments in Hydrobiology 9. Dr W. Junk Publishers, The Hague: 491–500. Reprinted from Hydrobiologia 91/92.

Korzeniewski, K., J. Trojanowski & C. Trojanowska, 1985. Hydrochemical study of Lake Szczytno Małe with trout cage culture. Pol. Arch. Hydrobiol. 32: 157–174.

Logan, T. J., T. C. Oloya & S. M. Yaksich, 1979. Phosphate characteristics and bioavailability of suspended sediments from streams draining into Lake Erie. J. Great Lakes Res. 5: 112–123.

Long, E. Jr. & G. D. Cooke, 1978. Phosphorus variability in three streams during storm runoff events: chemical analysis vs. algal assays. Verh. int. Ver. Limnol. 21: 441–452.

Marengo, G. & G. Premazzi, 1985. Biological availability of P-loads to Lake Lugano. Verh. int. Ver. Limnol. 22: 3351–3355.

Marsden, M. W., 1989. Lake restoration by reducing external phosphorus loading: the influence of sediment phosphorus release. Freshwat. Biol. 21: 139–163.

Mayer, T. & P. G. Manning, 1989. Variability of phosphorus forms in suspended solids at the Lake Erie-Grand River confluence. J. Great Lakes Res. 15: 687–699.

Miller, W. E., J. C. Greene & T. Shiroyama, 1978. The Selenastrum capriconutum Printz algal assay bottle test. USEPA Document No. EPA-600/9-78-018, 126 p.

Murphy, J. & J. P. Riley, 1962. A modified single solution method for the determination of phosphate in natural waters. Analyt. chim. Acta. 27: 31–

NCC, 1990. Fish Farming and the Scottish freshwater environment. Nature Conservancy Council, Edinburgh, 285 pp.

Penczak, T., A. Galicka, M. Molinski, E. Kusto & M. Zalewski, 1982. The enrichment of a mesotrophic lake by carbon, phosphorus and nitrogen from the cage aquaculture of rainbow trout, Salmo gairdneri. J. app. Ecol. 19: 371–393.

Persson, G., 1990. Utilization of phosphorus in suspended particulate matter as tested by algal bioassays. Verh. int. Ver. Limnol. 24: 242–246.

Phillips, M. J., 1985. The environmental impact of Scottish freshwater lochs. HIDB report, HIDB Inverness, U.K. 106 p.

Premazzi, G. & A. Provini, 1985. Internal P loading in lakes: a different approach to its evaluation. Hydrobiologia 120: 23–33.

Sinke, A. J. C. & T. E. Cappenberg, 1988. Influence of bacterial processes on the phosphorus release from sediments in the eutrophic Loosdrecht Lakes, The Netherlands. Arch. Hydrobiol. Beih. 30: 5–13.

Skogheim, O. K. S. Sanni & A. Bergheim, 1982. Some ecological effects of a fish-cage farm in a Norwegian lake. Report to 12th EIFAC session, Working Party on Fish Farm Effluents, Budapest, 31 st May–5th June 1982.

Trojanowski, J., C. Trojanowska & H. Ratajczyk, 1982. Effect of intensive trout culture in Lake Łetowo on its bottom sediments. Pol. Arch. Hydrobiol. 29: 659–670.

Trojanowski, J., C. Trojanowska & H. Ratajczyk, 1985. Primary production in lakes with cage trout culture. Pol. Arch. Hydrobiol. 32: 113–129.

van Liere, L., J. Peters, A. Montijn & L. R. Mur, 1982. Release of sediment phosphorus and the influence of algal growth on this process. Hydrobiol. Bull. 16: 191–200.

Weismann, D., H. Scheid & E. Pfeffer, 1988. Water pollution with phosphorus of dietary origin by intensively fed rainbow trout (Salmo gairdneri Rich.). Aquaculture 69: 263–270.

Weston, D. F., 1991. The effects of aquaculture on indigenous biota. In: Brune, D. E. & J. R. Tomasso (eds), Aquaculture and Water Quality, Advances in World Aquaculture, Vol. 3. pubs. World Aquaculture Society, Baton Rouge, Louisiana, pp. 534–567.

Hydrobiologia **253**: 373–376, 1993.
P.C.M. Boers, T.E. Cappenberg & W. van Raaphorst (eds),
Proceedings of the Third International Workshop on Phosphorus in Sediments.
© 1993 *Kluwer Academic Publishers.*

Abstracts

How to quantify the impact of internal phosphorus loading on the phosporus concentration in shallow lakes?

Paul Boers[1], Joop van der Does[2]
[1]*RIZA, P.O. Box 17, 8200 AA Lelystad, The Netherlands;* [2]*HHR, P.O. Box 156, 2300 AA Leiden, The Netherlands*

After a reduction of the external loading, a release of phosphorus from the sediments may delay the response of the phosphorus concentration on the reduction of the loading. Therefore, it is important to quantify this internal loading. One approach is the measurement of the flux in the laboratory. A completely different approach is to quantify the availability of the sediment phosphorus to support algal growth by bioassays or selective extractions. Comparisons between selective extractions and bioassays usually reveal that the two methods are incompatible.

As far as we know, no comparisons between measurements of fluxes and bio-availability have been made before.

We collected data on flux rates, bio-availability of sediment phosphorus and phosphorus balances for a number of Dutch shallow lakes. In all these lakes restoration measures have been carried out or are under consideration.

The specific characteristics of both types of measurements and their potentials to describe the internal loading in the different lakes will be discussed.

Phosphorus loading in the sediments of lake "Izvorul Muntelui-Bicaz" from salmonid fish culture

Baciu Dorin-Sorin
Statiunea Salmonicola Potoci-Bicaz R-5650, Romania

Sediments play an important role in the overall phosphorus metabolism of lakes acting both as a sink and a source of phosphorus.

The leakage of phosphorus from cage culture fish farming may play a significant role in nutrients loading of lakes and is regulated by the physical-chemical environment, the biological activity and by phosphorus composition in the food rests and the fecal faeces.

The objectives of our investigation are:

– the establishment qualitative and quantitative of the clay minerals from sediments in two control stations: POTOCI FARM and RUGINESTI;
– the fish food composition and quantity as compared to the sediment composition;
– fractional distribution of phosphorus in sediment from the control stations;
– the establishment of the physical-chemical environment.

Cage culture farming in fresh water systems has a eutrophicating effect.

Phosphorous release from inundated land of water supply reservoir

Anton Filip, Milan Ilić [1] and Dragan Manojlović [2]
[1] Energoprojekt Holding Corporation, Beograd, Yugoslavia; [2] Faculty of Science, Beograd University

Phosphorous release from land to be inundated by creation of two water supply reservoirs: Rovni and Barje is determined. The aim was to define the cost effective degree of land clearance works before the inundation. Such works should guarantee the best protection, i.e. minimize the water quality deterioration after the reservoir body formation.

The quantities of phosphorous which might be released from inundated land are determined on the basis of measurements of organic matter content being accumulated as trees and grass debris on the land surface, as well as in core samples of soil from the 10 and 20 cm depths. The kinetics of phosphorous release is determined on a laboratory scale, through the interaction of the samples with the river water within one month period. The amounts of possible total phosphorous to be released due to decay of the accumulated vegetation debris are estimated on the basis of the measurements at several locations with the results that the woodland contains 6000 g/sq.m and the meadows 600 g/sq.m. of organic matter, while in the arable land the organic matter content is considerably smaller.

Accordingly the estimated amount of phosphorous release from the inundated land is 11,000 kg P for Rovni reservoir inundation, with 86 percent participation from woodland, 12 percent from meadows and 2 percent from other land-use patterns. For Barje reservoir, potential phosphorous release from the inundated land is estimated to be 7,000 kg P, with 98 percent participation from woodland, 1 percent from meadows and 1 percent from the rest of inundated land.

The organic matters are, practically, absent from 10–20 cm depth of topsoil layer.

The investigation of phosphorous release kinetics, on a laboratory scale, shows that its rate is within a range 0,004–0,008 gP/sq.m.day.

Phosphorous load of the future reservoirs originating from the catchment area in the course of the average flow condition is 880 kg P/year from the Rovni catchment area of the total surface of 107 sq.km, and 1768 kg P/year from the Barje catchment area of the total surface of 236 sq.km. These quantities of total phosphorous are considerably smaller than those which might be released from the inundated land during the decay process even taking into account that it can last for several years.

It is, therefore, recommended to remove all organic debris and 10–20 cm topsoil from the woodlands and meadows, in order to prevent deterioration of water quality in the newly created reservoirs, both from the point of view of the prevention of dissolved oxygen decrease and the increase of the nutrients.

Phosphate immobilization in peaty sediment

Peer Keizer, Joanna Galas, Anja Sinke & Peter de Joode
Limnological Institute, Rijksstraatweg 6, 3631 AC Nieuwersluis, The Netherlands

On an annual basis, most sediments act as a permanent sink for phosphate, although during summer the sediment usually acts as a temporal source for phosphate in the overlying water. The phosphate cycle in the sediment may be influenced by the addition of aluminum and iron hydroxides or calcium salt. To what extent these additions suppress the phosphate release from the sediment and enhance the annual in lake phosphate retention depends on the external phosphate load, residence time, lake morphometry, hydrology, hydroxide dose, sediment composition, microbial activity and pore water chemistry. The mechanisms that contribute to the phosphate immobilization in Lake Loosdrecht, a shallow eutrophic lake with a peaty sediment, were investigated.

Adsorption of phosphate by aerobic and anaerobic sediment cores with various hydroxide treatments were measured. In addition adsorption was studied over a period of six months in anaerobic batch experiments with mixtures of iron or aluminum hydroxide and sediment material from different depths.

The results show that the effectiveness of iron additions is limited because the reduction of iron is a relatively fast process and accompanied by an almost complete loss of its ability to adsorb phosphate, even under methanogenic conditions. Furthermore, no decrease of the dissolved phosphate concentration was observed upon prolonged aging. This indicates that in this sediment the conversion of calcium carbonate into calcium phosphate and the formation of vivianite at natural pore water conditions is nonexistent or extremely small. The consequences of the insignificant diagenesis of phosphate minerals in peaty sediments for lake restoration perspectives will be discussed.

The use of vegetated ditches for water quality improvement

A.F.M. Meuleman & B. Beltman
Department of Plant Ecology and Evolutionary Biology, University of Utrecht, P.O. Box. 800.84, 3508 TB Utrecht, The Netherlands

One of the main problems of nature conservation in the Netherlands is the high nutrient and mineral load of river water. Supply of river water during periods of drought, leads to eutrophication and has detrimental effects on the vegetation. One option for water quality improvement in nature reserves is leading polluted river water through vegetated ditches or reed marshes. Removal of nutrients from the river water is due to biological processes (uptake by the vegetation, nitrification/denitrification) and physical and chemical processes (adsorption, precipitation).

In the Netherlands several ditch systems are in use for water quality improvement in nature reserves. Preliminary results of a research project on water quality improvement by vegetated ditches in a nature reserve (wet meadow) near Zegveld (Province of Utrecht) indicates a low removal capacity for nitrogen and a high removal capacity for phosphorus (90–95%). The vegetation at the part of the ditch system where removal of phosphorus took place shows a remarkable high cover, with a high abundance of the brown moss *Fontinalis antipyretica*. The mechanisms involved are probably plant uptake and chemisorption. The relevance of these processes for the removal of phosphorus will be studied more closely. Decreasing phosphorus concentrations along the ditch system are reflected in the vegetation zonation. Along the flow path, the eutrophic vegetation is replaced by a mesotrophic vegetation. The high removal efficiency for phosphorus is of utmost importance for the conservation of wet meadows, as plant growth and primary production of these ecosystems often seem to be phosphorus limited.

On the regulation of sediment phosphorus release by calcium–phosphorus minerals in Onondaga Lake, N.Y. (USA)

Steven W. Effler[1], Martin T. Auer and Michael R. Penn[2], Charles T. Driscoll[3]

[1] Upstate Freshwater Institute, Syracuse, New York, USA; [2] Michigan Technological University, Houghton, Michigan, USA; [3] Syracuse University, Syracuse, New York, USA

Redox conditions at the sediment–water interface are widely accepted as the principal regulator of sediment phosphorus (P) release in lakes. Under aerobic conditions, P is believed to be sorbed to oxidized forms of iron in a microzone at the sediment surface and the release of reduced species (e.g. Fe^{2+}) and associated dissolved P is prevented. With the onset of hypolimnetic anoxia, the microzone is lost and abrupt release of Fe^{2+} and dissolved P are observed.

Onondaga Lake is a hypereutrophic, industrially polluted, urban lake located in Syracuse, New York, USA. High hypolimnetic levels of H_2S that develop after anoxia limit Fe^{2+} concentrations. High rates of $CaCO_3$ deposition, promoted by Ca_{2+} discharges from an adjoining soda ash manufacturing plant, have resulted in unusually high sediment concentrations of $CaCO_3$ and dilution of other sedimentary constituents (including P). Substantial hypolimnetic P accumulations are observed and laboratory experiments indicate that elevated levels of sediment P release prevail in the lake.

The time course of concentrations of dissolved P and related chemical species is presented for the bottom waters of Onondaga Lake for 1990 and 1991 and contrasted with transformations generally accepted as reflecting redox controlled systems. The onset of sediment P release in Onondaga Lake precedes the development of anoxia and the initial release of Fe^{2+} by more than one month, but coincides with the increase in CO_2 and decrease in pH which occurs soon after the onset of stratification. This anomaly suggests that processes other than the dissolution of Fe precipitates and complexes are important in mobilizing sediment P.

It is hypothesized that sediment P release in Onondaga Lake is regulated by Ca–P minerals. The hypothesis is supported by thermodynamic calculations made with a chemical equilibrium model (MINEQL), and the results of preliminary laboratory experiments. This rather unique regulation of P release has important implications with respect to choices for remediation alternatives for the lake, e.g. hypolimnetic aeration.